Ecohydrology

This volume is devoted to the derivation and application of simplified bioclimatic boundary conditions at vegetated land surfaces using natural selection of vegetation characteristics driven by productivity maximization. It seeks to answer two questions:

- How are the physical characteristics of trees and their forest communities related, at equilibrium, to the climates and soils in which they are found?
- What is the maximum biological productivity to be expected from a given species in a given climate and soil?

The book begins with the small-scale physics of internal control of forest growth represented by the vertical fluxes of light, CO_2, water vapor, and sensible heat within the canopy. The scope then widens to consider the large-scale physics of the external control offered by the balances of thermal energy and water. This leads to the specification of the two state variables of a maximally productive canopy structure and completes the hydrologic surface boundary condition for given climate, soil, and plant species. Finally, by idealizing the stomatal responses to both light and available water, and assuming a preferred average state of zero plant stress, necessary relations are obtained among species, climate, and soil for maximum plant productivity. These new ideas are used to define the "climax" bounds of plant habitat and to estimate net primary productivity in the canopy.

Ecohydrology bridges the fields of hydrology and ecology and proposes new unifying principles derived from the concept of natural selection. It also has potential application in determining the response of vegetation to slow variations in climate. This book will therefore provide fascinating reading for graduate-level students and research scientists working in the fields of ecohydrology, hydroclimatology, forest ecology, and surface water hydrology.

PETER EAGLESON is Emeritus Professor of Civil and Environmental Engineering at the Massachusetts Institute of Technology. He was Head of the Department of Civil Engineering between 1969 and 1975, and has also held the positions of President of the American Geophysical Union (1986–8) and Chairman of the National Research Council Committee on Opportunities in the Hydrologic Sciences (1987–91). Professor Eagleson has been at the forefront of theoretical research into the physical interactions between climate, soil, and vegetation for more than twenty years. This volume brings together many of the advances in hydrology that he has pioneered during that time. Professor Eagleson has been honored many times for his outstanding contributions in this area including, more recently, the William Bowie Medal of the American Geophysical Union (1994), the Stockholm Water Prize (1997), and the John Dalton Medal of the European Geophysical Society (1999). His previous books include *Dynamic Hydrology* (1970) and *Land Surface Processes in Atmospheric General Circulation Models* (Cambridge University Press, 1982).

Nikko fir. (Photographed in the Arnold Arboretum, Boston by William D. Rich; Copyright © 2001 William D. Rich.)

Ecohydrology

Darwinian expression of vegetation form and function

Peter S. Eagleson
Massachusetts Institute of Technology

PUBLISHED BY THE PRESS SYNDICATE OF THE UNIVERSITY OF CAMBRIDGE
The Pitt Building, Trumpington Street, Cambridge, United Kingdom

CAMBRIDGE UNIVERSITY PRESS
The Edinburgh Building, Cambridge CB2 2RU, UK
40 West 20th Street, New York NY 10011–4211, USA
477 Williamstown Road, Port Melbourne, VIC 3207, Australia
Ruiz de Alarcón 13, 28014 Madrid, Spain
Dock House, The Waterfront, Cape Town 8001, South Africa

http://www.cambridge.org

First published 2002
First paperback edition 2005

Typefaces Times 10.25 / 13.5 pt and Joanna *System* LaTeX 2ε [TB]

A catalogue record for this book is available from the British Library

Library of Congress Cataloguing in Publication data

Eagleson, Peter S.
Ecohydrology : Darwinian expression of forest form and function / Peter S. Eagleson.
p. cm.
Includes bibliographical references (p.).
ISBN 0 521 77245 1 (hardback)
1. Forest ecology. 2. Forest productivity. 3. Forest plants – Ecophysiology. 4. Forest microclimatology. 5. Forest canopies. 6. Ecohydrology. I. Title.

QK938.F6 E24 2002
577.3–dc21 2002017395

ISBN 0 521 77245 1 hardback
ISBN 0 521 61991 2 paperback

To my dearest Bev who has taught me how to live and to love, and in so doing has inspired my work and enriched my life beyond measure.

To believe in something not yet proved and to underwrite it with our lives: it is the only way we can leave the future open.

Lilian Smith

The Journey, 1954

Contents

Foreword

Ecosystems are complex, evolving structures whose characteristics and properties depend on many interrelated links between climate, soil, and vegetation. The dynamics of these links are critically influenced by the scale at which the phenomena are studied, as well as by the physiological characteristics of the vegetation, the pedology of the soil, and the type of climate. The evolution of an ecosystem is crucially dependent on the need of its vegetation for light and water as external inputs that drive its productive mechanisms. These inputs are highly variable in time and space and their assimilation depends on the plant characteristics and the ecosystem structure. Thus, vegetation plays an active role in the space–time dynamics of soil water being both cause and effect of these dynamics. Similarly, plant structure and the canopy three-dimensional characteristics result from and at the same time control the use and impact of the radiative energy. Light and water exert key controls on the functioning of the leaf stomates and thus in the uptake and assimilation of carbon dioxide. Insolation and rainfall vary throughout the day and the seasons; moreover, and especially in the case of rainfall, the intermittent and unpredictable nature of the resource makes the study of its impact on ecosystem functioning an especially challenging field.

The hydrologic and ecologic mechanisms underlying the climate–soil–vegetation dynamics and thus controlling the most basic ecologic patterns and processes are one of the most exciting scientific frontiers at the start of the twenty-first century. It is a frontier full of challenging and unexplored questions which go to the heart of hydrology and ecology, embracing problems which are crucially related to biodiversity and ecosystem functioning. Its study is of fundamental importance for understanding the environment in which we live and the state in which it will be inherited by future generations.

This book is at the center of the above frontier undertaking a general analysis of many of the most crucial questions on the form and functioning of forests. It does so in a unique manner. The approach is analytical proceeding from the small-scale physics of the internal controls of forest growth to the large-scale physics which crucially control such growth. Professor Eagleson has been, throughout the last 40 years, a fundamental contributor to many of the key ideas and developments underlying the quantitative description and modeling of these physical processes. Nevertheless, this book is much, much more than an excellent synthesis of the state of the art. It in fact sets a new level for the science of ecohydrology.

The first part of the book presents a comprehensive and detailed description of the radiant and turbulent fluxes within the canopy and between the canopy and the atmospheric boundary layer. It continues with the study of the energy and water balances and the intimate links which these balances establish between soil, vegetation, and climate. The depth and focus of the presentation as well as the many novel contributions present in the first six chapters would be enough to make this book an outstanding contribution to the understanding of ecosystems. Nevertheless, this part just sets the stage for a *tour de force* of imagination and analytical prowess where the author develops optimality principles for the form and function of natural forests. Thus, the reader embarks on an exciting exploration of the links between soil, climate, and species characteristics, necessary to maximize plant productivity. Furthermore, Professor Eagleson proceeds to derive the criteria for optimal canopy structure in which canopy conductance of water vapor and carbon dioxide is maximum.

Chapters 7 through 11 present for the first time the new full theory. The driving general assumption is that evolutionary pressure drives forests towards a bioclimatic optimal state characterized by a maximum probability of reproductive success assumed to correspond to maximum biomass productivity. This principle of bioclimatic optimality is analytically structured around the following secondary principles: (1) optical optimality : foliage arrangement and climate insolation leading to maximum carbon assimilation; (2) mechanical optimality : leaf angle at which canopy conductance of carbon dioxide and water vapor is maximized; (3) thermal optimality : leaf temperature equals the photosynthetically optimum temperature; and (4) hydrologic optimality : moisture state at which plant is at incipient stress and the average insolation is a maximum, for open stomates. The author carries extensive comparisons of the consequences of the above principles against historical data collected in forests of different species and in different climates. These data cover a wide range of environmental conditions and community types and provide significant support for Professor Eagleson's theory, which I am sure will become a cornerstone in ecohydrology. Figure 10.3 is remarkably suggestive of bioclimatic optimality where the vast majority of the observed species lie inside the range of global maximum productivity.

This book contains the signatures of how the best science is created: it is a fascinating blend of knowledge and imagination which opens new and exciting avenues to fundamental problems in ecohydrology. Its profound originality, physical understanding, and analytical elegance will have a lasting and major impact in the field. Moreover, the importance of its ideas and the excitement transmitted by a truly creative mind at work will inspire many hydrologists, ecologists, and environmental scientists to dedicate their best efforts to continue the paths opened by Professor Eagleson.

Ignacio Rodriguez-Iturbe
Theodora Shelton Pitney Professor of Environmental Sciences
and Professor of Civil and Environmental Engineering
Princeton University

Preface

This is a research monograph in which I explore and evaluate the biophysical relationship between ambient climate and the form and function of the associated vegetation. I hope the results will be useful in anticipating the changes to be expected in vegetated surfaces under conditions of slowly changing climate.

At the 70 percent of Earth's surface which is water, the atmospheric boundary conditions are, under idealized circumstances, readily expressible in terms of the classical laws of thermodynamics and fluid mechanics. For the remaining 30 percent of the surface however, there is a greater or lesser presence of vegetation, and the boundary conditions are determined in part by biological behaviors which differently constrain the moisture and energy fluxes. In such cases the boundary conditions are termed "interactive" because the structure and type of vegetation help to determine the climate through their exchange of heat and moisture with the atmosphere, while the climate through its supply of moisture, carbon dioxide, and light to the surface helps to determine the type and structure of the vegetation.

This book is devoted to theoretical generation of these atmospheric boundary conditions at vegetated surfaces under conditions which are necessarily highly idealized. As John Monteith (1981, p. 753) said: "in a topic such as this, progress can be made only if the number of variables is held to a minimum at every stage in the analysis . . . ". Here I consider only monocultures; I neglect the activities of insects and other animals (including man), and omit the influence of disease and fire; I assume infinite nutrient and carbon dioxide reservoirs – leaving water and light as the limiting resources; and I analyze the climate–soil–vegetation system as though it operated in its long-term average state with a neutrally stable atmosphere and no lateral advection of energy or water from adjacent landscapes. Furthermore, I sever the feedback link from surface to atmosphere and thus consider the vegetation as a passive responder to climatic forcing at this level of approximation.

Ecologists, micrometeorologists, and plant scientists will likely find this approach to be naïve and may even be offended by my neglect of much important natural complexity. Engineers, on the other hand, will be familiar with and may welcome the reduction of an intricate multidisciplinary problem to a small set of simple, albeit approximate, rules. In either case, I ask the reader to withhold judgement pending the presentation of evidence.

Believing, after Darwin, that biology is an expression of physical optimality, my hypothesis holds natural selection responsible for both the form and the function of vegetation, and I seek its analytical formulation. However, selecting the modes of expression of natural selection from such complexity involves (at best) informed conjecture, and as G.K. Chesterton put it, "There are no rules of architecture for a castle in the clouds"!

Denying altruism in these organisms, I assume natural selection to be expressed through each individual tree rather than through the community of trees constituting the forest. In their classic paper, Parkhurst and Loucks (1972, p. 505) reasoned "Natural selection leads to organisms having a combination of form and function optimal for growth and reproduction in the environments in which they live." Accordingly, I imagine selection pressure to maximize reproductive success of the plant through maximization of its biomass (and hence seed) production, and I formulate this productivity maximum in terms of the physical variables defining the climate, the soil, and what are then called the "optimal" features of the vegetation.

This is *not* intended to be a micrometeorological textbook, an ecophysiological textbook, or even a hydrological textbook. It *is* an effort to draw out, from among the separate sciences of hydrology, micrometeorology, and ecophysiology, those behavioral characteristics that together may contribute to natural selection, and to connect them in a way that retains sufficient simplicity *and* scientific fidelity to be useful to those charged with forecasting the components of climate change. Accordingly I have tried to select models of the process physics and biology which are simple enough to permit their combination without losing sight of the goal (and without unnecessarily discouraging the reader!). I have tried to stick with the "traditional" models as best fitting this bill. In multidisciplinary endeavors, all the rich scientific detail of each contributing field can't be retained in their joining lest the resulting complexity negate the utility of the result.

The book is logically divided into two parts:

Part I contains "Biophysics". Following a summary in Chapter 1 of what lies ahead, it presents the small-scale physics of those *internal* controls on forest growth, the vertical fluxes within the canopy of light, carbon dioxide, water vapor, and sensible heat. Chapter 2 reports the observed distribution and orientation of leaf area in the crown and the canopy, and Chapter 3 examines the effect of solar altitude and foliage orientation on the penetration of solar radiation into the canopy. Chapter 4 gives considerable detail of the vertical turbulent flux of horizontal momentum within canopies of various geometry and, by analogy, the fluxes of sensible heat and mass as well. The canopy resistance controlling these vertical fluxes is derived by building a series–parallel resistive model which bridges the scales from stomate to canopy. The scope then widens to consider the large-scale physics of the *external* controls on forest growth offered by the thermal energy balance (Chapter 5) and the water balance (Chapter 6). While Part I is mostly familiar material several new ideas are developed.

Part II is best described as "Darwinian ecology". With all details now in hand, Chapter 7 develops an optimum canopy structure which specifies the two canopy state variables (i.e., the canopy conductance and canopy cover) completing the hydrologic surface boundary condition for given climate, soil, and plant species. Next, through idealization of (1) the species- and temperature-sensitive photosynthetic capacity of the leaf, and (2) the species-, soil- and climate-sensitive leaf water stress, Chapter 8 demonstrates the necessary relations among species, climate, and soil for maximization of plant productivity under the additional assumption of a preferred zero-stress average state. These ideas are then used in Chapter 9 to define the "climax" bounds of plant habitat, and in Chapter 10 to estimate net primary productivity at the canopy scale. The material of these four chapters is entirely new and represents the essence of the book. Chapter 11 presents a brief summary along with some speculation and identifies opportunities for useful extensions of this work.

The work is guided and evaluated along the way by comparisons of the optimal vegetal features with observations from the literature whenever possible. These comparisons are favorable for some features and less favorable (if not unfavorable) for others. The data are too few for useful statistical assessment, but I find the results encouraging considering the enormous complexity of this ecological system.

The exposition has two threads:

1. Description of the observed physical characteristics of natural (and some agricultural) vegetation canopies, and computation of the vertical fluxes therein of momentum, mass, sensible heat, and light. While much of this is a collection of material already present in the literature there is a revealing new representation of effective canopy resistance that is helpful in understanding possible processes of natural selection, and there is a new finding of optical optimality that allows quantitative comparison of canopy carbon dioxide supply and demand.
2. Search for the mechanisms by which natural selection is expressed in forest form and function. This material is new and has not appeared before in print.

This breadth of scope gives the book utility both as a graduate level text for the developing interdisciplinary field of *ecohydrology*, and as a reference for researchers in landscape ecology. In view of the former, the writing style is tutorial with attention given to completeness of detail.

My interest in the role of vegetation in climate arose from my water balance work in 1975 and has been advanced with the inspiration, advice, and assistance of many.

For awakening my interest in biomechanics I cite the pathbreaking work of Ralph Slatyer and of John Monteith, while for early advice concerning the climate connection I acknowledge Joseph Smagorinsky and Syukuro Manabe, as well as the late Jule Charney and Yale Mintz. Over the years I have been blessed with exceedingly capable and insightful graduate research assistants who have repeatedly pointed me in fruitful directions and rescued me from serious error; in particular I am endebted to P.C.D. Milly, Randal Koster, Pedro Restrepo-Posada, Tobin Tellers, Lelani Arris, Dara Entekhabi, and Guido Salvucci. Three special friends have been continuous sources of support during the nine years that this manuscript has been in preparation: Dara Entekhabi has unselfishly contributed both scientific advice and computational assistance; Andrea Rinaldo has acted as ex officio editor guiding me through the maze of modern manuscript preparation; and Ignacio Rodríguez-Iturbe has been inspirational through his scientific example, his unfailing encouragement to "go for the home run", and the constancy of his friendship. I thank them all and the many others who remain anonymous.

Peter S. Eagleson
Cambridge, MA
2002

Acknowledgments

Many have contributed to this work in important ways and deserve special mention at the outset: Dara Entekhabi translated the original data of Appendix F into accessible format; Andrea Rinaldo graciously edited an early draft of Chapter 4; Lelani Arris donated an exhaustive catalogued library of early ecological references that eased the literature search; David Benney assisted with mathematical complexities and gave continuing moral support; William D. Rich provided the artful digital images of many interesting trees from both the Arnold and Holden Arboreta; Beverly G. Eagleson and Peter S. Eagleson, Jr. contributed additional photographs; Robert H. Webb supplied the image of the creosote bush community and F. Eugene Hester that of the Okefenokee swamp; and John MacFarlane prepared the line drawings. The author thanks them each sincerely.

In addition, the author is indebted to: the MIT Department of Civil and Environmental Engineering for generous financial assistance with manuscript preparation through resources of the Edmund K. Turner Professorship; the Rockefeller Foundation for provision of a Bellagio Center Residency during a crucial early stage of this long project; and his colleagues at the Parsons Laboratory for keeping him eager for each new day at the office.

Author's note

The photographs of individual trees presented here were chosen, for the most part, as illustrations of their infinite variety and their intrinsic beauty rather than to illustrate particular ideas put forth in the text. Note that species gathered from afar and open-grown in an arboretum are likely to display different adaptive characteristics than they would were they part of a homogeneous stand within their natural habitat.

Notation

A = surface area, m^2
A = exponential variable (Eq. D.1), arbitrary dimensions
A_{cm} = surface area of conical monolayer (Eq. A.89), m^2
A_0 = reference value of exponential variable, same dimensions as A
AU = astronomical unit = 1.496×10^8 km
A_D = horizontal component of the effective drag-producing area of the solid matter, m^2
A_b = cross-sectional area of right circular cylinder circumscribing the crown, m^2
A_l = one-sided area of single leaf, cm^2
A_L = leaf area, m^2
A_0 = modified gravitational component of infiltration rate (Eq. 6.17), mm h^{-1}
A_s = horizontal shadow area per unit horizontal area of canopy (Eq. 3.18), m^2
A_s = lateral surface area of cone, m^2
A_{cm} = surface area of conical monolayer, m^2
A_{cr} = horizontal cross-sectional area of the crown, m^2
A_1 = single-sided surface area of needles lying in and comprising the monolayer crown surface, m^2
A_* = water balance parameter for quadratic soil (Eq. C.22), dimensionless
a = scattering dimension, m
$a(\xi) \equiv a_t(\xi)/L_t$ = single-sided foliage area homogeneity function, dimensionless
a_{cr} = crown one-sided leaf area density = (one-sided leaf area)/(crown volume), $m^2\ m^{-3}$
a_D = canopy drag area density, m^2(drag area) / m^3(crown)
a_d = extinction parameter for horizontal foliage shear stress (Eq. 4.5), dimensionless
a_L = canopy leaf area density = single-sided leaf area per unit of canopy volume, m^{-1}
a_m = extinction parameter for horizontal momentum flux (Eq. 4.6), dimensionless
a_t = canopy foliage area density = single-sided foliage area per unit of canopy volume (Eqs. 4.14, 4.15, A.3), m^{-1}

a_w = extinction parameter for horizontal wind velocity (Eq. 4.4), dimensionless
a_1, a_2 = coefficients, dimensionless
B = coefficient (Eq. C.2), dimensionless
B = rate of plant metabolic use of energy, cal cm^{-2} min^{-1}
B = proportionality constant in interference function (Eq. 4.87), dimensionless
B = additions to plant biomass, g$_s$ m^{-2} y^{-1}
B_0 = proportionality constant, dimensionless
B_o = evaporation proportionality coefficient, g cm^{-2} s^{-1} mb^{-1}
B_* = water balance parameter for quadratic soil (Eq. C.23), dimensionless
b = diameter of circular planform of conceptual crown, m
b = breadth of roughness element, m
b = scattering dimension, m
b_1, b_2 = coefficients, dimensionless
C = coefficient (Eq. C.3), dimensionless
C = chord of needle cross-section perpendicular to longitudinal axis, mm
C = plant matter consumed by grazers, g$_s$ m^{-2} y^{-1}
C_A = arc-chord of needle cross-section, m
C_D = fluid dynamic coefficient of form drag for the canopy solid matter, dimensionless
C_{D0} = foliage drag coefficient at high Reynolds number (Eq. 4.10), dimensionless
C_{D1} = foliage drag coefficient due to element shape (Eq. 4.10), dimensionless
C_{D2} = foliage drag coefficient due to foliage element interference (Eq. 4.11), dimensionless
C_f = foliage surface drag coefficient (Eq. 4.12), dimensionless
C_N = nitrogen solubility coefficient, g(N_2) g^{-1}(H_2O)
C_3 = pentose phosphate pathway for fixation of carbon during photosynthesis
C_* = water balance parameter for quadratic soil (Eq. C.24), dimensionless
CO_2 = carbon dioxide
c = scattering dimension, m
c = mass of carbon dioxide per unit mass of moist air, g(CO_2) g^{-1}, (O_2)
c = permeability index of soil (Eq. C.20), dimensionless
c_o = mass of carbon dioxide per unit mass of moist air at the uppermost leaf, g(CO_2) g^{-1}, (O_2)
c_p = specific heat of air at constant pressure, cal g^{-1} K^{-1}
c_s = mass of carbon dioxide per unit mass of moist air at the lowermost leaf, g(CO_2) g^{-1}, (O_2)
D = diffusivity, cm^2 s^{-1}
D = diameter of the right circular cylinder circumscribing the crown, m
D = spacing of roughness elements, m
D = plant matter shed as detritus, g$_s$ m^{-2} y^{-1}
D_e = effective desorption diffusivity (Eq. 6.37), cm^2 s^{-1}
D_H = horizontal flux divergence of sensible heat, cal cm^{-2} min^{-1}
D_i = sorption diffusivity (Eq. 6.61), cm^2 s^{-1}
D_p = drying power of atmosphere (Eq. B.2), cal cm^{-2} min^{-1}
D_{Qv} = horizontal flux divergence of latent heat, cal cm^{-2} min^{-1}
D_{dz} = zonal brightness = diffuse radiance on horizontal surface, cal cm^{-2} min^{-1}

d = diffusivity index of soil, dimensionless
d = diameter of cylindrical foliage element, mm
d = diameter of crown at elevation ξ, m
d_0 = zero-plane displacement height, m
d_{os} = zero-plane displacement height for the substrate, m
$d\sigma$ = element of surface area, m^2
$d\omega$ = elemental solid angle (Eq. 3.8), steradians
E = vertical flux density of water vapor (i.e., mass evaporation) (Eq. 4.115), $g\ cm^{-2}\ s^{-1}$
E = bare soil evaporation effectiveness (Eq. 6.38), dimensionless
$E[\ldots]$ = expected value of ...
$E_1(z)$ = exponential integral (Eq. 10.13)
E_a = component of E_v depending mainly upon the atmosphere, $g\ cm^{-2}\ s^{-1}$
E_o = dimensionless kernel of the arid climate asymptote of β_s (Eqs. C.9–C.10)
E_p = shorthand for E_{ps}, $g\ m^{-2}\ y^{-1}$ or $cm\ y^{-1}$
E_{ps} = potential rate of evaporation from a wet, simple surface (Eq. 5.20), $g\ m^{-2}\ y^{-1}$ or $cm\ y^{-1}$
E_{psd} = free water surface potential evaporation during dormant season, $g\ m^{-2}\ y^{-1}$ or $cm\ y^{-1}$
$E_{ps\tau}$ = free water surface potential evaporation during growing season, $g\ m^{-2}\ y^{-1}$ or $cm\ y^{-1}$
E_{pv} = potential rate of canopy transpiration, $g\ m^{-2} y^{-1}$ or $cm\ y^{-1}$
$E_{ps}^{\otimes}$ = potential rate of evaporation calculated for daylight hours only, $g\ m^{-2}\ y^{-1}$ or $cm\ y^{-1}$
E_r = component of E_v depending mainly upon the radiation, $g\ cm^{-2}\ s^{-1}$
E_r = storm surface retention, mm
E_{sj} = interstorm exfiltration, mm
E_T = rate of evapotranspiration, $g\ cm^{-2}\ s^{-1}$ which for water at 1 $g\ cm^{-3}$ is $cm\ s^{-1}$
E_{TA} = average annual evapotranspiration (Eq. 6.49), cm
E_v = rate of canopy transpiration (Eqs. 5.28, 8.14), $g\ m^{-2}\ y^{-1}$ or $cm\ y^{-1}$
E_{vj} = interstorm transpiration (Eq. C.8), mm
E_v^0 = normalized rate of canopy transpiration, dimensionless
$E_v^{\otimes}$ = rate of transpiration calculated for daylight hours only (Eq. 8.16), $g\ m^{-2}\ y^{-1}$ or $cm\ y^{-1}$
EGL = energy grade line
e = vapor pressure = partial pressure of water vapor, Pa or millibars (mb)
e_0 = vapor pressure at leaf surface, Pa or millibars (mb)
e_2 = vapor pressure at site number 2, Pa or millibars (mb)
e_s = saturation vapor pressure at the temperature of the evaporation site, Pa or millibars (mb)
e_s = instantaneous bare soil evaporation rate, $mm\ s^{-1}$
erf[...] = error function of ..., (Fig. C.2)
F = proportionality factor relating NPP to the productive equivalent of E_v (Eq. E.8), dimensionless
F_n = dimensionless factor, $n = 1, 2, 3, \ldots$, etc.

F_T = total canopy drag per unit of crown basal area, N m^{-2}
F_λ = flux of radiation through an arbitrarily oriented surface (Eq. 3.4), cal cm^{-2} min^{-1}
f_D = horizontal component of foliage resistance per unit volume of canopy (Eq. 4.9), N m^{-3}
f_I = shorthand for $\frac{1-e^{-\kappa L_t}}{\kappa L_t}$ (Eq. 8.27), dimensionless
f_e^* = exfiltration capacity (Eq. 6.32), mm h^{-1}
f_i^* = infiltration capacity (Eq. 6.16), mm h^{-1}
$f(A_L)$ = probability density function of leaf area, cm^{-2}
$f(S_r) \equiv q_b/D_p$ = empirical evaporation function (Eq. B.18), dimensionless
$f_D(\beta L_t)$ = carbon demand function = potential productive gain = $\frac{NPP}{p_D}$ (Eqs. 10.21, 10.22), dimensionless
$f_i(i)$ = probability density function of storm intensity (Eq. 6.11), h mm^{-1}
$f_h(h)$ = probability density function of storm depth (Eq. 6.2), cm^{-1}
$f_{h,t_b}(h, t_b)$ = joint probability density function of storm depth and immediately following interstorm interval (Eq. 6.45), h^{-1} cm^{-1}
$f_I(\beta L_t)$ = insolation averaging function (Eq. 8.30), dimensionless
$f_S(\beta L_t)$ = carbon supply function = limiting productive gain = $\frac{NPP}{p_S}$ (Eq. 10.30), dimensionless
$f_{t_b}(t_b)$ = probability density function of time between storms (Eq. 6.8), h^{-1}
$f_{t_r}(t_r)$ = probability density function of storm duration (Eq. 6.5), h^{-1}
$f_{i,t_r}(i, t_r)$ = joint probability density function of storm intensity and duration (Eq. 6.20), mm^{-1}
G = flux density of sensible heat transfer into the substrate, cal cm^{-2} min^{-1}
G = gravitational infiltration parameter (Eq. 6.23), dimensionless
$G(h_\otimes)$ = G-function = average over the canopy of the projection of a unit foliage area on a plane normal to the Sun direction, thus $G(h_\otimes)/\sin(h_\otimes)$ = shadow area per unit of foliage area (Eq. G.21)
$\widetilde{G_L}(\theta_L)$ = cumulative distribution function of leaf orientation angle (Eq. 2.11), dimensionless
GPP = gross primary productivity (Eq. 10.2), g(CO_2) m^{-2} y^{-1}
g = gravitational constant, m s^{-2}
g( ) = functional notation
g(CO_2) = grams of CO_2
g(N_2) = grams of nitrogen
g_L( ) = functional notation
$\widetilde{g_L}(\theta_L)$ = probability density function of leaf orientation angle (Eq. 2.10), deg^{-1}
g_s = grams of solid biomass
g_w = grams of water
$\widehat{g(T_l)}$ = generalized light-saturated photosynthetic rate as a function of leaf temperature, dimensionless
H = vertical flux density of sensible heat from canopy to atmosphere (Eq. 4.117), cal cm^{-2} min^{-1}
H_f = potential available to support flow resistance from soil to stomatal surface (Eq. 8.11), mb

H_G = gravitational potential due to height of trees, mb
H_L = potential drop across the spongy mesophyll cells from leaf vein to stomatal surface, mb
H_o = osmotic potential across root membranes, mb
h = height of tree from ground surface to top of crown, m
h = height of roughness element, m
h = storm depth, mm
h_l = potential drop due to flow resistance in the leaves, mb
h_c = depth of crown, m
h_o = surface retention depth, mm; depth of crown segment, m
h_r = potential drop due to flow resistance in the roots, mb
h_s = stem height (i.e., height of crown base above substrate), m
h_s = capillary potential (i.e., "head") loss to support soil moisture flow to the root surfaces, mb
h_x = potential drop due to flow resistance in the xylem, mb
$h_\otimes$ = solar altitude (Eq. 3.3), degrees
$\overline{h_o}$ = expected value of surface-retained precipitation (Eq. 6.52), mm
$\hbar$ = Euler's constant = 0.577 . . .
I = insolation, cal cm^{-2} min^{-1} or MJ m^{-2} h^{-1} or W m^{-2}
I_b = beam insolation (Eq. 3.7), cal cm^{-2} min^{-1} or MJ m^{-2} h^{-1} or W m^{-2}
I_d = diffuse insolation, cal cm^{-2} min^{-1} or MJ m^{-2} h^{-1} or W m^{-2}
I_{d0} = diffuse insolation at canopy top (Eq. 3.8), cal cm^{-2} min^{-1} or MJ m^{-2} h^{-1} or W m^{-2}
I_j = interstorm infiltration, mm
I_k = compensation insolation, cal cm^{-2} min^{-1} or MJ m^{-2} h^{-1} or W m^{-2}
$I_k^o \equiv I_k/I_0$ = normalized compensation insolation (Eq. 7.3), dimensionless
I_l = leaf insolation, cal cm^{-2} min^{-1} or MJ m^{-2} h^{-1} or W m^{-2}
I_{lc} = leaf insolation at which stomates become effectively fully open, cal cm^{-2} min^{-1} or MJ m^{-2} h^{-1} or W m^{-2}
I_{lw} = outgoing or "back" longwave radiation (Eq. B.8), cal cm^{-2} min^{-1} or MJ m^{-2} h^{-1} or W m^{-2}
I'_{lw} = longwave incoming insolation from a clear sky (Eqs. B.5, B.6), cal cm^{-2} min^{-1} or MJ m^{-2} h^{-1} or W m^{-2}
I^*_{lw} = net incoming longwave insolation (Eq. B.7), cal cm^{-2} min^{-1} or MJ m^{-2} h^{-1} or W m^{-2}
$\widehat{I_l}$ = crown-average insolation (Eq. 8.29), cal cm^{-2} min^{-1} or MJ m^{-2} h^{-1} or W m^{-2}
I_0 = actual climatic insolation at screen height, cal cm^{-2} min^{-1} or MJ m^{-2} h^{-1} or W m^{-2}
I_{sj} = storm infiltration, mm
I_{SL} = saturation or carbon-critical insolation, i.e., the insolation at intersection of the asymptotes of the photosynthetic capacity curve and thus the minimum insolation for fully-open stomates (water non-limiting) (Eq. 8.28), cal cm^{-2} min^{-1} or MJ m^{-2} h^{-1} or W m^{-2}
$\widehat{I_{SL}}$ = canopy average (i.e., "effective") saturation insolation (Eq. 8.28), cal cm^{-2} min^{-1} or MJ m^{-2} h^{-1} or W m^{-2}

I^o_{SL} = relative saturation insolation, dimensionless

I_{SW} = desiccation or water-critical insolation, i.e., the insolation at intersection of the asymptotes of the leaf transpiration-insolation curve and thus the maximum insolation for fully open stomates (light non-limiting) (Eq. 8.26), cal cm^{-2} min^{-1} or MJ m^{-2} h^{-1} or W m^{-2}

$\widehat{I_{SW}}$ = canopy average (i.e., "effective") leaf desiccation insolation (Eq. 8.32), cal cm^{-2} min^{-1} or MJ m^{-2} h^{-1} or W m^{-2}

$I^o \equiv I/I_0$, dimensionless

I_λ = insolation in wave length interval around λ (Eq. 3.6), cal cm^{-2} min^{-1} or MJ m^{-2} h^{-1} or W m^{-2}

i = storm intensity, mm h^{-1}

j = storm counting variable

K = turbulent transfer coefficient, cm^2 s^{-1}

$K(1)$ = effective saturated hydraulic conductivity of soil, cm s^{-1}

K_c = carbon dioxide eddy diffusivity, cm^2 s^{-1}

K_h = thermal eddy diffusivity, cm^2 s^{-1}

K_m = kinematic eddy viscosity (eddy momentum diffusivity) (Eq. 4.37), cm^2 s^{-1}

K_v = water vapor eddy diffusivity, cm^2 s^{-1}

$K_m(0)$ = kinematic eddy viscosity at canopy top, cm^2 s^{-1}

$K^o_m = K_m(\xi)/K_m(0)$, dimensionless

$\widehat{K^o_m} \equiv \widehat{K_m}(\xi)/K_m(0)$ = canopy-average eddy viscosity (Eq. 4.105), dimensionless

$(\widehat{K^o})_d$ = canopy-average eddy viscosity for dense monolayer crown, dimensionless

$(\widehat{K^o})_s$ = canopy-average eddy viscosity for sparse monolayer crown, dimensionless

k = von Kármán's constant = 0.40

k = decay parameter, dimensionless

k = empirical photosynthetic "binding" constant = I_{SL} (Eqs. 8.4, 8.5), cal cm^{-2} min^{-1} or MJ m^{-2} h^{-1} or W m^{-2}

$k^o \equiv k/I_o = I_{SL}/I_o \equiv I^o_{SL}$ = relative saturation radiance, dimensionless

k_c = proportionality coefficient for crown depth (Eq. 10.26), dimensionless

k_L = sum of geometrically scaled potential-loss coefficients (Eq. 8.13), s

k_l = empirical reduction factor (Eq. B.10), dimensionless

k_v = canopy conductance (Eq. 5.29), dimensionless

$k^o_v \equiv k_v/k^*_v$ = normalized canopy conductance (Eq. 8.12), dimensionless

$k^*_v \equiv E_v/E_{ps}$ = potential canopy conductance (Eqs. 5.30, 7.31), dimensionless

k^*_{vm} = monolayer conductance, dimensionless

$k^*_{v1} = k^*_v$ at $M = 1$, dimensionless

L = leaf area index ≡ one-sided leaf area per unit of basal area, dimensionless

L' = area index of foliage elements other than the leaves (Table 2.1), dimensionless

L_t = foliage area index = upper-sided area of all foliage elements per unit of basal area, dimensionless

L'_t = equivalent reduced foliage area index for scatterless radiation transmission (Eq. 3.41), approximated by L to relate diffusion and assimilation, dimensionless

l = mixing length, m

l_K = von Kármán's mixing length (Eq. 4.46), m

l_P = Prandtl's mixing length (Eq. 4.43), m
M = canopy cover (i.e., fraction of total ground surface covered by vegetation), dimensionless
M_{max} = maximum canopy cover for conceptual canopy of packed circular discs, dimensionless
M_o = canopy cover at which the soil moisture is maximum, dimensionless
MJ(tot.) = energy at all frequencies in the radiational spectrum, MJ
m = exponent relating shear stress on foliage to horizontal wind velocity and having the nominal value 0.5 for the foliage elements of trees (Eqs. 4.19, 4.20), dimensionless
m = soil pore size distribution index, dimensionless
m_d = mean length of the dry season, days
m_h = mean of storm depth (Eqs. 6.3, 6.21), cm
m_i = mean of storm intensity (Eq. 6.12), mm h^{-1}
$m_P \equiv P_A$ = mean annual or seasonal precipitation (Eq. 6.14), cm
m_{t_b} = mean time between storms, h
$m_{t_b'}$ = mean interstorm time available for transpiration (Eq. 6.63), h
$m_{t_b''}$ = mean interstorm time available for bare soil evaporation (Eqs. 6.8, 6.64), h
m_{t_r} = mean storm duration (Eq. 6.5), h
m_ν = number of independent storms per month or season, $time^{-1}$
m_τ = mean length of the growing season, daylight hours
N = fraction of sky covered by clouds, dimensionless
N_s = rate of nitrogen supply to plant (Eq. 10.39), $g(N_2)\ m^{-2}\ y^{-1}$
$N(L_t)$ = leaf area function (Eqs. 7.9, A.101), dimensionless
NPP = net primary productivity of canopy (Eq. 10.1), $g_s\ m^{-2}\ y^{-1}$
n = counting variable, dimensionless
n = number of sides of each foliage element producing surface resistance to wind and having the nominal value 2 for the foliage elements of trees
n_e = effective soil porosity, dimensionless
n_l = number of leaves in a crown
n_s = number of stomates per leaf
n_s = stomated leaf area / projected leaf area, dimensionless
n_s = rate of nitrogen supply to diffuse monolayer (Eq. 10.40), $g(N_2)\ m^{-2}\ y^{-1}$
n_s = number of parallel diffusion paths from leaf surface to atmosphere (Eq. 7.10), dimensionless
O(...) = order (...)
P = annual or seasonal precipitation, cm
P = assimilation rate, net photosynthetic rate, productivity, $g_s\ m^{-2}\ y^{-1}$ or $g(CO_2)\ m^{-2}\ h^{-1}$
P = point in space
P = needle perimeter, m
P_A = average annual or seasonal precipitation (Eq. 6.14), cm
P_D = rate of above-ground carbon demand (Eq. 10.5), $g_s\ h^{-1}$ / unit of one-sided leaf area, $g(CO_2)\ m^{-2}\ h^{-1}$

$\widehat{P_D}$ = canopy average above-ground assimilation capacity (Eq. 10.10), $g_s\ h^{-1}$ / unit of one-sided leaf area, $g(CO_2)\ m^{-2}\ h^{-1}$

P_L = actual biological productivity, $g(CO_2)\ m^{-2}\ h^{-1}$

P_r = rate of respiration (Eq. H.2), $g(CO_2)\ m^{-2}\ h^{-1}$

P_s = light-saturated rate of productivity (equal to P_{sm} when $T_l = T_m$) (Eq. E.9), $g_s\ m^{-2}\ y^{-1}$

$\widehat{P_T}$ = annual total above-ground carbon demand (Eq. 10.18), $g(CO_2)\ m^{-2}\ y^{-1}$

P_t = total rate of photosynthesis (Eqs. 8.2, H.1), $g(CO_2)\ m^{-2}\ h^{-1}$

P_W = climatic potential productivity as limited by available water, $g(CO_2)\ m^{-2}\ h^{-1}$

P_{sm} = saturation rate of CO_2 assimilation per unit of (one-sided) leaf area when $T_l = T_m$ (Eq. 8.1), $g(CO_2)\ m^{-2}\ h^{-1}$ or $g_s\ m^{-2}\ h^{-1}$, as noted

P_{wm} = desiccation rate of CO_2 assimilation as limited only by available water, $g(CO_2)\ m^{-2}\ h^{-1}$

$P^o = P/P_{sm}$ = normalized net photosynthesis (Eqs. 8.12, 8.15), dimensionless

p = atmospheric (total) pressure, Pa or millibars (mb)

p = net primary productivity of diffuse monolayer with $L_t = 1$, $g_s\ m^{-2}\ y^{-1}$

p_D = productive demand of a unit basal area of the groundcover monolayer, $\beta = 1$, $L_t = 1$ (Eq. 10.19), $g_s\ m^{-2}\ y^{-1}$

p_S = annual total above-ground supply of carbon downward from the atmosphere to a unit basal area of the diffuse monolayer, $L_t = 1$ (Eq. 10.28), $g_s\ m^{-2}\ y^{-1}$

Q = vertical flux density of CO_2 (Eqs. 4.116, 10.23), $g\ cm^{-2}\ s^{-1}$

$\widehat{Q_S}$ = annual total above-ground supply of carbon downward to the canopy, $g_s\ m^{-2}\ y^{-1}$

Q_v = vertical flux density of latent heat from canopy to atmosphere (Eq. 4.129), $cal\ cm^{-2}\ min^{-1}$

q_b = net flux of longwave back radiation at land surface (Eqs. B.10 *et seq.*), $cal\ cm^{-2}\ min^{-1}$

q_i = net flux of shortwave radiation at land surface, $cal\ cm^{-2}\ min^{-1}$

$q_b^* \equiv q_b\left[1 - \left(\frac{1-S_r}{0.85}\right)\right]$, $cal\ cm^{-2}\ min^{-1}$

q_v = specific humidity = mass of water vapor per unit mass of moist air (Eq. 4.122), $g(H_2O)\ g^{-1}$, (O_2)

R = radiant flux density or radiance, $cal\ cm^{-2}\ min^{-1}$

R = radius, m

R = respiration, $g(CO_2)\ m^{-2}\ y^{-1}$

R_a = absorbed radiance, $cal\ cm^{-2}\ min^{-1}$

R_b = beam radiance, $cal\ cm^{-2}\ min^{-1}$

R_{bo} = incident beam radiance at top of canopy, $cal\ cm^{-2}\ min^{-1}$

R_c = net longwave back radiation under clear sky (Eq. B.9), $cal\ cm^{-2}\ min^{-1}$ or $MJ\ m^{-2}\ h^{-1}$ or $W\ m^{-2}$

R_d = diffuse radiance, $cal\ cm^{-2}\ min^{-1}$

R_{gA} = average annual groundwater runoff, cm

R_i = incident (beam) radiance, $cal\ cm^{-2}\ min^{-1}$

R_n = net solar radiation (Eqs. 5.3, 5.4), $cal\ cm^{-2}\ min^{-1}$

R_r = reflected radiance, $cal\ cm^{-2}\ min^{-1}$

R_{sA} = average annual surface runoff, cm

R_{sj} = storm runoff, mm

R^*_{sj} = storm rainfall excess, mm
R_t = transmitted radiance, cal cm^{-2} min^{-1}
R_λ = radiance in wavelength interval around λ, cal cm^{-2} min^{-1}
$\boldsymbol{R}$ = Reynolds number, dimensionless
$\boldsymbol{R}_b$ = Bowen ratio $\equiv H/Q_v$ (Eqs. 4.131, 5.5, 5.40), dimensionless
$\boldsymbol{R}_c$ = Reynolds number of body with drag dimension c, dimensionless
$\boldsymbol{R}_d$ = Reynolds number of body with drag dimension d, dimensionless
$\boldsymbol{R}_D$ = Reynolds number of body with drag dimension D, dimensionless
$\boldsymbol{R}_i$ = Richardson number (Eq. 4.120), dimensionless
$\boldsymbol{R}_x$ = surface resistance Reynolds number with drag dimension x, dimensionless
r = radius of cone at elevation ξ, m
r_a = lumped atmospheric resistance over the 2 m above the canopy top (Eq. 4.143), s mm^{-1}
r_c = lumped resistance to flow through the canopy (Eqs. 4.156, 7.6, 7.11), s mm^{-1}
r_i = interleaf layer resistance (Eq. 4.165), s mm^{-1}
r_o = radius, m
r_o = equivalent stomatal cavity resistance (Eq. 4.144), s mm^{-1}
r_{cH} = equivalent interleaf resistance to heat flux, s mm^{-1}
r_{ci} = equivalent interleaf resistance to mass flux, s mm^{-1}
r_{cs} = equivalent canopy stomatal resistance, s mm^{-1}
r_{cu} = cuticular diffusion resistance of leaf, s mm^{-1}
r_{ic} = intercellular resistance within the stomatal cavity, s mm^{-1}
$r_{i,n}$ = nth interleaf layer resistance, s mm^{-1}
r_{la} = leaf boundary layer resistance, s mm^{-1}
r_{ls} = equivalent leaf stomatal resistance (Eq. 4.147), s mm^{-1}
r_{lls} = equivalent leaf layer stomatal resistance (Eq. 4.162), s mm^{-1}
$r_l = r_{ls} + r_{la}$ = equivalent leaf resistance (Eq. 4.148), s mm^{-1}
r_r = residual resistance of cell walls lining stomatal cavity, s mm^{-1}
r_s = stomatal resistance regulating plant water loss and CO_2 assimilation, s mm^{-1}
$(r_c/r_a)_d$ = resistance ratio for dense monolayer crown, dimensionless
$(r_c/r_a)_s$ = resistance ratio for sparse monolayer crown, dimensionless
$\boldsymbol{r}$ = directional vector, m
$\boldsymbol{r}_L$ = radius vector to leaf, m
$\widehat{\boldsymbol{r}_L\boldsymbol{r}}$ = angle between the vectors $\boldsymbol{r}_L$ and $\boldsymbol{r}$, radians
S = rate of heat storage within canopy, cal cm^{-2} min^{-1}
S_e = exfiltration sorptivity (Eq. 6.33), mm h$^{-1/2}$
S_i = infiltration sorptivity (Eq. 6.18), mm h$^{-1/2}$
S_r = atmospheric saturation ratio (i.e., fractional relative humidity) at screen height, dimensionless
S_W = annual moisture supply to the root zone (Eq. C.12), dimensionless
S_z = average net upward transport of s per unit horizontal area per unit time (Eqs. 4.1, 4.2), quantity cm^{-2} s^{-1}
S_{oo} = solar constant = 1.96 cal cm^{-2} min^{-1} = 1367 W m^{-2}

S'_{oo} = solar irradiance on horizontal plane at Earth's surface (Eq. 3.2), W m^{-2}
s = instantaneous amount of some fluid quantity (e.g., momentum, heat, or solute mass), quantity g^{-1}
s = soil moisture concentration ($s = 1$ at saturation), dimensionless
s_c = critical soil moisture concentration, dimensionless
s_o = space–time average soil moisture concentration in the root zone, dimensionless
s_{oc} = critical space–time average root zone soil moisture concentration (Eq. 6.58), dimensionless
s_5 = soil moisture concentration at $\psi_s = \psi_{sc} = 5$ bar, dimensionless
s_* = soil moisture for "quadratic soil" (Eq. C.21), dimensionless
$s(0)$ = soil moisture concentration (its maximum value) at end of average storm, dimensionless
$s(\xi)$ = interference (or shielding) function (Eq. 4.11), dimensionless
$\bar{s}$ = temporal mean of fluid quantity, s
s' = turbulent temporal fluctuation of fluid quantity, s
T = temperature, °C
T_l = leaf temperature, °C
$\widehat{T_l}$ = crown-average leaf temperature, °C
T_m = "optimal" leaf temperature = temperature at which P_{sm} is maximum for a given species, °C
T_0 = climatic atmospheric temperature at screen height above canopy, °C
T_{0K} = climatic atmospheric temperature at screen height above canopy, K
T_s = surface temperature, °C
T_{sK} = surface temperature, K
t = time, s
t = thickness of foliage layer, mm
t_b = time between storms, h
t'_b = interstorm time available for transpiration, h
t''_b = interstorm time available for bare soil evaporation, h
t_e = time between exhaustion of surface retention and cessastion of soil moisture evaporation, h
t_j = Julian day, dimensionless
t_o = time after exhaustion of surface retention ($t = t^*$) at which $f_e^* = E_{ps}$, h
t_r = storm duration, h
t_s = time to stress (Eq. C.6), h
t'_s = time during which the vegetation transpires (Eq. 6.43), h
t^* = time at which evaporation exhausts surface retention, h
u = horizontal wind velocity, m s^{-1}
u = integration variable (Eq. 10.11), dimensionless
u_0 = horizontal wind velocity at canopy top, m s^{-1}
$\bar{u}$ = average wind velocity, m s^{-1}
u_* = shear velocity $\equiv \sqrt{\tau_0/\rho}$ (Eq. 4.27), m s^{-1}
V_e = volume of soil moisture (per unit of surface area) available for exchange with atmosphere during average interstorm period (Eqs. 6.41, 6.59, 6.67, 6.68, 8.27), mm

$\forall$ = volume of the crown-enclosing right circular cylinder, m^3
$\forall_{co}$ = volume of conical crown (Eqs. A.14, A.75), m^3
$\forall_{cm}$ = foliage volume of conical monolayer (Eq. A.88), m^3
$\forall_{cr}$ = crown volume, m^3
$\forall_s$ = volume of spherical segment of height, h_c (Eqs. A.1, A.58, Fig. A.3), m^3
$\forall_{so}$ = volume of hemisphere (Eq. A.60), m^3
v = rate of percolation to water table during wet season (Eq. 6.50), mm h^{-1}
w = rate of capillary rise from water table to surface year round (Eq. 6.51), mm h^{-1}
w = vertical component of fluid velocity, m s^{-1}
w' = turbulent temporal fluctuation of vertical fluid velocity, m s^{-1}
x = horizontal coordinate distance, m
Y = long-term average water yield (Eq. 6.56), mm
y = horizontal coordinate distance, m
Z = diffusion penetration depth = 0.885 (Eq. C.28), dimensionless
z = vertical coordinate distance (usually the elevation above ground surface), m
z_e = average penetration depth of exfiltration (Eq. 6.36), m
z_g = elevation of roots, m
z_i = penetration depth of the soil wetting process (Eq. 6.60), m
z_l = elevation of leaf, m
z_0 = surface roughness length, m
z_r = rooting depth (Eq. 6.60), m
z_w = depth to water table, m

α = slope of C_D vs. $\boldsymbol{R}_D$ relationship in given $\boldsymbol{R}_D$ range (Eq. 4.10), dimensionless
$\alpha = 1/m_{t_b}$, h^{-1}
$\boldsymbol{\alpha}$ = Priestley–Taylor coefficient of potential evaporation (Eq. 5.36), dimensionless
α_T = coefficient of radiation absorption, dimensionless
β = momentum extinction coefficient = cosine of angle leaf surface makes with horizontal, dimensionless
β_s = bare soil evaporation efficiency (Eqs. 6.40, C.5, Fig. 6.8), dimensionless
β_v = canopy transpiration efficiency (Eqs. 6.46, 6.47, 8.17, Fig. 8.14), dimensionless
$\hat{\beta}$ = crown-average cosine of the leaf inclination angle (Eqs. 2.18, 2.21, 2.22)
$\widehat{\beta L_t}$ = critical canopy absorption index, separates regions of productivity control by carbon supply or carbon demand (Eq. 10.32), dimensionless
$\Gamma(\cdot)$ = factorial gamma function (Eq. 6.25, Table 6.1), dimensionless
$\gamma \equiv mn\beta$ = extinction parameter for horizontal wind velocity (Eq. 3.47), dimensionless
γ_o = base angle of cone having height, h_c, and base radius, r_o, degrees
γ_o = surface psychrometric constant (Eq. 4.132), Pa K^{-1}
γ_o^* = canopy psychrometric constant (Eq. 5.27), Pa K^{-1}
$\gamma[\ldots,\ldots]$ or $\gamma(\ldots,\ldots)$ = incomplete gamma function
$\gamma_*(\kappa_o, \lambda_o)$ = gamma function probability density function of storm depth (Eq. 6.2), cm^{-1}
Δ = angle between Sun direction and leaf normal, degrees

Δ = slope, $\delta e/\delta T$, of saturation vapor pressure vs. temperature curve (Eq. 4.134), Pa K^{-1}

$\Delta(\ )$ = finite increment of ( )

ΔS = average carryover (from dormant season to growing season) soil moisture storage (Eq. 6.69), mm

$\overline{\Delta z}$ = lumped vertical intercrown diffusion distance for canopy of conical crowns, m

Δz_L = thickness of interleaf atmospheric layer, m

δ = solar declination, degrees

$\delta = 1/m_{t_r}$, h^{-1}

$\delta_f \equiv \frac{d}{t} L_t$ = monolayer foliage volume density (Eq. 7.22), dimensionless

$\varepsilon = P_{sm}/I_{SL}$ = "dry matter–radiation quotient" (Eq. 8.9), an empirical constant with the approximate value $\langle\varepsilon\rangle \approx 0.81\ g_s\ MJ(tot.)^{-1}$

ε = infinitesimal radius, mm

ε = surface absolute roughness, mm

ε = characteristic of hemispherical segment, dimensionless

ε_o = longwave emissivity of the surface, dimensionless

$\varepsilon(z)$ = residual of exponential integral

$\zeta(\sigma)$ = surface runoff function (Eq. 6.27, Fig. 6.5), dimensionless

$\eta_o \equiv \frac{\text{stomated (i.e., assimilating) leaf area}}{\text{illuminated (i.e., projected) leaf area}} \approx \frac{\text{perimeter}}{\text{arc—chord}}$ (Eq. 8.8), dimensionless

Θ = normalized evapotranspiration (Eq. 6.57), dimensionless

θ = angle between the ray, $\boldsymbol{r}$, and an upward normal, $\boldsymbol{n}$, to an arbitrarily oriented surface

θ_L = leaf angle (i.e., the angle between the leaf surface and the horizontal), degrees

θ_o = particular leaf inclination, degrees

θ_{os} = supplement of a particular azimuthal angle, degrees

θ_z = Sun zenith angle = complement of $h_\otimes$ (cf. Fig. 3.2b), degrees

ϑ = empirical conversion factor $\equiv \frac{\text{g(solid matter)}}{\text{g(}CO_2\text{assimilated)}} \approx 0.5$ (Eqs. 8.7, 10.3)

κ = insolation extinction coefficient, dimensionless

$\kappa_b^{(1)}$ = extinction coefficient of an isolated leaf with no scattering = one-dimensional shadow of a solitary leaf (Eq. 3.20), dimensionless

$\kappa_b^{(2)}$ = extinction coefficient for a single layer of leaves with idealized scattering, dimensionless

κ_b = decay or extinction coefficient for direct radiation, dimensionless

κ_d = decay or extinction coefficient for diffuse radiation, dimensionless

κ_o = shape parameter or distribution index of storm depth, dimensionless

κ^* = scatterless extinction coefficient for total clear sky radiation, dimensionless

λ = wavelength of radiation, nm

λ = roughness density or frontal area index (Eq. 4.106), dimensionless

λ = latent heat of vaporization of water = 597.3 cal g^{-1} (at 0 °C)

λ_0 = scale parameter of probability density function of storm depth, cm^{-1}

λ_{NIR} = 700 to 4000 nm

λ_{PAR} = 400 to 700 nm

λ_{UV} = less than 400 nm

μ = beam ratio = direct radiance fraction of total radiance, dimensionless
ν = kinematic fluid viscosity, m^2 s^{-1}
ξ = vertical distance down from the crown top $\equiv \frac{h-z}{h-h_s}$ (Eqs. 2.1, A.15), dimensionless
$\Pi_L \equiv I_{SL}/f_I(\beta L_t)I_o$ = bioclimatic light parameter (Eq. 8.41), dimensionless
$\Pi_W \equiv \widehat{I_{SL}}/\widehat{I_{SW}}$ = bioclimatic water parameter (Eq. 8.42), dimensionless
π = 3.14159.....
ρ = fluid mass density, g cm^{-3}
ρ_l = longwave albedo of surface, dimensionless
ρ_{NIR} = reflection coefficient of NIR, dimensionless
ρ_o = Earth–Sun distance = 1 AU = 1.496 × 10^8 km
ρ_{PAR} = reflection coefficient of PAR, dimensionless
ρ_T = spectral reflectance (i.e., albedo) of surface or canopy, dimensionless
$\rho[i, t_r]$ = rainfall intensity–duration correlation
σ = Stefan–Boltzmann constant = 0.826 × 10^{-10} cal cm^{-2} min^{-1} K^{-4}
σ = capillary infiltration parameter (Eq. 6.24), dimensionless
σ_h = standard deviation of storm depth (Eq. 6.4), cm
σ_P = standard deviation of mean annual or seasonal precipitation (Eq. 6.15), cm
σ_{t_b} = standard deviation of time between storms (Eq. 6.10), h
σ_{t_r} = standard deviation of storm duration (Eq. 6.7), h
τ = solar hour angle, degrees
τ = time constant, same units as t
τ = growing season length, days
τ = momentum flux per unit horizontal area (i.e., shear stress) (Eq. 4.114), N m^{-2}
τ_0 = horizontal shear stress at canopy top (Eq. 4.36), N m^{-2}
τ_f = horizontal shear stress on the foliage, N m^{-2}
τ_s = horizontal shear stress on substrate, N m^{-2}
τ_T = coefficient of radiation transmission, dimensionless
$\tau_{zx} = \tau_f + \tau_s$, N m^{-2}
Φ = geographical latitude, degrees
ϕ = azimuthal angle, radians
ϕ_e = exfiltration or desorption diffusivity (Eq. 6.35, Fig. 6.7), dimensionless
ϕ_i = sorption diffusivity (Eq. 6.19, Fig. 6.3), dimensionless
ϕ_L = azimuthal angle to leaf, radians
ϕ_o = particular azimuthal angle (Eq. G.16), radians
ϕ_{os} = supplement of ϕ_o (Eq. G.17), radians
$\varphi = \tan^{-1}\gamma_o$ = angle of inclination of adiabatic line in e–T space (Eq. 5.7), tan(φ) in mb °C^{-1}
Ψ = proportionality constant in drag partition (Eq. 4.111), dimensionless
$\psi(1)$ = saturated matrix potential of soil (suction), cm
$\psi_l(s)$ = leaf moisture potential, cm
$\psi_{lc}(s)$ = critical leaf moisture potential, cm
$\psi_s(s)$ = soil moisture potential (Eqs. 6.30, 8.10), cm
$\psi_{sc}(s)$ = critical soil moisture potential, cm
Ω = evaporation decoupling factor, dimensionless

ω = central angle, radians
$\omega = 1/\sigma_i$, h mm^{-1}
ω_s = scattering coefficient, dimensionless
$\ldots_d$ = ... for the dry season
$\ldots_d$ = ... for a dense monolayer
$\ldots_s$ = ... for a sparse monolayer
$\ldots_j$ = ... for the jth storm
$\ldots_\tau$ = quantity, ..., averaged over the growing season τ
$(\widehat{\ldots})$ = average over the crown depth
$\langle \ldots \rangle$ = average over the sample
$\overline{(\ldots)}$ = time average

Units

IRRADIATION

(I.e., quantity of energy arriving at a surface during a given time interval)
(From Iqbal, M., *An Introduction to Solar Radiation*, Table B.1, p. 378, Copyright © 1983 Academic Press, with kind permission from Academic Press)

Units	$J\ m^{-2}$	$W\ h\ m^{-2}$	$cal\ cm^{-2}$	$Btu\ ft^{-2}$
1 $J\ m^{-2}$	1	2.778×10^{-4}	2.39×10^{-5}	8.81×10^{-5}
1 $W\ h\ m^{-2}$	3.60×10^{3}	1	0.0860	0.317
1 $cal\ cm^{-2}$	4.187×10^{4}	11.63	1	3.69
1 $Btu\ ft^{-2}$	1.136×10^{4}	3.155	0.271	1

IRRADIANCE

(I.e., rate of energy arriving at and normal to a surface)
(After Iqbal, 1983, Table B.2)

Units	$J\ m^{-2}\ h^{-1}$	$W\ m^{-2}$	$cal\ cm^{-2}\ min^{-1}$	$Btu\ ft^{-2}\ h^{-1}$
1 $J\ m^{-2}\ h^{-1}$	1	2.778×10^{-4}	3.983×10^{-7}	8.81×10^{-5}
1 $W\ m^{-2}$	3.600×10^{3}	1	1.433×10^{-3}	0.317
1 $cal\ cm^{-2}\ min^{-1}$	2.511×10^{6}	698	1	221.2
1 $Btu\ ft^{-2}\ h^{-1}$	1.135×10^{4}	3.155	4.521×10^{-3}	1
1 ft c	7.2×10^{-11}	2.0×10^{-14}	2.87×10^{-17}	6.34×10^{-15}

MISCELLANEOUS

1 Ångstrom = 10^{-10} meters
1 calorie = 4.187 Joule

1 μmol CO_2 m^{-2} s^{-1} = 0.044 mg CO_2 m^{-2} s^{-1}[†]
"full sunlight" = 10 000 ft c = 108 kilolux ("klx")[‡]
100 kilolux $\approx$ 1.5 cal cm^{-2} min^{-1} = 3.77 MJ m^{-2} h^{-1}
1 lux = 0.0929 ft c
1 E ("Einstein") = 2.39×10^5 J (at $\lambda = 500$ nm)[§]
1 mole of (quantum) radiation = 2.85×10^5 J (at $\lambda = 500$ nm)[††]
100 W m^{-2} = 4.18×10^{-4} E m^{-2} s^{-1} (at $\lambda = 500$ nm)[§]
1 dm^2 = 1 square decimeter of (single-sided) leaf area
1 metric ton per hectare = 10^{-5} grams per square centimeter

† Larcher (1983, p. XVII).
‡ Hicks and Chabot (1985, p. 262).
§ Gates (1980, p. 84).
†† Gates (1980, p. 85).

Common and scientific names

Abies Mill. spp.	fir
Abies balsamea (L.) Mill.	balsam fir
Abies homolepsis	Nikko fir
Abies lasiocarpa (Hook.) Nutt.	subalpine fir
Acer L. spp.	maple
Acer mono	painted maple
Acer negundo L.	box elder
Acer platanoides L.	Norway maple
Acer pseudoplatanus L.	sycamore maple
Acer rubrum L.	red maple
Acer saccharinum L.	silver maple
Acer saccharum Marsh.	sugar maple
Aesculus glabra Willd.	Ohio buckeye
Alnus B. Ehrh. spp.	alder
Alnus glutinosa (L.) Gaertn.	European black alder, black alder, European alder
Beta vulgaris	sugar beet
Betula L. spp.	birch
Betula alba L. (*B. pubescens* Ehrh.)	birch, European white
Betula papyrifera Marsh.	paper birch, white birch, canoe birch
Betula pendula Roth. (*B. verrucosa* J.F. Ehrh.)	silver birch, European silver birch
Betula verrucosa Roth. (see *B. pendula*)	European white birch

Brassica napus L.	rape
Cajanus cajan	pigeon pea
Carpinus betulus	European hornbeam
Cedris libani	cedar of Lebanon
Celtis occidentalis L.	common hackberry
Chamaecyparis thyoides	Northern white cedar
Coleogyne ramosissima	blackbrush
Eucalyptus L'Hér. spp.	eucalyptus
Fagus L. spp.	beech
Fagus grandifolia J.F. Ehrh.	American beech
Fagus orientalis	Oriental beech
Fagus sylvatica L.	European beech
Festuca	fescue
Ficus macrophylla	Moreton Bay fig
Ficus tetrameles nudiflora	spong
Fraxinus excelsior L.	European ash
Fraxinus pennsylvanica Marsh.	green ash, red ash
Glycine max (L.) Merr.	soybean
Goethalsia meiantha	rainforest species
Gossypium L. spp.	cotton
Helianthus L. spp.	sunflower
Hibiscus rosa sinensis	Chinese hibiscus
Hordeum	barley
Juglans cinera	butternut
Juglans regia L.	common walnut, English walnut
Juniperus L. spp.	juniper, red cedar
Juniperus virginiana	Eastern red cedar
Larix Mill. spp.	larch, tamarack
Larix leptolepis	Japanese larch
Larrea divaricata	creosote bush
Larrea tridentata	creosote bush
Lactuca sativa	lettuce
Liquidambar styraciflua L.	sweet gum, red gum
Liriodendron tulipifera L.	yellow poplar, tulip poplar, tulip tree
Magnolia L. spp.	magnolia
Malus domestica Borkh.	apple
Medicago sativa L.	alfalfa (lucerne)
Metasequoia glypostroboides	dawn redwood
Morus bombycis, Moraceae 'UNRYU'	mulberry cultivar
Nyssa sylvatica, Nyssaceae	tupelo or sourgum
Oryza L. spp.	rice
Palmaceae	palm
Picea abies	Norway spruce
Picea engelmannii Parry ex Engelm.	Engelmann spruce

Picea exelsa	Norway spruce
Picea glauca (Moench) Voss	white spruce
Picea mariana (Mill.) B.S.P.	black spruce
Picea pungens Engelm.	blue spruce
Picea pungens 'Glauca'	Colorado spruce
Picea rubens Sarg.	red spruce
Picea sitchensis (Bong.) Carr.	Sitka spruce
Pinus L. spp.	pine
Pinus banksiana Lamb.	jack pine
Pinus cembra L.	Swiss stone pine
Pinus contorta Dougl. ex Loud.	lodgepole pine
Pinus densiflora Sieb. & Zucc.	Japanese red pine
Pinus edulis Engelm.	Colorado pinyon pine
Pinus jeffreyi	Jeffrey pine
Pinus lambertiana Dougl.	sugar pine
Pinus nigra	Cevennes pine
Pinus nigra Arnold	Austrian pine
Pinus nigra ssp. Salzmanii	European black pine
Pinus palustris Mill.	longleaf pine
Pinus ponderosa Dougl. ex P. Laws. & C. Laws.	ponderosa pine, Western yellow pine
Pinus radiata D. Don	Monterey pine, radiata pine
Pinus resinosa Ait.	red pine
Pinus rigida Mill.	pitch pine
Pinus serotina	pond pine
Pinus strobus L.	Eastern white pine, white pine
Pinus sylvestris L.	Scots pine, Scotch pine
Pinus taeda L.	loblolly pine
Pinus virginiana	scrub pine
Pinus wallichiana	Himalayan pine
Pisum sativum	pea
Platanus × *acerifolia*	London plane tree
Polyalthia longifolia (Sonnerat) Thwaites	Asoka tree
Populus L. spp.	aspen, cottonwood, poplar
Populus alba var. *pyramidalis*	Bolleana poplar
Populus deltoides Bartr. ex Marsh.	Eastern cottonwood
Populus tremula L.	European aspen
Populus tremuloides Michx.	trembling aspen, quaking aspen
Prunus L. spp.	plum, prune, apricot, cherry, almond, peach
Prunus armeniaca L.	apricot
Prunus persica (L.) Batsch.	peach, nectarine
Prunus serotina J.F. Ehrh.	black cherry
Pseudolarix kaempferi	golden larch

Pseudotsuga menziesii	Douglas fir
Pyrus L. spp.	pear
Quercus L. spp.	oak
Quercus alba L.	white oak
Quercus borealis (*Q. rubra*)	red oak
Quercus coccinea	scarlet oak
Quercus ellipsoidalis E.J. Hill	Northern pin oak
Quercus macrocarpa Michx.	bur oak
Quercus palustris Muenchh.	pin oak
Quercus rubra L. [*Q. borealis* var. maxima (Marsh.) Ashe]	Northern red oak
Quercus suber L.	cork oak
Robinia L. spp.	locust
Roystonea regia	royal palm
Salix L.	willow
Salix alba 'Tristas'	golden weeping willow
Salix amygdaloides Anderss.	peachleaf willow
Salix babylonica	weeping willow
Secale cereale	rye
Semanae semane	umbrella tree
Sequoiadendron giganteum	giant sequoia
Solanum tuberosum L.	potato
Sorbus L.	mountain ash
Sorghum vulgare	sorghum
Tilia cordata Mill.	littleleaf linden
Trifolium	clover
Trifolium pratense L.	red clover
Triticum L. spp.	wheat
Tsuga canadensis (L.) Carr.	Eastern hemlock, Canadian hemlock
Ulmus americana L.	American elm
Ulmus fulva	slippery elm
Vicia faba	horse beans
Yucca angustissima	narrowleaf yucca
Zea L. spp.	corn, maize

1

Introduction and overview

This research monograph describes a search for the mechanisms by which natural selection influences vegetation form and function.

Here we summarize the work by describing, in non-mathematical terms, the discovered suboptimalities of the factors of net primary productivity, and how through increasing horizontal leaf area they lead to a productive gain which maximizes when the foliage density constrains the CO_2 supply to equal the light-stimulated CO_2 demand. We find a global maximum productivity over a range of horizontal leaf area that closely matches the range for observed species, thereby confirming our fundamental hypothesis that natural selection favors increased productivity.

Discovered dimensionless bioclimatic stability and optimality conditions are shown to yield the climax boundaries to the feasible habitat space.

A Introduction

The biophysical system

The physical interaction between a vegetation canopy and its atmospheric and soil environments is governed by both the plant structure and the biochemistry of the individual plants. The spacing of the individual plants; their height and diameter; the depth and shape of their crowns and root systems; the size, shape, number, color, texture and spatial arrangement of their leaves along with the associated pods, stems, twigs or shoots and branches, all contribute to the instantaneous vertical exchanges of momentum, mass and energy between canopy and atmosphere and/or to the extraction of moisture and nutrients from the soil. Plant biology modulates these fluxes through such transient mechanisms as stress-induced variability of leaf stomatal resistance to transpiration and to CO_2 assimilation, short-term changes in leaf attitude, and seasonal changes in the color and density of the foliage.

The question

Can we formulate this complex interaction in a way that is at once simple enough and yet sufficiently exact to reveal the principal natural selection pressures that determine the observed configurations and functionings of natural plant communities?

Background

The early work of Darwin (1859) on natural selection led to the concept of ecological optimality. This connection between natural selection and the principles of physics was recognized by Lotka (1922) who proposed that "natural selection tends to make the energy flux through the system a maximum, so far as is compatible with the constraints to which the system is subject . . .", and further that "in the struggle for existence, the advantage must go to those organisms whose energy-capturing devices are most efficient in directing available energy into channels favorable to the preservation of the species."

Rosen (1967, p. 7) proposed a more general connection between natural selection and the environment. In his words "On the basis of natural selection, then, it may be expected that biological organisms, placed for a sufficiently long time within a specific set of environmental circumstances, will tend to assume characteristics which are optimal with respect to these circumstances." Parkhurst and Loucks (1972, p. 505) refined this to the form commonly used today. That is "Natural selection leads to organisms having a combination of form and function optimal for growth and reproduction in the environments in which they live."

Methodology

We seek to define, in terms of the key structural and behavioral parameters of a monocultural plant community, those conditions under which the reproductive potential of the individual plants is maximum for a given climate and soil. We assume the resulting community will prevail in the given environment. While our interest is in natural systems, primarily forests, we will also examine the behavior of crops to the extent that this may aid in generalization. We formulate our optimization arguments mechanistically in the belief that quantification is the key to understanding.

B Overview

Objective

We wish to find a set of general biophysical relations that define the optimum natural habitat for a given vegetation species and that alternatively will define the maximally productive stable (i.e., *climax*) vegetation community for a given climate and soil.

Plate 1.1. Hemlock and fir forest. Western hemlock (foreground) and Douglas fir (background) in the Olympic National Park, Washington. (Photograph by William D. Rich; Copyright © 2001 William D. Rich.)

Assumptions and organization

Here we consider only monocultures. We neglect the activities of insects and other animals (including man), omit the influence of disease and fire, and assume infinite nutrient and CO_2 reservoirs – leaving water and light as the limiting resources. We analyze the climate–soil–vegetation system as though it operated in its long-term temporal average and local spatial average states with a neutrally stable atmosphere and no lateral advection of energy or water from adjacent landscapes. Finally, we sever the feedback link from surface to atmosphere and consider the vegetation as a passive responder to climatic forcing.

We assume that the individual plants have arrived at their particular characteristics through a process of natural selection driven by the competition to survive in a given average environment and fine-tuned to the environment's local variabilities by adaptation. We deal here only with the natural selection and to do so analytically we assume that seed productivity is a surrogate for survival probability and is proportional to the net primary productivity, NPP.[†] It is our thesis that the physical parameter values leading to maximum productivity in a given environment are those that we should find in a climax community[‡] and that the analytical expression of this optimum equilibrium may provide a useful means of coupling the vegetated land surface to the atmosphere in a slowly changing climate.

For pedagogical elegance the work is organized, beginning with Chapter 2, as the classical physical science textbook would be; understanding is developed analytically and in a reductionist manner from "first principles", building gradually toward NPP in Chapter 10. However in this overview, in order not to lose sight of the forest for the trees, we omit mathematical detail and *begin* with the discussion of productivity.

Factors of productivity

We define the environment of a forest in terms of: the average length, m_τ, of its growing season for particular species; the climatic time averages[§] over this season of the insolation,[††] I_0, the precipitation, P_τ, and the daylight-hour atmospheric temperature, T_0; under the assumption that the reservoir of nitrogen and other nutrients in the soil, and that of CO_2 in the free atmosphere, remain non-limiting to production.[§§] With these simplifications we recognize the principal productive needs for: *light* to keep the leaf stomates open for uptake of CO_2, and to fuel its assimilation; *nitrogen* to nourish the formation of plant tissue; *water* to keep leaf stomates open by maintaining plant turgor, to transport the nitrogen from soil to plant, and to regulate plant temperature through evaporative cooling; and finally, the crown's *turbulent flux capacity* to evacuate water vapor at a rate meeting the plant's demand for water-borne nitrogen, and to supply atmospheric CO_2 at a rate meeting that demand for even the lowest leaf. We find biophysical bases for optimizing each of these.

We begin by assuming for the moment that the capacity of the turbulent flux to supply CO_2 to the leaves exceeds the light-driven demand of the leaves for CO_2. Such demand-limited assimilation is illustrated by the classical experimentally-determined *photosynthetic capacity curve* for an isolated C_3 leaf at constant leaf temperature, T_l. This curve defines the rate of CO_2 assimilation, P, as a saturating function of the variable insolation, I, maximizing at the value P_s for large I. As sketched in Fig. 1.1, for the temperature $T_l = T_m$ producing maximum assimilation, the net photosynthesis is defined by two species-dependent parameters, the saturation rate of maximum carbon assimilation per unit of basal leaf area, P_{sm}, and the *saturation insolation*,

[†] The proportionality constant is likely to vary with species making uncertain those conclusions about interspecies competition that are based solely on NPP.

[‡] This neglects the fact that competitive advantage in a mixed community may shift with age leading possibly to a dominant stable species that is not "climax" in the globally optimum sense used here for the monoculture.

[§] Time-averaged quantities remain undifferentiated in notation because we write *all* our dynamic relationships in terms of climatic time-averages.

[††] Insolation is the flux density of solar radiation on a horizontal surface.

[§§] This crude assumption is an expedient that allows us to proceed and must be remembered when evaluating our results.

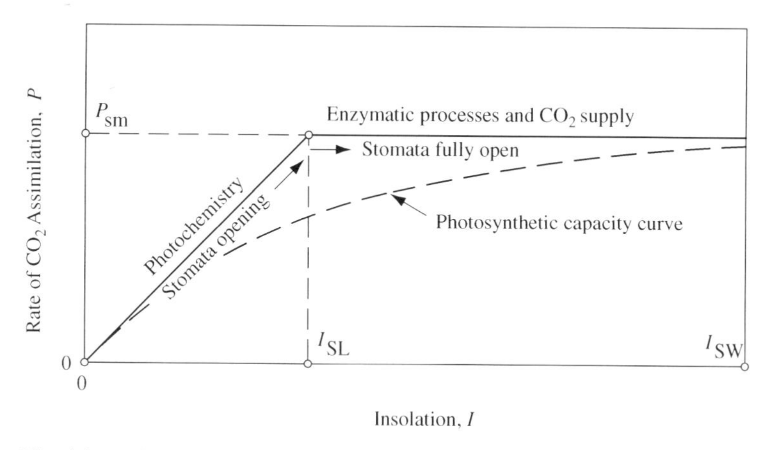

Fig. 1.1. Definition of the leaf light characteristic. Unit basal leaf area of C_3 species; water and nutrients non-limiting; optimal temperature; ambient CO_2.

I_{SL}, at which the asymptotes of the photosynthetic capacity curve intersect. We call these asymptotes the leaf *light characteristic*, and show that provided water is not limiting, I_{SL} marks the insolation at which the stomates are effectively fully open. For maximum productivity and maximum efficiency of light utilization, the optimal bioclimatic state for a single leaf of the given species sets the climatic value, I_0, of the insolation equal to I_{SL}, with $T_l = T_m$, as is shown in Fig. 1.2. We also show for isolated leaves that $I_{SL} = P_{sm}/\varepsilon$ with ε constant over a wide range of species as is shown in Fig. 1.3. The relationship $P_{sm} = \varepsilon I_{SL}$ is therefore the *biochemical assimilation capacity* for C_3 leaves.

Consider now an increasing *climatic insolation*, I_0, which causes decreasing average soil moisture concentration, s_0, when the other climatic variables remain fixed. With I_0 exceeding I_{SL} (and hence non-optimal), but not yet causing water-limitation due to a generous value of the fixed precipitation, the stomates remain fully open and the individual leaf continues to transpire at its climatically potential rate, $E_v = E_{ps}$.[†] However, at the particular climatic insolation, $I_0 = I_{SW}$ (which equals or exceeds I_{SL}), the declining average soil moisture concentration reaches a critical value, s_{oc}, at the end of the average interstorm period, causing the stomates to begin closing and transpiration to decline. This situation is illustrated in Fig. 1.4 where we define the evaporation function by its asymptotes which in this case we call the leaf *water characteristic* and which intersect at the *desiccation* or *water-critical* insolation, I_{SW}. We show how I_{SW} is estimated from the water balance (cf. Chapter 8). We also show that at constant temperature, the transpiration rate is an approximate surrogate for productivity of the same species (cf. Appendix E) allowing us to use the productivity, P, rather than the transpiration rate as the ordinate in Fig. 1.4, and making $P_{wm} = P_{sm}$. The range $I_{SL} \leq I_0 \leq I_{SW}$ is thus bioclimatically optimal for the isolated leaf because it maximizes leaf productivity (cf. Fig. 1.7).

[†] E_{ps} is the potential rate of evaporation from a wet, simple surface (cf. Appendix B).

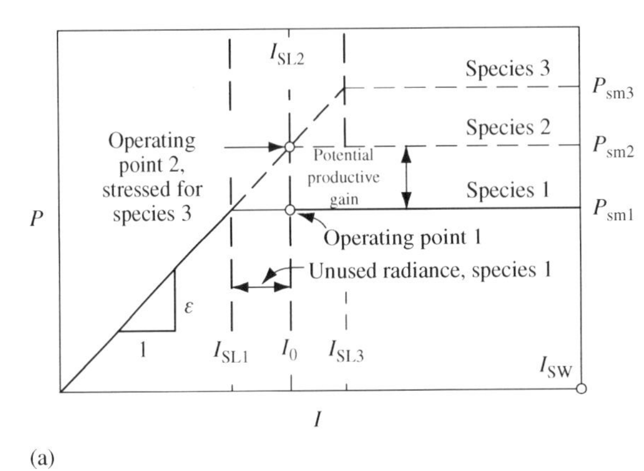

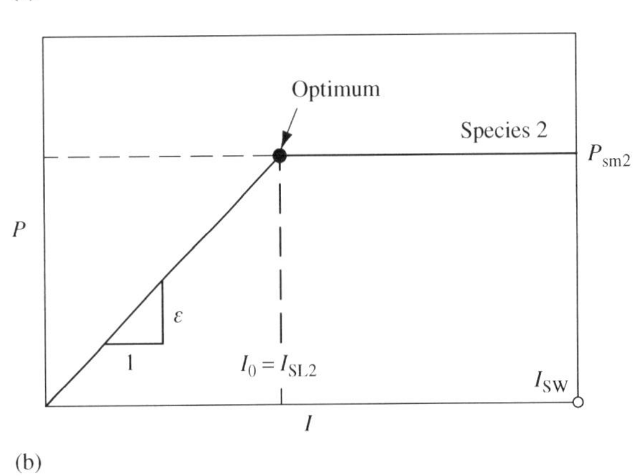

Fig. 1.2. Bioclimatic operating states under light control. Unit basal leaf area of C_3 species; water and nutrients non-limiting; optimal temperature; ambient CO_2.
(a) Suboptimal; (b) optimal.

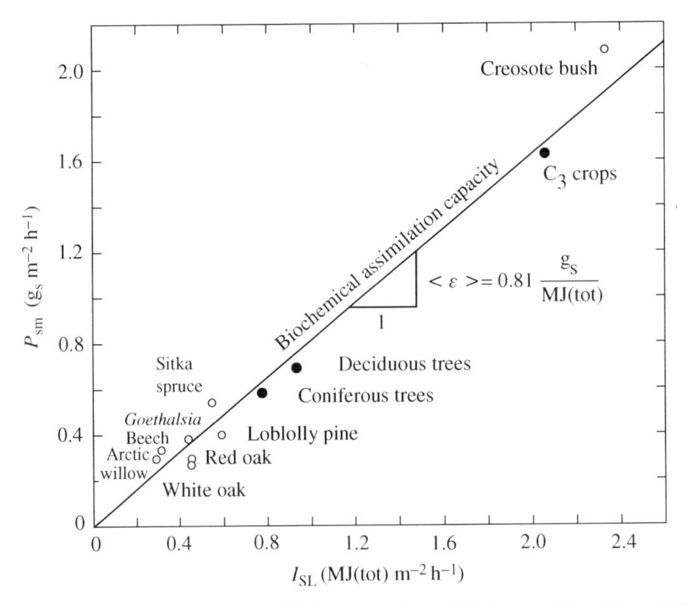

Fig. 1.3. Biochemical assimilation capacity of C_3 leaves. Data from Table 8.2.

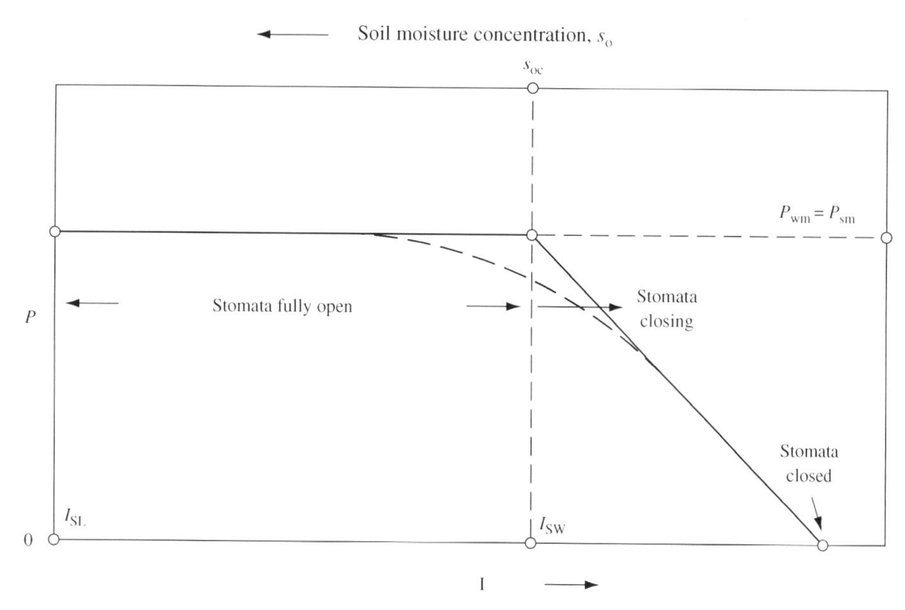

Fig. 1.4. Definition of the leaf water characteristic. Unit basal leaf area; light and nutrients non-limiting; ambient CO_2; fixed species, soil, precipitation and temperature.

To proceed toward canopy production, we must expand P_{sm} from isolated leaf to full canopy. Expanding in the vertical direction we multiply first by the leaf area index, L_t, to incorporate the total basal leaf area per unit of crown basal area, and next by the so-called *carbon demand function*, $f_D(\beta L_t)^\dagger$ (cf. Chapter 10) to average the leaf assimilation rate induced by a decaying insolation over the crown depth. This function is based upon our following proposed conditions for *optical optimality* in multilayers:

The reflection coefficient of the photosynthetically active component, PAR, of the incident radiation is small and nearly invariant over the range of vegetated latitudes, thus the incident light is used optimally for photosynthesis when *in the time average:*

(1) the upper leaf surfaces are in full sunlight and the crown basal area is in full shadow at the average solar altitude, $h_\otimes$; by including scattering we show (cf. Chapter 3) that this mandates an important geometrical relation between the solar altitude and the leaf angle, resulting in $\beta = \kappa,^\ddagger$ as is sketched for a pair of opaque leaves in Fig. 1.5 and as is verified from observations of full crowns of translucent leaves in Fig. 1.6; and

(2) the insolation at the lowest leaf is the minimum for leaf metabolism (i.e., the so-called *compensation intensity*, I_k); we show that its relative value, I_k/I_0, is a species constant (cf. Appendix H) that fixes the value of the *insolation*

† βL_t is called the (horizontal) *momentum absorption index*, or the *horizontal leaf area index*, since β is the cosine of the leaf angle, θ_L, with the horizontal.

‡ κ is the light extinction coefficient.

extinction index, κL_t, making it and the now-equal momentum absorption index, βL_t, species constants also.

In this vertical expansion from isolated leaf to whole crown, we average the insolation over the depth of the crown at optical optimality to get the crown-average

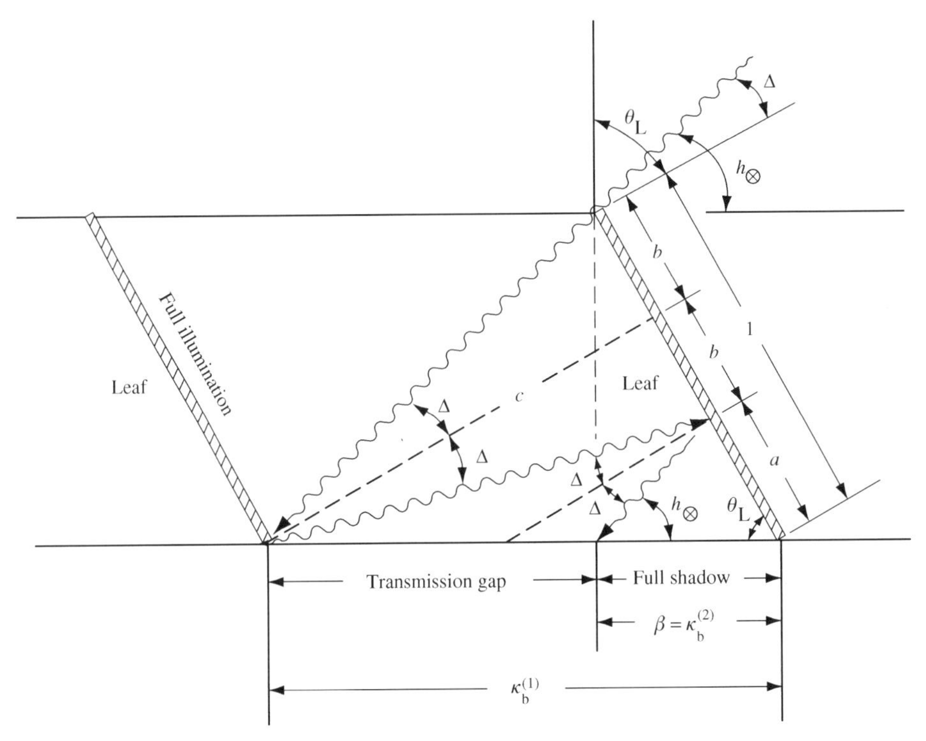

Fig. 1.5. Optically optimal beam-leaf geometry for opaque-leaved multilayers. Crown-average and daylight hour average conditions; exponent$^{(1)}$ signifies single leaf with no scattering; exponent$^{(2)}$ signifies two or more leaves with idealized scattering.

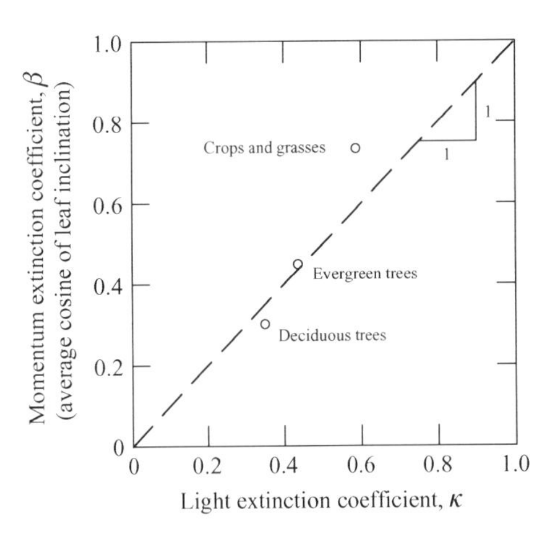

Fig. 1.6. Equality of observed extinction coefficients. Data from Table 3.8.

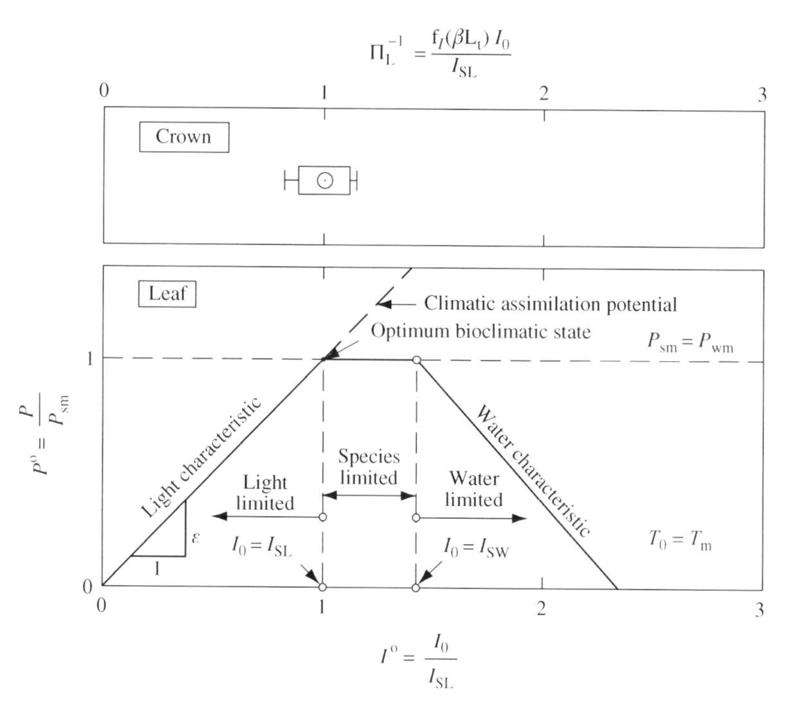

Fig. 1.7. Optimal bioclimatic state and the supporting evidence. Data from Table 9.1. Eight communities from desert shrub to rainforest. Plotted circle shows mean, box shows ± one standard deviation, and bar shows range (fixed C_3 species, soil and precipitation; ambient CO_2; nutrients non-limiting).

insolation, $\widehat{I_l} = f_I(\beta L_t)I_0$. We assume all leaves in the crown to have the same light characteristic making the crown-average $\widehat{I_{SL}}$ equal the isolated leaf I_{SL}, whereupon the *optimal bioclimatic state* for the canopy becomes $I_{SL} = \widehat{I_l} = f_I(\beta L_t)I_0$. This is supported in Fig. 1.7 by observations over a range of communities from desert shrub to rainforest (cf. Chapter 9). Together with Fig. 1.3, it defines the *climatic assimilation potential*, $P_{sm} = \varepsilon f_I(\beta L_t)I_0$, fixing the maximum CO_2 assimilation rate for given climate and species. The uppermost leaves in the canopy will reach their desiccation moisture state, I_{SW}, first and at the lower canopy-average radiance, $\widehat{I_{SW}} = f_I(\beta L_t)I_{SW}$. Comparison of the relative magnitudes of $\widehat{I_l}$, I_{SL}, and $\widehat{I_{SW}}$, in all their permutations, defines the range of natural habitats in water and light space and reveals the boundaries of this space as *climax conditions* which compare favorably in Fig. 1.8 with observations from the same wide range of communities (cf. Chapters 8 and 9).

Expanding crown productivity in the horizontal direction we multiply by the fraction of the ground surface covered by crown basal areas (i.e., the canopy cover, M). Being a canopy property, M depends upon the collective action of the individual plants in satisfying their individual water needs and is not maximized by Dawkins's "selfish gene". We show that the canopy demands water over the full growing season at the time-averaged rate ME_v, where $E_v \equiv k_v^* E_{ps}$ is the canopy transpiration rate (cf. Chapter 6),

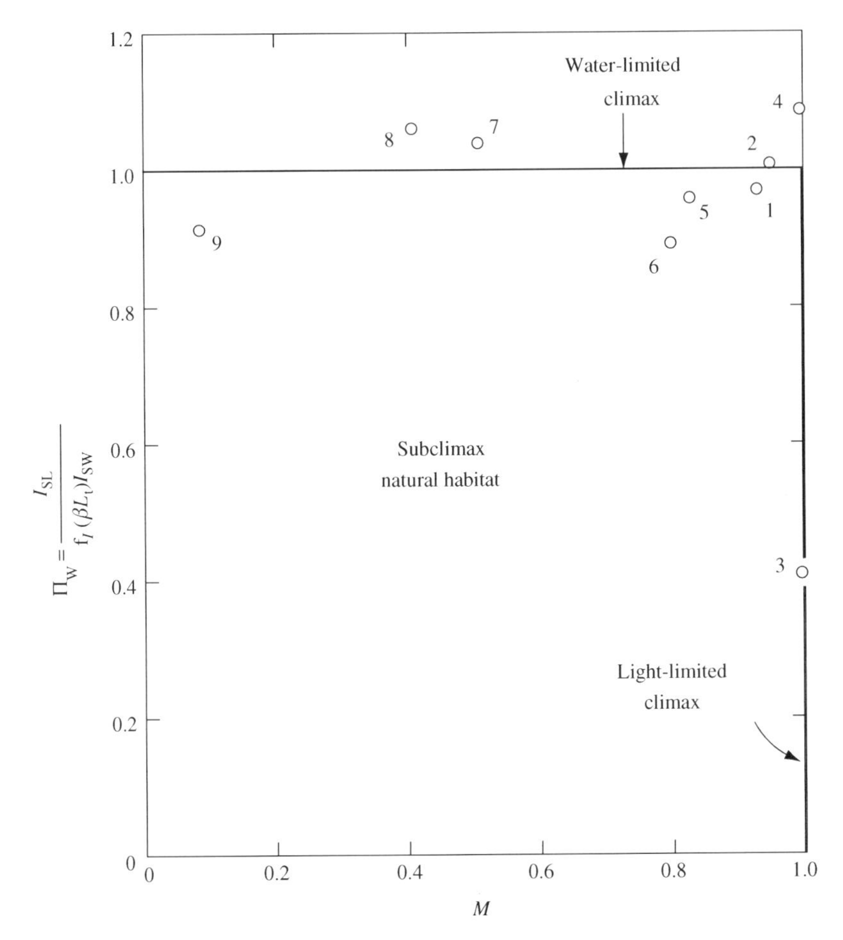

Fig. 1.8. Climax conditions as bounds to natural habitat. Data from Table 9.2. 1, Beech; 2, oak; 3, *Goethalsia* (rainforest); 4, Sitka spruce; 5, ponderosa pine; 6, loblolly pine; 7, pinyon-alligator juniper; 8, pinyon-Utah juniper; 9, creosote bush.

and k_v^* is the *canopy conductance* (cf. Chapter 6). In order to maintain the water balance, M will be less than 1 when E_v exceeds the climatically available water supply, and such a vegetation community is termed *water-limited*. Otherwise $M = 1$ and the community is *light-limited* with an excess of water. We recognize that the proportionality between soil-to-plant nitrogen flux and the flux of water, when coupled with the sensitivity of CO_2 assimilation to nitrogen availability, provides a degree of productivity-motivated selection pressure to maximize the plant transpiration rate, E_v. This may be accomplished by maximizing k_v^*, through minimizing the normalized canopy resistance, r_c/r_a (at given temperature) as is shown in Fig. 1.9 (cf. Chapter 7).[†] For given β we find this minimization to favor tapered crowns when L_t is large. For

[†] We show that the open-stomate condition of the optimal bioclimatic state allows approximation of the canopy flux resistance, r_c/r_a, as being independent of stomatal resistance. We then use the canopy-average eddy viscosity to derive r_c/r_a in terms of β and L_t.

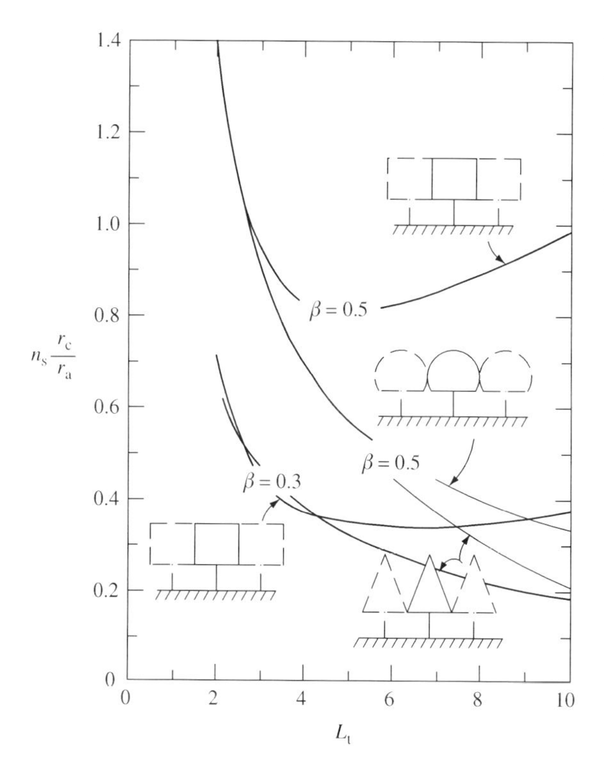

Fig. 1.9. Effect of crown shape on canopy resistance. Homogeneous multilayers with $M = 1$; r_c = canopy resistance; r_a = atmospheric resistance; $n_s = \frac{\text{stomated leaf area}}{\text{projected leaf area}}$.

water-limited canopies this transpiration maximization requires reduction of M and thus a larger investment in root mass by the individual plant with an associated loss in its seed productivity. This tradeoff of seed production for transpiration rate in order to gain seed production through added nitrogen flux is of indeterminate advantage to the plant. However with the plentiful water of light-limited situations no tradeoff is required, and rapid nitrogen recycling is fostered by this mechanism. Can this be the selection pressure leading to tapered crowns in nitrogen-poor soils?

There is much evidence (cf. Figs. 3.15 and 3.18) that β is heavily dependent upon $h_\otimes$. For a given deciduous species having cylindrical crowns, optical optimality in combination with the minimum resistance can determine both L_t and β as is shown in Fig. 1.10, and the range of L_t so determined is that observed (cf. Chapter 7). For evergreen species we show empirically that the needles are normal to the radiation beam (cf. Fig. 3.18). Under extreme temperature conditions these β vs. $h_\otimes$ relations appear to be biased by the need to control the reflection of the near infrared (NIR) component of radiation (cf. Fig. 3.15).

Returning now to the productivity, we next incorporate the temperature sensitivity of leaf photosynthesis by multiplying by the function $g(\widehat{T_l})$ where $\widehat{T_l}$ is the crown-average leaf temperature. We show in Fig. 1.11, albeit for only three communities, that the local growing season average atmospheric temperature, T_0, equals T_m, the species-dependent leaf temperature yielding maximum CO_2 assimilation (cf. Chapter 9). We infer that $\widehat{T_l}$ is approximately equal to T_0 by demonstrating growing season average

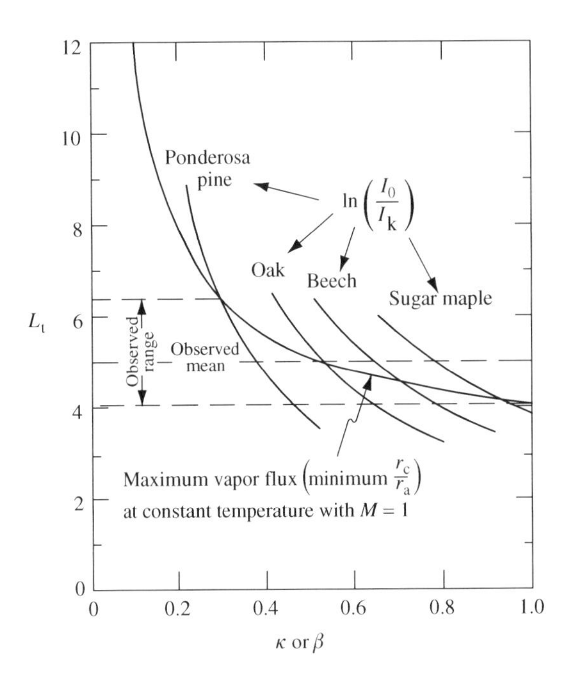

Fig. 1.10. Foliage parameters for maximum vapor flux. Homogeneous cylindrical multilayers with $M = 1$.

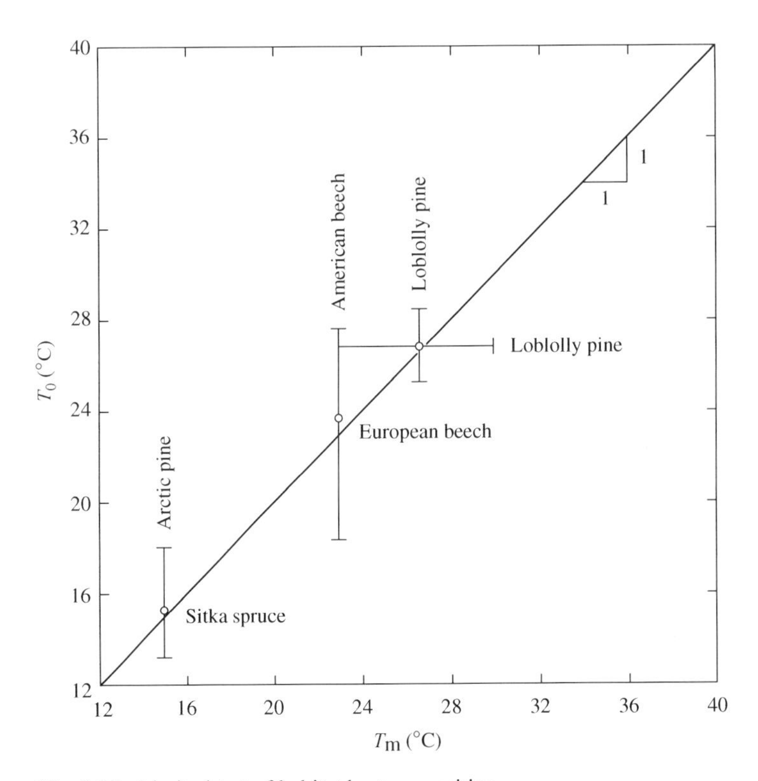

Fig. 1.11. Limited test of habitat heat proposition.

Bowen ratios on the order of 10^{-1} for a wide range of communities (cf. Table 9.2), and therefore that nominally $g(\widehat{T}_l) \approx g(T_0) = g(T_m) = 1$.

Finally, in order to get the potential (i.e., carbon demand-limited) annual production for a given species, we multiply by the average length of the growing season, m_τ, and to convert the *gross* primary productivity calculated in $g(CO_2)$ into *net* primary productivity, NPP, measured in grams of (above-ground) solid matter, g_s, we multiply by the widely accepted nominal empirical conversion factor, $\vartheta = 0.50\ g_s\ g^{-1}$.

Potential productive gain

We can now write the product of the above productivity factors in the convenient form (cf. Chapter 10)

$$\frac{\mathrm{NPP}}{p_D} = \frac{\mathrm{NPP}}{g(T_0)P_{sm}M m_\tau} = f_D(\beta L_t). \tag{1.1}$$

We show p_D to be the productive demand of a unit basal area of the monolayer, $L_t = 1$, for the given species, making $\frac{\mathrm{NPP}}{p_D}$ the potential (contingent upon adequate CO_2 supply) productive *gain* for this species resulting from the canopy structure. The independent variable, βL_t, is a species parameter (cf. Appendix H), and $\frac{\mathrm{NPP}}{p_D}$ is the maximum for a given species due to our use of both optical optimality and bioclimatic optimality in deriving the demand function, $f_D(\beta L_t)$. Equation 1.1 is plotted as Curve (a) in Fig. 1.12 where we see it to be monotonically increasing.

Picking a form for the vertical gradient of CO_2 concentration within the crown and calculating the eddy momentum diffusivity from a "big leaf" model of the crown, we use the flux–gradient relation to find the downward CO_2 supply function,

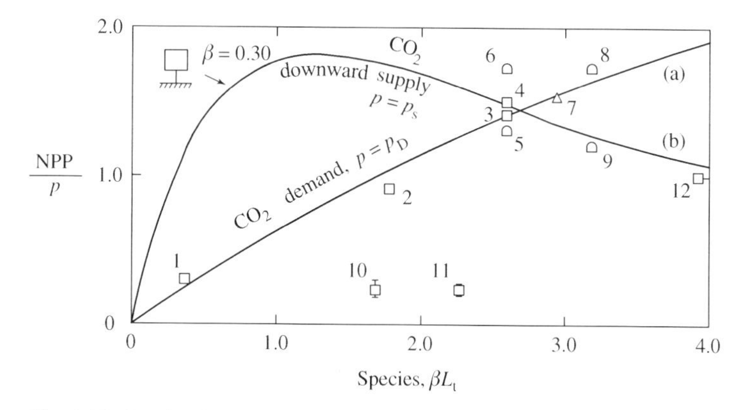

Fig. 1.12. Productive gain of C_3 multilayer canopies. Curve (a) = canopy CO_2 demand compared to monolayer demand; curve (b) = canopy downward CO_2 supply compared to monolayer supply. Data points: 1, Creosote bush; 2, ponderosa pine; 3, loblolly pine; 4, oak; 5, rainforest (Ghana); 6, rainforest (Congo); 7, red spruce; 8, beech (Tennessee); 9, beech (Eastern Europe); 10, pinyon-alligator juniper; 11, pinyon-Utah juniper; 12, sugar maple. Symbol shape indicates crown shape.

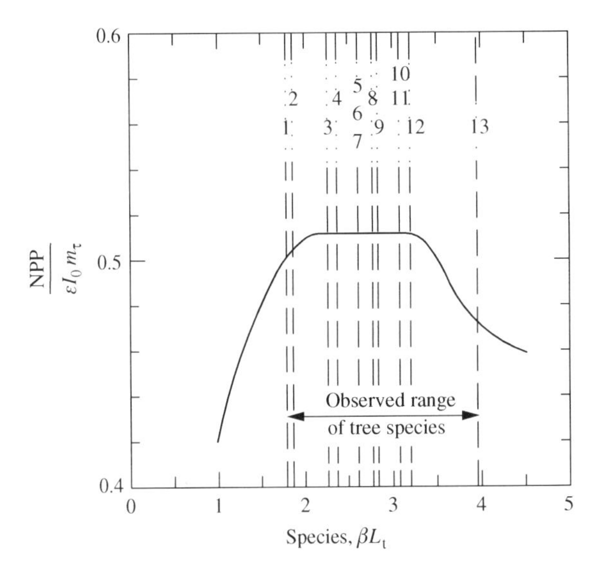

Fig. 1.13. Evidence for maximization of light-limited productivity of trees. Observed range from Baker's I_k/I_o (1950, Table 12, p. 143) using $\kappa = \beta$; Table 3.9. 1, Ponderosa pine; 2, Scots pine; 3, Northern white cedar; 4, tamarack; 5, lodgepole pine; 6, Douglas fir; 7, red oak; 8, hackberry; 9, Engelmann spruce; 10, Norway spruce; 11, Eastern hemlock; 12, beech; 13, sugar maple.

$\mathrm{NPP}/p_S = \mathrm{f_S}(\beta, L_t)$ for cylindrical crowns shown as Curve (b) in Fig. 1.12 for the commonly observed value $\beta = 0.30$. Here p_S is the carbon supply analog of p_D which we assume to equal p_D for the $L_t = 1$ monolayer. With this critical assumption, comparison of the supply and demand functions shows a *critical absorption index*, $\widehat{\beta L_t}$, separating the region ($\beta L_t \leq \widehat{\beta L_t}$) in which the atmosphere can supply the entire CO_2 demand from that ($\beta L_t > \widehat{\beta L_t}$) in which a growing fraction of the CO_2 demand must be met from below by decaying plant matter. We show that for tapered crowns (i.e., cones or hemispheres) the atmospheric CO_2 supply is monotonically increasing with βL_t so that these crown shapes can meet the CO_2 demand at all βL_t without local recycling. Limited observations seem to support our derivation of this optimum productivity as shown by the plotted points on Fig. 1.12.

Maximization of species potential productivity

Fundamental to all this work is the assumption that the dominant selective pressure is to maximize the individual plant's reproductive potential as expressed through maximization of annual biomass production with its proportional seed production. By substituting the climatic assimilation potential, $P_{sm} = \varepsilon \mathrm{f}_I(\beta L_t) I_0$,[†] in the denominator of Eq. 1.1[‡], for non-limiting water supply, $M = 1$, and with $g(T_0) = 1$, we write

$$\frac{\mathrm{NPP}}{\varepsilon I_0 m_\tau} = \mathrm{f_D}(\beta L_t)\,\mathrm{f}_I(\beta L_t), \tag{1.2}$$

[†] We find $\langle\varepsilon\rangle = 0.81\ \mathrm{g_s\ MJ(tot)^{-1}}$.

[‡] Without the need here to have the denominator on the left-hand side of Eq. 1.1 represent monolayer productivity we are free to extract from P_{sm} its dependence upon species through $\mathrm{f}_I(\beta L_t)$ and to move this function to the right-hand side in Eq. 1.2.

the species-dependence of the potential (i.e., light-limited) NPP is now all on the right-hand side. Equation 1.2 is plotted as the solid line in Fig. 1.13 and shows a broad global maximum over a particular range of βL_t. The range of observed tree species is indicated by the spread of the vertical dashed lines at the respective species-constant βL_t. Their clustering in the range of the global maximum NPP is taken as confirmation of our fundamental assumption.

Extreme climates call for extreme βL_t in order to conserve heat (large βL_t) or water (small βL_t) leading to locally optimal NPP that are smaller (due to CO_2 or water limitation respectively) than this global, light-limited maximum. While the above development of productivity and its underlying optical and bioclimatic optimalities is at the heart of this work, we also present and evaluate applications of these ideas to natural habitats and to ecotone location, and we discuss their potential use to assess some of the apparent consequences of global climate change.

Part I

Biophysics

2

Canopy structure

Basic physical features of the forest canopy are defined, observations are presented, and simplified models are proposed.

A Introduction

The diffusion of water vapor, CO_2, momentum, heat and light vertically through the canopy governs forest growth under the simplifying assumptions employed herein (see Chapter 1). In turn, these diffusions are ultimately controlled by the magnitude, distribution, and orientation of the solid matter comprising the forest canopy. This chapter is devoted to generalized quantitative description of these geometrical characteristics.

B Stand structure

We are interested in the collective physical interaction between the environment (atmosphere and soil), and a homogeneous community of plants called a *stand*. The stand is composed of individual plants, each consisting of a *crown* supported above the *substrate* (i.e., land surface) by a *trunk* or *main stem*. The collection of crowns in the stand constitutes the plant *canopy* which carries that name because it is often elevated, umbrella-like, above the ground. Each crown is composed of *crown elements* which include the *branches*, and the *foliage elements* such as leaves, pods, and stems.

Assuming the individual plants to be identical and with circular plan form, the plan view of the stand (see Fig. 2.1) shows circular disks of foliage nested at a greater or lesser density depending upon the availability of some growth-limiting resource. Letting that limiting resource be water and defining Rosen's (1967) "cost functional" as water-induced plant stress, Eagleson (1978f) assumed the optimal density of this nesting to be that which, for the given climate and soil, produced the maximum average soil moisture concentration, because this condition minimizes plant stress.[†] He defined

[†] As we will see in Chapter 6, Salvucci (1992) showed that a somewhat greater water use is achieved at a slightly greater vegetated fraction in return for somewhat higher risk of stress.

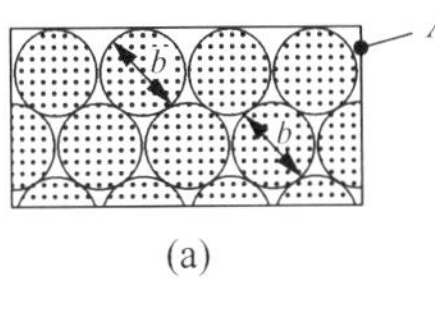

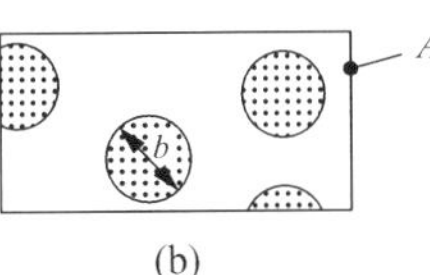

Fig. 2.1. Definition of canopy cover. (a) Closed canopy, $M = M_{max}$; (b) open canopy, $M < M_{max}$.

the fraction of surface area covered by the projected crown areas (i.e., the disks) as the *canopy cover*,† M, and showed its variability in water-limited situations. We use this definition here. When the disks are as close as possible (i.e., $M = M_{max}$) we refer to the canopy as "closed" as in Fig. 2.1a, and when the disks are less densely packed as in Fig. 2.1b we refer to the canopy as "open". From idealized geometrical considerations $M_{max} = 0.84$, however in discussion of this extreme state, we will refer to it as $M = 1$. In nature, the trees may depart from the circular planform assumed here, allowing a packing density approaching $M_{max} = 1$. The spaces between disks represent area that may be occupied by an understory and/or by bare soil. For simplicity here we consider it to be bare soil.

The individual crowns are described by their shape, solid density, leaf orientation, and leaf type. Each of these physical features of the crown influences its absorption of radiation, as well as the fluxes of water vapor, carbon dioxide, and heat to or from the leaves. Therefore, through selection and/or adaptation, we expect the density, arrangement, types, and orientation of the leaves to reflect the incident radiation regime (among other factors) in some way (Russell *et al.*, 1989a, p. 32).

C Crown shape

Trees have long been classified in part according to the shape of their crowns (see Baker, 1950, pp. 71 *et seq.*). Sinnot (1960) observed that plant species are often symmetrical in their external shape with respect to a central vertical axis which allows modeling individual plants as simple geometrical figures filled with plant elements. We follow this precedent limiting our geometrical classes to cylindrical (i.e., "flat top"), hemispherical (i.e., "rounded"), and conical (i.e., "pointed"). These are illustrated in Fig. 2.2 along with the defining measures of median size: b = diameter of crown, h = canopy height (i.e., height of tree), h_c = depth of crown, and h_s = stem height (i.e., height of crown bottom). Note that we show the geometry common for a single tree within an extensive closed canopy (i.e., $M = 1$) of identical trees and that the crown does not extend all the way to the ground. As the canopy opens (i.e., $M < 1$), the penetration of light increases and foliage may be found closer and closer to the ground.

† Eagleson (1978a) called M the "canopy density" which is ambiguous in three-dimensional studies such as this, so we adopt here the two-dimensional term "canopy cover".

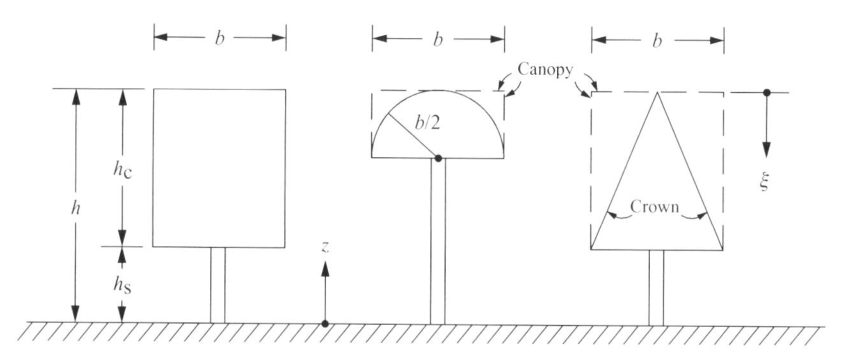

Fig. 2.2. Idealized crown shapes.

Plate 2.3. Typical cylindrical crown. Giant sequoia in Arnold Aboretum, Boston. (Photograph by William D. Rich; Copyright © 2001 William D. Rich.)

D Solid density

The diffusion of water vapor, CO_2, momentum, heat, and light through the canopy is ultimately controlled by the magnitude, distribution, and orientation of the solid matter comprising the canopy. The primary constituent of this solid matter is the leaves or needles.

Crown and canopy internal properties are assumed to be spatially variable in only the vertical direction. To study such distributed properties it is convenient to consider that portion of the crown or canopy lying above a given elevation, z. To do so we prefer to measure vertical distance downward from an origin at the top of the crown as is indicated by the variable ξ in Fig. 2.2. Taking into account the elevated crown bottom, we define this variable over the range $0 \leq \xi \leq 1$ in the dimensionless form:

$$\xi = \frac{h - z}{h - h_s}. \tag{2.1}$$

The crown's solid density is commonly expressed through the *leaf area index*, L, which is defined, after Ross (1975, p. 24), as the area of leaves[†] (upper side only) within that vertical circular cylinder which just encloses the crown divided by the horizontal cross-sectional area of that cylinder. The leaf area index is a dimensionless quantity which varies with season, age, and species, and is a critical parameter controlling both the physical and the biological processes of plant canopies (Chen and Black, 1992). For homogeneous leaf area density, the leaf area index is written:

$$L = \frac{4a_{cr}(h - h_s)}{\pi b^2} \int_0^1 A_{cr}(\xi)\, d\xi = a_{cr}(h - h_s)\frac{\forall_{cr}}{\forall}, \tag{2.2}$$

in which $A_{cr}(\xi)$ = horizontal cross-sectional area of the crown,
$\forall_{cr}$ = crown volume,
$\forall$ = canopy volume as given by the crown-enclosing circular cylinder, and
a_{cr} = *crown leaf area density* defined (Thom, 1975) as the area of leaves (one side only) per unit of crown volume ($m^2\ m^{-3}$).

The other crown elements (e.g., branches, twigs, stems, and pods) may be incorporated into a more inclusive *plant* (or *foliage*) *area index*, L_t, defined in the same manner as the leaf area index. If L' is the area index of the other crown elements, then $L_t = L + L'$ = plant or foliage area index, PAI, (e.g., Shaw and Pereira, 1982). These other crown elements also intercept light and contribute to the drag of the canopy. Their importance relative to that of the leaves is a function of species (as well as season and age) as can be seen in Table 2.1. From this table we note that for most deciduous

[†] Needle leaves have thickness on the order of the leaf width and thus one or more of the needle surfaces may have transverse curvature. In this case the term "projected" leaf area is conventionally used to describe the needle shadow area obtained when the light source is directly above the horizontal needle.

trees the other crown elements constitute less than 10 percent of the total area, but for agricultural crops the percentage rises markedly. Measurements for coniferous trees are few and hard to find, but Jarvis *et al.* (1976, p. 179) suggest that for radiation at least, the effective foliage units in coniferous canopies are the twigs rather than the individual needles. We assume that the energetic and dynamic importance of these crown elements is proportional to their relative magnitude. Our developments here will be in terms of the plant area index, L_t. However, because reported observations of this parameter are few, we will use available values of L as a first approximation to L_t wherever necessary.†

For simplicity we assume herein that the solid matter is distributed in one of two fashions (cf. Fig. 2.3): (1) *multilayer* – disbursed either homogeneously throughout the crown (i.e., a_{cr} is constant) forming what we call a homogeneous multilayer crown,‡ or non-homogeneously such that the foliage area index is constant with radius§ from the trunk; and (2) *monolayer* – concentrated in a thin layer of uniform density at the crown surface forming what we call a homogeneous monolayer crown.

The crown cross-section variation, $A_{cr}(\xi)$, is species-dependent of course and is idealized in Fig. 2.2. For $a_{cr} = \text{constant}$, $dL(z)/dz$ is proportional to $A_{cr}(\xi)$ from

Table 2.1. *Relative importance of crown element areas*

Species	L	L'	$L_t = L + L'$	L'/L_t	Reference
Leafy plants					
Oak	4.60	0.48	5.08	0.09	Rauner (1976)
Maple	5.02	0.16	5.18	0.03	Rauner (1976)
Aspen	4.73	0.17	4.90	0.03	Rauner (1976)
Linden	4.26	0.30	4.56	0.07	Rauner (1976)
Beans	6.25	0.85	7.10	0.12	Thom (1971)
Stemmy plants					
Maize				0.18	Den Hartog and Shaw (1975)
Maize				0.30	Ross (1981, p. 78)
Willow	4.50	1.50	6.00	0.25	Cannell *et al.* (1987)
Wheat				0.74	Ross (1981, p. 80)

† The leaf area index may be estimated in any one of several ways: (1) "destructive testing" involving cutting and measuring the individual leaves (e.g., Thom, 1971); (2) "inclined point quadrats" which involve statistical analysis of the contacts between the leaves and a thin probe passed at various angles through the crown (e.g., Levy and Madden, 1933; Warren Wilson, 1960, 1963; Philip, 1965); and (3) the quality of light reaching the canopy substrate (e.g., Evans and Coombe, 1959; Jordan, 1969; Lang *et al.*, 1985, 1991; Lang, 1986; Lang and Xiang, 1986; Lang and McMurtrie, 1992).

‡ We will see in Chapter 7 that this is precisely obtained only for cylindrical crowns due to the need for the same light intensity at the lowermost leaf.

§ This is the situation which must prevail for tapered crowns in order to achieve the same minimum light intensity at the base of the crown for all radii.

Eq. 2.2. Typical observed forms of $dL(\xi)/d\xi$ for multilayers are generalized by Ross (1981, Fig. 59b, p. 216) and are reproduced here as Fig. 2.4. According to Ross (1981), grasses and young conifers have a $dL(z)/dz$ that most closely resembles Type 1. He reports Type 2 to be quite common, while horse beans have the more unusual Type 4 distribution. There is ample additional observational evidence that canopy leaf area index has a Type 2 variation in a wide variety of crops and trees. For example see Uchijima (1976, Fig. 13), for rice and maize; Saugier (1976, Fig. 6b), for sunflower; Amiro (1990a, Fig. 2), for pine; Norman and Jarvis (1974, Fig. 8), for Sitka spruce; and Rauner (1976, Fig. 2), for oak, linden, and aspen. The last is reproduced here in Fig. 2.5.

In the present work we will approximate the commonly observed Type 2 distribution for multilayers by the uniform distribution designated Type 3 in Fig. 2.4.

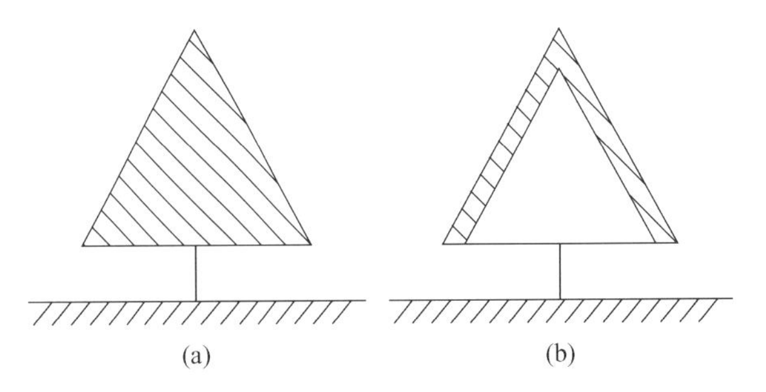

Fig. 2.3. Spatial distributions of crown solid matter. (a) Multilayer; (b) monolayer.

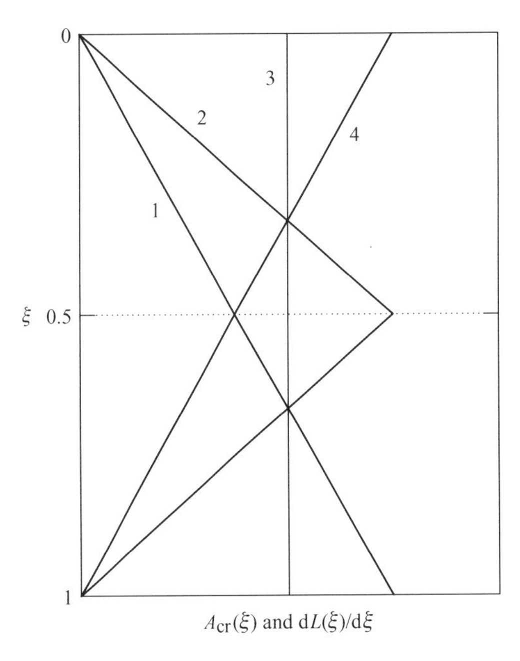

Fig. 2.4. Observed vertical distributions of leaf area. 1, Grasses and young conifers; 2, many crops and trees; 3, uniform; 4, horse beans. (After Ross, J., *The Radiation Regime and Architecture of Plant Stands*, 1981, Fig. 59b, p. 216, Copyright © 1981 Dr W. Junk Publishers, The Hague, with kind permission from Kluwer Academic Publishers.)

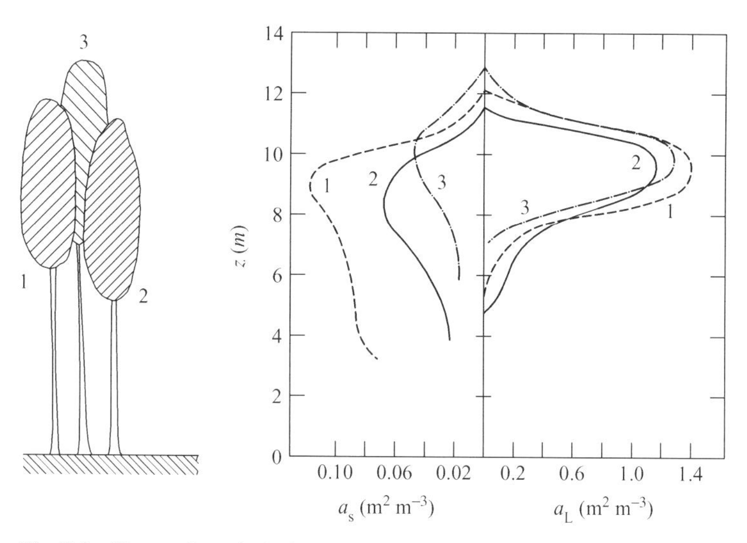

Fig. 2.5. Observed vertical distributions of solid matter in typical deciduous forests. a_L = Canopy leaf area density; a_s = canopy non-leaf area density. 1, Oak; 2, linden; 3, aspen. (From Rauner, J.L., 1976, Fig. 2, p. 244, Copyright © 1976 Academic Press Ltd, with kind permission from Academic Press.)

E Leaf orientation

The orientation of the leaves in space is also of great significance to plant behavior. Leaf projection on the horizontal plane defines the proportion of light flux that the leaves intercept (cf. Chapter 3), while the projection on the horizontal and vertical planes is a primary determinant of surface and form drag respectively (cf. Chapter 4). Leaf orientation is a species characteristic which may, through growth or heliotropism, adapt to ambient radiation conditions.

In a given species, the leaf distribution and orientation varies in the vertical direction (e.g., Grulois, 1967) but evidence supports our assumption of uniformity of orientation with respect to angular position (i.e., azimuth) about the stem at a given elevation (Nichiporovich, 1961a; de Wit, 1965). Therefore, the most common simplifying assumption is that at any elevation the leaves are distributed symmetrically about the vertical axis of the tree. Here we will extend the assumptions to include assumed homogeneity of both distribution and orientation throughout the crown in order to derive a simple analytical framework for leaf orientation which will be useful later when we take up the absorption of light by the canopy. These developments follow closely those of Ross (1981).

Distribution functions

We define the orientation of leaves in a volume by the probability density function (pdf) of normal directions to the top surfaces of the leaves. Consider this volume to

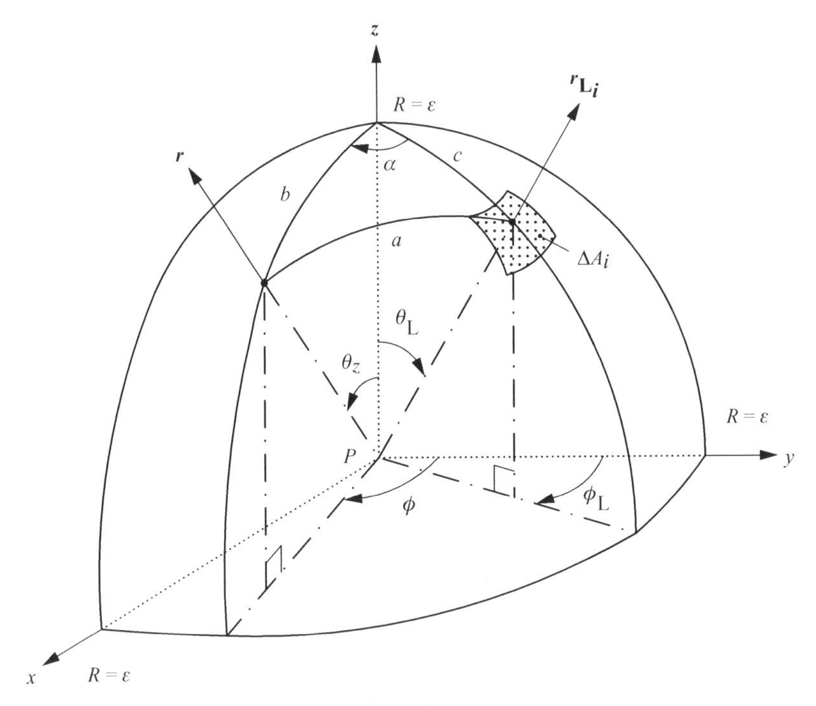

Fig. 2.6. Definition sketch for leaf orientation.

be a sphere of infinitesimal radius, $R = \varepsilon$, centered on any point, P, in the crown as is illustrated in the sketch of Fig. 2.6. We divide the surface of the sphere into i equal areas, ΔA_i, each of which therefore subtends the same central angle, $\Delta\omega_i$, centered on the direction indicated by the radius vector $\boldsymbol{r}_{\mathrm{L}_i}$.

From spherical coordinates

$$\Delta \mathrm{A} = \varepsilon^2 \Delta\omega = [(\varepsilon \sin\theta_{\mathrm{L}})\, \Delta\phi_{\mathrm{L}}] \cdot [\varepsilon \Delta\theta_{\mathrm{L}}]\,, \tag{2.3}$$

or

$$\Delta\omega = \sin\theta_{\mathrm{L}} \Delta\theta_{\mathrm{L}} \Delta\phi_{\mathrm{L}}. \tag{2.4}$$

Assuming, for computational simplicity, homogeneity of leaf distribution throughout the crown, and noting that $\sum_i \Delta\omega_i = 2\pi$, we see that the fraction of the total leaf area lying within the solid angle $\Delta\omega$ is $\frac{\Delta\omega}{2\pi}$. Let the fraction of this last area having normals in the $\boldsymbol{r}_{\mathrm{L}}$ direction be the function $\mathrm{g_L}(P, \boldsymbol{r}_{\mathrm{L}})$. If we sum over all i (that is, over all $\boldsymbol{r}_{\mathrm{L}_i}$) we obtain, in the limit

$$\int_{\omega} \frac{\mathrm{g_L}(P, \boldsymbol{r}_{\mathrm{L}})}{2\pi}\, \mathrm{d}\omega \equiv 1, \tag{2.5}$$

from which we see that

$$\frac{\mathrm{g_L}(P, \boldsymbol{r}_{\mathrm{L}})}{2\pi} = \text{pdf of foliage area orientation and of leaf normals.} \tag{2.6}$$

Since very few leaves have their "top" side facing downward, few normals lie in the lower hemisphere of our infinitesimal sphere, making $\frac{g_L}{2\pi} \approx 0$ therein. Noting that at any point, P, $\boldsymbol{r}_L = r_L(\theta_L, \phi_L)$, and assuming there to be no preferred azimuthal orientation, we use Eq. 2.4 to rewrite Eq. 2.5

$$\int_{\omega} \frac{g_L(P, \boldsymbol{r}_L)}{2\pi} \, d\omega = \int_0^{2\pi} d\phi_L \int_0^{\frac{\pi}{2}} \frac{1}{2\pi} g_L(P, \theta_L) \sin\theta_L \, d\theta_L = 1. \tag{2.7}$$

or simply

$$\int_0^{\frac{\pi}{2}} g_L(P, \theta_L) \sin\theta_L \, d\theta_L = 1. \tag{2.8}$$

For homogeneous foliage orientation, g_L is independent of position P in the crown, and Eq. 2.8 becomes

$$\int_0^{\frac{\pi}{2}} g_L(\theta_L) \sin\theta_L \, d\theta_L = 1, \tag{2.9}$$

in which

$$g_L(\theta_L) \sin\theta_L = \widetilde{g_L}(\theta_L) = \text{pdf of leaf orientation angle}, \tag{2.10}$$

and the cumulative distribution function (i.e., CDF) of leaf orientation angle is by

Plate 2.2. Typical hemispherical crown. Variety of fig, Agra, India. (Photograph by Peter S. Eagleson.)

definition

$$\widetilde{G_L} \equiv \int_0^{\theta_L} \widetilde{g_L}(\theta_L)\, d\theta_L. \tag{2.11}$$

De Wit (1965) distinguished four classes of cumulative distribution functions:

(1) *planophile* – in which the leaves tend toward the horizontal,

(2) *erectophile* – in which the leaves tend toward the vertical,

(3) *plagiophile* – in which the leaves tend toward a 45° inclination, and

(4) *extremophile* – in which both horizontal and vertical tendencies predominate.

Because it does not seem important in nature, we have replaced the last of these with the *uniform* distribution in which all leaf inclinations are equally likely. We have assigned simple analytical functions to represent these four classes:

(1) *planophile*: $\widetilde{g_L}(\theta_L) = \cos\theta_L$; $\widetilde{G_L}(\theta_L) = \sin\theta_L$ (2.12)

(2) *erectophile*: $\widetilde{g_L}(\theta_L) = \sin\theta_L$; $\widetilde{G_L}(\theta_L) = 1 - \cos\theta_L$ (2.13)

(3) *plagiophile*: $\widetilde{g_L}(\theta_L) = \sin 2\theta_L$; $\widetilde{G_L}(\theta_L) = \frac{1}{2}(1 - \cos 2\theta_L)$ (2.14)

(4) *uniform*: $\widetilde{g_L}(\theta_L) = \frac{2}{\pi}$. (2.15)

All these functions are plotted in Fig. 2.7.

De Wit (1965) made observations of leaf orientation angle† for a variety of agricultural species and presents them in the form of the cumulative distribution function, $\widetilde{G_L}(\theta_L)$. These are reproduced here in Fig. 2.8 (solid lines) with the addition, for comparative purposes, of the most relevant of Eqs. 2.12–2.14 (dashed lines).

Lang *et al.* (1985) and Ross (1981) also observed leaf orientation in certain field crops, but they presented their results in terms of the probability density function, $\widetilde{g_L}(\theta_L)$. These have been reproduced in Fig. 2.9 where they are compared with the pdfs of Eqs. 2.13 and 2.14. Note the difference in the pdf for sunflower in the morning as opposed to that for the afternoon, a clear demonstration of the well-known heliotropism of this species. From these observations it appears that grasses (i.e., rye, corn, and sorghum) tend to be erectophile, while other erect field crops (i.e., rape, sugar beet, and sunflower) tend to be plagiophile. It is no surprise that the recumbent crops (i.e., potato and clover) appear to be planophile. According to Ross (1981, p. 115), "Data concerning leaf orientation in tree stands are exceedingly scarce." However, we will see later that given the leaf area index, the mean of θ_L can be inferred from observations of the vertical decay of *either* light *or* momentum in the canopy.

† Leaf angles may be estimated in the field either through the methods of inclined point quadrats (e.g., Warren Wilson, 1963, Philip, 1965, Miller, 1967), of gap frequencies (Lang *et al.*, 1985), from transmission of direct sunlight (Lang, 1986), or by painstaking use of protractors to measure the angle of a statistically significant sample of leaves (e.g., Nichiporovich, 1961b).

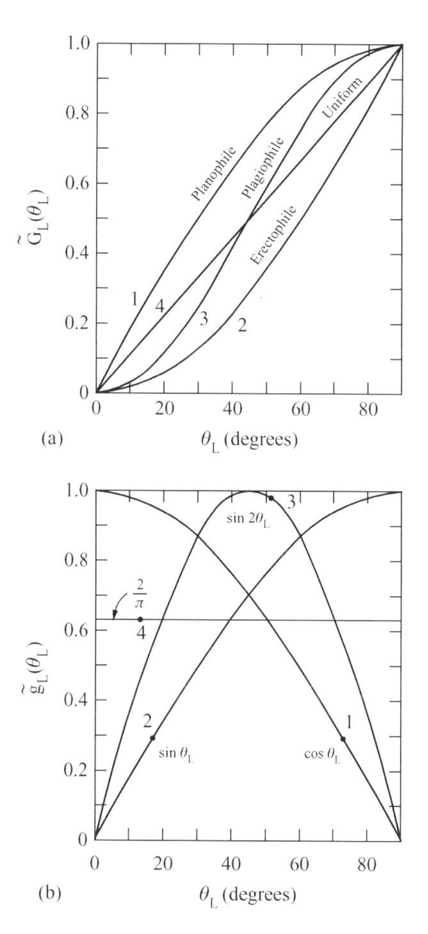

Fig. 2.7. Distribution functions of leaf inclination. (a) Cumulative distribution classes defined by de Wit (1965); (b) Simple probability density functions corresponding to the de Wit (1965) classes.

Mean leaf inclination angle

For both drag and phytometric calculations it is important to know the leaf area projected on the vertical and/or horizontal planes. Since for homogeneous canopies (Eq. 2.6)

$$\frac{g_L(\boldsymbol{r}_L)}{2\pi} = \text{distribution function for the projection of leaf area on to a plane perpendicular to the leaf normal} \tag{2.16}$$

then,

$$\frac{g_L(\boldsymbol{r}_L)}{2\pi}|\cos \widehat{\boldsymbol{r}_L \boldsymbol{r}}| = \text{distribution function for the projection of leaf area on to a plane perpendicular to the vector } \boldsymbol{r}(\theta, \phi). \tag{2.17}$$

Integrating as before (Eq. 2.7), and assigning the symbol β

$$\hat{\beta}(\boldsymbol{r}) = \int_{\omega} \frac{g_L(\boldsymbol{r}_L)}{2\pi} |\cos \widehat{\boldsymbol{r}_L \boldsymbol{r}}| \, d\omega \equiv \langle \cos \widehat{\boldsymbol{r}_L \boldsymbol{r}} \rangle, \tag{2.18}$$

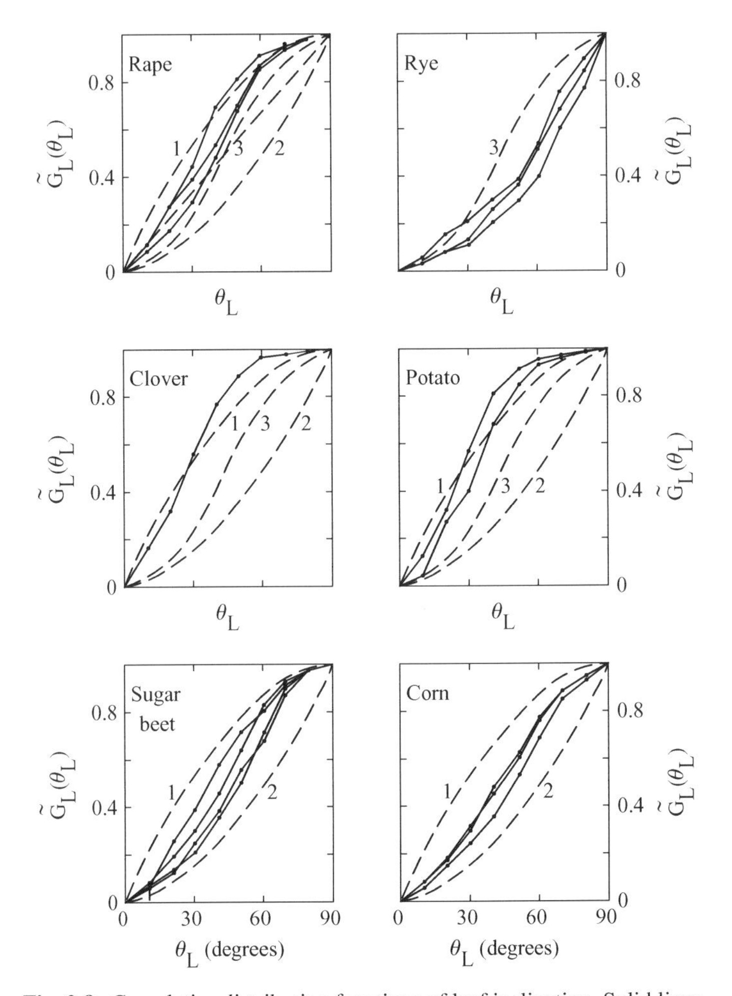

Fig. 2.8. Cumulative distribution functions of leaf inclination. Solid lines = observations of de Wit, 1965, at different stages of plant maturation; dashed lines 1, 2, 3 = Eqs. 2.12, 2.13, 2.14, respectively. (Adapted from de Wit, 1965, with kind permission from the Wageningen University and Research Centre, Wageningen, The Netherlands.)

giving the spatial average, throughout the crown, of the cosine of the angle between the leaf normal and the directional vector $\boldsymbol{r}$.

The radius vectors $\boldsymbol{z}$, $\boldsymbol{r}_L$, and $\boldsymbol{r}$ pierce the surface of the sphere of Fig. 2.6 at three points defining the vertices of a spherical triangle whose opposite sides we have labeled respectively a, b, and c. From the spherical law of cosines (see mathematical tables, e.g., Abramowitz and Stegun, 1964, p. 79)

$$\cos \widehat{\boldsymbol{r}_L \boldsymbol{r}} = \cos b \cos c + \sin b \sin c \cos \alpha, \tag{2.19}$$

or

$$\cos \widehat{\boldsymbol{r}_L \boldsymbol{r}} = \cos \theta \cos \theta_L + \sin \theta \sin \theta_L \cos (\phi - \phi_L). \tag{2.20}$$

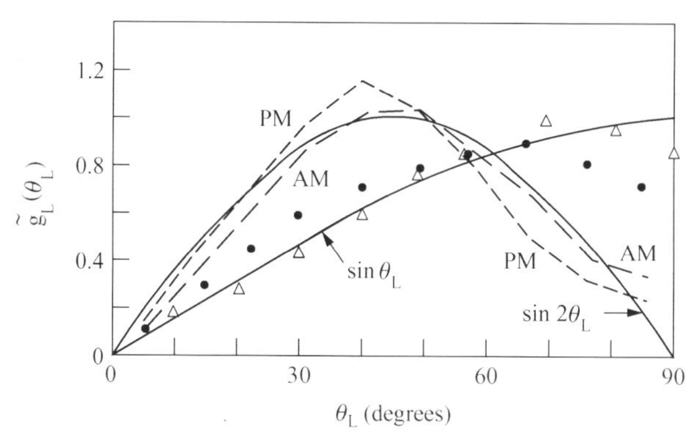

Fig. 2.9. Probability distribution functions of leaf inclination. Observations:- - - sunflower; AM, morning, PM, afternoon (reprinted from Lang, A. R. G., Yueqin Xiang, and J. M. Norman, Crop structure and the penetration of direct sunlight, Fig. 10, p. 93, *Agricultural and Forest Meteorology*, 35, Copyright © 1985 Elsevier, with kind permission from Elsevier Science); • sorghum (Lang *et al.*, 1985, Fig. 5, p. 91); △ corn (from Ross, J., *The Radiation Regime and Architecture of Plant Stands*, 1981, Fig. 32d, curve 3, p. 117, Copyright © 1981 Dr W. Junk Publishers, The Hague, with kind permission from Kluwer Academic Publishers); fitted distributions: —.

Plate 2.3. Typical spherical sector crown. Umbrella tree, Ayutthaya, Thailand. (Photograph by Peter S. Eagleson.)

For the special case in which $\boldsymbol{r}$ is oriented vertically upward (i.e., $\boldsymbol{r} \equiv z$), Eqs. 2.18 and 2.20 give

$$\hat{\beta} = \langle \cos \theta_L \rangle, \tag{2.21}$$

which is the spatial average, throughout the crown, of the cosine of the leaf inclination angle, θ_L. In the general case, this expectation is evaluated analytically by

$$\hat{\beta} = \int_0^{\frac{\pi}{2}} \widetilde{g_L}(\theta_L) \cos \theta_L \, d\theta_L. \tag{2.22}$$

Monteith (1969) gives values of $\hat{\beta}$ as low as 0.24 for the more-or-less vertical leaves of ryegrass, and as high as $\hat{\beta} \approx 1$ for the horizontal leaves of clover. Broadleaved and conifer trees normally have values in the range $0.30 < \hat{\beta} < 0.50$ (see Jarvis *et al.*, 1976, and Rauner, 1976). A collection of field estimates is included herein (see Tables 2.2, 2.3, and 2.4).[†]

Estimates of β derived from the distributions of Eqs. 2.12–2.15 corresponding to the de Wit (1965) classifications are given in Table 2.5. We see that these distributions give values of β which are generally larger than found for trees but agree well with the distributions observed for crops, as proposed by de Wit (1965) and as seen in Figs. 2.8 and 2.9. Also in Table 2.5 we give the range of observed β for crops and grasses, and the average β for deciduous and coniferous trees as assembled from the literature and listed in Tables 2.2–2.4. Note the higher values of observed β for crops and grasses than for trees, and note the general correspondence of the range of observed β for crops and grasses with the range of β for the de Wit (1965) distribution types. We conclude from this that the de Wit distributions are not applicable to trees, but are well suited for crops and grasses.

Determinants of leaf inclination

What determines β? Is it related to the direction of light? As we will see in Chapter 3, diffuse (i.e., cloudy sky) radiation may be assumed isotropic because on an element of surface it is incident with equal intensity from all sectors of the (hemispherical) sky. Cloudy skies are characteristic of wet climates in which available light rather than available water may be expected to limit plant productivity. In such situations, maximum productivity will require, among other things, maximum interception of available photosynthetically active (i.e., shortwave) radiation which can only happen when the reflection of this radiation is minimized. Although the reflection coefficient is relatively insensitive to incidence angle in these wave lengths, it is minimized when the radiation is perpendicular to the leaf surface. For isotropic radiation intensity, this condition would favor leaf areas whose normals are distributed as are those on the surface of a hemisphere.

[†] Henceforth we will drop the crown average notation, $\hat{\beta}$, and use β in which the averaging is understood.

Table 2.2. *Observed canopy parameters for evergreen vegetation*

Type	β	$a_w = \gamma L_t$	L_t	κ	Reference
Douglas fir			8.9		Fritschen *et al.* (1977)
			5.4		Lee and Black (1993)
Eucalyptus			1.05	0.37	Whitford *et al.* (1995)
(*E. maculata* Hook.)			2.8	0.37	Denmead *et al.* (1993)
(*E. grandis*)			2.9		Lang and McMurtrie (1992)
Larch		1.00	9.4		Allen (1968)
Sourgum		4.42			Cionco (1978)
Pine			6.9	0.41	Jarvis *et al.* (1976)
			2.3		Jacquemin and Noilhan (1990)
			9.5		Walter (1973)
Jack pine		2.6	2.0		Amiro (1990a)
Pinus radiata			1.7		Whitehead *et al.* (1994)
			4.38–9.23		Lang (1991)
Pinus taeda			2.6	0.43	Jarvis *et al.* (1976)
Pinus sylvestris			4.3	0.46	Oliver (1971)
			2.6		Lindroth (1985)
	0.45				Norman and Jarvis (1974)
Pinus ponderosa			4.0		Denmead and Bradley (1985)
Pinus resinosa/strobus			3.1[a]	0.42[a]	Jarvis *et al.* (1976)
Pinus resinosa			2.6	0.40	Jarvis *et al.* (1976)
Spruce		2.74			Cionco (1978)
Sitka spruce				0.4–0.6	Ross (1981)
		1.16, 1.67, 2.74	9.6[b], 3.3[c]		Landsberg and Jarvis (1973)
			9.8	0.49–0.56	Jarvis *et al.* (1976)
Black spruce		4.8	10.1		Amiro (1990a)
Jungle forest		3.84			Cionco (1978)
Tropical rainforest					
Ivory Coast			8.5		Walter and Breckle (1986)
Nigeria			9.5		Walter and Breckle (1986)
Puerto Rico			6.4		Odum *et al.* (1963)
Shrub					
Chaparral			2.5		Mooney *et al.* (1977)
Desert shrub			1.0		Larcher (1983)
Average	0.45	2.77	5.4	0.44	

[a] Average of 5 values.
[b] Includes entire crown.
[c] Dynamically active crown only.

If we take the leaf area, A_L, as statistically distributed with respect to angle in the same way as is the surface area of a hemisphere (Nichiporovich, 1961b), the foliage area is oriented equally in all hemispheric directions. That is, the distribution of foliage orientation is "hemispheric". Further assuming axial symmetry, we can use the sketch of Fig. 2.10 to express the θ-dependence of leaf area in this case as

$$\mathrm{d}A_L = (\pi D \sin\theta)\frac{D}{2}\,\mathrm{d}\theta. \tag{2.23}$$

Scaling by the total surface area of the hemisphere, we obtain the probability density function of leaf area, $\mathrm{f}(A_L)$, which is identical to that of leaf angle, $\widetilde{g_L}(\theta_L)$, as

$$\widetilde{g_L}(\theta_L) \equiv \mathrm{f}(A_L) = A_L^{-1}\frac{\mathrm{d}A_L}{\mathrm{d}\theta} = \sin\theta \equiv \sin\theta_L, \tag{2.24}$$

which we recognize as the pdf of leaf inclination for *erectophile* plants. Equations 2.22 and 2.24 then give, for the average leaf angle

$$\beta = \int_0^{\frac{\pi}{2}} \sin\theta_L \cos\theta_L \,\mathrm{d}\theta_L = 0.5. \tag{2.25}$$

Table 2.3. *Observed canopy parameters for deciduous vegetation*

Type	β	$a_w = \gamma L_t$	L_t	κ	Reference
Deciduous forest	0.25–0.35				Rauner (1976)
average			4.0		Uchijima (1976)
Oak		2.68			Cionco (1978)
			5.1	0.28	Rauner (1976)
			3.5	0.27	Rauner (1976)
			5.5, 4.6	0.42	Rauner (1976)
Ash			4.2–4.4		Vertessy *et al.* (1994)
Aspen	0.30		4.9	0.38	Rauner (1976)
			3.5	0.29	Rauner (1976)
			4.0	0.26	Rauner (1976)
			7.4		Rauner (1976)
			4.0		Amiro (1990a)
Birch	0.25		6.04		Rauner (1976)
Maple	0.33		5.18		Rauner (1976)
		4.03			Cionco (1978)
Linden	0.35		5.10, 4.0		Rauner (1976)
Willow	0.30		6.0	0.40	Cannell *et al.* (1987)
			7.5	0.50	Lindroth (1993)
Average	0.31	3.36	4.68	0.35	

Table 2.4. *Observed canopy parameters for crops and grasses*

Type	β	$a_w = \gamma L_t$	L_t	κ	Reference
Grassland		2.60	4.1	0.47	Ripley and Redmann (1976)
			5.0	0.31	Rauner (1976)
Rye				0.24	Monteith (1969)
				0.49	Ross (1981)
	0.50				de Wit (1965)
Barley	0.34				Mägi and Ross (1969)
Lucerne	0.67				Warren Wilson (1965)
Sorghum	0.64				Lang *et al.* (1985)
			4.4		Ross (1981)
Fescue			3.8		Aylor *et al.* (1993)
Clover				1.05	Monteith (1969)
	0.87				de Wit (1965)
Red clover		2.5			Lemon (1965)
Wheat	0.50				Nichiporovich (1961b)
		2.2–3.3		0.55	Denmead (1976)
				0.32–0.50	Ross (1975)
			3.6		Denmead (1976)
		2.45			Cionco (1972)
		2.8			Uchijima (1976)
Oats		2.8			Cionco (1972)
Horse beans	0.78				Ross (1981)
Soybeans		2.79			Cionco (1978)
Beans		2.92, 2.17, 1.36, 1.17	3.6		Thom (1971)
Corn			4.25		Allen *et al.* (1964)
	0.71				de Wit (1965)
	0.59				Ross (1981)
Maize	0.71, 0.89				Fakorede and Mock (1977)
			4.45		Udagawa *et al.* (1968)
		2.88			Wright and Brown (1967)
		2.0			Hicks and Sheih (1977)
		2.6	3.7	0.56	Brown and Covey (1966)
		2.46–2.88			Uchijima *et al.* (1970)
		3.0	4.16		Inoue and Uchijima (1979)
			5.6	0.28	McCaughey and Davies (1974)
			3.0		Jacquemin and Noilhan (1990)
		1.97			Cionco (1972)
			1.05	0.37	Whitford *et al.* (1995)
			4.0–4.2		Lemon (1963)
Potato	0.87				de Wit (1965)
	1.00				Ross (1981)
	1.00				Nichiporovich (1961b)
Rape	0.84				de Wit (1965)
Sugar beet	0.78				de Wit (1965)
Rice		2.5–3.0	4.01	0.45–0.65	Uchijima (1976)
Sunflower			1.7	0.72, 0.90, 0.66, 0.71, 0.80, 0.92	Saugier (1976)
			1.2		Ross (1981)
		1.61			Cionco (1978)
Average	0.73	2.41	3.63	0.59	

This value can be compared with observations in Table 2.5. It appears that coniferous trees, having $\beta = 0.45$, and being common to moist, cloudy climates, may exhibit this dependence of β on the direction of light.

For direct (as opposed to diffuse) light, the proposition that the leaf is perpendicular to the incident light leads directly to the relation

$$\beta = \cos\left(\frac{\pi}{2} - h_{\otimes}\right) = \sin(h_{\otimes}), \tag{2.26}$$

in which $h_{\otimes}$ is the effective solar altitude. In Chapter 3 (Fig. 3.9) we will see that over the latitude range $0^\circ \leq \Phi \leq 60^\circ$, the daylight-hour average solar altitude at summer solstice is $27^\circ \leq h_{\otimes} \leq 40^\circ$ whence, from Eq. 2.26, $0.46 \leq \beta \leq 0.65$. The observed average for deciduous trees in Table 2.2 is $\beta \approx 0.31$. We conclude from this that the leaf surface of deciduous trees is approximately 26 degrees more vertical than is the normal to the direction of light. In Chapter 3 we will consider these important questions in more detail.

F Leaf type

There are basically two types of leaves, broad leaves and needle leaves.

Broad leaves

So-called "broad leaves" are the thin flat leaves, anywhere from 2 cm to 25 cm in characteristic dimension depending upon species. In plan view, again depending upon species, they may be star-shaped with a varied number of points, or they may be heart-shaped with serrated edge, or have multiple lobes. Their stomates are predominantly on the underside of the leaf giving them the name *hypostomatous*. However, the leaves

Table 2.5. *Estimates of average leaf inclination*

Form of distribution	$\langle\beta\rangle$ from Eq. 2.22	$\langle\beta\rangle$ from observation[a]
Hemispherical (Eq. 2.24)	0.50	Deciduous trees $\langle\beta\rangle \approx 0.31$
Erectophile (Eq. 2.13)	0.50	Coniferous trees $\langle\beta\rangle \approx 0.45$
Uniform (Eq. 2.15)	0.64	
Plagiophile (Eq. 2.14)	0.67	Crops and grasses $0.50 < \langle\beta\rangle < 0.89$
Planophile (Eq. 2.12)	0.79	

[a] See Tables 2.2–2.4.

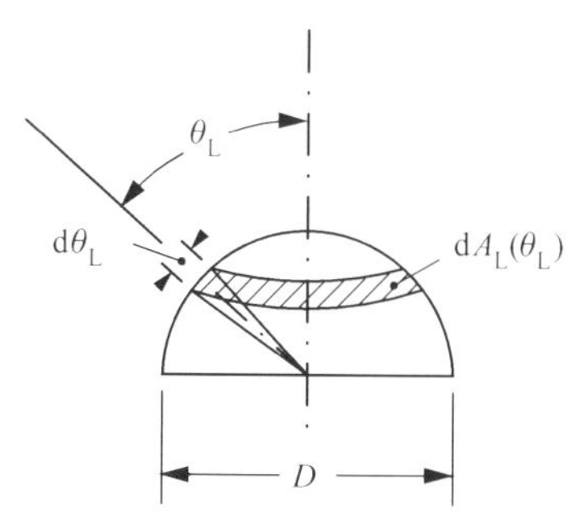

Fig. 2.10. Definition sketch for hemispheric distribution of crown surface area.

Plate 2.4. Typical conical crown. Eastern red cedar (foreground). (Photographed by William D. Rich in Holden Arboretum, Cleveland. Copyright © 2001 William D. Rich.)

of grasses and herbaceous broadleaved plants have stomates on both the upper and lower surfaces and are called *amphistomatous*.

Needle leaves

There is a wide range of geometries within the group of needle-leaved plants, but we will identify and treat only the two extremes:

"Individual needle" varieties

Most *Pinus* needles emerge from the stem in a tight circular bundle (called a *fascicle*) of multisided individuals which opens progressively along the needle axis, thereby providing a varying degree of mutual interference to the transport of momentum and to the penetration of light. Those varieties with very long needles (8–45 cm) in which the individuals become separated to the point of relative independence we call "individual needle" varieties. As is pictured by Zim and Martin (1987), these are exemplified by the longleaf pine (*Pinus palustris*) and the loblolly pine (*Pinus taeda*), and (to a somewhat lesser degree) include others such as sugar pine (*Pinus lambertiana*), pitch pine (*Pinus rigida*), ponderosa pine (*Pinus ponderosa*), and Eastern white pine (*Pinus strobus*).

There are three needles to the circular bundle of *Pinus taeda* (Kozlowski and Schumacher, 1943) and of Monterey pine (*Pinus radiata*) (Raison *et al.*, 1992), so each needle of these varieties has a cross-section consisting of a 120° circular sector. The important ratio of total needle perimeter (P) to arc-chord (C_a) of these varieties is $P/C_a = 2.36$ (Raison *et al.*, 1992). All three sides of these needles are stomated (Kramer and Decker, 1944, p. 352).

The cross-section of *Pinus strobus* needles is a 72° circular sector (i.e., five to the bundle) with stomates only on the straight sides (Kozlowski and Schumacher, 1943), and Scots pine (*Pinus sylvestris*) needles have a semicircular cross-section with stomates on only the straight side. Over a range of needle varieties, Waring (1983) found the average perimeter-to-chord ratio, $\langle P/C_a \rangle = 2.5$.

"Clustered needle" varieties

Particularly spruce, but to a lesser degree fir, larch, pinyon pine (*Pinus edulis*) and other varieties of needle-leaved plants have short (2–3 cm), stiff, relatively thick needles[†] that are clustered closely together on long twigs forming what are called "shoots". There is a high degree of interference among these needles so that the characteristic physical dimension is that of the cluster rather than that of the individual needle. Needle diameters are about 2 mm, and cluster widths are about one order of magnitude larger.

These leaf characteristics will be important later when we consider momentum transport and rainfall interception in the canopy.

[†] Spruce needles have a "+" cross-section such as a square with four deeply fluted quadrants (cf. Zim and Martin, 1987, p. 29).

3

Radiant fluxes

Characteristics of incident solar radiation and its partition are described, and the average insolation during the nominal growing season is determined as a function of latitude. Penetration of radiation into the canopy is considered theoretically as a function of solar altitude and leaf angle, and its extinction coefficient is compared with observation.

Maximum canopy absorption of solar radiation is reasoned to occur when the light at the lowest leaves is at the biological limit called *compensation intensity*, and when by virtue of scattering the coefficient of light absorption equals the cosine of the leaf angle as is shown by the observations. These conditions define the state of *optical optimality* that appears to govern those systems that are productively limited by available photosynthetically active light. Systems limited by water and apparently by leaf temperature are observed to regulate leaf angle in ways that accept or reject heat-producing long-wave radiation.

A Introduction

The average rate at which radiant solar energy passes through a plane unit area normal to the Sun's rays is referred to here as *radiant flux density*. At the outer limit of the Earth's atmosphere this quantity is termed the *solar constant*; it is given the symbol S_{oo} and has the value (Fröhlich and Brusa, 1981):

$$S_{oo} = \text{solar constant} = 1.96\,\text{cal cm}^{-2}\,\text{min}^{-1} = 1367\,\text{W m}^{-2}. \tag{3.1}$$

It is distributed spectrally according to Planck's distribution function for a blackbody (e.g., Liou, 1980, p. 9) evaluated at the Sun's surface temperature, 5700 to 5800 K.

In the absence of an atmosphere, the radiant flux density on a horizontal plane at the Earth's surface is given by the astronomical relation

$$S'_{oo} = \frac{S_{oo}}{\rho_o} \sin h_{\otimes}, \tag{3.2}$$

in which ρ_o is the distance between the Sun and Earth in astronomical units, AU (1 AU = mean Sun–Earth distance = 1.496×10^8 km), and $h_\otimes$ is the solar elevation. Spherical trigonometry gives

$$\sin(h_\otimes) = \sin\Phi \sin\delta + \cos\Phi \cos\delta \cos\tau, \tag{3.3}$$

where Φ is the geographical latitude, δ is the solar declination (see List, 1951), and τ is the solar hour angle (with zero at apparent noon and positive values in the afternoon). For more details of this celestial geometry see the illustration and caption of Fig. 3.1 as taken from Eagleson (1970, p. 404).

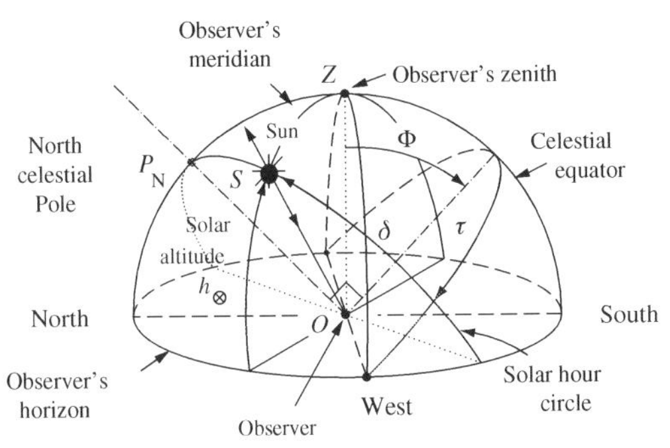

Fig. 3.1. Great circles on the celestial sphere. Consider an *observer* at point O on the Earth's surface. A coordinate system is established having its vertical perpendicular to the tangent plane at this point. This coordinate system is defined as *topocentric*, and the measures of solar position seen from point O are called *apparent* measures. Transferring our tangent plane down the radius until point O coincides with the Earth's center we have the *geocentric* coordinate system shown in Fig. 3.1. The *celestial sphere* (in this case the *solar sphere*) is defined by the imaginary spherical shell passing through the Sun and having its center at point O. The observer's *horizon* will then be the great circle defined by the intersection of a plane, tangent to the Earth's surface at point O, with the celestial sphere. The *celestial equator* is another great circle given by the intersection of the celestial sphere with a plane through the Earth's equator. A perpendicular to the equator plane, erected at point O, pierces the celestial sphere at the *celestial pole* P; and a perpendicular to the observer's horizon plane, erected at point O, pierces the celestial sphere at the observer's *zenith* Z. A great circle on the celestial sphere and including both P and Z is called the observer's *meridian.* The angle Φ between the equator plane and an east–west plane containing Z and O is the *astronomical latitude* and is positive for points O in the northern (terrestrial) hemisphere. A great circle on the celestial sphere containing the Sun S and both celestial poles P_N and P_S is known as the solar *hour circle*, and the angular distance between the planes of the meridian and the Sun's hour circle is known as the Sun's *hour angle* τ. It is positive for Sun positions to the west of the meridian. A plane containing S, O, and Z defines another great circle along the celestial sphere. This plane is perpendicular to the observer's horizon plane. The solar *altitude* is defined by the angle $h_\otimes$ measured along this great circle, between the observer's horizon plane and the plane containing the solar hour circle. The angular distance between the celestial equator plane and the Sun, measured from the former (and positive when the Sun lies north of the Earth's equator) and along the solar hour circle, is known as the *solar declination* δ.

The presence of an atmosphere causes attenuation in the intensity of this so-called *direct* radiation and modification in its spectral distribution. Molecular and particulate scattering, which are wavelength-selective refraction and reflection processes, produce what is called *diffuse sky radiation*. Exitation of molecular resonances on the other hand produces energy absorption and subsequent long-wave radiation by the atmosphere.

Computation of the attenuation of radiation by the atmosphere under specific conditions is a complex and still highly empirical task (e.g., Kondratyev, 1969; Iqbal, 1983). We will avoid this calculation herein by using nominal values of incident light intensity at the top of the vegetation canopy calculated by de Wit (1965) as a function of solar altitude for the two limiting conditions: very clear sky with only direct radiation, and overcast sky where all radiation is diffuse.

Our primary interests here are in the total quantity and average incidence angle of radiation intercepted by the surfaces of the vegetation during the growing season. It is likely that certain structural properties of the vegetation are adaptations to the radiation regime (Ross, 1981, p. 91), particularly its angle of leaf incidence, and its ability to shed or retain heat.

We begin with a look at the characteristics of the incident radiation regime following the work of Kondratyev (1969) and Ross (1981).

B Definitions

We define the incident radiation field in terms of its energy, in which case the primary characteristic is its *radiant flux density*, R, measured in cal cm^{-2} min^{-1}. The radiant flux density or *radiance* for short, is a function of the wavelength, λ, of the radiation; the time, t; the location, P, in space; and the direction, $\boldsymbol{r}$, of the ray. Accordingly, we write, formally

> *radiance* $= R_\lambda(t, P, \boldsymbol{r}) =$ rate at which energy in the wavelength interval around λ, and in the increment of solid angle, $d\omega$, around the ray, $\boldsymbol{r}$, passes through a unit of surface area, $d\sigma$, perpendicular to the ray at the particular time, t, and location, P.

This is illustrated in the definition sketch of Fig. 3.2a where θ is the angle between the ray, $\boldsymbol{r}$, and an (upward) normal, $\boldsymbol{n}$, to an arbitrarily oriented surface. If R_λ is independent of direction, $\boldsymbol{r}$, the radiation field is *isotropic*. Only blackbody radiation is strictly isotropic, but we will see that diffuse sky radiation under overcast skies approximates this condition.

While the flux of radiation through a unit area perpendicular to the ray, $\boldsymbol{r}$, has been termed *radiance*, and is independent of the surface on which it is incident, the flux of radiation through an arbitrarily oriented surface (such as a leaf) is known as the *irradiance* of the surface and depends upon the surface orientation. That is, *radiance* refers to the ray, and *irradiance* refers to the surface. The flux of radiation, F_λ, through

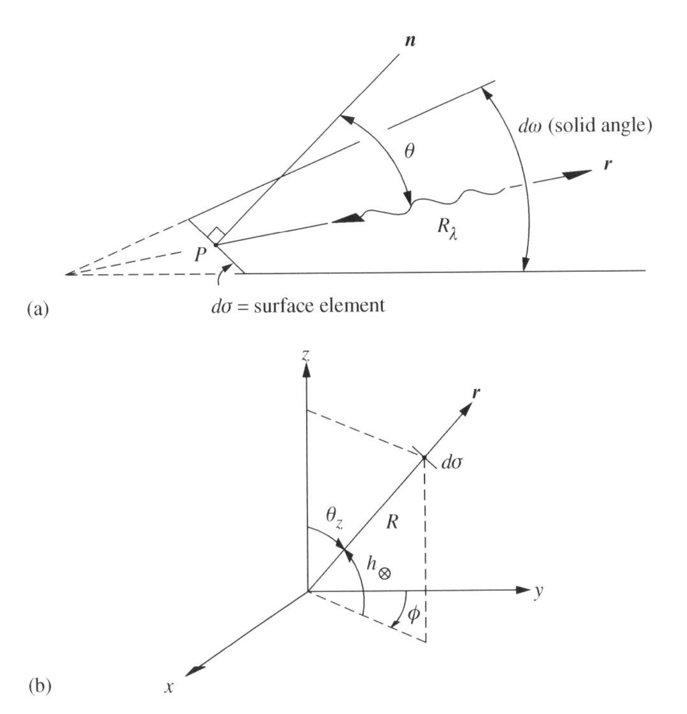

Fig. 3.2. Definition sketches. (a) Elementary cone for radiance. (After Kondratyev, K.Ya., *Radiation in the Atmosphere*, Fig. 1.2, p. 10, Copyright © 1969 Academic Press, with kind permission from Academic Press, Orlando, FL.) (b) Spherical coordinates for radiant flux.

the arbitrarily oriented surface of Fig 3.2a is simply

$$F_\lambda = R_\lambda \langle \cos \widehat{\boldsymbol{nr}} \rangle, \tag{3.4}$$

in which $\widehat{\boldsymbol{nr}}$ is the angle between the vectors $\boldsymbol{n}$ and $\boldsymbol{r}$, and the operator $\langle \cdot \cdot \rangle$ signifies local spatial averaging over the many surface orientations (e.g., leaves) present at location, *P.*

In this work we are interested only in climatic issues. Therefore, we will describe climatic radiance solely in terms of its first-order dependent variables:

(1) latitude, Φ, and
(2) growing season timing, t, and length, τ.

Thus, we will deal with radiances that are heavily averaged, both in space at a given latitude, and in time over an ensemble of growing season months. The characteristic radiances at a given latitude which result from this process are therefore "nominal" and may be expected at best to explain only first-order spatial variations in vegetation form and function. To indicate that the radiances have been so averaged we give them an overbar. The time-averaged form of Eq. 3.4 is then

$$\overline{F_\lambda} = \overline{R_\lambda \langle \cos \widehat{\boldsymbol{nr}} \rangle}. \tag{3.5}$$

Plate 3.1. Giant *Ficus* walk toward the light. At Angkor, Cambodia, spong trees (*Ficus tetrameles nudiflora*) step over the wall surrounding the ruins of Ta Prom seeking access to light available where the temple created a gap in the canopy. (Photograph by Peter S. Eagleson.)

A special case of particular interest here is when the surface of Fig. 3.2a has a horizontal orientation. The irradiance of this unit horizontal area is known as the *insolation*, I_λ. In terms of averaged quantities we use Eq. 3.5 and Fig 3.2b to write

$$\overline{I_\lambda} = \overline{R_\lambda \langle \cos \widehat{zr} \rangle}. \tag{3.6}$$

The total local surface irradiance is the *climatic insolation*, I_0.

Direct radiation

Radiation coming directly from the sun without modification by the absorption and scattering effects of intervening clouds is called *direct* or *beam* radiation, R_b, and is in the spectral range 280 nm $< \lambda <$ 4000 nm (Ross, 1981, p. 159). About half of this (van Wijk, 1963) is in the region of photosynthetically active radiation, PAR, where 400 nm $< \lambda_{PAR} \equiv \lambda_P <$ 700 nm, and about half is in the near-infrared region, NIR, where 700 nm $< \lambda_{NIR} \equiv \lambda_N <$ 4000 nm. A small amount occurs in the ultra-violet region, UV, where $\lambda_{UV} <$ 400 nm; however we neglect that here and assume PAR = NIR = $R_b/2$. We will concern ourselves primarily with PAR in this work, thus all radiation symbols refer to PAR unless stated otherwise. The PAR absorbed by the atmosphere and by the surface heats these receptors to earthly rather than solar temperatures at which they reradiate in correspondingly longer wavelengths. The net, q_b, of this bi-directional longwave flux is important only to the surface energy balance of later chapters and its estimation is considered in Appendix E.

Direct solar radiation is zero outside the small solid angle $\Delta\omega$ corresponding to the angular diameter of the Sun located at position $\boldsymbol{r}$. Thus we can use Eq. 3.6 to approximate the seasonal and zonal average beam insolation, $\overline{I_b}$, as

$$\overline{I_b}(\overline{h_\otimes}) = \overline{R_b}(h_\otimes)\sin(\overline{h_\otimes}), \tag{3.7}$$

in which $\overline{R_b}$ is the (seasonal and zonal) average solar (i.e., direct) radiance in the PAR band at the top of the canopy for solar altitude, $h_\otimes$, and $\overline{h_\otimes}$ is the average (daylight-hour) solar altitude during the growing season.

Diffuse radiation

That part of solar radiation which reaches the vegetation from above after multiple scattering from atmospheric particles and clouds as well as after reflection from the vegetation (and ground) and additional atmospheric scattering is called the *diffuse* sky radiation, R_d (Ross, 1981, p. 161). Its spectral distribution is different from that of direct radiation. In general, the angular distribution of diffuse radiation is non-isotropic. In this case, radiation is incident upon our unit surface area from all sectors of the sky, and the seasonal and zonal average diffuse insolation, $\overline{I_d}$, is

$$\overline{I_d}(h_\otimes) = \overline{\langle R_d(\boldsymbol{r}:\ h_\otimes)\cos\widehat{\boldsymbol{zr}}\rangle}. \tag{3.8}$$

Noting that the elemental solid angle, $d\omega$, surrounding $\boldsymbol{r}$ is written

$$d\omega = \sin\theta_z\, d\theta_z\, d\phi$$

in which θ_z is the complement of $h_\otimes$. Using Fig. 3.2b, Eq. 3.8 can be written

$$\begin{aligned}\overline{I_d}(h_\otimes) &= \int_0^{2\pi} d\phi \int_0^{\frac{\pi}{2}} \overline{R_d}(\boldsymbol{r}:\ h_\otimes)\sin\theta_z\cos\theta_z\, d\theta_z \\ &= \pi\int_0^{\frac{\pi}{2}} \overline{R_d}(\boldsymbol{r}:\ h_\otimes)\sin 2\theta_z\, d\theta_z.\end{aligned} \tag{3.9}$$

The integrand of Eq. 3.9 defines the diffuse radiance on a horizontal surface. This radiance is also called the *zonal brightness* of the sky, $\overline{D_{dz}}$, and its angular distribution is symmetrical with respect to the solar vertical plane.

Magnitudes

On average, the direct and diffuse components of the total radiation are distributed among the three primary spectral bands (i.e., UV, PAR, and NIR) as is shown in Table 3.1 taken from Ross (1975, Table II, p. 19). The UV component is negligible for our purposes. Only a small amount of the total incident energy, less than 10 percent, is used *directly* by the plants in photosynthesis (cf. Gates, 1980, p. 26; Larcher, 1983, p. 20) where it is stored chemically in the high-energy organic compounds of biomass. This is from the PAR component. The remaining majority is converted into heat where it powers transpiration, catalyzes photosynthesis, and exchanges convectively with the atmosphere. The mixture of direct and diffuse radiation varies with place and time. Without distinguishing among the spectral components, meteorologists define this situation by the *beam ratio*, μ, which is the direct radiance fraction of total radiance. Values of μ under conditions of clear atmospere and industrial cloudiness are presented in Fig. 3.3 as a function of solar altitude, $h_{\otimes}$, using data from Gates (1980, Table 6.1, p. 114).

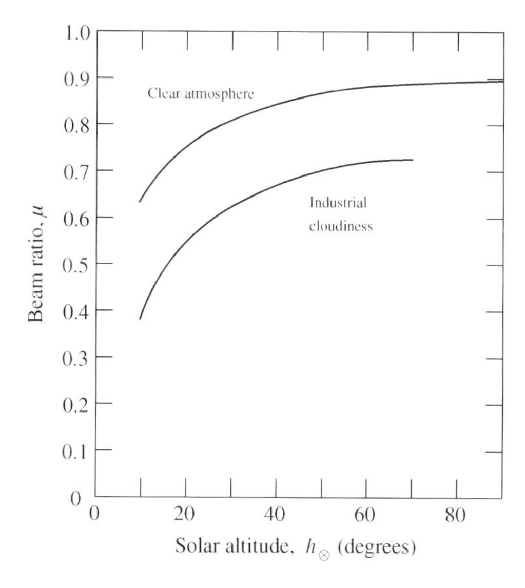

Fig. 3.3. Typical values of the beam ratio. (Data from Gates, 1980, Fig. 6.1, p. 114.)

Table 3.1. *Mean radiation fraction*

	Waveband (μm)		
	UV (0.28–0.38)	PAR (0.38–0.71)	NIR (0.71–4.0)
Direct	0.02	0.42	0.56
Diffuse	0.10	0.65	0.25
Total (cloudless)	0.03	0.50	0.47

Source: Ross, J., Radiative transfer in plant communities, Table II, p. 19, in *Vegetation and the Atmosphere*, Vol. 1, *Principles*, edited by J.L. Monteith, Copyright © 1975, Academic Press, with kind permission from Academic Press Ltd, London.

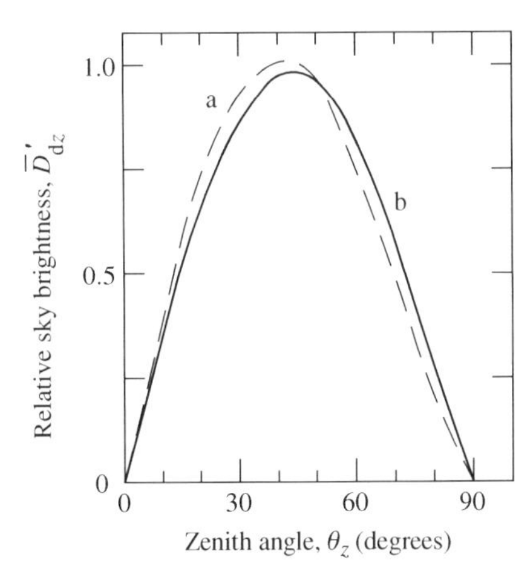

Fig. 3.4. Verification of isotropic radiance for overcast sky. (a) Overcast sky observations (Põldmaa, 1963); (b) isotropic radiance. (After Ross, J., *The Radiation Regime and Architecture of Plant Stands*, Fig. 41, p. 164, Copyright © 1981 Dr W. Junk Publishers, The Hague, with kind permission from Kluwer Academic Publishers.)

Limiting cases

As stated earlier, we will recognize two limiting radiation cases:

(1) *Very clear sky* – in which there is no diffuse radiation (i.e., beam ratio = 1), the Sun's rays are everywhere at the same inclination, and the averaged insolation is given by Eq. 3.7.

(2) *Overcast sky* – in which there is no direct radiation (i.e., beam ratio = 0), and the averaged insolation is given by Eq. 3.9.

Observations show (Põldmaa, 1963, reproduced in Ross, 1981, Fig. 41, p. 164) that under overcast skies, the diffuse radiation is closely isotropic. Under isotropic conditions, $\overline{R}_d$ is independent of $\boldsymbol{r}$ and Eq. 3.9 reduces to

$$\overline{I}_d = \pi \overline{R}_d, \tag{3.10}$$

which is Lambert's law (Kondratyev, 1969, p. 13). The zonal brightness (cf. Eq. 3.9) may be normalized to give the *relative brightness* of the sky, $\overline{D}'_{dz}$, under isotropic conditions, as

$$\overline{D}'_{dz}(\theta) = \frac{\overline{D}_{dz}}{\overline{R}_d} = \sin 2\theta_z. \tag{3.11}$$

Equation 3.11 is compared with Põldmaa's (1963) observations in Fig. 3.4 as adapted from Ross (1981, Fig. 41, p. 164). The excellent agreement confirms the isotropic assumption for the radiance of an overcast sky.

C Partition of radiance

Total beam radiance, R_i, incident upon the canopy is partitioned by the foliage elements, into three components: absorption (R_a), reflection (R_r), and transmission (R_t)

as is indicated schematically in Fig. 3.5. (The sum of reflection and transmission is known as *scattering*.) Normalizing by the incident total insolation, this partition is expressed mathematically

$$\frac{R_a}{R_i} + \frac{R_r}{R_i} + \frac{R_t}{R_i} = \alpha_T + \rho_T + \tau_T = 1 \tag{3.12}$$

in which α_T, ρ_T, and τ_T are the coefficients of absorption, reflection, and transmission respectively. For average green leaves the irradiance[†] partition coefficients have been calculated separately for each of the three major spectral bands by Ross (1975, Table V, p. 22) and have the values presented here in Table 3.2. Note the low reflection and transmission of PAR and hence its high absorption. This important point is made

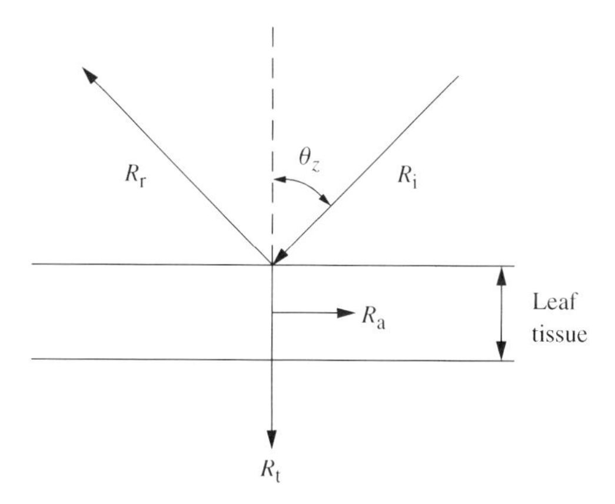

Fig. 3.5. Partition of incident radiance by green leaf. (After Kriedeman, P. E., *et al.*, Photosynthesis in relation to leaf orientation and light interception, Fig. 4, p. 597, *Australian Journal of Biological Science*, 17, 591–600, Copyright © 1964 CSIRO Publishing, with kind permission from CSIRO Publishing, Collingwood, Australia.)

Table 3.2. *Partition of irradiance for average green leaves*[a]

Coefficient	UV	PAR	NIR
Reflection, ρ_T	0.30	0.09	0.51
Transmission, τ_T	0.20	0.06	0.34
Absorption, α_T	0.50	0.85	0.15

[a] Note that, by definition, irradiance is the flux density experienced by a surface perpendicular to the beam and when the surface under consideration is horizontal the flux density is often referred to as "insolation".
Source: Ross, J., Radiative transfer in plant communities, Table V, p. 22, in *Vegetation and the Atmosphere*, Vol. 1, *Principles*, edited by J.L. Monteith, Copyright © 1975 Academic Press, with kind permission from Academic Press Ltd, London.

[†] Remember that by definition irradiance is perpendicular to the surface in question.

graphically in Fig. 3.6 (Larcher, 1983, Fig. 2.4, p. 10, after Gates, 1965) where the full spectral distribution of the coefficients is shown for a poplar (*Populus deltoides*) leaf.

Consider next the sensitivity of radiance partition to the angle of radiation incidence. Ross (1981, Table II.3.2, p. 185) has examined this for isolated leaves of lettuce (thin and pale) and Chinese hibiscus (thick and glossy) under incident white light, and his results are presented here in Table 3.3. Note:

Table 3.3. *Sensitivity of partition of white light to angle of radiation incidence*

	Angle of radiation incidence (zenith angle)						
	0°	30°	40°	50°	60°	70°	75°
Lettuce							
τ	0.175	0.17	0.16	0.145	0.12	0.11	0.10
ρ	0.14	0.145	0.155	0.165	0.19	0.23	0.25
α	0.685	0.685	0.685	0.69	0.69	0.66	0.66
Chinese hibiscus							
τ	0.06	0.055	0.05	0.045	0.04	0.04	0.04
ρ	0.065	0.075	0.08	0.085	0.095	0.11	0.13
α	0.875	0.87	0.87	0.87	0.865	0.85	0.83

Note: Leaves of lettuce (*Lactuca sativa*) were thin and pale green in color, while those of Chinese hibiscus (*Hibiscus rosa sinensis*) were thick and glossy.
Source: Ross, J., *The Radiation Regime and Architecture of Plant Stands*, Table II.3.2, p. 185, Copyright © 1981 Dr W. Junk Publishers, The Hague, with kind permission from Kluwer Academic Publishers.

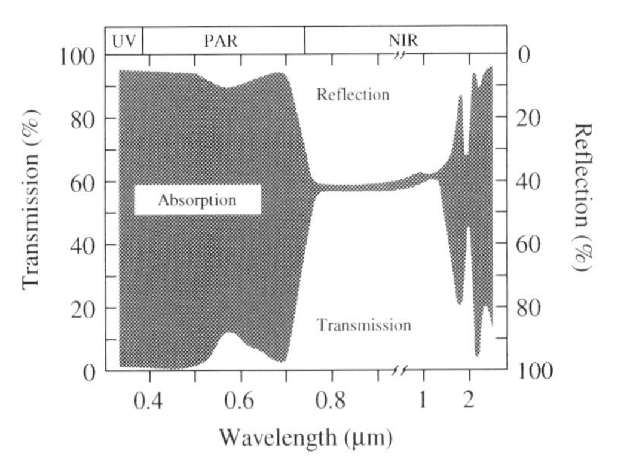

Fig. 3.6. Spectral partition of radiance by leaf of poplar *(Populus deltoides)*. (From Larcher, W., *Physiological Plant Ecology*, corrected printing of the 2nd edn, Fig. 2.4, p. 10, Copyright © 1983 Springer-Verlag Publishers; after Gates, D. M., Energy, plants and ecology, Fig. 3, p. 5, *Ecology*, 46/1, 2, 1–13, Copyright © 1965 Ecological Society of America, with kind permission from Springer-Verlag GmbH & Co. KG, W. Larcher, and the Ecological Society of America.)

(1) absorption is insensitive to angle of radiation incidence,
(2) reflection increases beyond a zenith angle of incidence of about 45°, and consequently, from Eq. 3.12, and
(3) transmission decreases beyond a zenith angle of incidence of about 45°.

Plant structures which minimize leaf *upward* reflection will therefore maximize downward scattering and thus maximize the potential canopy depth, a condition which should lead to maximum biological productivity, all else remaining constant.

Partition of total irradiance for the leaves of a wide selection of tree species is given by Birkebak and Birkebak (1964, Table 1, p. 648) and because of its importance to our later work it is reproduced here in Table 3.4.

Table 3.4. *Partition of total beam irradiance of upper surface of tree leaves*[a]

Common name[b]	Reflectance, ρ_T	Transmittance, τ_T	Absorptance, α_T
Bolleana poplar	0.285	0.18	0.535
Quaking aspen	0.32	0.195	0.485
Eastern cottonwood	0.280	0.275	0.445
Peachleaf willow	0.27	0.18	0.55
Weeping willow	0.285	0.22	0.515
Butternut	0.28	0.24	0.48
Common white birch	0.32	0.24	0.44
European white birch	0.30	0.22	0.48
Common alder	0.23	0.21	0.56
Northern red oak	0.27	0.24	0.49
Northern pin oak	0.235	0.17	0.595
White oak	0.28	0.235	0.485
Bur oak	0.24	0.23	0.53
American elm	0.235	0.18	0.585
Slippery elm	0.24	0.275	0.485
Black cherry	0.24	0.19	0.57
Silver maple	0.30	0.21	0.49
Norway maple	0.25	0.24	0.51
Box elder	0.31	0.22	0.50
Ohio buckeye	0.27	0.19	0.54
Common locust	0.325	0.255	0.42
Green ash	0.29	0.21	0.50
Average	0.27	0.22	0.51

[a] Values are measured hemispherical reflectances and transmittances for normal incidence with absorptances calculated from Eq. 3.12.
[b] See listing of "Common and scientific names used", p. xxxvi.
Source: Birkebak, R., and R. Birkebak, Solar radiation characteristics of tree leaves, Table I, p. 648, in *Ecology*, 45/3, 646–649, Copyright © 1964 Ecological Society of America, with kind permission from the Ecological Society of America, Washington, DC.

Moving now from individual leaves to the optical properties of a dense stand of trees, Ross (1975, Fig. 11, p. 42) demonstrates theoretically the important sensitivity of the reflection coefficient to the angle of incidence of the beam radiation as is shown here in Fig. 3.7. Using the series approximation suggested by Goel (1989, p. 209) we have added the following empirical approximations of the NIR and PAR curves for use in estimation:

$$\rho_{\mathrm{NIR}} = 0.60 - 0.31\, h_{\otimes} - 0.03\, (h_{\otimes})^2, \tag{3.13}$$

and

$$\rho_{\mathrm{PAR}} = 0.08 + 0.01\, h_{\otimes} - 0.03\, (h_{\otimes})^2, \tag{3.14}$$

in which the solar altitude, $h_{\otimes}$, is in radians. Note that ρ_{NIR} is also sensitive to β, and therefore Eq. 3.13 is a good approximation only for a uniform distribution of leaf normals.

Neglecting the small UV component we see, as above, that the reflection of the PAR is small and is relatively insensitive to incidence angle, while the reflection of NIR is large and is highly sensitive to incidence angle. In principal, this NIR behavior provides a primary means of plant temperature regulation, climatically through natural selection and adaptation for those average leaf angles that allow for spilling or retention of NIR, and transiently through either heliotropism to follow the sun or turgor loss (i.e., wilting) to spill the radiation. However, Fig. 3.7 shows that over the practical range $27^{\circ} \leq h_{\otimes} \leq 40^{\circ}$,[†] a 90° change in θ_{L} will change ρ_{NIR} by at most 10 percent.

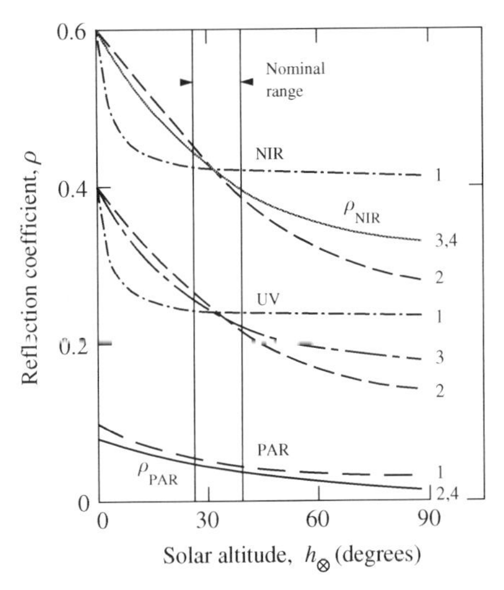

Fig. 3.7. Theoretical reflection coefficients of dense stands for direct radiation. 1, Horizontal leaves; 2, vertical leaves; 3, uniform distribution of leaf normals; 4, empirical approximations, Eqs. 3.13, 3.14. (From Ross, J., Radiative transfer in plant communities, Fig. 11, p. 42, in *Vegetation and the Atmosphere*, Vol. 1, *Principles*, edited by J. L. Monteith, Copyright © 1975 Academic Press Ltd, with kind permission from Academic Press Ltd, London.)

[†] To be shown later in Fig. 3.9.

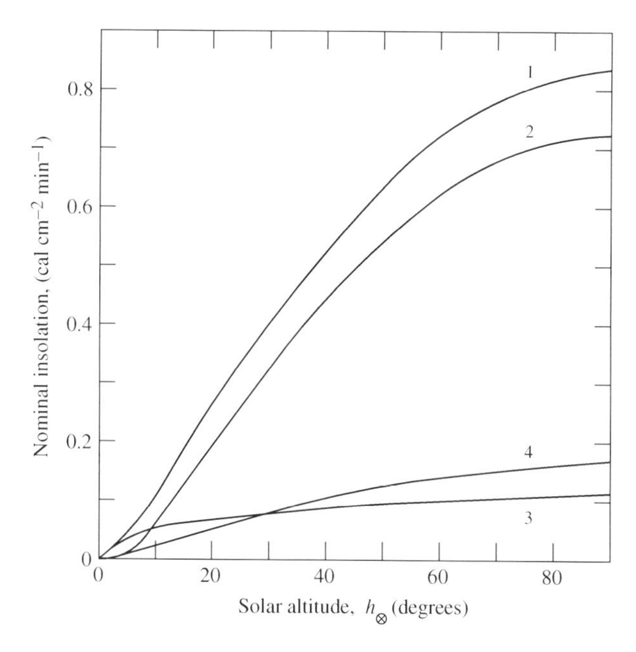

Fig. 3.8. Incident radiance (PAR) on a horizontal surface (i.e., insolation). 1, Total light with very clear sky; 2, direct light with very clear sky; 3, diffuse light with very clear sky; 4, total (same as diffuse) light with overcast sky. (From de Wit, C. T., *Photosynthesis of Leaf Canopies*, Fig. 2, p. 7, Copyright © 1965 Wageningen University, with kind permission from Wageningen UR, Wageningen, The Netherlands.)

Finally, to calculate *quantities* of radiant energy received by the vegetation during the growing season we will need to multiply the averaged radiances at a given latitude by the similarly averaged growing season length at that latitude.

D Nominal average insolation over the growing season†

Assumptions

In later chapters dealing with the biological response of vegetation canopies to solar radiation it will be helpful to express this relation approximately in terms of long-term averaged rates. Accordingly, we will need representative values of the local insolation averaged over the growing season and hence over the range of cloud covers and beam ratios characteristic of local climate in the growing season. Although such generalizations are certainly problematic, de Wit (1965) has offered one for the limiting cases of "very clear" and "overcast" skies, and we make use of his results here.

De Wit (1965) used the heating and ventilating nomograms of Fritz (1949), based on the earlier work of Kimball and Hand (1921) and of Klein (1948), to estimate the insolation as a function of solar altitude, $h_{\otimes}$, for a "very clear sky" (i.e., cloudless and dustless) in which the precipitable water was 10 mm. He multiplied these values by 0.5 to yield the PAR (van Wijk, 1963) which are plotted here as Curve no. 1 of Figure 3.8. He further used the calculations of Jones and Condit (1948) to estimate the direct and diffuse components of this PAR and these are reproduced here as Curves nos. 2 and 3 of

† Henceforth, we will work solely with climatic variables averaged over the growing season and for simplicity will discontinue the overbar notation for such time averages.

Fig 3.8. For overcast skies de Wit (1965) estimated the insolation at $h_{\otimes} = 90°$ to be 20 percent of that for very clear sky and distributed it with solar altitude according to the isotropic assumption to be discussed in a later section of this chapter. This is shown as Curve no. 4 of Fig. 3.8. We can replace the variable $h_{\otimes}$ of Fig. 3.8 with latitude, Φ, through Eq. 3.3 provided we assume representative values for the solar declination, δ, and the solar hour angle, τ.

Solar declination

We fix the solar declination at a value representative of the vegetation growing season. As we will see, the growing season changes in length with latitude and varies somewhat in timing according to the climate. However, the first-order accuracy of these radiation estimates suggests using a single average value of the declination as being representative of the growing season irrespective of latitude and climate; we choose the summer solstice, $\delta = 23°$. Northern hemisphere daylengths for this declination are given by Trewartha (1954, p. 12) and are listed here in Table 3.5.

Solar hour angle

Photosynthetic efficiency is extremely sensitive to leaf temperature, falling off sharply on either side of a species-dependent optimum temperature, T_m, as we will see in Chapter 8. To a considerable extent, the struggle for productivity hinges on the ability of the plant to maintain optimal heat for photosynthesis. Therefore, we hypothesize that natural selection and adaptation fix T_m as the "operating thermal state"

Table 3.5. *Properties of northern hemisphere daylight at summer solstice*

Latitude (°N)	Daylength[a] (hr–min)	Daylength (radians)	τ_o[b] (radians)	τ_o[b] (degrees)	$\overline{\cos\tau}$[c]
0	12–00	π	0.50 π	90.0	0.637
10	12–35	1.05 π	0.53 π	95.4	0.600
20	13–12	1.10 π	0.55 π	99.0	0.572
30	13–56	1.16 π	0.58 π	104.4	0.532
40	14–52	1.24 π	0.62 π	111.6	0.477
50	16–18	1.36 π	0.68 π	122.4	0.395
60	18–27	1.54 π	0.77 π	138.6	0.273
66.5	24–00	2 π	π	180.0	0
70	24–00	2 π	π	180.0	0
80	24–00	2 π	π	180.0	0
90	24–00	2 π	π	180.0	0

[a]From Trewartha (1954, p. 12).

[b]$\tau_o = \frac{\tau_2 - \tau_1}{2}$

[c]$\overline{\cos\tau} = \frac{1}{\tau_2 - \tau_1}\int_{\tau_1=\text{sunrise}}^{\tau_2=\text{sunset}} \cos\tau \, d\tau \approx \frac{\sin\varepsilon_o}{\tau_o}$

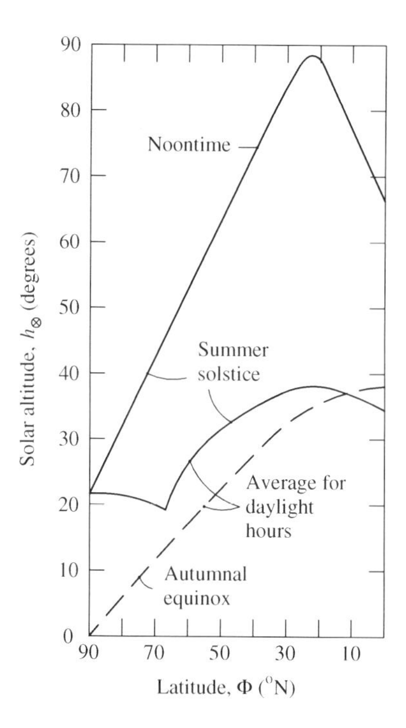

Fig. 3.9. Effective solar altitude as a function of latitude (effective solar altitude is defined as the daylight-hour average solar altitude).

of the vegetation. Given the effective solar altitude, the tree appears to achieve this temperature control by setting the leaf angle at a value which compromises between reflecting heat-producing NIR while absorbing biomass-producing PAR.

We begin to examine this relationship by setting the effective solar hour angle at the daylight-hour average.[†]

Daylight average hour angle

A fixed declination gives daylight hours which are symmetrical about noon, leading to the average, $\overline{\cos\tau}$, of those daylight hours as is presented in the last column of Table 3.5. Letting the daylight-hour average solar altitude be $h_\otimes$,[‡] Eq. 3.3 is then written

$$\sin(h_\otimes) = \sin\Phi\sin\delta + \cos\Phi\cos\delta\,\overline{\cos\tau}, \tag{3.15}$$

which gives the desired $h_\otimes$ vs. Φ relation for the summer solstice plotted in Fig. 3.9. For comparison purposes, the relation for the autumnal equinox, $\delta = 0°$, $\overline{\cos\tau} = \frac{2}{\pi}$, is also given in Figure 3.9. For further comparison, the $h_\otimes$ vs. Φ relation for noontime at the summer solstice is obtained from Eq. 3.3 with $\tau = 0°$ and is also plotted in Fig. 3.9.

Finally, Fig. 3.9 allows transformation of Fig. 3.8 into the nominal growing season *daylight hour average* PAR (on a horizontal surface), I_{0p}, as a function of latitude

[†] Note that this choice minimizes the daily extremes of the angle of incidence of solar radiation on the leaves and hence minimizes scattering thereby maximizing the radiation available for photosynthesis.

[‡] Remember, we are now omitting the overbar indicator of temporal averaging.

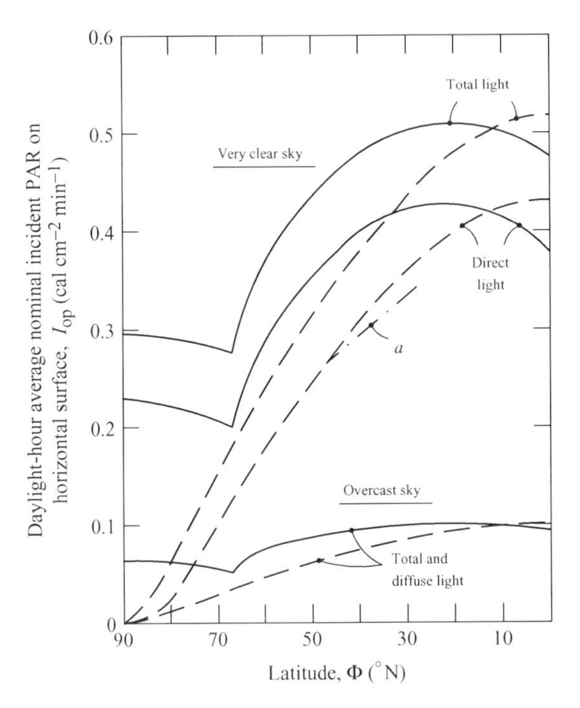

Fig. 3.10. Daylight-hour average nominal insolation at canopy top. Solid line, summer solstice; dashed line, autumnal equinox; *a*, observed average total light (Jensen and Haise, 1963).

(cf. Fig. 3.10). Curves for the autumnal equinox ($\delta = 0°$) are added to the figure for comparison. At any given location the actual average insolation will lie somewhere in between the bounding "very clear" and "overcast" conditions. Jensen and Haise (1963) compiled observations of average annual total insolation in the latitudinal range $25°\text{ N} < \Phi < 48°\text{ N}$, and the regression of these values (PAR component only) is added to Fig. 3.10 for reference.

E Length of growing season

While the intensity of solar radiation controls the rate of biomass production for a stand of trees, its annual productivity depends also upon the duration of photosynthetic activity, that is, the length of the growing season, τ. Seasonal initiation and termination of growth in trees is forced by such factors as: (1) photoperiod (i.e., day length), (2) the strong temperature dependence of photosynthetic biochemistry in woody plants (cf. Fig. 8.4), or (3) moisture stress (Addicott and Lyon, 1973). For autumnally deciduous trees, season change is bounded by bud break in the spring and leaf-color change in the fall.† For evergreen trees, the external signs of the growing season are less apparent and observations are inconclusive;‡ observers usually resort to repeated

† "Summer" deciduous trees (e.g., Mediterranean and tropical climates) shed all or part of their leaves in the dry season, whenever it occurs during the calendar year.

‡ Ahlgren (1957) finds that red pine, white spruce, and black spruce begin activity when the atmospheric temperature rises above freezing. Fraser (1966) concludes that photoperiod controls the growth cessation of black spruce. Baldwin (1931) finds the initiation of evergreen growth to be temperature dependent and its cessation of growth to be undetermined. However, Romell (1925) finds evergreen growth initiation insensitive to both latitude and altitude.

detailed measurements of the dimensions of saplings. In neither case is there an extensive literature to support estimation of season length in a specific case.† As a substitute we have collected a set of observations from the literature (cf. Table 3.6), and we will procede empirically from this admittedly sparse sample.

The data of Table 3.6 are correlated in Fig. 3.11 where we observe several things:

(1) The season of deciduous multilayers shortens monotonically with latitude in temperate climates as does that of the evergreen ponderosa and white pines as well as pinyon–juniper. This common behavior is generally accepted to be due to shortening of the photoperiod with increasing latitude.

(2) The season of monolayer evergreens such as spruce and fir in temperate climates *lengthens* with latitude.

(3) The season of multilayer ponderosa and white pines as well as white cedar appears also to lengthen with latitude in cooler climates.

The unexpected observed increase in evergreen season length with increasing latitude (i.e., decreasing photoperiod and decreasing temperature) may possibly be due to the criterion used for its estimation. The bulk of evergreen observations in Table 3.06 use measured growth of the uppermost shoot or "leader" to define the season while various investigators note that needle growth continues for two weeks (Marie-Victorin, 1927) to two months (Tolsky, 1913) after cessation of leader growth. Furthermore, Gail (1926) found that photosynthesis continued after growth ceased with the carbohydrates produced being stored for later use.‡ Despite these caveats, we will accept

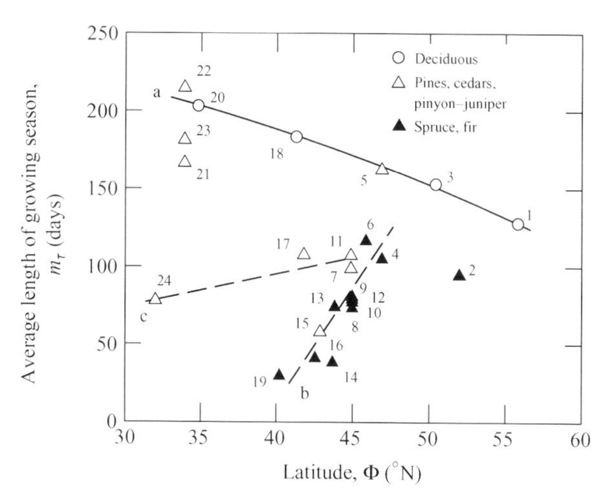

Fig. 3.11. Observations of the growing season of trees. Data points keyed to Table 3.6. (a) multilayer, temperate climate; (b) monolayer, temperate climate; (c) multilayer, cool climate.

† For background on bud break and leaf color change see Valentine (1983), Lechowicz (1984), Taylor (1974), Gee and Federer (1972), and Mitchell (1936) among others. For models of season length see Britton (1878a, 1878b), Nemeth (1973), and Federer and Lash (1978).

‡ Various investigators report low-level conifer productivity to super-cooled temperatures (cf. Arris and Eagleson, 1989b).

Table 3.6. *Observations of the growing season of trees*

Observation	Location	Latitude °N	Species	Growing season			Method of estimation	Reference
				Start (Julian day)	Stop (Julian day)	Length (days)		
1	Russian plain	56	birch, aspen	130	258	128	Foliation to color change	Rauner (1976)
2	Canada (Boreal)	52	black spruce	142	236	94	Bud break to terminal bud set	Morganstern (1978)
3	Russian steppe	50.5	oak	120	274	154	Foliation to color change	Walter and Breckle (1985)
4	Canada (Great Lakes)	47	black spruce	144	250	106	Bud break to terminal bud set	Morganstern (1978)
5	Southern Quebec	47	white pine			165	Unknown	Marie-Victorin (1927) [a]
6	Canada (Acadia)	46	black spruce	144	258	114	Bud break to terminal bud set	Morganstern (1978)
7	New England	45	white pine	133	235	102	1–99% growth of sapling leader	Baldwin (1931) [b]
8	New England	45	red spruce	162	234	73	1–99% growth of sapling leader	Baldwin (1931) [b]
9	New England	45	balsam fir	157	237	81	1–99% growth of sapling leader	Baldwin (1931) [b]
10	New England	45	white spruce	157	231	75	1–99% growth of sapling leader	Baldwin (1931) [b]
11	Berlin, NH	45	white cedar	138	247	110	1–99% growth of sapling leader	Baldwin (1931) [b]

12	Cupsuptic Lake, ME	45	black spruce	163	242	79	1–99% growth of sapling leader	Baldwin (1931)[b]
13	Mt. Desert Isle, ME	44	red spruce, white pine	156	232	76	Growth of leader	Moore (1917)[a]
14	Mt. Desert Isle, ME	44	balsam fir	152	191	39	Growth of leader	Moore (1917)[a]
15	Austria	43	Austrian pine	141	200	59	Growth of leader	Nakashima (1929)[a]
16	Adirondack Mts., NY	43	red spruce	166	212	46	Growth of leader	Rees (1929)[a]
17	Ithaca, NY	42	white pine	120	227	107	Growth of leader	Brown (1912, 1915)[a]
18	Wauseon, OH	41.5	mixed deciduous	105	288	183	Bud break to leaf fall	Smith (1915)[b]
19	Pennsylvania	40.7	white pine, Norway spruce	152	182	30	Growth of leader	Illick (1923)[a]
20	North Carolina	35	tulip poplar, red maple	95	297	202	Leaf emergence to leaf fall	Reader *et al.* (1974)[b]
21	Flagstaff, AZ[c]	34	ponderosa pine			169	Avg. date of latest spring frost to avg. date of earliest fall frost	Williams and Anderson (1967)
22	Flagstaff, AZ[c]	34	pinyon–Utah juniper			215	Avg. date of latest spring frost to avg. date of earliest fall frost	Williams and Anderson (1967)
23	Flagstaff, AZ[c]	34	pinyon–alligator juniper			182	Avg. date of latest spring frost to avg. date of earliest fall frost	Williams and Anderson (1967)
24	Arizona	32	ponderosa pine	135	213	78	Growth of leader	Pearson (1924)[a]

[a] As quoted by Baldwin (1931).
[b] Extensive observations at many sites.
[c] 2100 m elevation.

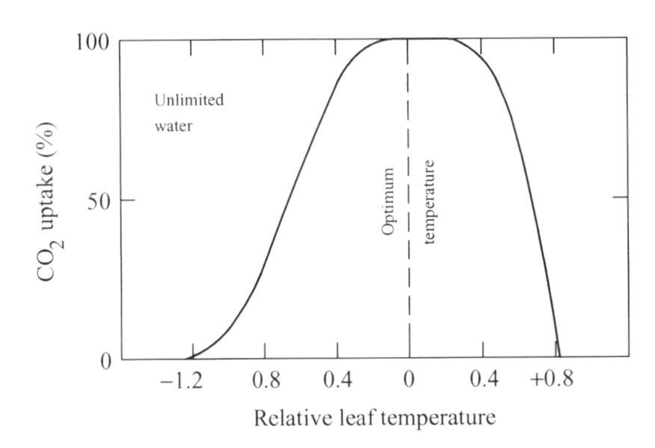

Fig. 3.12. Temperature dependence of photosynthesis in evergreens. (Adapted from Larcher, 1983, Fig. 3.35, p. 114.)

the data of Table 3.6 as observations of the complete growing season, and we will use their empirical correlations with latitude (cf. Fig. 3.11) in later chapters where season length plays a role.

Analyzing the data of Table 3.6, we find that the average Julian day at which the spruce–fir season begins is 157 (i.e., early in June) with a standard deviation (sample size of 24) of only 11 days. This supports the earlier observation of Romell (1925) that evergreen growth initiation is insensitive to latitude and altitude. In contrast, the season length is highly sensitive to latitude having an average value of 75 days with standard deviation of 21 days. This puts the mean cessation of spruce–fir growth at Julian day 232 which is mid-August, the peak of summer heat, suggesting that *high* temperature may control termination of growth in these species.

A possible mechanism for this control is the asymmetric sensitivity of evergreen photosynthetic capacity with variations of the deviation in leaf temperature from its optimum value.† This is sketched in Fig. 3.12 where we see the abrupt shutdown of photosynthesis when the leaf temperature reaches a critically *high* value. Note in this figure that the range of productive temperatures less than the optimum is about twice that of temperatures greater than the optimum. With this in mind we see that increasing latitude brings an increasing photosynthetic season as measured from initiation in early June until cessation when the temperature reaches 0.7× the optimum leaf temperature for the given spruce–fir species.

F Penetration of direct radiation into the canopy‡

The penetration function

The attenuation of direct radiation penetrating a homogeneous absorbing medium such as a dense plant canopy is given by the classical Boguer–Lambert law (Gates, 1980, p. 244; Ross, 1981, p. 313). Following the work of Monsi and Saeki (1953), Ross

† We discuss this biochemical mechanism further in Chapter 8.

‡ For an exhaustive analytical analysis of this issue see Ross (1981, Chapter II.4, pp. 188–238).

and Nilson (1965), Warren Wilson (1965, 1967), and Anderson (1966), Cowan (1968) derived this attenuation as follows:

Let R_{bo} be the incident beam radiance at the top of the vegetation canopy for solar altitude $h_{\otimes}$. This radiation is intercepted and partitioned progressively by the leaves as the beam penetrates the stand, leaving R_b unintercepted at vertical depth ξ. In a cascading process, there is at each leaf surface reflection, transmission, and absorption, and some of the reflected and transmitted energy is partitioned anew at the bottom side of other leaves. The sum of the multiple reflections and transmissions is called *scattering*, and it results in additional absorption of radiation.† Forward (i.e., downward) scattering results in a reduced decay of the penetrating beam. Bookkeeping these many internal canopy reflections and transmissions is a tedious and approximate task that is at a much higher level of detail than is called for in this work.‡ Cowan (1968) ignores these details and assumes the change in R_b per unit length of optical path to be separably proportional to R_b and to the foliage area density in the manner

$$\frac{dR_b}{d\xi/\sin(h_{\otimes})} = G\, L_t R_b, \tag{3.16}$$

in which

$G \equiv G(h_{\otimes}) =$ dimensionless proportionality function representing the spatial arrangement and orientation of the leaves in relation to the solar altitude,
$\frac{L_t}{1} = L_t =$ homogeneous canopy foliage area density index (i.e., one-sided leaf area per unit horizontal area per unit of dimensionless crown depth), and
$\frac{\xi}{\sin(h_{\otimes})} =$ distance in the direction of the penetrating beam.

Equation 3.16 gives

$$\frac{R_b}{R_{b0}} \equiv \frac{I_b(\xi)}{I_{b0}} = \exp(-\kappa_b L_t \xi), \tag{3.17}$$

in which I_{b0} is the incident insolation, and $I_b(\xi)$ is the insolation transmitted past depth $\xi \equiv z/h_s$ in the absence of scattering (h_s is the stem height). In this form Eq. 3.17 is often called the *penetration function* by plant scientists, and κ_b is called the *decay* or *extinction coefficient* for direct radiation. However, the total extinction effect of the canopy structure is given by the important product, $\kappa_b L_t$, called here the direct *radiation absorption index* after the custom in physics (cf. Feynman, 1963, Vol. I, pp. 31–38).

The extinction coefficient without scattering

It is easiest to understand the physical significance of $G(h_{\otimes})$ by noting that the rate at which energy is intercepted by a unit area of a thin horizontal layer of the canopy equals the area of shadow cast by the leaves on a horizontal plane times the insolation

† The fraction of incoming energy escaping back to the sky constitutes the *albedo* of the canopy.
‡ For more details of the formulation of this process see Sellers (1989).

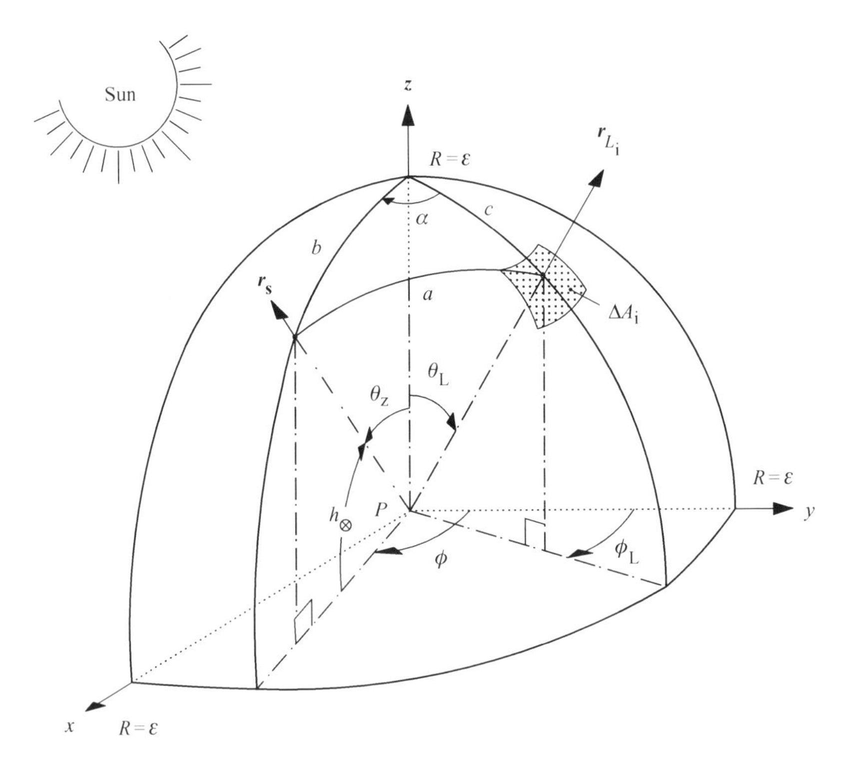

Fig. 3.13. Definition sketch for radiant flux.

(Monteith, 1973, p. 52). Then, using Eqs. 3.16 and 3.17, the shadow area, A_s, per unit horizontal area of canopy, is

$$A_s = \frac{dR_b}{R_b d\xi} = \frac{G(h_\otimes)L_t}{\sin(h_\otimes)} \equiv \kappa_b L_t, \tag{3.18}$$

in which $G(h_\otimes)$ is called the *G-function* (Ross and Nilson, 1965). In the above:

$$\frac{G(h_\otimes)}{\sin(h_\otimes)} \equiv \kappa_b = \text{shadow area per unit of foliage area, and}$$

$G(h_\otimes)$ is the average, over the canopy, of the projection of a unit foliage area on a plane normal to the sun direction, $\boldsymbol{r}_s$, at arbitrary point $\boldsymbol{P}$ (see Fig. 3.13 for definitions), and as such it is the average cosine of the angle between the direction $\boldsymbol{r}_s$, and the effective leaf normal, $\boldsymbol{r}_L$. The nature of κ_b is further clarified by exploring its relation to solar altitude and leaf angle using an idealized simple geometry.

Two-dimensional beam–leaf geometry

The geometry of beam radiation incident upon a single opaque leaf of unit dimension is sketched in Fig. 3.14 for the separate cases of upward and downward specular

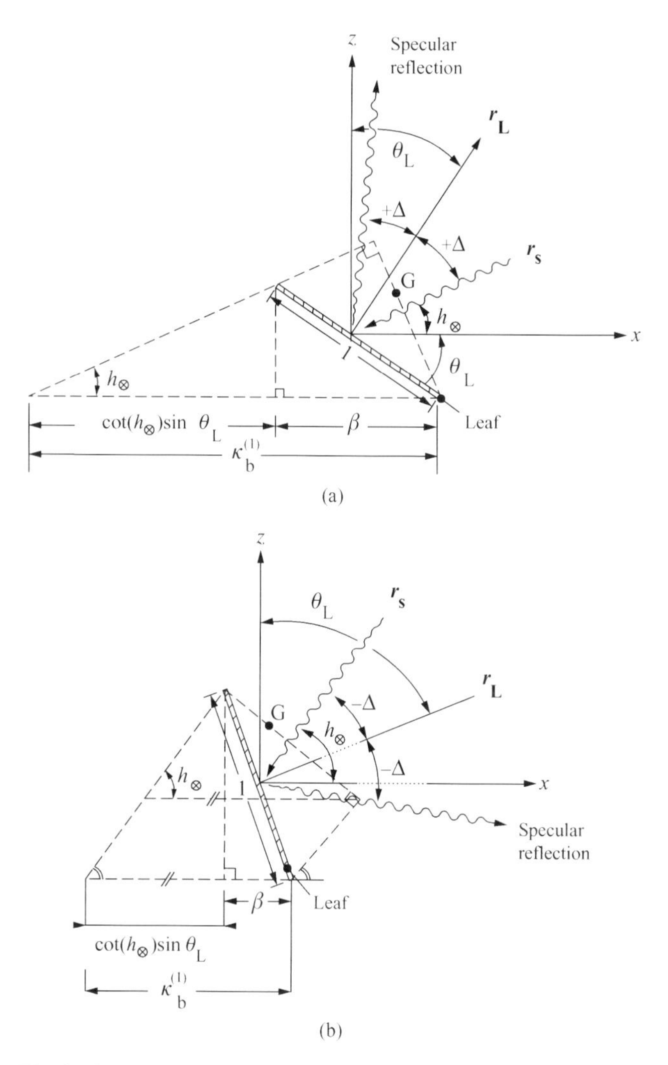

Fig. 3.14. Geometry of specular reflection from opaque leaf in two dimensions. (a) Upward specular reflection; (b) downward specular reflection.

reflection.[†] From either case we use the above definition of the G-function to write

$$G \equiv \cos\Delta = \cos\left[\frac{\pi}{2} - (h_\otimes + \theta_L)\right] = \sin(h_\otimes + \theta_L), \tag{3.19}$$

so that from Eq. 3.18, the extinction coefficient, $\kappa_b^{(1)}$, of an isolated leaf is

$$\kappa_b^{(1)} = \cos\theta_L + \cot(h_\otimes)\sin\theta_L = \beta + \cot(h_\otimes)\sin\theta_L, \tag{3.20}$$

in which $\beta \equiv \cos\theta_L$. Equation 3.20 shows that without scattering, $\kappa_b^{(1)} \geq \beta$.[‡]

[†] Specular reflection is not the primary reflective mechanism on any but the smoothest leaves; however it provides a convenient reference by defining the principal scattering directions (cf. Ross, 1981, Chapter II.6).

[‡] We use the superscript on κ_b (i.e., $\kappa_b^{(n)}$) to signify its applicability to a canopy consisting of a limited number, n, of leaves, in contrast to the unsuperscripted notation for the decay coefficient of the full canopy.

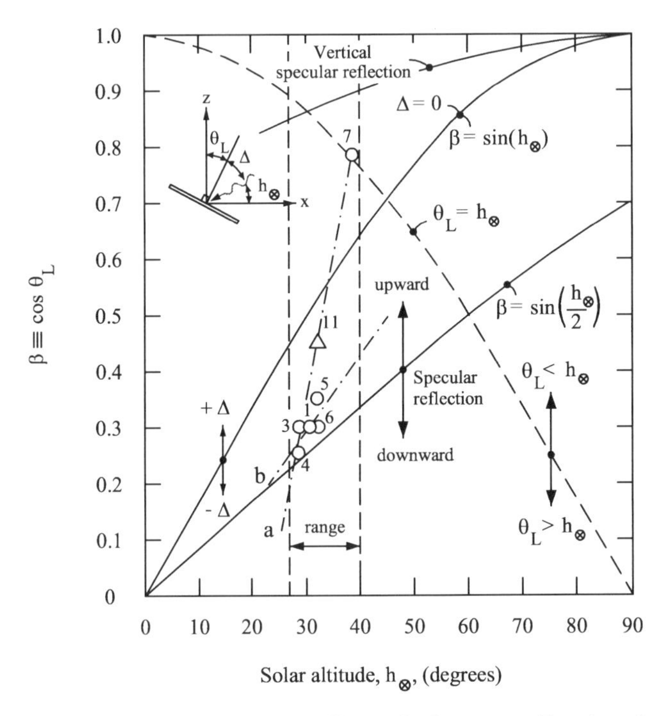

Fig. 3.15. Graphical summary of beam–leaf geometry. Data keyed to Table 3.7. ○, Deciduous; △, evergreen; Curve (a), empirical correlation; Curve (b), optical optimality for multilayers (Eq. 3.51).

From Fig. 3.14 we see that κ_b is the horizontal shadow area per unit of (opaque) leaf area (Monteith, 1973, p. 52). It is easy to visualize how in reality diffraction at the leaf edges, coupled with re-reflections from the underside of other leaves and transmittance through the leaf and the interleaf gaps all act to reduce this shadow and hence to bring $\kappa_b \rightarrow \beta$.

In Fig. 3.15 the beam–leaf geometry is summarized in $\beta - h_\otimes$ space. The function $\beta = \sin(h_\otimes)$ divides the space into regions of positive and negative Δ, while the function $\beta = \sin(h_\otimes/2)$ separates regions of upward and downward specular reflection. Assuming vegetation growth is limited to latitudes between the Equator and 60°, the effective solar altitude, $h_\otimes$, at summer solstice is $27^\circ \leq h_\otimes \leq 40^\circ$ as given by Fig. 3.9. These limits on $h_\otimes$ are plotted as the vertical dashed lines on Fig. 3.15. A few observations of β from the literature are listed in Table 3.7 and are plotted (with identifying numbers) on Fig. 3.15 where we note:

(1) specular reflection is upward in all cases,

(2) leaf angles in the temperate deciduous forest (i.e., points nos. 1, 3, 4, 5, 6) are larger than that for normality with the beam,

(3) leaf angles at the coldest latitude (i.e., no. 4) produce nearly horizontal specular reflection,

(4) leaf angles at the hottest latitude (i.e., tropical rainforest, no. 7) produce nearly vertical ($h_\otimes + 2\,\theta_L = 90°$) specular reflection, and

(5) the single evergreen observation (i.e., no. 11) has leaves almost normal to the beam.

The *primary* variability in κ_b resides in the three-dimensional angular relationship between the solar beam and the leaf surface since this determines the distribution of "gaps" between the leaves, and it is through these gaps that the beam propagates. The frequency of gaps along the path of a given solar ray has been studied extensively in the field by the method of *point quadrats* which involves passing very thin, long needles through a vegetation stand in a fixed direction and recording all contacts with the foliage. These observations have guided theoretical study of the significance of gap distributions under the leadership of Warren Wilson and Reeve (1959, 1960), Warren Wilson (1960, 1963), Anderson (1966), and Nilson (1971) and they lead to the following improved prediction of κ_b using more realistic geometry albeit still without consideration of scattering.

Beam–leaf geometry of three-dimensional, axially symmetric, multilayers

We now explore the effects of foliage orientation as well as solar elevation on the coefficient κ_b, in a homogeneous, multilayer, axially symmetric closed canopy in

Table 3.7. *Observations of extinction coefficients*

Observation	Species	Latitude, Φ (degrees)	$h_\otimes$ (degrees)[a]	β	κ	Reference
1	Deciduous forest	54	31	0.30		Rauner (1976)
2	Oak	52	32		0.32	Rauner (1976)
3	Trembling aspen	56	29	0.30	0.31	Rauner (1976)
4	Paper birch	56	29	0.25		Rauner (1976)
5	Maple	52	32	0.35		Rauner (1976)
6	Linden	52	32	0.30		Rauner (1976)
7	*Goethalsia*	10	39	0.79		Allen and Lemon (1976)
8	Eucalyptus	−35	39		0.37	Denmead *et al.* (1993)
9	Sitka spruce	57	29		0.50	Jarvis *et al.* (1976)
10	Red pine	46	34		0.42	Jarvis *et al.* (1976)
11	Scots pine	52	32	0.45	0.46	Jarvis *et al.* (1976)

[a] Effective solar altitude as given by daylight-hour average at summer solstice.

which scattering is neglected. Derivation of the important G-function is presented in Appendix G in considerable mathematical detail. The result is

$$\frac{G(h_\otimes)}{\sin(h_\otimes)} \equiv \kappa_b = \beta + \frac{2}{\pi}\int_{h_\otimes}^{\frac{\pi}{2}} \tilde{g}(\theta_L)\,[\tan(\phi_{os}) - \phi_{os}]\cos(\theta_L)\,d\theta_L, \quad 0 \le \theta_L \le \frac{\pi}{2}, \tag{3.21}$$

in which $\tilde{g}(\theta_L)$ is the pdf of leaf orientation angle (cf. Chapter 2). Monteith (1973, p. 48) shows 2 ϕ_{os} to be the shadowed sector of the axially symmetric tree.† From the geometry

$$\phi_{os} \equiv \cos^{-1}[\tan(h_\otimes)\cot(\theta_L)]. \tag{3.22}$$

Beam extinction coefficients for constant leaf inclination

Equation 3.21 is difficult to evaluate analytically except for the simplest distributions, $\tilde{g}(\theta_L)$. Anderson (1966) has done so for the special case of *constant leaf inclination*, θ_o (maintaining azimuthal symmetry of course), in which

$$\tilde{g}(\theta_L) = \delta(\theta_L - \theta_o), \tag{3.23}$$

where $\delta(\cdots)$ is the Dirac delta-function. Using Eq. 3.23, Eq. 3.21 reduces to

$$\frac{G(h_\otimes)}{\sin(h_\otimes)} \equiv \kappa_b = \left\{ \begin{array}{ll} \beta, & \theta_L \le h_\otimes \\ \beta\{1 + \frac{2}{\pi}[\tan(\phi_{os}) - \phi_{os}]\}, & \theta_L > h_\otimes \end{array} \right\}, \tag{3.24}$$

in which $\theta_L = \theta_o$. Eq. 3.24 is plotted in Fig. 3.16 as taken from Anderson (1966, Fig. 1, p. 44). Several specific cases of constant leaf inclination are of particular interest:

Horizontal leaves ($\theta_o = 0$)

$$\frac{G(h_\otimes)}{\sin(h_\otimes)} \equiv \kappa_b = \beta = 1, \tag{3.25}$$

Vertical leaves ($\theta_o = \frac{\pi}{2}$)

With $\theta_o = \frac{\pi}{2}$, Eq. 3.24 is indeterminate for $h_\otimes < \theta_L$, so we return to Eqs. G.05 and G.07 to find (as did Ross, 1981, p. 22) that $\beta = 0$ and

$$\frac{G(h_\otimes)}{\sin(h_\otimes)} \equiv \kappa_b = \frac{2}{\pi}\cot(h_\otimes), \quad 0 \le h_\otimes \le \frac{\pi}{2}. \tag{3.26}$$

Leaves perpendicular to solar beam ($\theta_o = \frac{\pi}{2} - h_\otimes$)

$$\cos(\phi_{os}) = \cot^2(\theta_L) = \tan^2(h_\otimes). \tag{3.27}$$

† Mathematically, ϕ_{os} is the supplement of the azimuthal angle (cf. Fig. 3.13) at which sign change occurs.

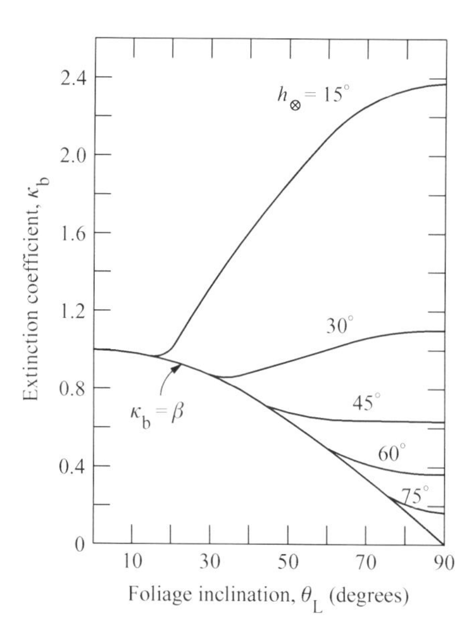

Fig. 3.16. Direct radiation extinction coefficient for invariant foliage inclination and without scattering. (From Anderson, M. C., Stand structure and light penetration, II. A theoretical analysis, Fig. 1, p. 44, *Journal of Applied Ecology*, 3, 41–54, Copyright © 1966 Blackwell Science Ltd, with kind permission from Blackwell Science Ltd, Oxford, UK.)

Plate 3.2. Typical equatorial crown shape. Trees of this equatorial Kenyan savanna elevate their monolayer leaf surface above the shrubs for cooling and display a spherical-sector crown of large radius in keeping with the near-zenith position of the Sun during the growing season. (Photograph by Peter S. Eagleson.)

Beam extinction coefficient for uniform distribution of leaf inclinations

For hemispherical distribution[†] of leaf inclinations, $\beta = 0.5$. For simplicity, we again work directly from the geometry to obtain $G(h_\otimes) = 1/2$, and therefore

$$\kappa_b = \frac{1}{2\sin(h_\otimes)}. \tag{3.28}$$

Beam–leaf geometry of three-dimensional, axially symmetric, monolayers

Opaque vertical right circular cylinders

This case covers species in which the reflecting surface may be approximated by an opaque vertical right circular cylinder giving $\beta = 0$, and includes dense vertical monolayers such as cypress, and some spruce, as well as sparsely foliated trees where the trunk is the primary scatterer. Calculation of κ_b is simpler in such cases because we need not include any leaf under-surfaces. We follow Monteith (1973, Chapter 4) in which for solid bodies he directly calculates the shadow area per unit of surface area. In this case with D = cylinder diameter and h = cylinder height, we have for $h \gg D$ (i.e., neglecting the shadow of the circular cross-section)

$$\kappa_b = \frac{\text{shadow area}}{\text{surface area}} = \frac{D\,h\cot(h_\otimes)}{\pi\,D\,h} = \frac{1}{\pi}\cot(h_\otimes). \tag{3.29}$$

Opaque right circular cones

Working from the geometry as described above, Monteith (1973, Eq. 4.6, p. 49) gives

$$\kappa_b = \frac{\text{shadow area}}{\text{surface area}} = \frac{(\pi - \phi_{os})\cos(\gamma_o) + \sin(\gamma_o)\cot(h_\otimes)\sin(\phi_{os})}{\pi[1 + \cos(\gamma_o)]}, \tag{3.30}$$

in which γ_o is the base angle of a cone having height, h_o, and base radius, r_o. That is

$$\tan(\gamma_o) = \frac{h_o}{r_o}. \tag{3.31}$$

For the special case in which the beam is normal to the slant side of the cone, $\tan(\gamma_o) = \cot(h_\otimes)$, and Eq. 3.30 becomes

$$\kappa_b = \frac{(\pi - \phi_{os})\sin(h_\otimes) + \cos(h_\otimes)\cot(h_\otimes)\sin(\phi_{os})}{\pi[1 + \sin(h_\otimes)]}. \tag{3.32}$$

[†] The term "hemispherical" distribution is used interchangeably here with "uniform" distribution to describe the leaves having a spatial distribution identical to the surface of a hemisphere. That is, the distribution of area with hemispherical radius vector is constant, i.e., the fractional area is the same in all vector directions on the hemisphere.

Opaque hemispheres

Once again, the geometry dictates

$$\kappa_b \equiv \frac{\text{shadow area}}{\text{surface area}} = \frac{\frac{\pi R^2}{2}[1+\sin(h_\otimes)]}{2\pi R^2} = \frac{1}{4}\left[1+\frac{1}{\sin(h_\otimes)}\right]. \tag{3.33}$$

Plate 3.3. There is variety in foliage arrangement. In temperate climates such as Rome, trees have many alternative successful foliage arrangements. The spiral branching pattern of this evergreen in the Borghese Gardens allows vertical branch separation for leaf cooling and azimuthal branch separation for light penetration. (Photograph by Peter S. Eagleson.)

G Penetration of diffuse radiation into the canopy

We assume (after Ross, 1975, p. 34) that penetration of diffuse radiation into the homogeneous canopy follows the same laws as does the direct radiation for a particular value of θ and of ϕ.† For simplicity we once again neglect scattering. However, in the diffuse case there will be contributions from each segment of the sky so it will be necessary to integrate Eq. 3.16 over the entire celestial hemisphere. For isotropic diffuse radiation (i.e., "uniform" sky) this integral takes the form (Ross, 1975, Eq. 28)

$$\frac{I_d(\xi)}{I_{d0}} = 2\int_0^{\frac{\pi}{2}} \exp\left[-L_t \frac{G(h_\otimes)}{\sin(h_\otimes)}\right] \sin(h_\otimes)\cos(h_\otimes)\,dh_\otimes. \tag{3.34}$$

Note that if $\dfrac{G(h_\otimes)}{\sin(h_\otimes)} \equiv \kappa_d$ is assumed to be independent of $h_\otimes$, we get

$$\frac{I_d(\xi)}{I_{d0}} = \exp(-\kappa_d L_t \xi). \tag{3.35}$$

Ross's numerical solutions of Eq. 3.34 (Ross, 1975, Table VII, p. 34) for a uniform sky and for the common value $L_t = 5$ are plotted as the solid lines in Fig. 3.17 for

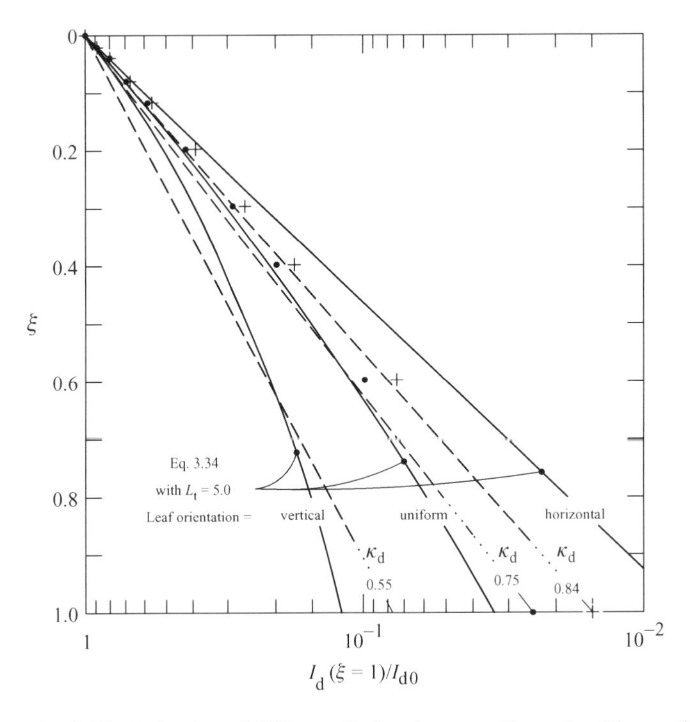

Fig. 3.17. Extinction of diffuse radiation from a uniform sky. Observations: •, maize; +, cotton. Solid lines = theory, Eq. 3.34 with $L_t = 5.0$; dashed lines = exponential approximation, Eq. 3.35. (From computations of Ross, 1975, Table VII, p. 34.)

† This is strictly true only for horizontal leaves (Ross, 1981, p. 247) as is seen for a special case in Fig. 3.17.

horizontal, vertical, and uniform leaf orientations in comparison with his observations in maize and cotton canopies. In that figure we see:

(1) the agreement between the observation and the uniform leaf orientation is fairly good for maize but is poor for cotton,
(2) the observations agree well with Eq. 3.35 (dashed lines) in which the fitted values of the extinction coefficient, κ_d, for diffuse radiation are about 0.75 and 0.84 for maize and cotton respectively, and
(3) Equation 3.35 fits the vertical leaf distribution approximately with $\kappa_d = 0.55$, and fits the uniform distribution fairly well with $\kappa_d = 0.75$.

We assume that Eq. 3.35 is equally applicable for trees.

H Penetration of total radiation into the canopy

Theoretical scatterless extinction coefficients for total radiation

Clear sky

Adding the direct and diffuse components (Eqs. 3.17 and 3.35 respectively) in the proportion given by the beam ratio, μ,

$$\frac{I(\xi)}{I_0} = \mu \exp(-\kappa_b L_t \xi) + (1 - \mu) \exp(-\kappa_d L_t \xi). \tag{3.36}$$

At large effective solar altitudes, and for a clear atmosphere, Fig. 3.3 shows that

$$\frac{1 - \mu}{\mu} = O(10^{-1}), \tag{3.37}$$

whereupon, if $\kappa_b \approx \kappa_d$, Eq. 3.36 is approximately

$$\frac{I(\xi)}{I_0} = \exp(-\kappa^* L_t \xi), \tag{3.38}$$

in which κ^* is the scatterless extinction coefficient for total clear sky radiation, and where for clear sky

$$\kappa^* \approx \kappa_b. \tag{3.39}$$

Overcast sky

For overcast skies, we neglect the beam radiation whereupon the extinction is again governed by Eq. 3.35 in which

$$\kappa^* = \kappa_d. \tag{3.40}$$

Comparison of theoretical and observed extinction coefficients

The observed momentum and light extinction coefficients collected in Table 3.7 are compared with the above separate scatterless theories for clear sky (i.e., beam radiation) and cloudy sky (i.e., diffuse radiation) in Figs. 3.18a and 3.18b respectively.

In Fig. 3.18a we plot $\kappa_b(h_\otimes)$ for uniform leaf distribution (Eq. 3.28, Curve (b)), and three fixed θ_L solutions of Eq. 3.24: $\theta_L = 0°$ (Curve (a)), $\theta_L = 90°$ (Curve (c)), and $\theta_L = 90° - h_\otimes$ (Curve (f)). Note that Curves (c) and (f) are essentially the same in the realistic range of $h_\otimes$ (bounded by the vertical dashed lines), but are higher than the observations by about a factor of two. Opaque right circular cylinders, right circular cones and hemispheres are represented by Curves (d), (e), and (h) respectively.

In Fig. 3.18b we show Ross's (1975, Table VII, p. 34) solutions to Eq. 3.34 for horizontal leaves (Curve (a), $\beta = 1$ and $\kappa_d = 1$), for uniformly distributed leaves (Curve (b), $\beta = 0.50$ and $\kappa_d = 0.75$), and for vertical leaves (Curve (c), $\beta = 0.5$ and $\kappa_d = 0.55$). In Fig. 3.15 the sample of deciduous trees showed $\beta \approx 0.30$. Interpolating linearly gives an estimated value of $\kappa_d = 0.67$ for the observed deciduous trees in diffuse light (Curve (d)).

To help extract order from these observations we next consider scattered light.

Estimated scattering

Absent an appropriately simple formulation of the scattering phenomenon[†] we will estimate the sense and the approximate magnitude of its modification to the scatterless extinction coefficients derived above. We begin by recalling (cf. Fig. 3.14) that the effect of scattering is to reduce the size of the shadow. It may be thought of crudely as scatterless radiation in a crown with reduced foliage area index, L_t', where we let

$$L_t' = (1 - \omega_s)L_t, \tag{3.41}$$

with ω_s being a scattering coefficient assumed here to be the sum of the reflectance and transmittance since these are the partition components that contribute to reduction of canopy shadow. Using Eq. 3.12 that is

$$\omega_s \approx \rho_T + \tau_T = 1 - \alpha_T. \tag{3.42}$$

Equation 3.38 and 3.41 then give

$$\frac{I(\xi)}{I_0} = \exp\left[-\kappa^* (1 - \omega_s)\,\mathrm{L}_t\xi\right] = \exp\left(-\kappa \mathrm{L}_t\,\xi\right), \tag{3.43}$$

in which the estimated true (i.e., including scattering) light extinction coefficients, κ, are defined from Eq. 3.42 for beam and diffuse radiation respectively

$$\kappa = (1 - \omega_s)\,\kappa_b = \alpha_T\,\kappa_b,$$

[†] Refer to Sellers (1989) for a description of the complex "two stream" approximation.

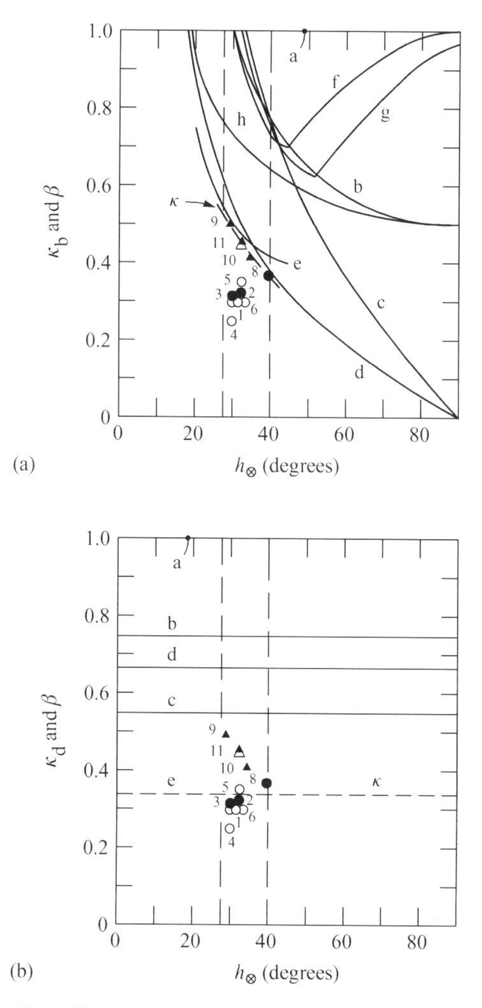

Fig. 3.18. Extinction coefficients: theory vs. observation. Solid lines are theoretical κ neglecting scattering. Dashed lines are κ adjusted for estimated scattering. Data are keyed to Table 3.7: solid points, κ; open points, β; ▲, evergreen; ○, deciduous. Vertical dashed lines enclose daylight-hour average solar altitude at summer solstice for vegetation zone: $0^\circ \leq \Phi \leq 60^\circ$. (a) Beam radiation. a, horizontal leaves, Eq. 3.25; b, uniform leaf distribution, Eq. 3.28; c, vertical leaves, Eq. 3.26; d, opaque vertical right circular cylinders, Eq. 3.29; e, opaque vertical right circular cones with beam normal to the slant surface, Eq. 3.32; f, $\theta_L = 90^\circ - h_\otimes$, Eq. 3.24; g, Curve (f) incorporating estimated scattering as described in text; h, opaque hemisphere, Eq. 3.33. (b) Diffuse radiation. a, Horizontal leaves, $L_t = 5$, Eq. 3.34; b, uniform leaf distribution, $L_t = 5$, Eq. 3.34; c, vertical leaves, $L_t = 5$, Eq. 3.34 (a–c: Numerical solution at $L_t = 5$ (Ross, 1975, Table VII, p. 34), fitted with exponential decay in Fig. 3.17); d, Linearly interpolated κ_d for observed $\beta = 0.30$ (cf. Fig. 3.15); e, Curve (d) incorporating estimated scattering as described in text.

and

$$\kappa = (1 - \omega_s)\kappa_d = \alpha_T \kappa_d. \tag{3.44}$$

For mat (i.e., dull surface) leaves, scattering is independent of light direction (Ross, 1981, p. 137) and we find from Table 3.4 that for trees $\alpha_T = 0.51$. Hence, with scattering

$$\kappa \approx 0.51\,\kappa^*. \tag{3.45}$$

We now use the estimate of Eq. 3.45 to transform Curve (d) of Fig. 3.18b into Curve (e). The resulting scattering $\kappa(h_\otimes)$ only approximates the observations. Note also that applying this transformation to Curve (f) of Fig 3.18a ($\theta_L = 90° - h_\otimes$) reduces it to an excellent fit (dashed Curve (g)) with the evergreen observations.

Note that the observations of β from Table 3.7 are also plotted in Fig. 3.18 as open symbols, one for the evergreens and five for the deciduous. Considering their values in comparison with the κs corrected for scattering (dashed curves), we get the very interesting empirical result

$$\beta \approx \kappa. \tag{3.46}$$

Equation 3.46 is based on a very limited sample and deserves further investigation; we now assemble additional support for this finding

Additional observations

The literature contains many other estimates of κ ranging in value from 0.3 to 1.5 (Ross, 1975). The higher values of κ are associated with large, horizontal leaves and the lower values with more vertical leaves (Ross, 1975). Without reference to their geographical location (and hence their $h_\otimes$) we have collected observations of κ and of the momentum decay parameter, γ, for various plant types in Tables 2.2, 2.3, and 2.4. As we will see in Chapter 4

$$\gamma = mn\,\beta, \tag{3.47}$$

Table 3.8. *Comparison of extinction coefficients for total radiation and momentum*

	κ^a	γ^a	mn^b	β^c	β^a
Deciduous trees	0.35	0.72	1.00	0.72	0.31
Evergreen trees	0.44	0.52	1.00	0.52	0.45
Crops and grasses	0.59	0.66	1.17	0.57	0.73

[a]Observed: Tables 2.2, 2.3, and 2.4.
[b]Fig. 4.14.
[c]Calculated: $\beta = \gamma/(mn) = a_w/(mnL_t)$.

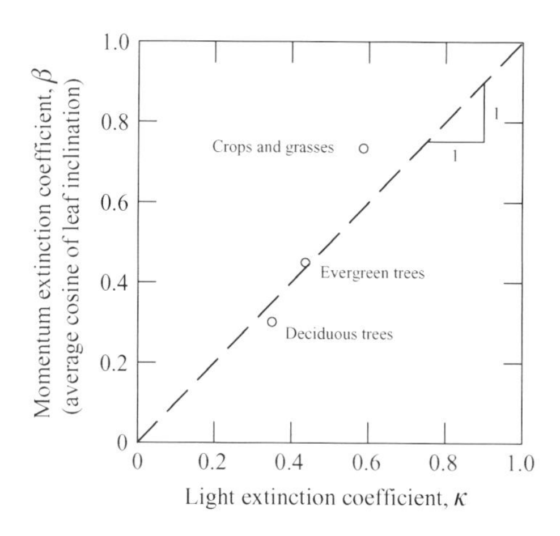

Fig. 3.19. Equality of observed extinction coefficients. Data from Table 3.8.

where

m = exponent governing the relationship between velocity and shear stress and having the nominal value 0.5 for the foliage elements of trees (Table 4.1), and
n = the number of sides of each foliage element experiencing drag and having the nominal value 2 for the foliage elements of trees.

The product, mn, has the nominal value of unity for trees but for crops and grasses $1 \leq mn \leq 1.35$ (Fig. 4.14). A typical tabulated field experiment (Tables 2.2, 2.3, and 2.4) measured *either* κ *or* γ but not both, and did not distinguish between clear and cloudy skies. Thus the only possible test of Eq. 3.46 is to compare the averages of the two parameters across broad classes of vegetation and cloudiness, and to use the above nominal values of mn to estimate β from γ. This is done in Table 3.8 and Fig. 3.19 and confirms Eq. 3.46 across the unknown range of cloudiness encompassed by the data. This finding is central to the arguments of this monograph. Equation 3.46 also increases our confidence in the conclusions arising from Fig. 3.18 by effectively doubling the sample size for either κ or β.

The equality of Eq. 3.46 was noted by Ross (1981, p. 313) at $h_{\otimes} = \pi/2$, but its more general applicability seems to have escaped prior notice. As we shall see in Chapters 7 and 8, this result has far-reaching physical significance. Is there selection pressure that leads κ naturally toward β?

I Optical optimality

Solar radiation is the energy source for production, and as we assume here, selection pressure is toward maximizing *plant* productivity. Therefore, whenever energy is the resource limiting production it will be used "optimally". As we will see in Chapter 6, *canopy* water demand will adjust to limited supply by reducing the canopy cover, M, while leaving the *plant* water use and productivity relatively unchanged. It is the availability of light, carbon, or nutrients that will limit plant production. In Chapter 10 we will consider the carbon and nutrients while in this chapter we deal only with the light.

Plate 3.4. Excess energy loads must be shed. In tropical climates such as Jodhpur, India (latitude = 26.5° N), leaves of this asoka tree have a mean angle of about 65° assuring that at the growing season solar altitude of over 30° (cf. Fig. 3.9) excess NIR will be reflectively dumped. (Photograph by Peter S. Eagleson.)

To ensure optimal productive use of the available solar energy there exist at least three simple mechanisms through which selection may operate:

(1) *Maximize the productive solar energy entering the canopy*. This is accomplished by minimizing the canopy reflectance, ρ_{PAR}. Figure 3.7 shows ρ_{PAR} to be insensitive to β over the practical range of $h_{\otimes}$, so the primary mechanism for control of the reflection of PAR is the nature of the leaf surface† rather than leaf angle optics.

† Larcher (1983, p. 10) suggests that a dense covering of hairs can reduce PAR reflection by a factor of two or three.

(2) *Maintain the optimum photosynthetic leaf temperature*. NIR is a significant source of plant heat, and in Fig. 3.7 we see that $\rho_{\rm NIR}$ is insensitive to leaf angle at $h_\otimes = 30°$. On the other hand, at $h_\otimes = 40°$, $\rho_{\rm NIR}$ increases with leaf horizontality (i.e., with increasing β). Note the strong empirical correlation between β and $h_\otimes$ indicated by the dashed Curve (a) on Fig. 3.15; specular reflection varies, from the vertical at the lowest latitude (i.e., highest $h_\otimes$) where heat must be shed,[†] to the horizontal at the highest latitude (i.e., lowest $h_\otimes$) where energy must be contained within the canopy. However, Fig. 3.12 raises sensitivity questions concerning leaf angle optics as a primary control of photosynthetic efficiency at least for temperate climate trees.[‡] Note in this figure that it takes a ± 40 percent change in leaf temperature to produce a change of only 10 percent in carbon assimilation.

(3) *Minimize the productive energy transmitted to the forest floor*. For a closed multilayer canopy this implies two conditions: (a) the upper leaf surfaces, where the chloroplasts are concentrated, must be in full light (i.e., unshadowed), and (b) the *crown* basal area[§] must be in full shadow (but at *incipient illumination*) since partial shadow indicates useful energy escaping the crown.

We will return to the important condition of incipient illumination, but first imagine the two-dimensional simplification of specular reflection from a unit of opaque leaf surface shown in Fig. 3.14 to be extended to a single-layer array of such leaves as sketched in Fig. 3.20. Considering only downward scattering, the above two conditions are met in this simplification by the geometrical arrangement sketched which, *on the daylight-hour average*, puts the basal area of the *crown-average* leaf in full shadow and transmits light downward to the next leaf layer through the unshadowed gap.[††] Thus in tandem, reflection from the adjacent leaf reduces the one-dimensional leaf shadow to the optimal group condition[‡‡]

$$\kappa_{\rm b}^{(2)} = \beta, \tag{3.48}$$

as long as $h_\otimes > 90 - \theta_{\rm L}$.[§§] From the sketch of Fig. 3.20, at optimality the solar altitude and leaf angle are related geometrically through dimensions a, b, and c according to

$$2b + a = 1, \tag{3.49}$$

[†] In sparse canopies (i.e., low $L_{\rm t}$), NIR may best be shed by low β which reflects to the forest floor, while for dense canopies (i.e., high $L_{\rm t}$) high β are needed to reflect NIR back to space from the canopy top.

[‡] Additional evidence is presented in Fig. 8.4.

[§] This is defined as the horizontal projection of crown area on the forest floor.

[††] Because of the low leaf reflection coefficient for PAR (cf. Fig. 3.7), we cannot rely upon multiple reflections to supply the lowest leaf with even the compensation PAR under these average geometrical conditions. Note however that Fig. 3.20 is for the time-averaged daylight-hour solar altitude so that for some hours near midday, particularly near the summer solstice, deep crown penetration of the incident radiation will occur.

[‡‡] Geometrical considerations show that for downward scattering, the leaf spacing drawn in Fig. 3.20 is optimal by providing maximum insolation absorption per unit of basal area through minimum leaf spacing at maximum energy absorption efficiency.

[§§] This weak dependence of light direction makes our argument approximately applicable also to diffuse radiation making $\kappa_{\rm d} = \beta$ as well.

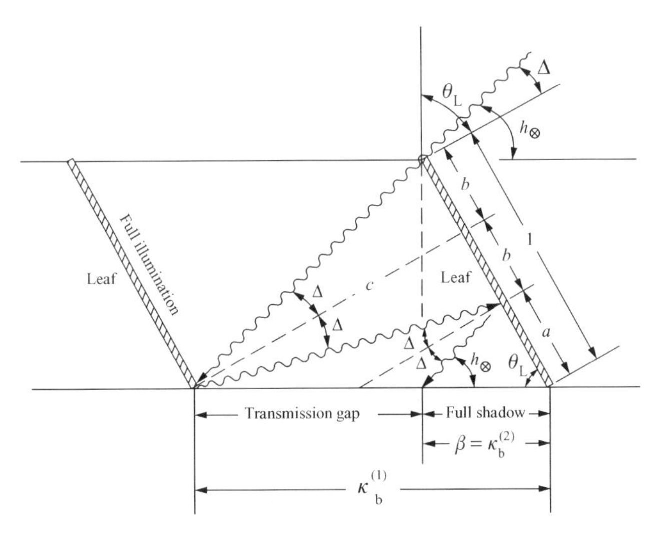

Fig. 3.20. Proposed optically optimal geometry for opaque leaves (crown-average and daylight hour average conditions).

in which

$$a = \frac{\beta}{\kappa_b^{(1)}} = \frac{\beta \sin h_\otimes}{\sin(\pi - h_\otimes - \theta_L)}; b = c \tan\left(h_\otimes - \frac{\pi}{2} + \theta_L\right); \text{ and}$$

$$c = \kappa_b^{(1)} \sin \theta_L \quad (3.50)$$

where, as earlier, $\kappa_b^{(1)} \equiv$ the one-dimensional shadow of a solitary leaf. Equations 3.49 and 3.50 give the idealized geometrical constraint on optical optimality

$$2\frac{\sin(\pi - h_\otimes - \theta_L)}{\sin h_\otimes} \sin \theta_L \tan\left(h_\otimes - \frac{\pi}{2} + \theta_L\right) + \frac{\beta \sin h_\otimes}{\sin(\pi - h_\otimes - \theta_L)} = 1, \quad (3.51)$$

which is plotted over the practical range of $h_\otimes$ as Curve (b) in Fig. 3.15.

In evaluating the mechanisms leading to Eq. 3.48 as the source of the observed $h_\otimes$ vs. β correlation, remember that the natural translucent case will include transmission, additional specular reflection from an array of neighboring leaves, and scattering in a variety of directions from irregularities in the reflecting surfaces.

We now postulate that through natural selection, *whenever light limits productivity*, leaf surface textures and the spatial arrangements of foliage elements of closed canopies scatter the incident light so as to put leaf surfaces in full light and the crown basal area in full shadow and thereby to reduce κ to the value[†]

$$\kappa = \beta, \quad (3.52)$$

while at extreme temperatures, *where photosynthetic temperature limits productivity*, Eq. 3.51 is biased by a need to control heat through regulation of ρ_{NIR}. For open canopies we expect $\kappa > \beta$ as is shown for individual leaves in Fig. 3.14.

[†] For cases in which plant productivity is limited by carbon or nutrient supply rather than light, the condition of Eq. 3.52 is relaxed. For cases in which excess heat is limiting plant productivity rather than insufficient light, the selection pressure leading to Eq. 3.52 is also relaxed and will be driven by the needs of leaf temperature control as determined by the coupled land–canopy–atmosphere heat balance. We do not consider this case here.

Compensation light intensity

Returning now to the corollary optimality condition at the base of the crown which we have called *incipient illumination*: to minimize the transmittance of useful energy to the forest floor, the light available to the lowest leaf in each crown must be exactly the minimum intensity, I_k, needed for leaf growth in the given species. That is, the gross photosynthesis is just balanced by the total respiration. This minimum light intensity is known as the *compensation light intensity* (Horn, 1971), and its achievement at the lowest leaf is just sufficient to put the crown basal area in full shadow but in a state of incipient illumination. Compensation light intensity, $I(\xi) \equiv I_k$, at the canopy base, $\xi = 1$, allows us to use Eq. 3.52 with Eq. 3.43 to write the important relation

$$\frac{I_k}{I_0} = \exp(-\kappa L_t) = \exp(-\beta L_t). \tag{3.53}$$

Baker (1950, Table 11, p. 143) presents values of I_k, observed under controlled artificial light, as a percentage of so-called "full sunlight". These are reproduced here as I_k^o in column 2 of Table 3.9. Baker (1950, p. 144) notes that "The compensation point is by no means fixed for individual species", but the meaning of his term "full sunlight" is somewhat ambiguous. Gates (1980, p. 94) gives it the constant value 107.6 klx for a clear summer day independent of species or habitat. Indeed, Baker (1950, Fig. 43, p. 142) sketches light–photosynthesis curves for various species in terms

Table 3.9. *Light intensity at the compensation point*

(1) Species	(2) Compensation point I_k^o (% of full sunlight[a])	(3) κL_t (from column 2 by Eq. 3.53)
Ponderosa pine	17.0	1.77
Scots pine	15.9	1.84
Northern white cedar	10.3	2.27
Tamarack	9.6	2.34
Lodgepole pine	7.6	2.58
Douglas fir	7.6	2.58
Red oak	7.4	2.60
Hackberry	6.4	2.75
Englemann spruce	6.0	2.81
Norway spruce	4.9	3.02
Eastern hemlock	4.7	3.06
Beech	4.1	3.19
Sugar maple	2.0	3.91

[a] "Full sunlight" is interpreted here as the average growing-season daylight-hour insolation at canopy top. See discussion following Eq. 3.53.

Source: Baker, F.S., *Principles of Silviculture*, Table 12, p. 143, Copyright © 1950 McGraw-Hill Book Co. Inc., with kind permission from the McGraw-Hill Companies.

Plate 3.5. Tree height often wins the battle. Elevation of the leaf surfaces above those of competing species captures the light at the metabolic cost of additional supporting trunk and branches. This higher elevation in the boundary layer feeds back positively into lower canopy resistance and thus to the higher nutrient fluxes needed to support this structure. (Photographed by Peter S. Eagleson in the Borghese Gardens, Rome.)

of a constant "full sunlight". However, using the Gates (1980, p. 94) value of full sunlight, Baker's percentages yield absolute values of I_k which are as much as two orders of magnitude larger than those listed by Larcher (1983, Table 3.6, p. 104); for example $I_k = 18.3$ klx (Baker) vs. 0.2 klx (Larcher) for the shade leaves of conifers, and $I_k = 2.15$ klx (Baker) vs. 0.45 klx (Larcher) for the shade leaves of deciduous trees.

The value of I_k is known to reflect habitat light conditions (e.g., Larcher, 1983, p. 103), thus we interpret the Baker values in column 2 of Table 3.9 as $I_k^o \equiv I_k/I_0$, where I_0 is the daylight-hour, growing-season average, canopy-top insolation of the particular habitat.

J Structural response to the radiation regime

Leaf angle

We draw the following tentative conclusions from the small samples of Fig. 3.15:

(1) For deciduous multilayer crowns, leaf angle is imposed by solar altitude through Eq. 3.51 in temperate climates where neither carbon or nutrients limit productivity.

(2) For climates in which temperature extremes limit photosynthetic efficiency, leaf angle is biased by the need to shed or retain NIR; vertical reflection from the top of hot climate multilayer rainforest canopies, horizontal reflection in cold climate multilayers, and normal to the leaf suface (i.e., minimum reflection) in cold climate monolayers.

From the small samples of Fig. 3.18 we further tentatively conclude:

(3) By virtue of their clear sensitivity to $h_\otimes$ (cf. Fig. 3.18a), the three evergreen κs seem to be governed by beam radiation in a manner that has the form of the theoretical curves for radiation normal to the leaves. The data follow closely Eq. 3.32 for opaque right circular cones at the higher latitudes (i.e., lower $h_\otimes$) and Eq. 3.29 for opaque vertical circular cylinders at the lower latitudes (i.e., higher $h_\otimes$). The opacity likely results from cold-climate-induced monolayer foliage structure, which in the idealized case has all leaves lying in the crown surface and results in no downward scatter of radiation. These observed κ_bs also lie below the theoretical value for translucent canopies having $\theta_L = 90^\circ - h_\otimes$ (Curve (f)) by a relatively constant factor (see dashed Curve (g)) that is consistent with the effect of downward scattering and confirms the previous conclusion from Fig. 3.15.

(4) The two deciduous observations of κ must include an effect of scattering due to the open structure of multilayer deciduous crowns. It is *possible* to explain these observations if the leaf angle is controlled by diffuse light as displayed in Fig. 3.18b because here the observations are smaller than the theoretical κ_ds by an amount that is approximated by downward scattering as we have seen. However, our postulated optical optimality condition leads to both $\kappa_b = \beta$, and $\kappa_d = \beta$, and to a relationship $\kappa_b(h_\otimes)$ (cf. Eq. 3.51) which has the magnitude of the observed $\beta(h_\otimes)$ as is shown by Curve (b) in Fig. 3.15. This confirms the first conclusion that leaf angle is determined by the solar altitude of beam radiation in temperate climate multilayers.

We conclude that solar elevation plays an important and perhaps dominating role in selecting the leaf angle in a given climate.

Crown shape

To absorb maximum solar energy the cold-climate conifer† has darkly pigmented needles which we have just seen to be oriented *on average* nearly normal to the radiation. To conserve this heat, the conifer may restrict sensible heat diffusion by packing the needles in a dense "monolayer" shell at the surface of the crown. This strategy is particularly successful in situations with large amounts of scattered light (e.g., high latitudes, overcast skies, or snow-covered surfaces) since light can then reach the side of the tree which is shaded by the monolayer in direct light. As a limiting case

† In contrast to "warm-climate" conifers such as loblolly pine.

of this monolayer we can imagine the needles in a single planar layer coincident with a crown surface that is perpendicular to the beam radiation. In such an idealized case we can derive the relationships between:

(1) solar altitude and the proportions of the conical crown, and
(2) solar altitude and leaf area index.

Proportions of the conical crown of cold climate conifers

With the leaves (i.e., needles) lying in the lateral surface of the cone, that surface must be normal to the solar beam at the effective azimuth and altitude of the sun. This readily yields the geometrical relationship for leaf inclination

$$\beta = \frac{r_o}{h_c}\left[1 + \left(\frac{r_o}{h_c}\right)\right]^{-1/2} = \cos\left[\tan^{-1}\left(\frac{h_o}{r_o}\right)\right], \tag{3.54}$$

in which $r_o = b/2$ is the radius of the cone at its base of diameter, b, and as earlier, h_c is its height. We define the crown aspect ratio as

$$\frac{h_c}{r_o} = \cot(h_\otimes). \tag{3.55}$$

The lateral surface area of the cone, A_s, is

$$\frac{A_s}{\pi r_o^2} = \left[1 + \left(\frac{h_c}{r_o}\right)^2\right]^{1/2}, \tag{3.56}$$

but the (single-sided) surface area, A_l, of the needles lying in and comprising the conical surface is larger than given by Eq. 3.56. Lang (1991) gives this as

$$A_l = \frac{\pi}{2} A_s. \tag{3.57}$$

By definition of the leaf area index, we can now use the last two equations to write

$$L_t \equiv \frac{A_l}{\pi r_o^2} = \frac{\pi}{2}\frac{A_s}{\pi r_o^2} = \frac{\pi}{2}\left[1 + \left(\frac{h_c}{r_o}\right)^2\right]^{1/2}. \tag{3.58}$$

Jarvis *et al.* (1976, Table XVI, pp. 234–235) give the latitudinal locations of leaf area indices measured for various conifer species. Selecting the pines because of the many varieties measured over a wide range of latitudes, we present the observations of L_t vs. Φ in Fig. 3.21. Combining Eqs. 3.55 and 3.58 and using Fig. 3.9 as before to translate $h_\otimes$ into Φ, we can plot Curve (a) in Fig. 3.21 in comparison with the observations. Except for the Japanese red pine the agreement is good and supports the notion that the aspect ratio of the conical cold-climate conifers reflects the solar altitude.

For three varieties of pine (Scots, ponderosa, and lodgepole) the averaged values of I_k^o reported by Baker (cf. Table 3.9, column 2) give $I_k/I_0 = 0.135$ which makes Eq. 3.53

$$L_t = \frac{1}{\beta}\ln\left(\frac{1}{0.135}\right). \tag{3.59}$$

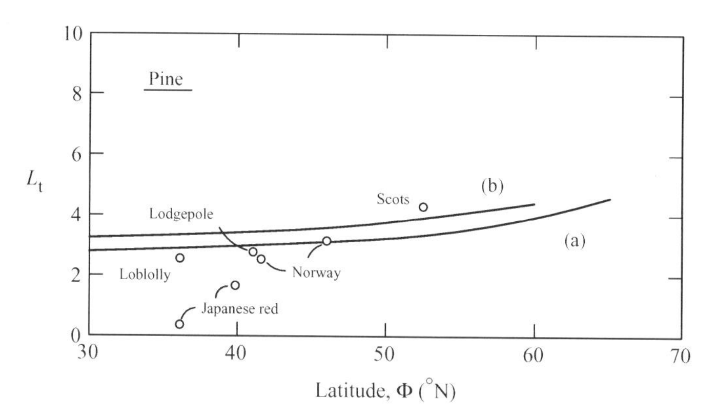

Fig. 3.21. Sensitivity of crown shape to solar altitude. Data from Jarvis *et al.* (1976, Table XVI, pp. 234–235). For explanation of Curves (a) and (b) see text.

With the needle surface perpendicular to the solar beam, $\theta_L = \frac{\pi}{2} - h_\otimes$, and Eq. 3.59 is, for pine

$$L_t = 2\csc(h_\otimes), \tag{3.60}$$

which, with the help of Fig. 3.9, is plotted as Curve (b) in Fig. 3.21. It serves both as a verification of Baker's (1950) compensation light intensity (as interpreted here) for these pine varieties, and as confirmation of the observed slight increase of leaf area index with latitude.

4

Turbulent fluxes

The momentum flux in a vegetation canopy is expressed theoretically in terms of the crown shape and the drag properties of the foliage elements for both multilayer and monolayer crowns. The vertical distribution of derived kinematic eddy viscosity has an anomalous maximum high in a canopy of tapered multilayer crowns that is verified by observations in Sitka spruce and which leads to superior vertical mixing.

The Penman–Monteith equation for transpiration from a unit area of dry leaf surface is applied to the entire dry forest canopy using the "big leaf" model. Transfer from leaf scale to big leaf canopy scale uses a stack of leaf layers with stomata fully open to minimize stress. Total canopy resistance is closely approximated by a series-parallel network of interleaf-layer atmospheric resistances only. The interleaf-layer resistance is calculated from the kinematic eddy viscosity using a lumped analogy with Ohm's law.

A Background

Under the assumptions of this work, the essential controls on forest growth are the vertical fluxes within the canopy of light, carbon dioxide, water vapor, and heat. We have considered the first of these in the previous chapter. Here we seek analytical expression of the turbulent fluxes of mass and heat in terms of the canopy structure and climate. To reach this goal we represent the three-dimensional canopy by a two-dimensional approximation.

We begin by reviewing the turbulent transfer of any conservative fluid quantity which does not change by virtue of its displacement in the direction of transfer. We let "s" be the instantaneous amount of this quantity per unit mass of fluid, being the sum of a temporal mean value, $\bar{s}$, and a turbulent fluctuation therefrom, s', and we assume that s varies in the z-direction only as is shown in Fig. 4.1.

Because of zero-mean vertical velocities, $w = w'$, a vertical exchange of fluid packets takes place through any horizontal area, A, even though the mean fluid velocity is in the horizontal direction, x. The vertical distance traveled by these packets before taking on the character of their new surroundings is assumed to be a random quantity,

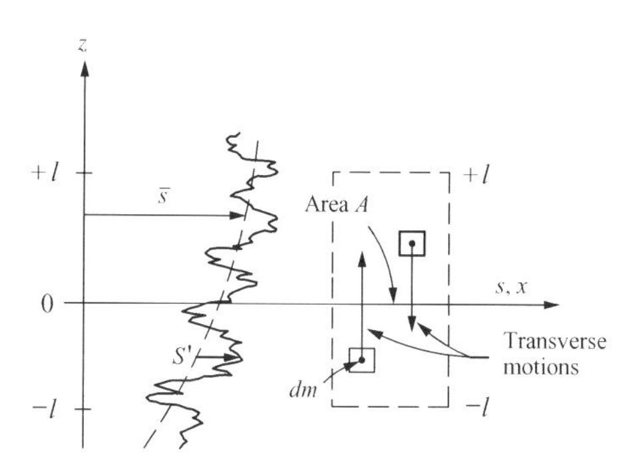

Fig. 4.1. Definition sketch for turbulent transport. (After Eagleson, 1970, Fig. 3-2, p. 17.)

the mean value of which we shall call the *mixing length*, l;† we will return to this quantity later in the chapter. The amount of the quantity s carried by a parcel of mass dm and passing through A will depend upon the point of origin of motion of the parcel. This is due to the assumed gradient of the quantity $\bar{s}$ in the direction of the random motion. Statistically speaking, no particle lying outside the limits $+l > z > -l$ will pass through A, and the average time needed for the parcel to travel the distance l will be called "t". The average rate of net upward transport of the quantity s per unit area is called S_z and may be written

$$S_z = \frac{l}{At}\left[\int_A \int_{-l}^{0} \rho\, s(z)\, \mathrm{d}z\, \mathrm{d}A - \int_A \int_{0}^{+l} \rho\, s(z)\, \mathrm{d}z\, \mathrm{d}A\right], \tag{4.1}$$

in which ρ is the fluid mass density (g cm^{-3}). Expanding $s(z)$ in a Taylor series about $z = 0$, retaining only the linear terms, and integrating gives the general form of the so-called *flux–gradient* relationship

$$S_z = -\frac{\rho l^2}{t}\left(\frac{\mathrm{d}s}{\mathrm{d}z}\right)_{z=0} = -\rho K\left(\frac{\mathrm{d}\bar{s}}{\mathrm{d}z}\right)_{z=0}, \tag{4.2}$$

in which K is a turbulent transfer coefficient (cm^2 s^{-1}) dependent upon the flow conditions. Alternatively, we can write this vertical transport as

$$S_z = \rho\, \overline{ws} = -\rho \overline{w's'}. \tag{4.3}$$

Estimation of the coefficient, K, is key to understanding S_z, and we will pursue it through first looking at the easiest case, the vertical transfer of horizontal momentum, in which case $K \equiv K_{\mathrm{m}}$ is customarily referred to as the *kinematic eddy viscosity*. Then for s = mass or heat we will apply the Reynolds analogy (Schmidt, 1917) which allows, *in stable atmospheres*, use of the K_{m} of momentum transfer to approximate the *eddy diffusivities*, K, governing the transfer of mass and heat as well (Wright and Brown, 1967; Denmead and McIlroy, 1970).

The use of flux–gradient relationships to estimate vertical transport within forest canopies has come into question (e.g., Raupach and Thom, 1981; Finnigan, 1985;

† Introduced by Prandtl (1926) as the "mixture length", l is the average distance traversed by a fluctuating fluid element before it aquires the velocity of its new region.

Finnigan and Raupach, 1987; Raupach, 1988; Raupach and Finnigan, 1988; Shuttleworth, 1989) on the basis of "frequent and pronounced" countergradient fluxes of heat, water vapor, and carbon dioxide observed at times in certain canopies (e.g., Denmead and Bradley, 1985, 1987). The countergradient fluxes are attributed to length scales of the vertical mixing (i.e., mixing lengths, l) that are related to the crown dimension, probably through eddies shed in the wake of individual crowns in a canopy with an "open" top (e.g., conical crowns), and thus violate the "fine grain" mixing implied in the linearization of Eq. 4.1. A more correct, higher-order formulation has been proposed by Wilson and Shaw (1977), and by Raupach and Shaw (1982), but its complexity is mismatched to the other gross simplifications of this work. In spite of the demonstrated inadequacies of the flux–gradient approach, Raupach and Thom (1981, p. 98) acknowledge that the flux–gradient methods have "given approximate but useful insight into the way in which physical and biological factors combine to govern the transpiration and photosynthesis rates of a plant canopy...".

General insight is the goal of this work rather than precision of detail, so we will use the flux–gradient approach with all its flaws as a suitable approximation. To reduce the resulting error, we will consider primarily "dense" canopies in which l is most likely to have the scale of the interleaf distance, and will concern ourselves with only the long-term average state of atmosphere–canopy–soil systems, a state under which (continuous) countergradient transport within the crown will be minimized.

B Introduction to momentum flux

The canopy and its substrate constitute a sink for the horizontal momentum that is characteristic of the ambient mean airflow. Turbulent diffusion carries horizontal momentum vertically downward within the canopy and at each level the resistance offered to this flow (i.e., the "drag") of the solid canopy elements extracts a portion of that momentum and converts it into turbulence and thence into heat. This continuous supply from the ambient atmosphere to the drag on the canopy elements defines what is called the vertical flux of momentum. Its formulation is the beginning point of our analysis.

During the second half of the twentieth century agriculturalists and later foresters studied momentum flux in various types of plant canopies; the literature is vast, but a useful summary is provided by Brutsaert (1982). A mixture of theory and empiricism has led the field to expressions of the form

$$\frac{u(\xi)}{u_0} = \exp(-a_w \xi), \tag{4.4}$$

$$\frac{\tau_f(\xi)}{\tau_0} = \exp(-a_d \xi), \tag{4.5}$$

and

$$\frac{K_m(\xi)}{K_m(0)} = \exp(-a_m \xi), \tag{4.6}$$

Plate 4.1. European hornbeam. (a) Bare; (b) leaved. (Photographed bare by William D. Rich; Copyright © 2001 William D. Rich, and photographed leaved by Peter S. Eagleson; Arnold Arboretum, Boston.)

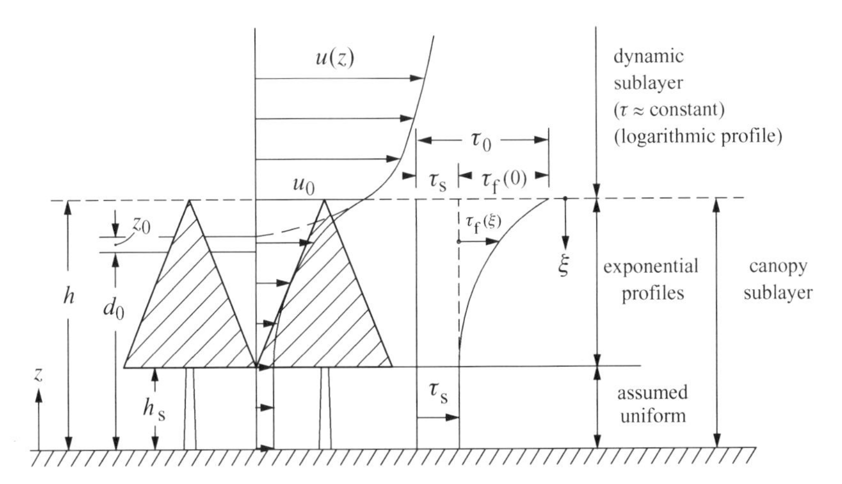

Fig. 4.2. Definition sketch for canopy kinematics. (After Brutsaert, W.H., *Evaporation into the Atmosphere*, Fig. 3.1, p. 54, Copyright © 1982 D. Reidel Publishing Company, with kind permission from Kluwer Academic Publishers, Dordrecht, The Netherlands.)

in which u, τ_{f}, and K_{m} are, respectively, the horizontal wind velocity, the horizontal shear stress on the foliage, and the kinematic eddy viscosity (see Eq. 4.2), while as before (see Fig. 4.2)

$$\xi \equiv \frac{h-z}{h-h_{\mathrm{s}}}. \tag{4.7}$$

The reference values, u_0, τ_0, and $K_{\mathrm{m}}(0)$, are those in the free stream at $\xi = 0$.

The coefficients a_{w}, a_{d}, and a_{m} are called *canopy indices* or *extinction parameters* and are estimated empirically to be $0.4 < a_{\mathrm{w}} < 4$ (Cionco, 1972), $a_{\mathrm{d}} \approx 3$ (Kaimal and Finnigan, 1994, p. 78), and $2.2 < a_{\mathrm{m}} < 4.25$ (Brutsaert, 1982, p. 106). To distinguish the behavior of different plant types, it is important that we understand the physical reasons for the variation in these indices. We will attempt to derive them from first principles of fluid dynamics, keeping in mind the cautionary words of Landsberg and Jarvis (1973, p. 645):

> The analysis of the momentum balance of a plant community is fraught with difficulties, . . . , and there is no certainty that the exchange coefficients derived from momentum balance calculations will be sufficiently accurate to provide good estimates of fluxes between various levels of the stand . . . . Nevertheless, such analysis can provide insights into effects of canopy architecture on the transfer processes . . . . Furthermore, . . . models of the momentum balance of canopies can provide a quick and relatively simple estimate of the ventilation conditions in the stand which, in association with models of the radiation regime and information on the photosynthetic characteristics, contribute to the refinement of estimates of the effects of weather conditions and management on productivity.

C General formulation of momentum flux

The equation of horizontal motion

Figure 4.2 provides a definition sketch for the kinematics of flow in a closed canopy (i.e., canopy cover, $M = 1$). As done by Brutsaert (1982, p. 54), we divide the atmospheric boundary layer at canopy height, $z = h$, into an outer, dynamic sublayer, and an inner portion closest to the surface which is called the canopy layer. In the general case, the horizontal resistance, τ_0, to airflow through the canopy is due both to horizontal fluid dynamic drag, τ_f, on the internal solid matter of the crown, and to horizontal resistance, τ_s, imposed "externally" by the substrate. The former is normally the principal component as we indicate by the large reduction in horizontal momentum through the crown in Fig. 4.2; the residual horizontal momentum below the crown is removed by substrate shear. The accompanying fluid shear stress profile is also shown in the figure.

The drag forces exerted on the solid canopy elements by the fluid flowing through the canopy are shown in Fig. 4.3 under the assumption of zero horizontal pressure gradient. The canopy chosen for these two definition sketches is the general, non-homogeneous case in that the crowns do not occupy the entire canopy space. We will call this *canopy scale non-homogeneity*. Non-homogeneity will also result from spatially variable leaf area density within the crown regardless of whether the crowns are space-filling or not. We will call this *crown scale non-homogeneity*. In our mathematical developments we will use either vertical coordinate, the dimensional z or the dimensionless ξ depending upon which seems the most convenient (see Eq. 4.7).

Canopies for which the momentum reaching the substrate is negligible will be called *dense*, and their boundary shear stress is $\tau_s = 0$. To be dense a canopy must be *closed*, $M = 1$, and also the foliage element drag area density and distribution must be such as to absorb the entire downward momentum flux. The canopy drag will be called *foliage resistance* and is expressed in terms of the equivalent horizontal shear stress, τ_f.

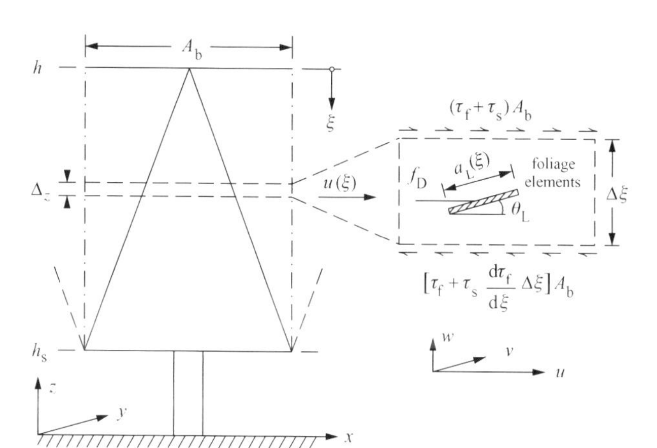

Fig. 4.3. Definition sketch for canopy fluid dynamics.

Table 4.1. *Discrete drag regimes for idealized foliage elements*

	$\boldsymbol{R}_D < 10^{-1}$ (two-dimensional circular cylinders)	$10^{-1} \leq \boldsymbol{R}_D < 10^0$ (two-dimensional circular cylinders)	$10^0 \leq \boldsymbol{R}_D < 10$ (two-dimensional circular cylinders)	$10 \leq \boldsymbol{R}_D < 10^3$ (two-dimensional circular cylinders)	$10^3 \leq \boldsymbol{R}_D$ and all practical $\boldsymbol{R}_x$ (two-dimensional circular cylinders and flat plates)
$C_D \sim u_0^{-\alpha}$	$\alpha = 1$	$\alpha = 2/3$	$\alpha = 1/2$	$\alpha = 1/4$	$\alpha = 0$
$\tau \sim u_0^{2-\alpha}$	$2 - \alpha = 1$	$2 - \alpha = 4/3$	$2 - \alpha = 3/2$	$2 - \alpha = 7/4$	$2 - \alpha = 2$
$u_0 \sim \tau^m$	$m = 1$	$m = 3/4$	$m = 2/3$	$m = 4/7$	$m = 1/2$
Wake separation	None	None	Laminar	Laminar	Laminar
Plant regimes (based upon C_D)		Inoue and Uchijima (1979) (rice and maize)	Thom (1971) (beans)	Landsberg and Jarvis (1973) (spruce shoots)	Amiro (1990b) (pine and aspen) Jarvis *et al.* (1976) (coniferous forest)

In this chapter we will work primarily with closed canopies where we will assume the boundary shear stress to be negligible. In the manner of Brutsaert (1982, p. 97), we assume the foliage resistance to be "diffusely distributed" throughout the flow field. That is, we assume there to be an infinite number of small, unconnected resistive elements behaving as a continuum-like sink for momentum. However, for illustrative purposes, in the definition sketches of Fig. 4.3 we have lumped this distributed resistance so that the isolated fluid element contains a single leaf. The horizontal drag force, f_D, which this leaf exerts on the fluid is balanced (in the absence of both a horizontal pressure gradient and horizontal velocity gradients) by the forces due to fluid shear stress, $\tau_{zx} = \tau_f + \tau_s$, on the top and bottom horizontal surfaces of the fluid element. The difference in these shearing forces, $\Delta\tau_f A_b$, is the increment of horizontal fluid momentum extinguished by the solid matter per unit of time. The horizontal equation of motion is then written in the form

$$-\frac{d\tau_f}{dz} + f_D = 0, \tag{4.8}$$

in which, using the meteorological convention,

$$f_D = \frac{\text{horizontal component of foliage resistance}}{\text{unit volume of canopy}} = \frac{C_D \rho u^2 A_D(z)}{A_b \Delta z}. \tag{4.9}$$

In Eq. 4.9, C_D is the fluid dynamic drag coefficient of the canopy solid matter, ρ is the air density, A_D is the horizontal component of the effective drag-producing area of the solid matter, and A_b is the cross-sectional area of the right circular cylinder that just circumscribes the crown.

Drag coefficients

In the general case illustrated in Fig. 4.3, the drag coefficient of the foliage elements may be influenced by one or more of three factors:

1. Shape (C_{D1})

Here we are concerned with the surface and form resistance of the foliage element shape in an infinite flow field. This is the principal component of the resistance and is given by the familiar empirically determined drag coefficient (C_D or C_f) vs. Reynolds number (***R***) curves. These as are given for flat plates ("leaves") by Schlichting (1955, Fig. 21.9, p. 448), and for two-dimensional circular cylinders ("stems, twigs, and shoots") by the generic curve found in most fluid dynamics texts (e.g., Rouse, 1946; Daily and Harleman, 1966) and presented here as Fig. 4.4, all for steady flow and stationary bodies. We will represent the latter curve, in the Reynolds number range of interest, by

$$C_{D1} = \frac{C_{D0}}{\boldsymbol{R}_D^{\alpha}}, \tag{4.10}$$

as is shown by the straight line segments in Fig. 4.4. The values of α are summarized with their Reynolds number ranges in Table 4.1.

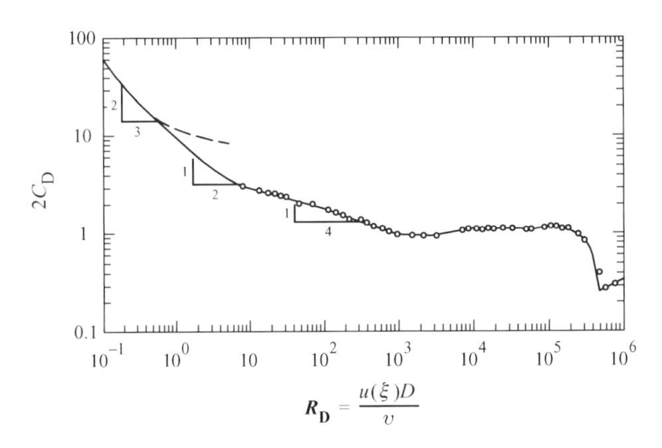

Fig. 4.4. Drag characteristics of two-dimensional circular cylinders (steady flow and infinite fluid).

2. Interference (C_{D2})

Here we are interested in the fact that proximity of multiple, two-dimensional drag-producing elements reduces the total drag on the set of elements to less than that on the sum of individuals each governed by the infinite fluid observations of Fig. 4.4. This effect is due to reduction in the form drag of body (a) due to the stagnation pressure bulb upstream of body (b) in the wake of body (a). This is illustrated in Fig. 4.5 for stationary two-dimensional circular cylinders and for stationary streamlined struts as given by Hoerner (1958). We will assume that the shape and two-dimensional interference effects are separable so that

$$C_{D2} = C_{D1} e^{-s(\xi)}, \tag{4.11}$$

in which $s(\xi)$ is the *interference function* or *shielding function* and is dependent upon the vertical distribution of the canopy solid matter. For homogeneous canopies $s(\xi)$ is a constant. With non-homogeneity, we encounter interference and hence variable $s(\xi)$ in two distinct fashions according to the volume density of the foliage elements: In the case of low foliage volume density, such as with crown-scale non-homogeneity of leafy plants, the individual foliage elements have a variable spacing and hence variable mutual interference as a function of vertical position in the crown. In the case of high foliage volume density, in plants having the canopy-scale non-homogeneity such as the conical crowns in Fig. 4.2, the crown shape itself deflects part of the flow around the crown. This causes crown form drag (as distinguished from foliage element form drag) which is subject to vertically variable interference from adjacent crowns. We assume that this interference manifests itself as in the first case, by reducing the drag on the individual foliage elements.

3. Roughness

In the case of canopy-scale non-homogeneity, the "open" canopy top causes a streamline pattern which is variable with free stream velocity. In the idealized case of two-dimensional roughness strips illustrated in Fig. 4.6, we show the free stream tending to follow the crown contours at low u_0, and to skim over them at high u_0. This causes

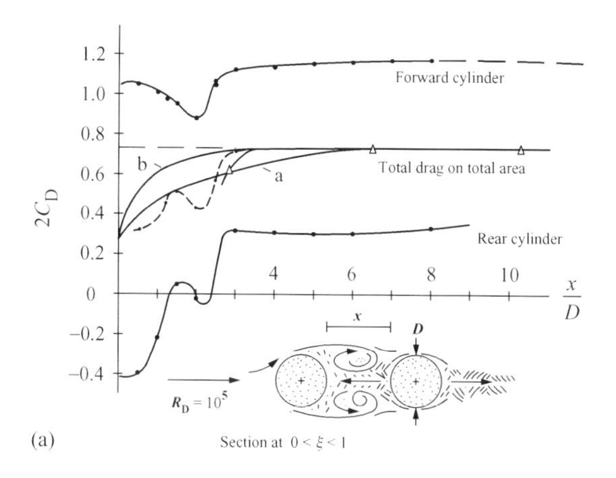

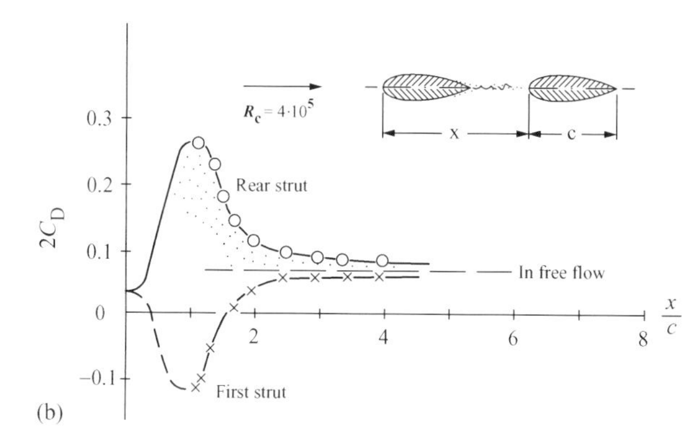

Fig. 4.5. Interference drag. (a) Drag coefficients of two interfering circular cylinders. Approximations for: (a) interfering cones: $\frac{x}{D} = \frac{1}{\xi} - 1$, $B_o = -1$, $\gamma L_t = 1.5$; and (b) interfering hemispheres: $\frac{x}{D} = [\xi(2-\xi)]^{-1/2} - 1$, $B_o = -1$, $\gamma L_t = 2.0$. (b) Drag coefficients of two interfering thin struts. (From Hoerner, S.F., *Fluid-Dynamic Drag*, Fig. 8.2, p. 1, and Fig. 8.3, p. 2, Copyright © 1958 Hoerner Fluid Dynamics; with kind permission from Mrs S.F. Hoerner, Bakersfield, CA.)

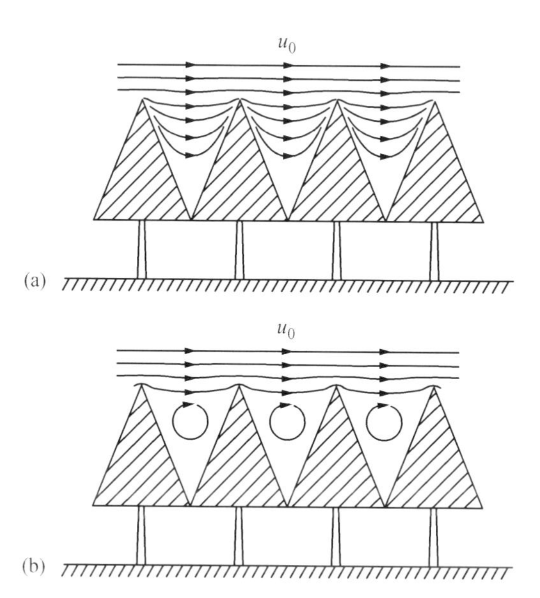

Fig. 4.6. Variable roughness of open canopy top. (a) Low free-stream velocity; (b) high free-stream velocity.

a decrease in the aerodynamic roughness of the canopy with increasing free stream velocity and is expressed through the canopy surface resistance coefficient, C_f, that is

$$\tau_0 = C_f(u_0)\rho u_0^2. \tag{4.12}$$

In any particular case, one or more of these influences may be present depending upon the crown shape, foliage volume density, leaf area density, and leaf area distribution; leafy plants may be areally dense yet volumetrically sparse.

The drag–shear stress relation for low foliage volume density

We continue the analysis with a further major simplification. In order to relate the foliage resistance and fluid shear, we replace the vertically variable crown cross-section with a circumscribing right circular cylinder of constant cross-sectional area, A_b, within which the canopy *drag area density*, $a_D(z)$, is a function of vertical position, z, but is diffusely distributed (i.e., uniform) over the full cylinder cross-section at each elevation. Then Eq. 4.9 can be written (Brutsaert, 1982)

$$f_D = \frac{C_D(u, z)\,\rho u^2\,A_D(z)}{A_b\,\Delta z} = C_D(u, z)\,\rho u^2 a_D(z) = \tau_f\,a_D(z). \tag{4.13}$$

The general momentum transfer equation in canopies with low foliage volume density†

Expressing the horizontal component of the drag-producing area per unit of canopy volume as

$$a_D(z) \equiv \frac{A_D(z)}{A_b \Delta z} = n\widehat{\cos\theta_L}\,a_t(z), \tag{4.14}$$

in which n = number of element sides producing skin-friction,
$\widehat{\cos\theta_L}$ = spatial average cosine of the angle between element's drag force and the horizontal plane (given the notation "β" in Chapter 2), and
$a_t(z)$ = *foliage area density* = single-sided foliage area per unit of canopy volume (m^{-1}), which must satisfy the condition

$$L_t = \int_{h_s}^{h} a_t(z)\,dz = \int_0^1 a_t(\xi)\,d\xi, \tag{4.15}$$

in which L_t = single-sided *foliage area index* (dimensionless).

† In sparse canopies where the substrate shear stress cannot be neglected the following development is inexact.

Using Eqs. 4.8, 4.13, and 4.14, and noting that by definition

$$a_t(\xi) \equiv (h - h_s)a_t(z), \text{ and } dz \equiv -(h - h_s)\,d\xi,$$

Eq. 4.8 can be rewritten

$$\frac{d\tau_f}{\tau_f} = -n\beta a_t(z)\,dz = -n\beta a_t(\xi)\,d\xi. \quad (4.16)$$

Defining

$$a(\xi) \equiv \frac{a_t(\xi)}{L_t} = \text{dimensionless, single-sided foliage area } \textit{homogeneity function},$$

and assuming β and n to be invariant with elevation, Eq. 4.16 integrates to give

$$\frac{\tau_f(\xi)}{\tau_f(0)} = \exp\left(-n\beta L_t \int_0^{\xi} a(\xi)\,d\xi\right). \quad (4.17)$$

The variation of fluid velocity is then obtained from Eqs. 4.13 and 4.17 depending upon the drag coefficient vs. Reynolds number relationship.

We will now carry out the detailed calculations for a number of canopies beginning with the limiting cases posed by low volume density homogeneous canopies of (1) "leafy" plants such as deciduous and needle-leafed trees (where the surfaces have a significant horizontal projection and the drag is primarily skin-friction); using the terminology of Horn (1971) we refer to such plants as *multilayer* plants due to the layered appearance of their leaves resulting from the branching structure, and (2) "stemmy" plants such as rice and wheat (where the surfaces are largely vertical and it seems a priori that form drag may predominate).

D Features of momentum flux in homogeneous canopies

The nature of homogeneity

The crowns of leafy plants typically have their foliage area density distributed vertically in a "bell" curve as shown in Fig. 2.5 for aspen (*Populus tremula*), linden (*Tilia cordata*), and oak (*Quercus robur*) taken from Rauner (1976, Fig. 2, p. 244). For truly homogeneous canopies, the foliage area density would be distributed uniformly through crowns that fill the entire canopy space (i.e., right cylinders with canopy cover, $M = 1$), and these vertical distributions would be rectangles demonstrating $a(\xi) = 1$. The degree to which this homogeneity condition is met by actual canopies is illustrated in Fig. 4.7 for the crowns of aspen (Rauner, 1976, Fig. 1, p. 243) which we will consider typical of "homogeneous" leafy plants.

For homogeneous canopies we use $a(\xi) = 1$ in Eq. 4.17 to obtain the desired expansion of Eq. 4.5

$$\frac{\tau_f(\xi)}{\tau_f(0)} = \exp(-n\beta L_t\,\xi). \quad (4.18)$$

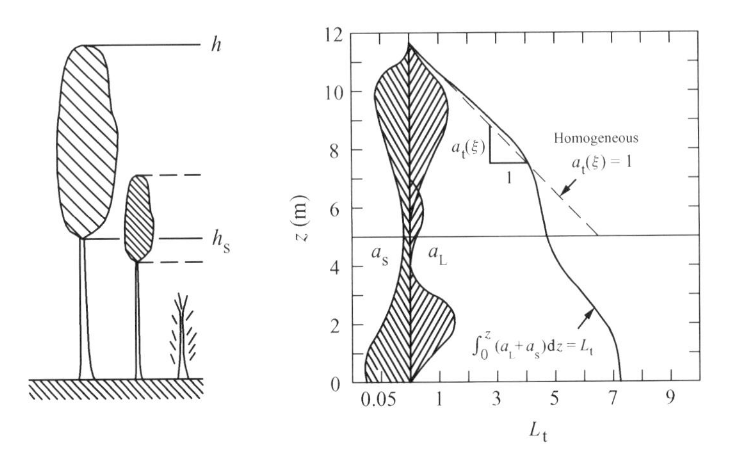

Fig. 4.7. Typical foliage area distribution of leafy canopy and its homogeneity approximation. a_s = non-leaf area (one-sided); a_L = leaf area (one-sided). (From Rauner, J.L., Deciduous forests, in *Vegetation and the Atmosphere*, Vol. 2, *Case Studies*, edited by J.L. Monteith, Fig. 1, p. 243; Copyright © 1976 Academic Press Ltd, with kind permission from Academic Press Ltd, London.)

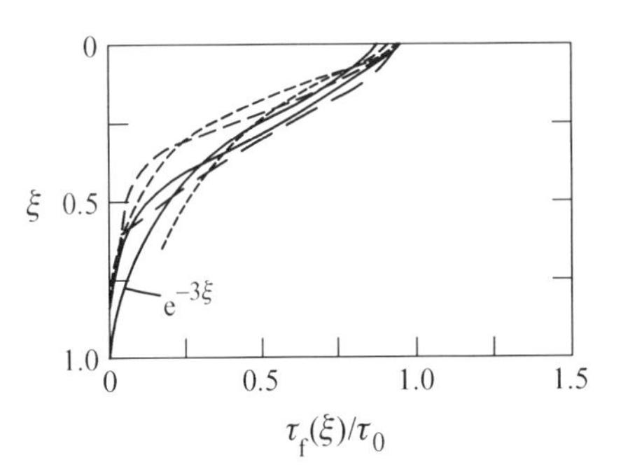

Fig. 4.8. Observed shear distributtion in various canopies. (Data from Kaimal and Finnigan, 1994, Fig. 3.3, p. 78.)

This simple exponential nature of the momentum flux decay is supported by observations of, among others, Hicks and Sheih (1977) in maize, Seginer *et al.* (1976) in an artificial canopy, and Kaimal and Finnigan (1994, p. 78) in a variety of natural and artificial canopies. The Kaimal and Finnigan (1994, Fig. 3.3b, p. 78) observations embrace artificial canopies (rods), as well as crops (wheat, corn), and forest (eucalyptus, pine), and they all cluster around an extinction parameter, $a_d = n\beta L_t = 3$ (see Fig. 4.8), which we take to represent the typical leafy canopy.

Kinematics and dynamics of flow in the homogeneous canopy

From the usual meteorological definition of shear stress as expressed in Eq. 4.13, we see that the inverse, $u(\tau_f)$, depends upon the nature of the drag coefficient. That is, velocity is proportional to shear stress according to

$$u \propto \tau_f^m, \tag{4.19}$$

where, referring to Eq. 4.10,

$$m = \frac{1}{2 - \alpha}. \tag{4.20}$$

Strictly speaking, the proportionality of Eq. 4.19 is variable with elevation even in homogeneous canopies due to α increasing with ξ as the velocity variation leads into a new Reynolds number regime (see Table 4.1). However, this variation is confined to the lower part of the canopy which contributes little to the dynamics. Consequently, we will assume the proportionality of Eq. 4.19 to be invariant and use it with Eq. 4.18 to write the general relation for horizontal windspeed

$$\frac{u(\xi)}{u_0} = \exp(-mn\beta L_t\xi), \tag{4.21}$$

where u_0 is the horizontal wind speed at the top of the canopy, $z = h$, or $\xi = 0$. This exponential form of the velocity extinction is confirmed by observations over a wide range of leafy plants (e.g., Rauner, 1976, p. 256; Cionco, 1978, p. 83; and Kaimal and Finnigan, 1994, p. 78). We reproduce the Rauner (1976, Fig. 11, p. 256) observations here in Fig. 4.9. We will continue to separate the effects of solid density (i.e., L_t), from those of geometry (i.e., $mn\beta$) by writing Eq. 4.21

$$\frac{u(\xi)}{u_0} = \exp(-\gamma L_t\xi), \tag{4.22}$$

in which

$$\gamma \equiv mn\beta. \tag{4.23}$$

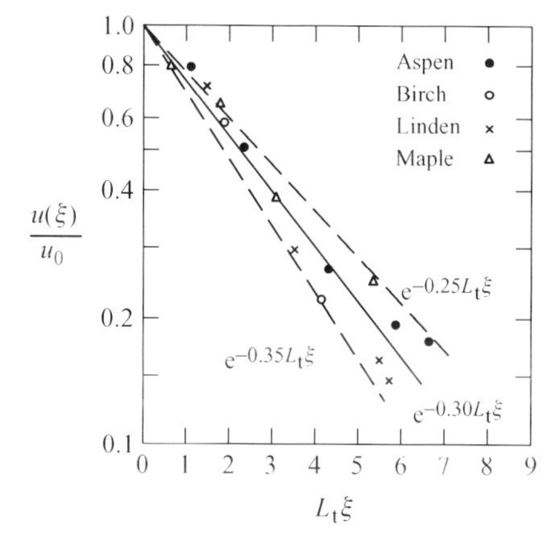

Fig. 4.9. Observed velocity distribution in leafy trees. (From Rauner, J.L., Deciduous forests, in *Vegetation and the Atmosphere*, Vol. 2, *Case Studies*, edited by J.L. Monteith, Fig. 11, p. 256; Copyright © 1976 Academic Press Ltd, with kind permission from Academic Press Ltd, London.)

As was shown in Chapter 3, optical optimality relates β to the light extinction parameter, κ, by

$$\beta = \kappa. \tag{4.24}$$

Thus the extinction parameters for both horizontal wind velocity and light are related by

$$\gamma = mn\kappa. \tag{4.25}$$

To evaluate m and n we must examine the drag characteristics of idealized foliage elements having specific form and orientation. First, however, we will complete the general formulation for homogeneous canopies having low foliage volume density.

Continuing our earlier assumption that the solid elements are diffusely distributed throughout the canopy, we can expect the kinematics and dynamics of the air flow to be continuous across the interface between the dynamic sublayer and the canopy sublayer (Fig. 4.2). In such cases we can reduce the number of free canopy parameters still further by introducing a set of "matching conditions" at the canopy top $h = z(\xi = 0)$.

Matching of velocity gradients

The vertical distribution of the mean horizontal wind velocity in the dynamic sublayer is assumed to follow the well-accepted logarithmic law attributed (in its earliest form) to von Kármán (1930) and written

$$u(z) = \frac{u_*}{k} \ln\left(\frac{z - d_0}{z_0}\right), \quad z \geq h, \tag{4.26}$$

in which $u(z)$ is the temporal mean wind velocity at elevation z, k is von Kármán's "constant", u_* is the *shear velocity* defined by

$$u_* = \sqrt{\tau_0/\rho} - C_f^{1/2} u_0, \tag{4.27}$$

where d_0 is the *zero-plane displacement height*, z_0 is the *surface roughness length*, and τ_0 is the shear stress at $z = h$. Various derivations of Eq. 4.26 are given by Monin and Yaglom (1971, pp. 274 *et seq.*), and a definition sketch is given here in Fig. 4.10. At the top of the canopy Eq. 4.26 gives

$$u_0 = \frac{u_*}{k} \ln\left(\frac{h - d_0}{z_0}\right), \quad z = h. \tag{4.28}$$

Differentiating Eq. 4.26 we obtain

$$\frac{du(z)}{dz} = \frac{u_*}{k(z - d_0)}, \tag{4.29}$$

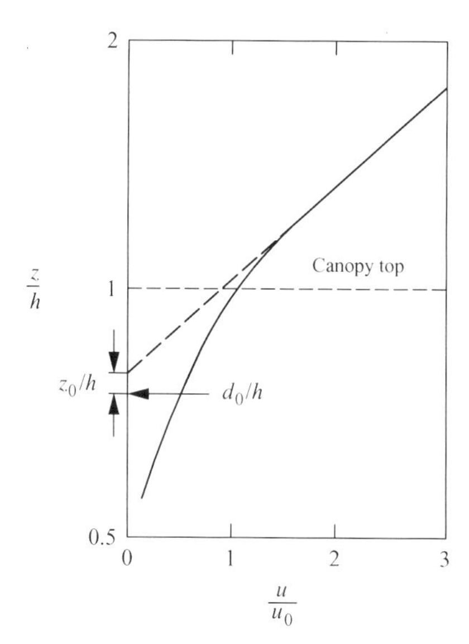

Fig. 4.10. Definition of boundary layer parameters. (Adapted from Kaimal and Finnigan, 1994, Fig. 3.1, p. 68.)

and introducing Eq. 4.28, gives

$$\frac{\mathrm{d}u(z)}{\mathrm{d}z} = \frac{u_0}{(h - d_0)\ln\left(\dfrac{z - d_0}{z_0}\right)}. \tag{4.30}$$

Within the canopy sublayer we have found (Eqs. 4.7 and 4.22)

$$u(z) = u_0 \exp\left[-\gamma L_t\left(\frac{h - z}{h - h_s}\right)\right], \quad h_s \le z \le h. \tag{4.31}$$

For $0 \le z \le h_s$ we will assume the velocity constant at $u(z) = u(h_s)$. From Eq. 4.31 this is

$$u(h_s) = u_0 \exp(-\gamma L_t), \quad 0 \le z < h_s, \tag{4.32}$$

which neglects the shear at $z = 0$ as being small with respect to that, τ_0, at $z = h$. This assumption is supported by Kaimal and Finnigan (1994, p. 78) who show, for a wide range of canopies having $L_t > 1$, that "the shearing stress transmitted to the ground surface is essentially zero". We have seen in Fig. 4.9 that the form of Eq. 4.32 fits observations in maple, aspen, linden, and birch forests.

Differentiating Eq. 4.31 gives the velocity gradient within the canopy sublayer:

$$\frac{\mathrm{d}u(z)}{\mathrm{d}z} = u_0\left(\frac{\gamma L_t}{h - h_s}\right)\exp\left[-\gamma L_t\left(\frac{h - z}{h - h_s}\right)\right]. \tag{4.33}$$

Finally, setting Eq. 4.30 equal to Eq. 4.33, gives, at the canopy top $z = h$, the *velocity gradient matching condition*:

$$\gamma L_t \left(\frac{1 - \frac{d_0}{h}}{1 - \frac{h_s}{h}} \right) \ln \left(\frac{1 - \frac{d_0}{h}}{\frac{z_0}{h}} \right) = 1, \qquad (4.34)$$

which was first presented by Kondo (1971) for the special case $h_s = 0$. Using Eq. 4.28, this velocity gradient matching condition can be written in the useful alternative form

$$\frac{u_*}{ku_0} = \left(\frac{h - d_0}{h - h_s} \right) \gamma L_t. \qquad (4.35)$$

Plate 4.2. Dawn redwood. (Photographed by William D. Rich in Arnold Arboretum, Boston; Copyright © 2001 William D. Rich.)

In Eq. 4.35, the left-hand side is called the *bulk drag parameter*, while the right-hand side is the *normalized canopy drag area density*, or more simply, the *crown density*.

Matching of momentum flux density

Immediately above the canopy the shear stress is independent of z with the value

$$\tau(z) = \tau_0 = \text{ constant}, \tag{4.36}$$

as is shown over a variety of canopies through the observations assembled by Kaimal and Finnigan (1994, p. 78). Inside the canopy sublayer the shear stress decays from this value according to Eq. 4.18. Of course the gradients, $\mathrm{d}\tau/\mathrm{d}z$, of momentum flux density cannot match across the sublayer interface, $z = h$, because of the absorption of momentum in drag forces on the solid matter of the canopy as formulated in Eq. 4.8.

Eddy viscosity

Introducing the usual assumption that the shear stress is proportional to the velocity gradient through the kinematic eddy viscosity,[†] K_m, we have, as a special case of Eq. 4.2:

$$K_\mathrm{m} \equiv \frac{\tau/\rho}{\mathrm{d}u/\mathrm{d}z} = \frac{(\tau_\mathrm{f} + \tau_\mathrm{s})/\rho}{\mathrm{d}u/\mathrm{d}z}. \tag{4.37}$$

Above the canopy, Eqs. 4.29 and 4.36 allow us to evaluate this quantity as

$$K_\mathrm{m}(z) = ku_*(z - d_0), \tag{4.38}$$

while within the areally dense canopy, τ_s is negligible, and Eqs. 4.07, 4.18, 4.27, and 4.33 give

$$K_\mathrm{m}(z) = \frac{u_*^2(h - h_\mathrm{s})}{u_0\gamma L_\mathrm{t}} \exp\left[-n(1-m)\beta L_\mathrm{t}\left(\frac{h - z}{h - h_\mathrm{s}}\right)\right]. \tag{4.39}$$

Since K_m is defined in terms of quantities that we are matching at $z = h$, K_m will automatically match there also without generating an independent matching condition. From Eq. 4.38 at $z = h$, $K_\mathrm{m}(h) = ku_*(h - d_0)$, and with Eqs. 4.07, 4.23, 4.35, and 4.40, Eq. 4.39 becomes finally

$$K_\mathrm{m}^\mathrm{o} \equiv \frac{K_\mathrm{m}(\xi)}{K_\mathrm{m}(0)} = \exp\left[-\frac{(1-m)}{m}\gamma L_\mathrm{t}\xi\right]. \tag{4.40}$$

Equation 4.40 has been found to match observed eddy viscosity decay in maize (e.g., Hicks and Sheih, 1977), pine forest (Lemon, 1965), rice (Uchijima, 1962), and wheat (Denmead, 1976). For aspen, Rauner (1976, Fig. 10, p. 255) finds a variation somewhat different from Eq. 4.40 as is shown here in Fig. 4.11. Such a variation is more like that which we will derive later for non-homogeneous canopies.

[†] Alternatively we may refer to K_m as the *eddy momentum diffusivity*.

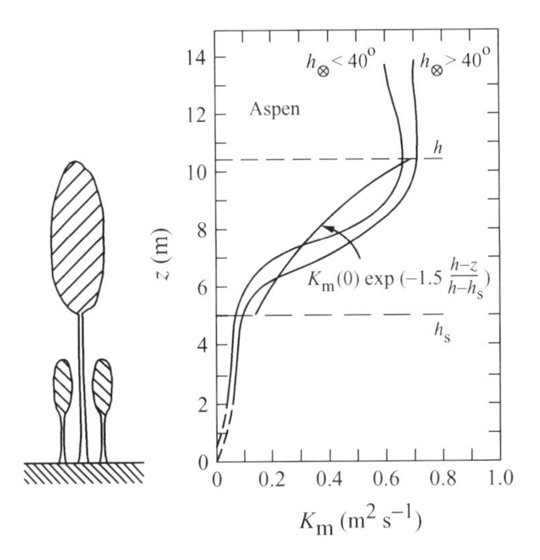

Fig. 4.11. Eddy viscosity distributtion in aspen. (From Rauner, J.L. Deciduous forests, in *Vegetation and the Atmosphere*, Vol. 2, *Case Studies*, edited by J.L. Monteith, Fig. 10, p. 255; Copyright © 1976 Academic Press Ltd, with kind permission from Academic Press Ltd, London.)

Averaging Eq. 4.40 over the depth of the homogeneous cylindrical crown, we obtain

$$\widehat{K^{\circ}_{m}} \equiv \int_0^1 \frac{K_m(\xi)}{K_m(0)} d\xi = \frac{1 - \exp\left[-\frac{(1-m)}{m}\gamma L_t\right]}{\left(\frac{1-m}{m}\right)\gamma L_t}. \tag{4.41}$$

The function $\widehat{K^{\circ}_{m}}$ is derived for non-homogeneous hemispherical and conical crowns in Appendix A.

Mixing length

In describing turbulent transport in fluids a length scale is needed, and to meet this need Prandtl (1926, cf. 1952, pp. 117–118) introduced the *mixing length*, l. Conceptually, l is the average distance traversed by a fluctuating fluid element before it acquires the velocity of its new region. He assumed the mixing length to be dominated by the geometrical characteristics of the flow; namely, in the presence of a solid boundary he took the size of the turbulent eddies to be proportional to the distance of the eddy from that boundary. This led him to relate the shear stress to the mean velocity gradient in turbulent flow in a manner similar to that for molecular shear but with a proportionality coefficient dependent upon the flow features. Subscripting l with Prandtl's initial we then write

$$\tau = \rho l_P^2 \left|\frac{du}{dz}\right| \frac{du}{dz}, \tag{4.42}$$

from which, using Eq. 4.38, we see that the kinematic eddy viscosity is defined by

$$K_m = l_P^2 \left|\frac{du}{dz}\right|. \tag{4.43}$$

Using Eqs. 4.33 and 4.39, Eq. 4.43 gives, within the canopy,

$$l_{\mathrm{P}}(z) = \frac{u_*}{u_0}\left(\frac{h - h_{\mathrm{s}}}{\gamma L_{\mathrm{t}}}\right)\exp\left[-\left(\frac{1-2m}{2m}\right)\gamma L_{\mathrm{t}}\left(\frac{h-z}{h-h_{\mathrm{s}}}\right)\right], \quad z \leq h. \tag{4.44}$$

Using Eqs. 4.29 and 4.38, Eq. 4.43 gives, above the canopy,

$$l_{\mathrm{P}}(z) = k(z - d_0), \quad z > h. \tag{4.45}$$

Von Kármán (1934) approached the need for a length scale differently than did Prandtl, basing his development upon a physical picture of the turbulent mechanisms. He assumed the mixing length to be governed by a local turbulence structure which is similar from point-to-point differing only in its length and time-scales and without any geometrical reference to the source of the turbulence. This led him to the expression for the (Kármán) mixing length

$$l_{\mathrm{K}} = k\left|\frac{\dfrac{\mathrm{d}u}{\mathrm{d}z}}{\dfrac{\mathrm{d}^2u}{\mathrm{d}z^2}}\right|. \tag{4.46}$$

With the velocity distribution of Eq. 4.31, Eq. 4.46 gives the constant mixing length

$$l_{\mathrm{K}} = \frac{k(h - h_{\mathrm{s}})}{\gamma L_{\mathrm{t}}}, \quad z \leq h, \tag{4.47}$$

while for $z > h$, Eq. 4.45 again results.

Zero-plane displacement

In experiments on an artificial canopy composed of a dense array of thin vertical rods, Thom (1971) found the zero-plane displacement to be "indistinguishable from the level of action of the drag on the component parts" of the canopy. Hicks and Sheih (1977) confirmed this finding for a maize canopy. If we accept this to be a general result as Thom proposes, we can calculate the zero-plane displacement in terms of the drag-producing characteristics of the canopy. Referring to Fig. 4.12, we take moments about an axis perpendicular to the flow and in the plane of the surface to write

$$d_0 F_{\mathrm{T}} = \int_{h_{\mathrm{s}}}^{h} z f_{\mathrm{D}}(z)\,\mathrm{d}z, \tag{4.48}$$

in which

$$f_{\mathrm{D}}(z) = \tau_{\mathrm{f}}(z) A_{\mathrm{D}}(z), \tag{4.49}$$

and the total drag, F_{T}, is given by the integral

$$F_{\mathrm{T}} = \int_{h_{\mathrm{s}}}^{h} f_{\mathrm{D}}(z)\,\mathrm{d}z. \tag{4.50}$$

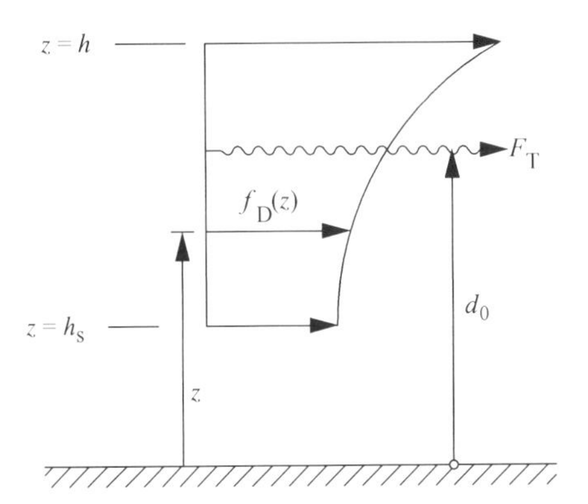

Fig. 4.12. Definition sketch for displacement thickness calculation.

Plate 4.3. Golden weeping willow. (Photographed by William D. Rich in Holden Arboretum, Cleveland; Copyright © 2001 William D. Rich.)

Using Eq. 4.18 with Eq. 4.14 in Eq. 4.49, we can integrate Eq. 4.48 with Eq. 4.50 for areally dense homogeneous canopies (i.e., $a(\xi) = 1$; see Eq. 4.17) to obtain

$$\frac{h - d_0}{h - h_s} = \frac{1}{n\beta L_t} - \frac{\exp(-n\beta L_t)}{1 - \exp(-n\beta L_t)}, \tag{4.51}$$

which is plotted in Fig. 4.13 in comparison with observations from the literature. Most observations in this figure do not include h_s, hence we assume h_s/h is negligible for purposes of the comparison; this certainly accounts for some of the scatter. Nevertheless, from the limited data available, Eq. 4.51 seems for trees to be a decided improvement over the constant value of $d_0/h = 2/3$ (dashed line) proposed by

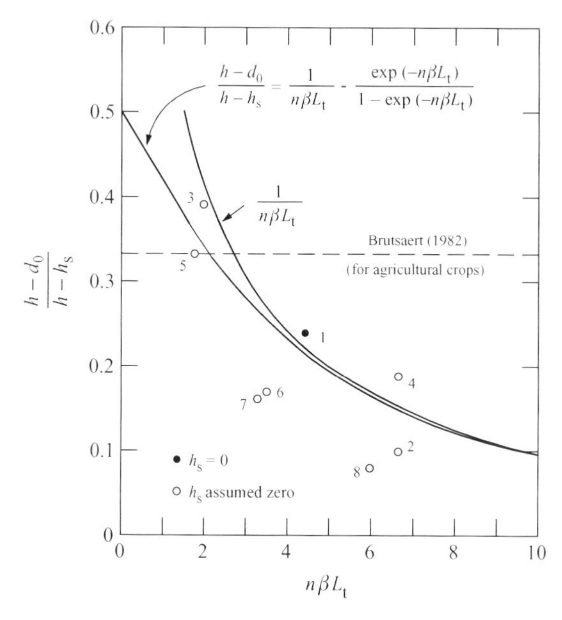

Fig. 4.13. Displacement thickness for forest canopies.

Point	Species	Quantities	Source
1	Vertical rods	$h = 14.3$ cm; $d_0 = 10.9$ cm; $h_s = 0$; $m = 2/3$; $\gamma L_t = 3.31$	Thom (1971)
2	*Pinus resinosa strobus*	$L_t = 7.97$; $\beta = 0.42$; $n = 2$	Martin (1971)
3	*Larix leptopolis*	$\gamma = 1$; $m = 1/2$; $d_0/h = 0.61$	Cionco (1978) Allen (1968)
4	*Pinus resinosa*	$d_0/h = 0.81$; $L_t = 7.97$	Leonard and Federer (1973) Martin (1971)
5	*Pinus densiflora*	$L_t = 2.06$; $\beta \approx 0.4$; $n = 2$; $d_0/h = 0.67$	Jarvis *et al.* (1976)
6	*Pinus densiflora*	$L_t = 4.37$; $\beta \approx 0.4$; $n = 2$; $d_0/h = 0.83$	Jarvis *et al.* (1976) Jarvis *et al.* (1976)
7	*Picea sitchensis*	$L_t = 3.32$; $\beta \approx 0.5$; $n = 2$	Landsberg and Jarvis (1973)
8	*Pinus taeda*	$L_t = 6.68$; $\beta \approx 0.45$; $n = 2$	Jarvis *et al.* (1976)

Brutsaert (1982, p. 116) from observations, largely on agricultural crops, made by Stanhill (1969) and others. Other early theoretical predictions of the zero-plane displacement (e.g., Shaw and Pereira, 1982) do not include the important parameters h_s, n, and β.

For canopies which are sparse due to $M < 1$, Raupach (1994) derives d_0 in terms of the single parameter, "roughness density", or "frontal area index" and obtains excellent agreement with observations.

To proceed with the areally dense homogeneous canopy we must determine the coefficients m and n which reflect the form and orientation of the foliage elements.

E Momentum flux in homogeneous leafy canopies: M = 1

Drag characteristics of idealized foliage elements

The drag characteristics of isolated, stationary, rough, flat plates in two-dimensional steady flow are presented by Schlichting (1955, Fig. 21.9, p. 448), and of fixed single cylinders in two-dimensional steady flow here in Fig. 4.4. Certainly, the leaves and other thin foliage elements of trees are not isolated nor are they stationary under average wind conditions. While interference and unsteadiness will affect the magnitude of the drag coefficient at a given Reynolds number, we do not seek these values; we are concerned only with the slope of the C_D vs. $\boldsymbol{R}$ curve which we assume to be insensitive to the gentle swaying motion of most leaves† under average wind conditions, and will use the drag on isolated stationary bodies in steady flow as a first approximation. As mentioned earlier, we have divided the Reynolds number range of Fig. 4.4 into discrete intervals in order to determine the appropriate value of the exponent m of Eq. 4.20. These intervals and the associated values of m are given in Table 4.1.

Broadleaved plants

The characteristic dimension of the leaves of broadleaved plants is about $x = 4$ cm (e.g., Murphy and Knoerr, 1977), and a nominal average wind speed in the canopy, $\bar{u}(\xi)$, will be taken as 2 m s^{-1}. The nominal drag-defining Reynolds number for broadleaved plants is then:

$$\boldsymbol{R}_x = \frac{\bar{u}(\xi)x}{\nu} = \frac{(2 \cdot \text{m s}^{-1})(4 \cdot 10^{-2}\ \text{m})}{1.5 \cdot 10^{-5}\ \text{m}^2\ \text{s}^{-1}} = \text{O}(10^3).$$

It seems reasonable to assume that the flat leaves of broadleaved plants offer surface resistance over at least a portion of both sides of the leaves, thus setting $n = 2$ in Eq. 4.25. At Reynolds numbers $\boldsymbol{R}_x = \text{O}(10^3)$ such as given above, and with an assumed leaf surface absolute roughness, $\varepsilon \approx 1$ mm, it is assumed that the surface is aerodynamically

† An obvious exception is the aspen (*Populus tremuloides*) the leaves of which vibrate rapidly in the wind.

"rough" and therefore has a constant surface drag coefficient, $C_D = C_f$, as is shown for flat plates by Schlichting (1955). This has been shown to be the case by Amiro (1990b) for aspen, as well as for the long, soft needles of jack pine.

With this nominal Reynolds number, C_D is constant, $m = 1/2$ (Table 4.1), and as argued above, $n = 2$. Thus for all leafy plants, $m = 1/2$, and

$$mn = 1. \tag{4.52}$$

Needle-leaved plants

There is a wide range of drag geometries within the group of needle-leaved plants, but we will identify and treat only the two extremes.

1. *"Individual-needle" varieties*

Most *Pinus* needles emerge from the stem in a tight circular bundle (called a "fascicle") of multi-sided individuals which opens progressively along the needle axis, thereby providing a varying degree of mutual drag interference. Those varieties with very long needles in which the individuals become separated to the point of relative independence with respect to drag we call "individual needle" varieties. As is pictured by Zim and Martin (1987), these are exemplified by the longleaf pine (*Pinus palustris*) and the loblolly pine (*Pinus taeda*), and (to a somewhat lesser degree) include others such as sugar pine (*Pinus lambertiana*), pitch pine (*Pinus rigida*), ponderosa pine (*Pinus ponderosa*), and Eastern white pine (*Pinus strobus*).

There are three needles to the circular bundle of *Pinus taeda* (Kozlowski and Schumacher, 1943) and of *Pinus radiata* (Raison *et al.*, 1992), so each needle of these varieties has a cross-section consisting of a 120° circular sector. The important ratio of total needle perimeter (P) to arc-chord (C_a) of these varieties is $P/C_a = 2.36$ (Raison *et al.*, 1992). All three sides of these needles are stomated.

The cross-section of *Pinus strobus* needles is a 72° circular sector (i.e., five to the bundle) with stomata only on the straight sides (Kozlowski and Schumacher, 1943), and Scots pine (*Pinus sylvestris*) needles have a semicircular cross-section with stomates on only the straight side. Over a range of needle varieties, Waring (1983) found the average perimeter-to-chord ratio, $\langle P/C_a \rangle = 2.5$.

Realizing that varying needle orientation exposes an average silhouette area for either drag or light interception, Lang (1991) used one of the Cauchy (1832) theorems to find $P/\widehat{C} = \pi$ where $\widehat{C}$ is the average chord of the cross-section when the needle is rotated about its longitudinal axis. Finally, Chen and Black (1992) averaged over all needle positions in three dimensions to find $\langle P/C_a \rangle = 4.0$.

Pinus needle "diameters" are about 2 millimeters, and assuming a nominal average wind speed, $\bar{u}(\xi)$, equal to 2 m s^{-1}, the nominal drag-defining Reynolds number for the "individual needle" varieties is

$$\boldsymbol{R}_x = \frac{\bar{u}(\xi)x}{\nu} = \frac{(2 \cdot \mathrm{m\,s^{-1}})(2 \cdot 10^{-3}\mathrm{m})}{1.5 \cdot 10^{-5}\ \mathrm{m^2\,s^{-1}}} = O(10^2).$$

With this Reynolds number Table 4.1 gives $m = 4/7$, and assuming surface resistance over the full perimeter, $n = 2$. Thus for the "individual needle" varieties, $m = 4/7$, and

$$mn = 8/7 = 1.15. \tag{4.53}$$

2. *"Clustered needle" varieties*

Particularly spruce, but to a lesser degreee fir, larch, pinyon pine (*Pinus edulis*) and other varieties of needle-leaved plants have short, stiff, relatively flat needles that are clustered closely together on long twigs forming what are called "shoots". There is a high degree of drag interference among these needles so that the characteristic drag dimension is that of the cluster rather than that of the individual needle. For example, Landsberg and Jarvis (1973) find the dimension of the shoot rather than the needle to govern the foliage element drag of spruce trees. In such cases we will assume that the drag arises from drag on the cluster rather than from drag on the individual needle. Needle diameters are about 2 millimeters, and cluster widths are about one order of magnitude larger. Thus with a nominal average wind speed, $\bar{u}(\xi)$, again equal to 2 m s^{-1}, the nominal drag-defining Reynolds number for needle-leaved plants is again

$$\boldsymbol{R}_D = \frac{\bar{u}(\xi)D}{\nu} = \frac{(2 \cdot \text{m s}^{-1})(2 \cdot 10^{-2}\ \text{m})}{1.5 \cdot 10^{-5}\ \text{m}^2\ \text{s}^{-1}} = \text{O}(10^3).$$

Just as with the leafy plants at this Reynolds number, C_D is constant and $m = 1/2$ (Table 4.1). Assuming the drag to be surface resistance on the composite shoot surface, $n = 2$. Thus for "clustered needle" varieties as well as for leafy plants, $m = 1/2$, and

$$mn = 1. \tag{4.54}$$

Finally, in conjunction with Eqs. 4.24 and 4.25, Eqs. 4.52–4.54 give:

$$\left.\begin{array}{l}\text{for leafy and clustered needle plants: } \gamma = \beta = \kappa \\ \text{and for individual needle plants: } \gamma = 1.15\beta = 1.15\,\kappa\end{array}\right\}. \tag{4.55}$$

This tie between the vertical extinction of solar radiation and of horizontal momentum flux within the canopy was derived and verified, apparently for the first time, in Chapter 3. We will show it to be a critical link in climatic determination of canopy structure.

Observational support for Eq. 4.55 is obtained from our collection of field estimates of various parameters of leafy canopies which we have attempted to convert to common definition and have included in Tables 2.2 (evergreen forest), 2.3 (deciduous forest), and 2.4 (crops and grasses). Wherever, in these tables, estimates of both γ and κ are available for the same species, these values are plotted in Fig. 4.14. In these coordinates, the slope of the line from the origin to the plotted point is the value of the product mn. For leafy canopies and for spruce, the limited data clearly cluster around the line

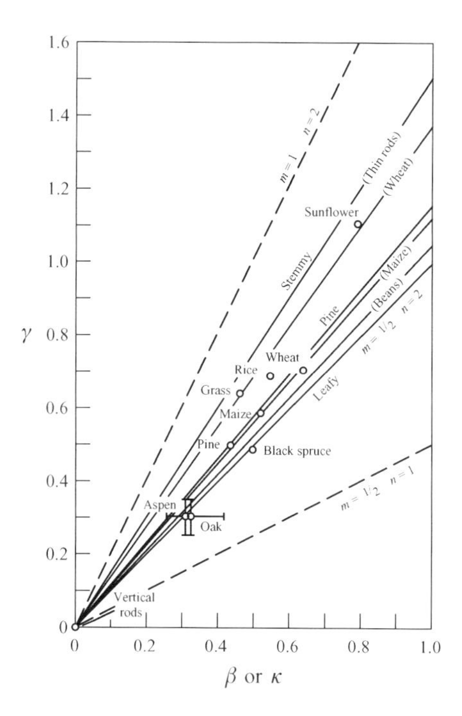

Fig. 4.14. Observed extinction coefficients ($\gamma \equiv mn\beta$).

$mn = 1$ which represents both broad-leaved and clustered needle trees. For the various *Pinus* varieties given in Table 2.2, the averaged data give $mn = 1.14$ which agrees well with that for individual needled plants.

Other data points on Fig. 4.14 are for those crops in which there is a significant component of canopy drag from the vertical stems which do not obey the relation $\beta = \kappa$. We will return to these "stemmy" plants shortly.

For very low Reynolds numbers, $C_D \propto u^{-1}$ (Lamb, 1932, p. 616), whereupon $m = 1$ (see Table 4.1), and with $n = 2$, $mn = 2$. This is the upper limit of mn shown by the upper dashed line in Fig. 4.14. The lower limit of $mn = 1/2$ is given by high Reynolds number flow with $m = 1/2$ over ground-hugging leafy plants (such as plantain) for which $n = 1$. It is shown as the lower dashed line in Fig. 4.14.

Mixing length

Considering its conceptual definition as "the average distance traversed by a fluctuating fluid element before it acquires the velocity of its new region", we expect the Prandtl mixing length to be related to the vertical spacing of the solid elements in the canopy. Thus for the uniform vertical spacing of the solid elements accompanying the homogeneous leafy canopies being considered here, it is reasonable to assume that within the crown

$$l_P = \text{constant}, \quad h_s \leq z \leq h, \tag{4.56}$$

as was proposed by Inoue (1963). From Eq. 4.44, this can be seen to require

$$m = 1/2, \tag{4.57}$$

as was reasoned above from classical C_D vs. $\boldsymbol{R}$ behavior. With the assumption that $n = 2$, our earlier observational evidence that $mn = 1$ for leafy trees is a demonstration that the mixing length really is constant within these canopies. From Eq. 4.44 this constant value within the crown is

$$l_P = \frac{u_*}{u_0}\left(\frac{h - h_s}{\beta L_t}\right), \quad h_s \le z \le h. \tag{4.58}$$

When the eddy-generating solid surfaces are parallel to the flow as for the leafy plants now under discussion, the Prandtl and von Kármán definitions of mixing length are equivalent, and we equate Eqs. 4.47 and 4.58 to obtain the theoretical value of the bulk drag parameter

$$\frac{u_*}{k u_0} = 1, \tag{4.59}$$

which can be shown to require a reversal in curvature of the velocity profile at $\xi = 0$. The bulk drag parameter was suggested as a constant for "tall, dense plant communities" by Kondo (1971) and has been shown empirically by Kondo (1972) to have the value of unity through collected observations in pine and Japanese larch canopies. The generally accepted value of the Kármán constant is $k = 0.4$ from which Eqs. 4.12, 4.27, and 4.59 imply

$$\frac{u_*}{u_0} = C_f^{1/2} \approx 0.4. \tag{4.60}$$

Kondo and Akashi (1976, Fig. 11) show, both analytically and from wind tunnel experiments, that Eq. 4.60 is satisfied in an areally dense canopy. Together, Eqs. 4.58–4.60 confirm our major initial assumption (see discussion at end of Section 4A):

$$\frac{\text{mixing length}}{\text{average vertical interleaf distance}} = \frac{l_P}{(h - h_s)/(\beta L_t)} = k < 1, \quad z \le h.$$

Using Eqs. 4.54 and 4.59 in Eq. 4.35, we have for areally dense, homogeneous, leafy canopies

$$\gamma L_t\left(\frac{h - d_0}{h - h_s}\right) \equiv \beta L_t\left(\frac{h - d_0}{h - h_s}\right) = 1, \tag{4.61}$$

which gives finally, using Eq. 4.34

$$\ln\left(\frac{1 - \frac{d_0}{h}}{\frac{z_0}{h}}\right) = 1, \tag{4.62}$$

or

$$1 - \frac{d_0}{h} = 2.718\left(\frac{z_0}{h}\right). \tag{4.63}$$

In Fig. 4.15, Eq. 4.63 is compared with observations compiled by Jarvis *et al.* (1976, Table V, pp. 188–189). When it is realized that the plotted points numbered 10 to 14 are for trees with conical crowns and thus non-homogeneous canopies, we can say that Eq. 4.63 represents homogeneous canopies quite satisfactorily. Note that Eq. 4.63

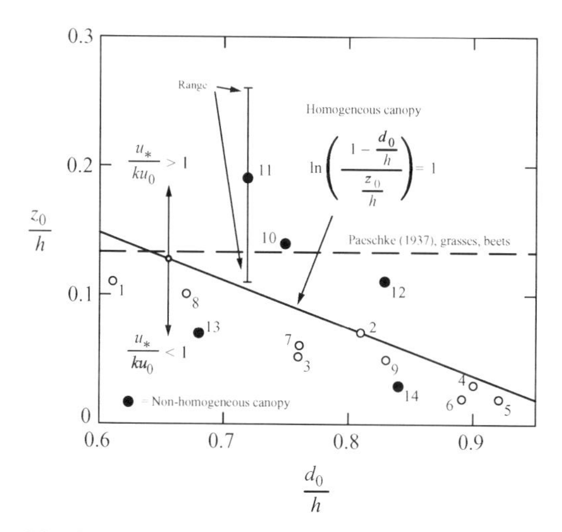

Fig. 4.15. Boundary layer parameters for needle-leaved canopies (assumed $\tau(z = h_s) = 0$).

Number	Common Name	h (meters)	L_t	z_0/h	d_0/h
1	Japanese larch	10.4	9.3	0.11	0.61
2	Norway pine	11.8	—	0.07	0.81
3	Lodgepole pine	10.0	2.8	0.05	0.76
4	Norway and eastern white pine	22.0	3.1	0.03	0.90
5	Loblolly pine	14.0	6.7	0.02	0.92
6	Loblolly pine	23.4	—	0.02	0.89
7	Scotch pine	15.5	4.3	0.06	0.76
8	Japanese red pine	4.5	0.8	0.10	0.67
9	Japanese red pine	23.0	1.7	0.05	0.83
10	Douglas fir	28.0	—	0.14	0.75
11	Norway spruce	27.2	21.6	0.26–0.11	0.72
12	Norway spruce	27.5	21.6	0.11	0.83
13	Norway spruce	22.0	—	0.07	0.68
14	Sitka spruce	11.5	3.3	0.03	0.84
			Average =	0.08	0.78

Data from Jarvis, P.G., G.B. James and J.J. Landsberg, Coniferous forest, in *Vegetation and the Atmosphere*, Vol. 2, *Case Studies*, edited by J.L. Monteith, Table V, pp. 188–189;

follows directly from the turbulent velocity distribution, Eq. 4.28, as soon as Eq. 4.59 is established.

F Momentum flux in homogeneous stemmy canopies

Definition

Erectophile crops such as the grasses (i.e., rye and sorghum), as well as wheat, rice, and sunflower, have a total foliage area made up primarily of stems and only secondarily of leaves (see Table 2.1). In such cases the majority of the foliage resistance arises from the drag of vertical "cylinders". We will refer to such plants as "stemmy" plants. They constitute another limit to the general dynamic behavior of plants in which both leaves and stems play a role. Maize, even though it has prominent stalks, has a total foliage area composed heavily of leaves as we see in Table 2.1, and we expect its behavior to be intermediate between the leafy and stemmy plants. We begin our analysis here by examining the limiting case of pure, stemmy behavior.

Artificial canopy of vertical rods

We next consider experiments on artificial canopies composed of thin circular cylinders which are tall, vertical, rigid, and smooth. The surface area of these rods is constant in the vertical direction so the canopies are precisely homogeneous and our earlier analysis is applicable.

1. Experiments of Thom (1971)

This set of observations was performed in a wind tunnel by Thom (1971) and used cylinders 1 millimeter in diameter and 14.3 centimeters high, placed on 1 centimeter centers in a diamond pattern. The average (they are velocity-dependent) zero-plane displacement and roughness height were calculated by Thom to be $d_0 = 11$ cm and $z_0 = 1.4$ cm respectively. At the average free-stream wind speed used, $u_0 = 0.5\ \mathrm{m\,s^{-1}}$, the cylinder Reynolds number is $\boldsymbol{R}_D = 33$, and for a single, two-dimensional circular cylinder in an infinite fluid (Table 7.1), $C_\mathrm{D} \propto u_0^{-1/4}$. However, Thom's (1971) wind tunnel measurements on a single, three-dimensional circular cylinder show that $C_\mathrm{D} \propto u_0^{-1/2}$ and these are reproduced here in Fig. 4.16.

Goldstein (1938, Fig. 154, p. 425) presents the separate skin friction and form (i.e., normal pressure) drag components for single, two-dimensional circular cylinders and this information is reproduced here in Fig. 4.17. From this figure, at $\boldsymbol{R}_D = 33$ (i.e., log $\boldsymbol{R}_D = 1.5$), we see that the drag of single, two-dimensional cylinders is about equally divided between the two components, and that the skin friction variation is $C_\mathrm{D} \propto \boldsymbol{R}_D^{-1/2}$ (i.e., $C_\mathrm{D} \propto u_0^{-1/2}$). It is thus the form drag which changes α from 1/2 to 1/4.

From a comparison of these two findings we conclude that Thom's (1971) single, three-dimensional cylinders experienced only skin friction, probably because the

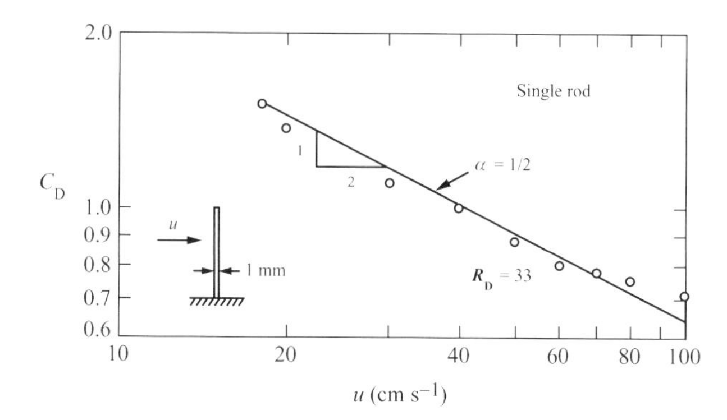

Fig. 4.16. Drag coefficient of single rod. (From Thom, A.S., Momentum absorption by vegetation, Fig. 3, p. 417, *Quarterly Journal of the Royal Meteorological Society*, 97; Copyright © 1971 The Royal Meteorological Society; with kind permission from the Royal Meteorological Society.)

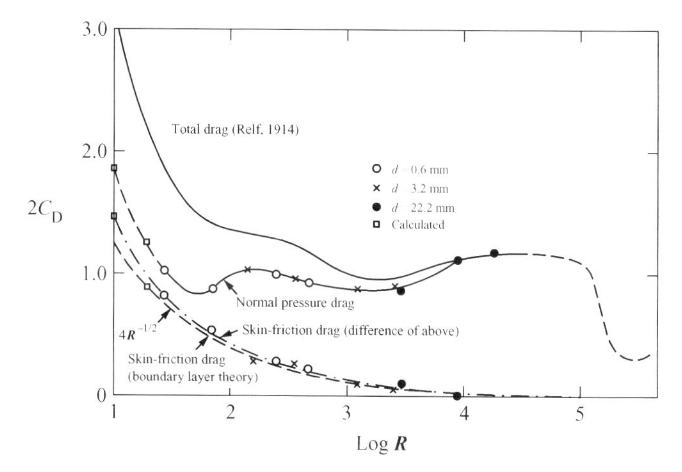

Fig. 4.17. Components of drag in flow around two-dimensional circular cylinders. (From Thom, A., An investigation of fluid flow in two dimensions, *Aeronautical Research Committee, Reports and Memoranda*, No. 1194; Copyright © 1929 HMSO London, as given in S. Goldstein, editor, *Modern Developments in Fluid Dynamics*, Fig. 154, p. 425, Copyright © 1938 Oxford at the Clarendon Press; with kind permission from both HMSO and Oxford University Press.)

ventilation effect of the three-dimensionality eliminates the laminar separation expected at about 90° for smooth surfaces at this $\boldsymbol{R}_D$. What effect will an array of cylinders have on this behavior and hence on α?

As we have seen in Fig. 4.5, arrays of two-dimensional cylinders create interference which may further change the drag. Their primary effect will be reduction of the effective fluid velocity and hence of the Reynolds number regime. To assess the extent of this change we must turn to experiments. Thom's (1971) observed velocity distribution is plotted in Fig. 4.18a where it is fitted over the upper half of the canopy using Eq. 4.21, with the dashed line

$$a_{\mathrm{w}} = mn\beta L_{\mathrm{t}} = 3.64. \tag{4.64}$$

Assuming the rods to be tall enough and their density to be high enough that no significant shear reaches the substrate, we can consider the individual rods to be foliage

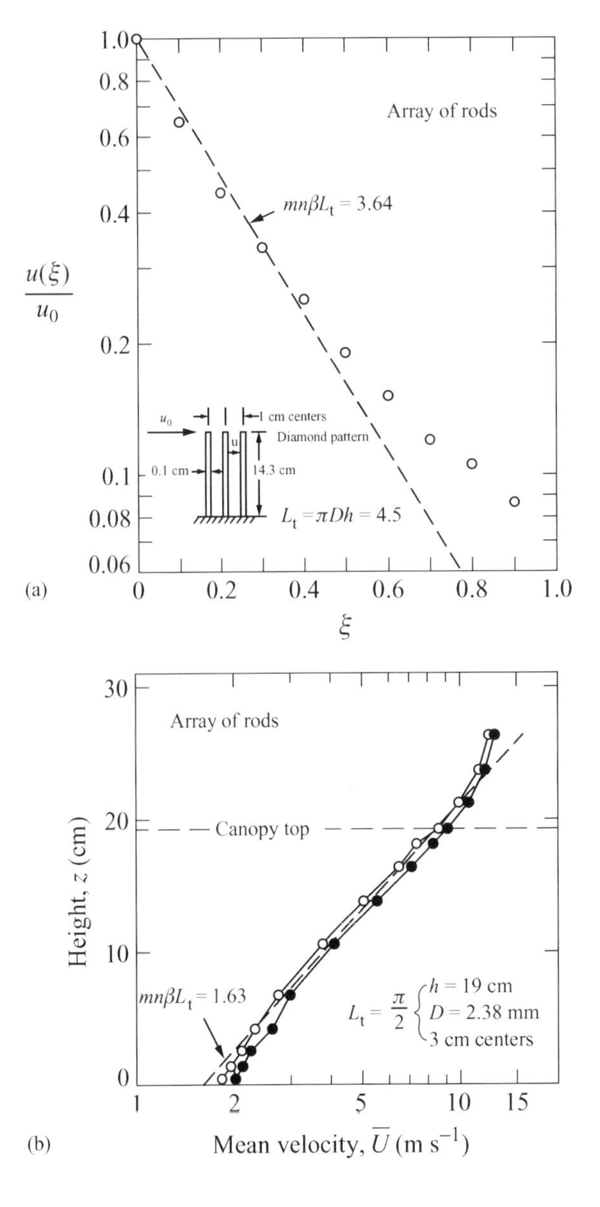

Fig. 4.18. Observed velocity profiles in arrays of rods. (a) Experiments of Thom (1971). (Data from Thom, A.S., Momentum absorption by vegetation, Fig. 6, p. 421, *Quarterly Journal of the Royal Meteorological Society*, 97; Copyright © 1971 The Royal Meteorological Society; with kind permission from the Royal Meteorological Society.) (b) Experiments of Seginer *et al.* (1976). (From Seginer, I., P.J. Mulhearn, E.F. Bradley and J.J. Finnigan, Turbulent flow in a model plant canopy, Fig. 9, p. 440, *Boundary Layer Meteorology*, 10; Copyright © 1976 D. Reidel Publishing Company, Dordrecht, The Netherlands; with kind permission from Kluwer Academic Publishers.)

elements in a dense canopy rather than entire plants in a sparse canopy. The basal area is then given by the rod spacing rather than by its diameter, and the single-sided foliage area index is

$$L_t = \frac{\pi Dh/2}{0.5\ \mathrm{cm}^2} = \pi(0.1)(14.3) = 4.50, \tag{4.65}$$

and for vertical cylindrical friction surfaces in a horizontal wind

$$\beta = \frac{2}{\pi}. \tag{4.66}$$

Equations 4.64–4.66 give $mn = 1.27$, and with no separation, $n = 2$; thus $m = 0.73$.

2. Experiments of Seginer *et al.* (1976)

A set of wind tunnel experiments was conducted by Seginer *et al.* (1976). In this case the rods were 19 cm high, with a diameter of 2.38 mm, and installed on 30-cm centers in a diamond pattern array. Observed velocity distributions are reproduced here in Fig. 4.18b where they give, following Eq. 4.28,

$$a_w = mn\beta L_t = 1.63. \tag{4.67}$$

With D and h in centimeters, the single-sided foliage area index for these rods is

$$L_t = \frac{\pi Dh/2}{4.5} = \pi(0.238)(19)/9 \cong \frac{\pi}{2}. \tag{4.68}$$

Once again $\beta = \frac{2}{\pi}$, and $n = 2$, giving $m = 0.82$.

Averaging the results of the two studies, we have for arrays of vertical cylindrical rods

$$\langle m \rangle = 0.78, \tag{4.69}$$

which is almost identical to the value $m = 0.75$ (Table 4.1) found on the basis of drag measurements for rice and maize. We conclude that $m = 0.75$ and $n = 2$ are the nominal values for thin, vertical rods and hence for the stems of stemmy plants. Their product, $mn = 1.5$, is drawn on Fig. 4.14, where along with the line $mn = 1$ for leafy plants, it provides a limit for those plants whose drag has significant components from both stems and leaves.

Stemmy crops†

We can now use the above bounding relations:

$$mn = 1.5 \quad \text{for stems}, \tag{4.70}$$

and

$$mn = 1.0 \quad \text{for leaves}, \tag{4.71}$$

to estimate mn for stemmy crops where both stems and leaves are important to drag. The relative proportion of the one-sided area of leaves and stems for wheat, maize, and beans are listed in Table 2.1. These can be used to estimate the average value of mn by simple proportion

$$\langle mn \rangle = \frac{\%_{\text{stems}}}{100}(mn)_{\text{stems}} + \frac{\%_{\text{leaves}}}{100}(mn)_{\text{leaves}}. \tag{4.72}$$

† Stems are assumed to be photosynthetically active.

1. Wheat

The relative foliage areas for wheat (Table 2.1) are 74% stems and 26% leaves. From Eq. 4.72

$$\langle mn\rangle = 0.74\left(\frac{3}{4}\cdot 2\right) + 0.26\left(\frac{1}{2}\cdot 2\right) = 1.37, \tag{4.73}$$

which is plotted on Fig. 4.14. The values of γ and κ obtained from the literature (Table 2.4) for wheat and for sunflower, rice, and grasses as well agree fairly well with Eq. 4.73.

2. Maize

The average relative foliage areas for maize (Table 2.1) are 24% stems and 76% leaves. From Eq. 4.72

$$\langle mn\rangle = 0.24\left(\frac{3}{2}\cdot 2\right) + 0.76\left(\frac{1}{2}\cdot 2\right) = 1.12, \tag{4.74}$$

which is also plotted on Fig. 4.14. The values of γ and κ obtained from the literature (Table 2.5) agree very well with Eq. 4.74.

Ross (1981, Fig. 28, p. 103) presents the probability density function for leaf inclination angle, $\theta_{\mathrm{L}} = \cos^{-1}(\beta)$, for various elevations in the maize canopy and at two stages of its growth. From these we conclude that for maize

$$\beta \approx 0.5. \tag{4.75}$$

The foliage area of maize is uniformly distributed over the typical mature plant height (see Fig. 4.19 taken from Allen *et al.*, 1964,) and thus our earlier analysis of homogeneous canopies is again applicable. However, the foliage area index, L_t, is highly variable both varietally and seasonally; the observations collected in Table 2.4 show $1.05 < L_t < 5.6$ with an average value of $L_t = 3.72$. These parameters allow us to

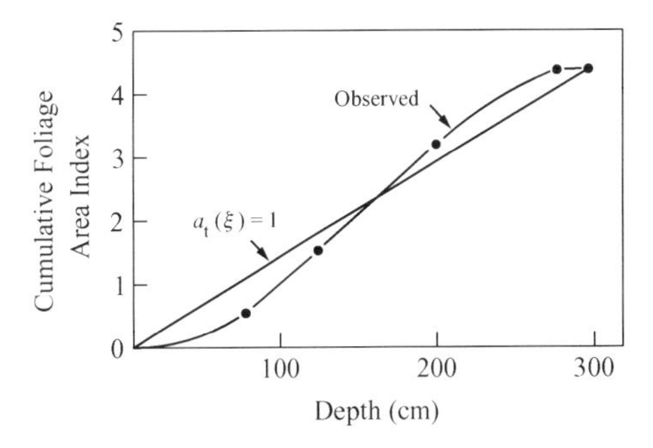

Fig. 4.19. Homogeneity of typical maize canopy. (From Allen, L.H., C.S. Yocum and E.R. Lemon, Photosynthesis under field conditions, VII, Radient energy exchanges within a corn crop canopy and implications in water use efficiency, Fig. 5, p. 255, *Agronomy Journal*, 56; Copyright © 1964 American Society of Agronomy; with kind permission from the American Society of Agronomy, Madison, WI.)

estimate the average extinction coefficients for maize in the manner we did for mn in Eq. 4.72:

(a) *Shear (Eq. 4.18)*

$$\langle a_{\rm d}\rangle = \langle n\beta L_{\rm t}\rangle = [0.24(2)(1) + 0.76(2)(0.50)]3.72 = 4.61, \quad (4.76)$$

which can be compared with the value, $a_{\rm d} = 5.6$, reported by Uchijima and Wright (1964) from drag experiments on a particular maize crop. Note that the range of observed variation in $L_{\rm t}$ gives a range of estimated $a_{\rm d}$, $1.30 < a_{\rm d} < 6.94$, which easily includes Uchijima and Wright's observation.

(b) *Velocity (Eq. 4.21)*

$$\langle a_{\rm w}\rangle = \langle mn\beta L_{\rm t}\rangle = \left[0.24\left(\frac{3}{4}\right)(2)(1) + 0.76\left(\frac{1}{2}\right)(2)(0.50)\right]3.72 = 2.75, \quad (4.77)$$

which compares favorably with the value, $a_{\rm w} = 3.0$, found in experiments by Inoue and Uchijima (1979). The range of observed variation in the leaf area index, $L_{\rm t}$, gives a range of estimated $a_{\rm w}$, $0.74 < a_{\rm w} < 4.14$, which readily encompasses Inoue and Uchijima's (1979) observation.

(c) *Eddy viscosity (Eq. 4.41)*

$$\langle a_{\rm m}\rangle = \langle (1-m)n\beta L_{\rm t}\rangle = \left[0.24\left(\frac{1}{4}\right)(2)(1) + 0.76\left(\frac{1}{2}\right)(2)(0.50)\right]3.72 = 1.86, \quad (4.78)$$

which is somewhat smaller than the value, $a_{\rm m} = 2.88$, determined by Wright and Brown (1967) from experimentally based momentum balance and energy balance analyses. In this case however, the range of observed variation in the leaf area index gives an estimated range $0.53 < a_{\rm m} < 2.80$ which does not quite include the observation of Wright and Brown (1967) unless the uncertainty in β (for the leaves) is also included.

We conclude that with reasonable estimates for β and for $L_{\rm t}$ along with the proportion of stem area to total area, we can make credible estimates of the extinction coefficients for stemmy crops.

G Momentum flux in non-homogeneous leafy canopies: M = 1

Assumptions

Given the biological requirement for compensation light intensity at the lowest leaf level (cf. Chapters 3 and 8), *truly* homogeneous foliage area distribution is possible only for crowns having the same depth at all radii, that is, for the cylindrical crowns considered above. For tapered crowns such as cones and hemispherical segments, the fraction of canopy space occupied by the crown varies with elevation. To meet compensation light intensity at the lowest leaf for all radii, the crown foliage area density at any elevation must increase with radial distance from the trunk. To compare this reasoning with observation we refer to Fig. 4.20b where the plotted points are observations from the upper, *photosynthetically active* region of Sitka spruce crowns. At the base of this region the leaf area density is represented fairly well by a homogeneous crown foliage area density. Light transmitted below this elevation ($z > 8$ m) is used to support thinning foliage in either an understory, or on dying lower branches as is shown in Fig. 4.20a. The foliage requirements of compensation light intensity are satisfied fairly well in this region (again see Fig. 4.20b). Therefore, for the purposes of estimating canopy resistance for tapered crowns we not only assume the crown to be homogeneous in β as we did in Chapter 3, but also we take the crown foliage

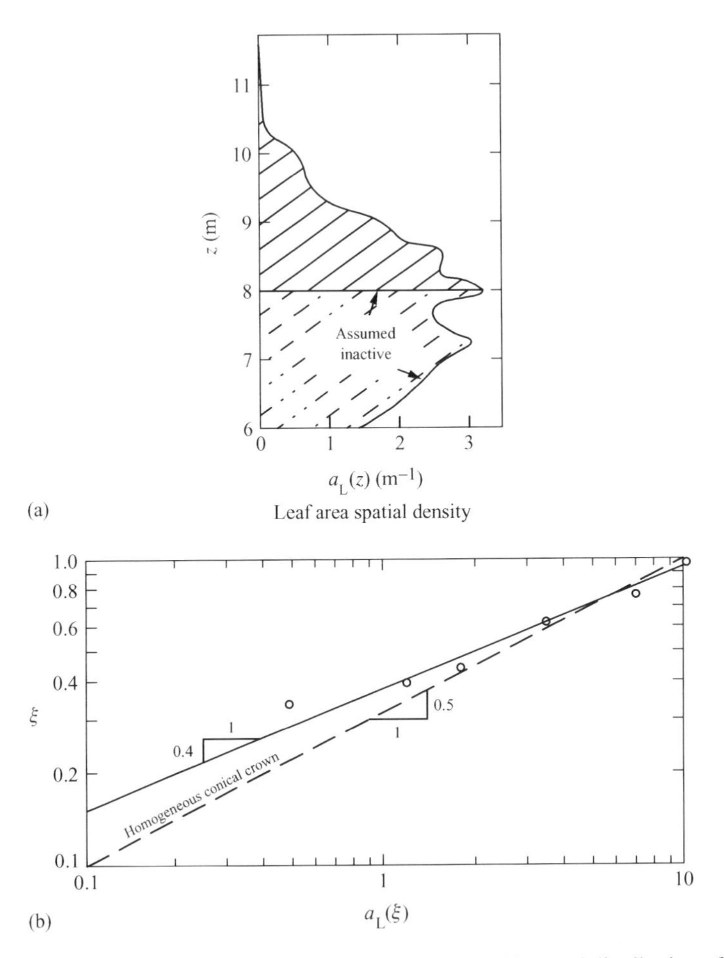

Fig. 4.20. Non-homogeneous Sitka spruce canopy. Observed distribution of leaf area density. (From Landsberg, J.J., and P.G. Jarvis, A numerical investigation of the momentum balance of a spruce forest, Fig. 1, p. 648, *Journal of Applied Ecology*, 10; Copyright © 1973 Blackwell Science Ltd, with kind permission from Blackwell Science Ltd. (b) Fitted leaf area distribution. (Data from Landsberg and Jarvis, 1973.)

area density, a_{cr}, to be homogeneous throughout. To further facilitate the calculations we next distribute the foliage at any elevation in the tapered crown over the entire cross-section of the crown's circumscribing right circular cylinder, thereby creating a fictitious cylindrical crown (cf. Fig. A.1b) having a conical or hemispherical vertical distribution of foliage density. The calculations are detailed in Appendix A.

We make the assumption that non-homogeneity enters canopy dynamics solely through vertical variation in the horizontal component of the drag area density, a_D. This was defined in Eqs. 4.14 *et seq.*, and is expanded here as

$$a_D(z) \equiv \frac{A_D(z)}{A_b \Delta z} = n\,\widehat{\cos\theta_L}\, a_t(\xi) \equiv n\beta L_t \cdot a(\xi), \tag{4.79}$$

where $a(\xi)$ is again the *homogeneity function* which is equal to unity in the homogeneous case. Here we drop the restriction that the canopies be areally dense in recognition of the fact that non-homogeneity of the foliage area density is a common cause of higher velocity penetration of the canopy and hence of significant substrate shear. We incorporate the substrate shear only to the first approximation by making the approximation

$$\frac{\tau_f(\xi)}{\tau_0} \approx \frac{\tau_f(\xi)}{\tau_f(0)}, \tag{4.80}$$

which, in conjunction with Eq. 4.17, introduces the total downward flux of momentum in the manner

$$\frac{\tau_f(\xi)}{\tau_0} = \exp\left[-n\beta L_t \int_0^{\xi} a(\xi)\,d(\xi)\right]. \tag{4.81}$$

We will consider only one non-homogeneous case, conical crowns such as are displayed by many conifers (Landsberg and Jarvis, 1973), and which are non-homogeneous both because $a(\xi) \neq 1$ (cf. Eq. 4.17), and because their crown shape does not fill the canopy space.

A particular non-homogeneous conical crown

Landsberg and Jarvis (1973) studied a Sitka spruce (*Picea sitchensis*) forest at Fetteresso near Aberdeen in northeast Scotland. We assume that their reported "understory" is dynamically unimportant and we will neglect it here. The physical parameters of this spruce canopy (neglecting the understory) are: $h = 11.5$ m; $h_s = 8$ m; $d_0 = 9.63$ m; and $z_0 = 0.34$ m.

The homogeneity function

Landsberg and Jarvis (1973) measured the spatial distribution of only the leaf area density, $a_L(\xi)$, in the spruce as is shown in Fig. 4.20a. We are interested in the total foliage area density, $a_t(\xi)$ and will assume the two are equal.[†] The observed average leaf area density at each elevation is fitted through the canopy in Fig. 4.20b by the solid line

$$a_t(\xi) \approx a_L(\xi) = 11.6\xi^{2.5}. \tag{4.82}$$

From the above physical parameters for this canopy

$$L_t = \frac{11.6}{3.5} = 3.32, \tag{4.83}$$

so that Eq. 4.82 can be written

$$a_t(\xi) = L_t(3.5\,\xi^{2.5}), \tag{4.84}$$

in which the homogeneity function (cf. Section C of this chapter) appears as

$$a(\xi) = 3.5\,\xi^{2.5}, \tag{4.85}$$

confirming by its departure from unity that the canopy has a non-homogeneous leaf area density, and by the departure of the exponent from "2" that the crown is non-homogeneous as well (cf. Appendix A).

No information is available on the volume density of the other foliage elements. We will assume it to be low and that the momentum absorption is as for the leafy plants studied so far.

Momentum flux

Using Eq. 4.85 in Eq. 4.17 yields the momentum flux variation

$$\frac{\tau_f(\xi)}{\tau_0} = \exp(-n\beta L_t \xi^{3.5}). \tag{4.86}$$

With the high degree of crown-scale non-homogeneity evidenced in Fig. 4.20 and in the resulting Eq. 4.85, we expect the foliage element drag coefficient to vary with elevation due to changing foliage geometry and density (i.e., changing interference), as well as to changing Reynolds number. The true form of the *interference function*, $s(\xi)$, is probably quite complicated and is certainly unclear. However, expressing interference by the exponential of Eq. 4.11, $s(\xi)$ must be linear in ξ in order to match velocity gradients at the canopy top. To maintain reasonable agreement with Fig. 4.05a, we assume the simple linear form

$$s(\xi) = B\xi, \tag{4.87}$$

and use Eqs. 4.10–4.13 to relate the canopy wind velocity and the foliage element drag according to

$$\tau_f(\xi) \propto u^{2-\alpha} \exp(-B\xi), \tag{4.88}$$

[†] The difference is on the order of 10% for most trees as we see in Table 2.1.

from which using Eq. 4.20

$$u(\xi) \propto \tau_f^m \exp(mB\xi). \tag{4.89}$$

Assuming, as in the homogeneous case (but with somewhat less justification), that the proportionality of Eq. 4.88 is now invariant with elevation, for this non-homogeneous case we set $B_0 = \frac{mB}{\gamma L_t}$, and using Eqs. 4.86 and 4.89, we have

$$\frac{u(\xi)}{u_0} = \exp[-\gamma L_t(\xi^{3.5} - B_0\xi)]. \tag{4.90}$$

Estimation of coefficients

Using Eq. 4.7, we differentiate Eq. 4.90 to obtain

$$\frac{du}{dz} = -\frac{u_0}{h - h_s} \cdot (-3.5\gamma L_t\xi^{2.5} + B_0\gamma L_t) \cdot \exp[-\gamma L_t(\xi^{3.5} - B_0\xi)], \tag{4.91}$$

which at the canopy top ($\xi = 0$) is

$$\left.\frac{du}{dz}\right|_{\xi=0} = -\frac{u_0}{h - h_s} \cdot (B_0\gamma L_t). \tag{4.92}$$

Using Eq. 4.92 with Eq. 4.29 (at $z = h$), we match the velocity gradients to get

$$B_0 = -\frac{\dfrac{u_*}{ku_0}}{\gamma L_t\left(\dfrac{h - d_0}{h - h_s}\right)}. \tag{4.93}$$

To evaluate B_0 we need estimates of mn, β, and the bulk drag parameter, u_*/ku_0.

We have seen in Fig. 4.14 that for leafy plants (both needle-leaved and broadleaved), $m = 1/2$ and $n = 2$, while at the latitude of Fetteresso (about 57° N), Figs. 3.9 and 3.18 give $\beta \approx 0.50$. With these parameters

$$\gamma L_t \equiv mn\beta L_t = 1.66, \tag{4.94}$$

and using the physical dimensions of this canopy given earlier

$$\frac{h - d_0}{h - h_s} = 0.53. \tag{4.95}$$

We estimate the bulk drag parameter, u_*/ku_0, by taking its average value over the canopy depth under the assumption that the von Kármán and Prandtl mixing lengths are equal at each depth. Using Eqs. 4.42, 4.86, and 4.91, the dimensionless Prandtl mixing length is

$$\frac{l_P}{h - h_s} = \left(\frac{u_*}{ku_0}\right)\frac{(k)\exp(-\gamma L_t B_0\xi)}{|-3.5\gamma L_t\xi^{2.5} + B_0\gamma L_t|}, \tag{4.96}$$

while from Eqs. 4.46 and 4.91, the dimensionless von Kármán mixing length is

$$\frac{l_K}{h - h_s} = k\left|\frac{-3.5\gamma L_t\xi^{2.5} + B_0\gamma L_t}{(-3.5\gamma L_t\xi^{2.5} + B_0\gamma L_t)^2 - 8.75\gamma L_t\xi^{1.5}}\right|. \tag{4.97}$$

Equating these two mixing lengths and using Eq. 4.93 gives

$$\frac{u_*}{ku_0} = \frac{\exp\left[-\left(\frac{h-h_s}{h-d_0}\right)\frac{u_*}{ku_0}\xi\right]}{\left|1 - \frac{8.75\gamma L_t\xi^{1.5}}{\left[3.5\gamma L_t\xi^{2.5} + \left(\frac{h-h_s}{h-d_0}\right)\frac{u_*}{ku_0}\right]^2}\right|}. \tag{4.98}$$

We solve Eq. 4.98 along with Eqs. 4.94 and 4.95 for u_*/ku_0 at various values of ξ to get the function shown in Fig. 4.21. From this figure, the average of the bulk drag parameter over the crown depth is

$$\left\langle\frac{u_*}{ku_0}\right\rangle = 0.89, \tag{4.99}$$

which with Eqs. 4.93, 4.94 and 4.95 gives

$$B_0 = -1. \tag{4.100}$$

Verification of theoretical momentum decay

With the parameters now defined we can verify the theoretical decay of horizontal momentum. We do this by comparing Eq. 4.90 with the observations of canopy wind speed reported by Landsberg and Jarvis (1973). This comparison is given in Fig. 4.22 where once again we see good agreement at low free stream wind speeds, and a

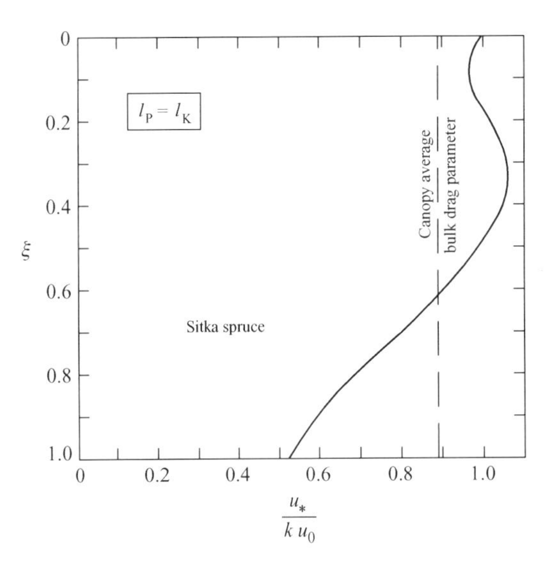

Fig. 4.21. Calculated bulk drag parameter for Sitka spruce ($l_P = l_K$).

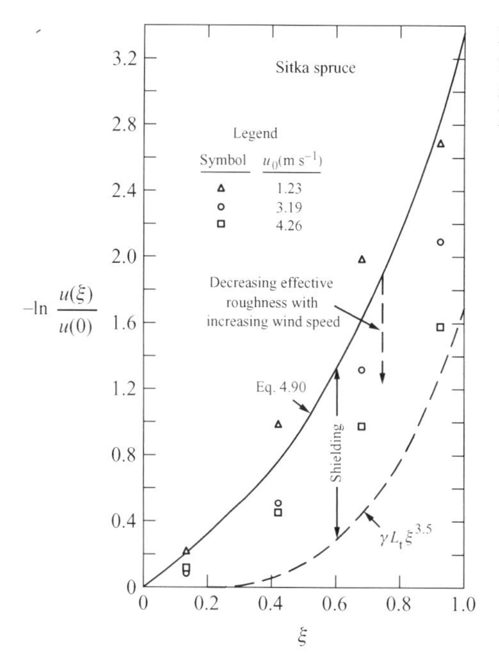

Fig. 4.22. Test of theoretical momentum decay for Sitka spruce. (Data from Landsberg and Jarvis, 1973.)

systematic underestimation with increasing wind speed due probably to reduction in effective roughness.

Eddy viscosity

With the definition of Eq. 4.37 and using Eqs. 4.40, 4.86, and 4.90, the theoretical spatial variation of the atmospheric eddy viscosity within the crown is

$$K_{\mathrm{m}}^{\mathrm{o}} \equiv \frac{K_{\mathrm{m}}(\xi)}{K_{\mathrm{m}}(0)} = \frac{\dfrac{u_*}{ku_0}}{\gamma L_{\mathrm{t}}\left(\dfrac{h-d_0}{h-h_{\mathrm{s}}}\right)} \cdot \frac{\exp\left\{-\gamma L_{\mathrm{t}}\left[\left(\dfrac{1-m}{m}\right)\xi^{3.5} - B_0\xi\right]\right\}}{1+3.5\gamma L_{\mathrm{t}}\xi^{2.5}}. \tag{4.101}$$

With $m = 1/2$, and using Eqs. 4.93, 4.94 and 4.100, this particular case becomes

$$K_{\mathrm{m}}^{\mathrm{o}} = \frac{\exp\left[1.5(\xi - \xi^{3.5})\right]}{1+3.5\xi^{2.5}}, \tag{4.102}$$

which is plotted in Fig. 4.23 in comparison with values calculated by Landsberg and Jarvis (1973) using observed velocities and laboratory-determined drag coefficients. The agreement is excellent lending strong support to the "observation-free" method of analysis introduced here. We have confidence now in our ability to estimate the effect of crown shape on the eddy viscosity distribution. The shape of the eddy viscosity distribution for this particular non-homogeneous conical crown is qualitatively different from the exponential decay found for the homogeneous circular cylinder in Eq. 4.40.

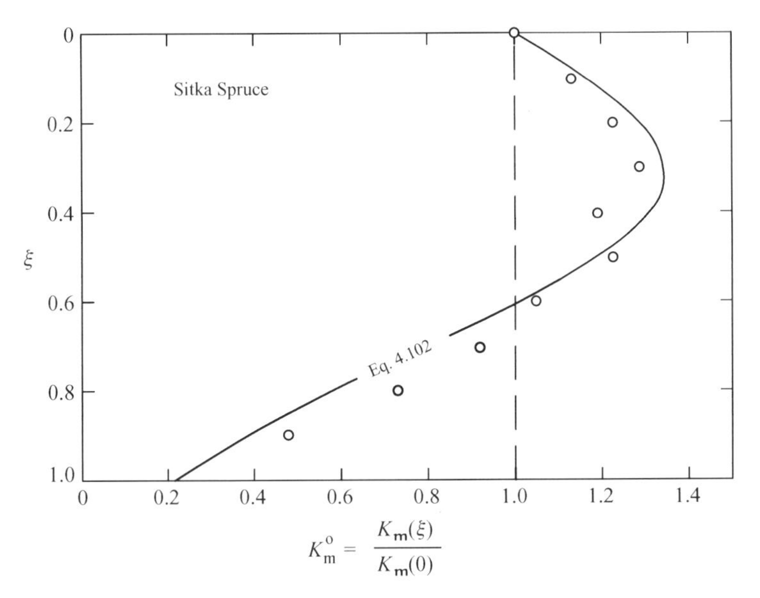

Fig. 4.23. Test of theoretical eddy viscocity distribution for Sitka spruce. (Data from Landsberg and Jarvis, 1973 for $u_0 = 1\ \mathrm{m\ s^{-1}}$.)

A second non-homogeneous crown

Thom (1971) studied a bean (*Vicia faba*) crop having the following physical characteristics:

$$h = 118\ \text{cm}$$
$$h_s = 18\ \text{cm}$$
$$d_0 = 88.5\ \text{cm}$$
$$z_0 = 7.67\ \text{cm}$$
$$L_t = 3.55.$$

Unpublished analysis of the bean data by the author shows, in a manner similar to that used above for the Sitka spruce, that $\gamma = 0.53$, $\langle \frac{u_*}{ku_0} \rangle = 0.59$, and $B_0 = 0.20$. However, in this case our predicted eddy viscosity distribution is a poor representation of that found from observations.

The homogeneous conical crown

A simpler case of the conical crown has homogeneous foliage volume density within the crown and all canopy heterogeneity arises by virtue of the non-space-filling crown shape.† This case is considered in Appendix A where the normalized eddy viscosity

† Strictly speaking this is an infeasible foliage arrangement since it precludes satisfaction of the compensation light intensity for all radii at the crown base.

distribution is (cf. Eq. A.22)

$$K_{\mathrm{m}}^{\mathrm{o}} = \frac{\exp\left[\gamma L_{\mathrm{t}}(\xi - \xi^3)\right]}{1 + 3\xi^2}, \tag{4.103}$$

showing the same increase of $K_{\mathrm{m}}^{\mathrm{o}}$ with ξ for small ξ found above for the Sitka spruce.

The homogeneous hemispherical crown

Similarly, homogeneous foliage volume density within a hemispherical crown is treated in Appendix A where the normalized eddy viscosity distribution for a hemispherical segment is (cf. Fig. A.1a and Eq. A.11)

$$K_{\mathrm{m}}^{\mathrm{o}} = \frac{\exp\left[-\gamma L_{\mathrm{t}}\left(\dfrac{3\varepsilon\xi^2 - \xi^3}{3\varepsilon - 1} - \xi\right)\right]}{1 + \dfrac{6\varepsilon\xi - 3\xi^2}{3\varepsilon - 1}}, \tag{4.104}$$

in which $\varepsilon = \frac{r_{\mathrm{o}}}{h - h_{\mathrm{s}}}$ and r_{o} = radius of the sphere. The limiting condition $\varepsilon = 1$ gives the full hemisphere showing a modified exponential decay similar to that of the right circular cylinder which the hemisphere resembles. The limiting condition $\varepsilon \gg 1$ gives the umbrella-like palm or savanna tree.

The eddy viscosity distribution for the cylinder, cone and limiting hemispherical segments are compared in Fig. 4.24a using Eqs. 4.40, 4.103, and 4.104 evaluated for the typical values $m = 1/2$, $n = 2$, and $\beta L_{\mathrm{t}} = 1.5$. The cone is seen to maintain large eddy viscosities throughout the upper canopy as has been found experimentally by others (e.g., Druilhet *et al.*, 1972). These differences in eddy viscosity distribution due to canopy non-homogeneity give theoretical support to the conjecture of Brutsaert (1982, p. 106) that the exponential variation is "probably valid only for canopies with a fairly uniform leaf area distribution". This qualitative and important difference in eddy viscosity distribution with crown shape is due to canopy-scale heterogeneity and its consequent effect is expressed through the interference function. The higher eddy viscosity reflects the degree to which the openness of the non-homogeneous canopy near the crown top admits higher velocities and hence promotes turbulent mixing. This has been noted previously by others; for example, Shuttleworth (1989, p. 147) when speaking of dense, tall tropical forests says, "The characteristically high efficiency of radiation and momentum capture is more directly a function of the relative depth to separation of depressions in the canopy top."

Canopy-average eddy viscosity for multilayers

It will be useful later to have general expressions for the vertical spatial average atmospheric eddy viscosity within multilayer canopies of differing crown shape. We have given this spatial average value the notation,

$$\widehat{K_{\mathrm{m}}^{\mathrm{o}}} \equiv \frac{\widehat{K_{\mathrm{m}}(\xi)}}{K_{\mathrm{m}}(0)} \equiv \int_0^1 \frac{K_{\mathrm{m}}(\xi)}{K_{\mathrm{m}}(0)} \mathrm{d}\xi, \tag{4.105}$$

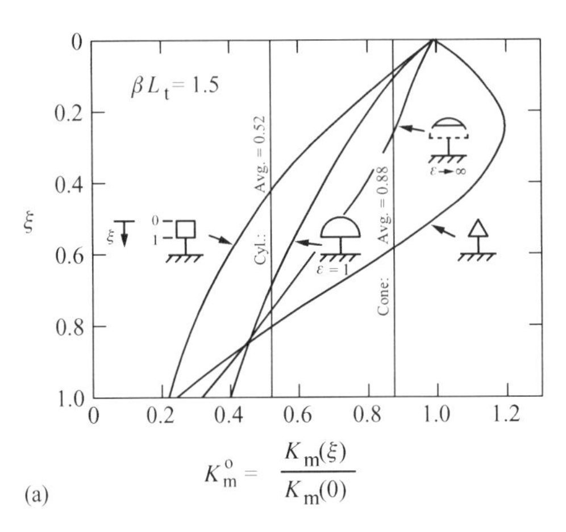

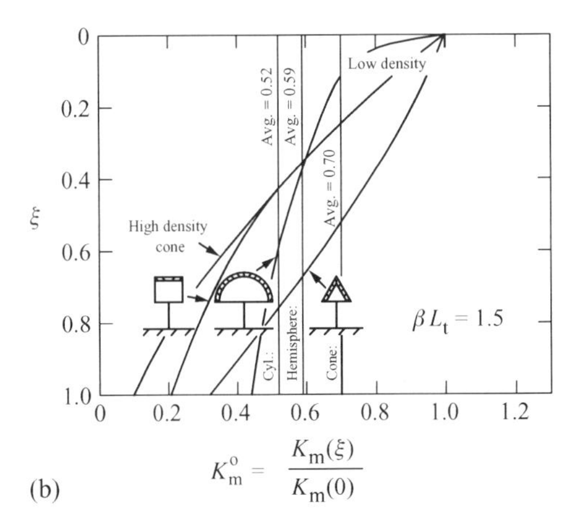

Fig. 4.24. The role of crown shape in turbulent mixing. (a) Homogeneous multilayer crowns. (b) Homogeneous monolayer crowns.

in Eq. 4.41 where it is evaluated for the homogeneous right circular cylindrical crown. In Appendix A we present the numerically integrated results for the cone and hemispherical segment in Eqs. A.25–A.27. They are compared graphically in Fig. 4.25a where we see again that for the homogeneous canopy of cylindrical crowns the average eddy viscosity declines with increasing horizontal leaf area, βL_t, while for the non-homogeneous canopies of conical and hemispherical crowns the average eddy viscosity increases with βL_t.

Generally speaking, non-homogeneities which produce open canopy tops admit higher wind velocities at a given depth than do homogeneous canopies or non-homogeneous canopies which become less dense with depth. The open canopy tops produce canopy-average eddy viscosities which increase with the degree of non-homogeneity at a given βL_t (greater than some minimum), and which *increase* with increasing βL_t for a given non-homogeneous crown shape. This is in marked contrast to the canopy-average eddy viscosities for homogeneous cylindrical crowns which *decrease* with increasing βL_t. In the following chapter we will show that these

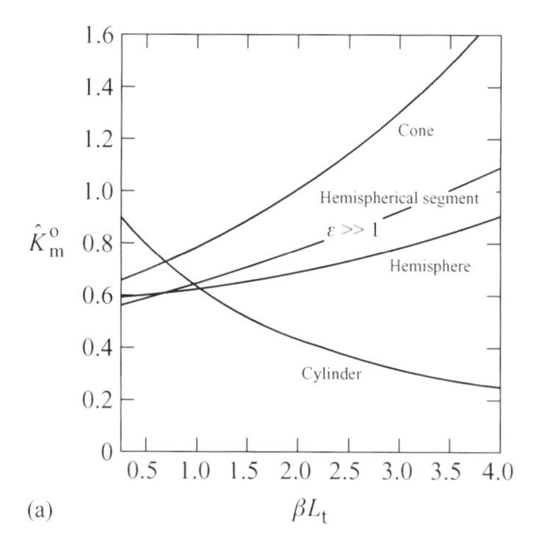

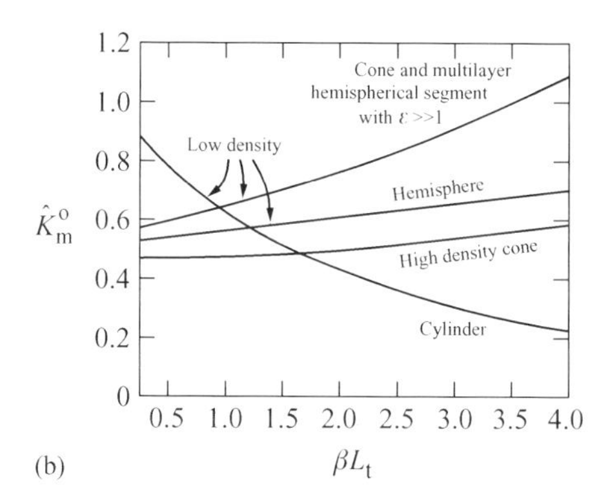

Fig. 4.25. Canopy-average eddy viscocity. (a) Homogeneous multilayer crowns. (b) Homogeneous monolayer crowns.

differences in turbulent transport within the canopy may help to explain the observed prevalence of certain crown shapes in particular climatic regimes. Suppose for example that survival favors those trees that have the highest ability to transfer heat and mass into and/or out of the canopy. Then Fig. 4.25a suggests that trees with low βL_t should be homogeneous while those with high βL_t should be non-homogeneous. The range $0.75 < \beta L_t < 1.0$ would seem to be more or less indifferent to crown shape. Of course the reality is not this simple as we will see!

Canopy-average eddy viscosity for monolayers

Horn (1971) identifies a second classification of crowns in which the foliage elements are concentrated (presumably for reasons of temperature control) in a relatively thin layer of thickness, t, where $\frac{t}{h-h_s} \ll 1$. He calls these plants *monolayers* even though their foliage area index may exceed unity. Use of the term herein identifies an extreme crown-scale non-homogeneity in which the foliage is packed closely together at the crown surface. We consider them here because, at moderate foliage volume densities, the closeness of their foliage elements brings about a fundamental change in the dominant mechanism of momentum absorption by the canopy, and hence in the sensitivity of the eddy viscosity to increasing leaf area.

Definition sketches for monolayer hemispherical and conical crowns are given in Appendix A along with derivations of the distribution and spatial average of the eddy viscosity for the cylinder, cone, and hemispherical segment. Graphical comparison of the crown shapes is presented in Figs. 4.24b and 4.25b, while multilayer and monolayer are compared between Figs. 4.24 and 4.25. Distributions of monolayer eddy viscosity (cf. Fig. 4.24b) are qualitatively similar to those of the multilayer but less variable over the crown depth.

The monolayer values of $\widehat{K_m^o}$ (cf. Fig. 4.25b) display only quantitative differences with the multilayer curves (cf. Fig. 4.25a); sensitivity to βL_t is reduced in the monolayer case. The magnitude of $\widehat{K_m^o}$ for the high-density conical monolayer is the smallest of all the shapes tested. It seems clear from this finding that the occurrence of high-density conical monolayers in nature is not due to any capability for high vertical transport rates. On the contrary, it is likely that they have adapted to situations in which low transport, probably of sensible heat, is the key to survival.

H Drag partition

Raupach (1992) presented an analysis of the drag on rough surfaces composed of differing densities of surface-mounted roughness elements, on the one hand, and of natural forests of differing canopy cover, M, on the other; his results are of great interest here. In Fig. 4.26a we reproduce his findings for k times the bulk drag parameter as a function of the *roughness density* or *frontal area index*, λ, defined as

$$\lambda = \frac{bh}{D^2} = \frac{4M}{\pi(b/h)}, \tag{4.106}$$

in which D is the roughness element spacing, while b and h are the roughness element breadth and height respectively. In this figure, the data (closed circles) of Raupach *et al.* (1980) and (open circles) of O'Loughlin (1965) are for geometrical roughness elements in a wind tunnel, while the data (open triangles) of Garratt (1977), Jarvis *et al.* (1976), and Raupach *et al.* (1991) are for natural vegetation canopies and for vegetation canopies modeled in a wind tunnel.

The Raupach (1992) work deals with what we have called canopy-scale nonhomogeneity due to "open" stand structure. For small roughness density, the drag parameter is small, approaching its minimum value, $\frac{u_*}{u_0} \equiv C_f^{1/2} = 0.055$, or

$$\frac{u_*}{ku_0} = 0.137, \tag{4.107}$$

which represents the drag on the average rough substrate as determined empirically (Raupach, 1992). For large roughness density (i.e., $M = 1$), this drag parameter

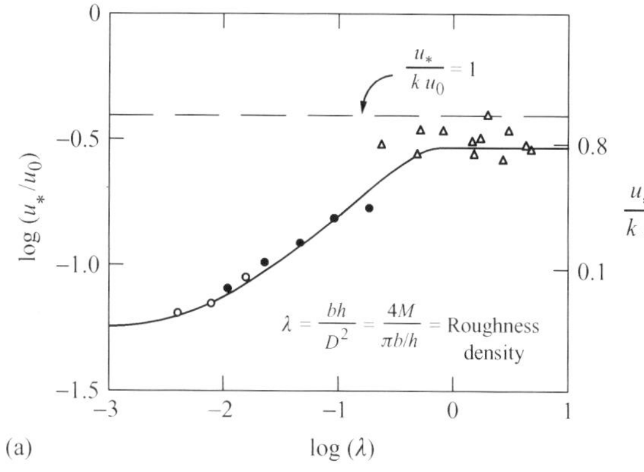

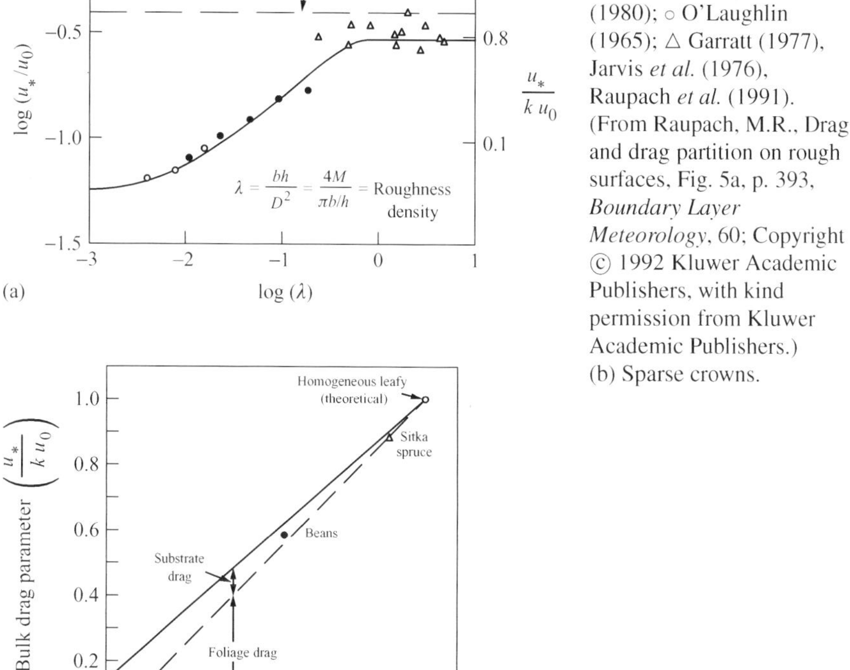

Fig. 4.26. Drag partition in canopies. (a) Open canopies. • Raupach *et al.* (1980); ○ O'Laughlin (1965); △ Garratt (1977), Jarvis *et al.* (1976), Raupach *et al.* (1991). (From Raupach, M.R., Drag and drag partition on rough surfaces, Fig. 5a, p. 393, *Boundary Layer Meteorology*, 60; Copyright © 1992 Kluwer Academic Publishers, with kind permission from Kluwer Academic Publishers.) (b) Sparse crowns.

approaches its maximum value, $\frac{u_*}{u_0} = 0.32$, or

$$\frac{u_*}{ku_0} = 0.80, \tag{4.108}$$

which indicates that closed natural vegetation canopies do not achieve the foliage density and/or homogeneity necessary for a bulk drag parameter of unity as we predicted in the analysis leading to Eq. 4.59.

In this chapter we have been dealing with closed canopies ($M = 1$), but we have encountered some sparse canopies in which substrate shear plays a role because the crown densities are insufficient to absorb all the downward momentum flux. In the previous section we examined one such case, Sitka spruce (*Picea sitchensis*), and by matching the von Kármán and Prandtl mixing lengths, we estimated its average bulk drag parameter. From Eqs. 4.96 and 4.93 we see that small ξ gives the first approximation

$$\frac{l_P}{h - h_s} \propto \gamma L_t \left(\frac{h - d_0}{h - h_s}\right), \tag{4.109}$$

and putting Eq. 4.109 back into the left-hand side of Eq. 4.96 gives the second approximation

$$\frac{u_*}{ku_0} \propto \gamma L_t\left(\frac{h-d_0}{h-h_s}\right). \tag{4.110}$$

Remembering that $\frac{\gamma L_t}{h-h_s}$ represents the average density of drag-producing foliage area in the crown (m^{-1}), and using the quantity, $h-d_0$, as a normalizing factor, the right-hand side of Eq. 4.110 is a dimensionless measure of canopy sparseness that we call *crown density*. For closed but sparse canopies crown density apparently governs the partition of drag between the foliage elements and the substrate as the *roughness density*, λ, does for the dense but open canopies studied by Raupach (1992).

We test this conclusion by comparing our limited data with Eq. 4.110 as is shown in Fig. 4.26b. Equation 4.110 was derived from relations containing only the foliage drag (i.e., with substrate shear neglected), so when the crown density is zero, Eq. 4.110 shows the bulk drag parameter to be zero. At the other end of the scale, we showed for homogeneous, leafy canopies with zero substrate shear stress, that the bulk drag parameter was unity. The dashed line in Fig. 4.26b connects these two points. To get the bulk drag parameter for the case in which substrate shear contributes to the total

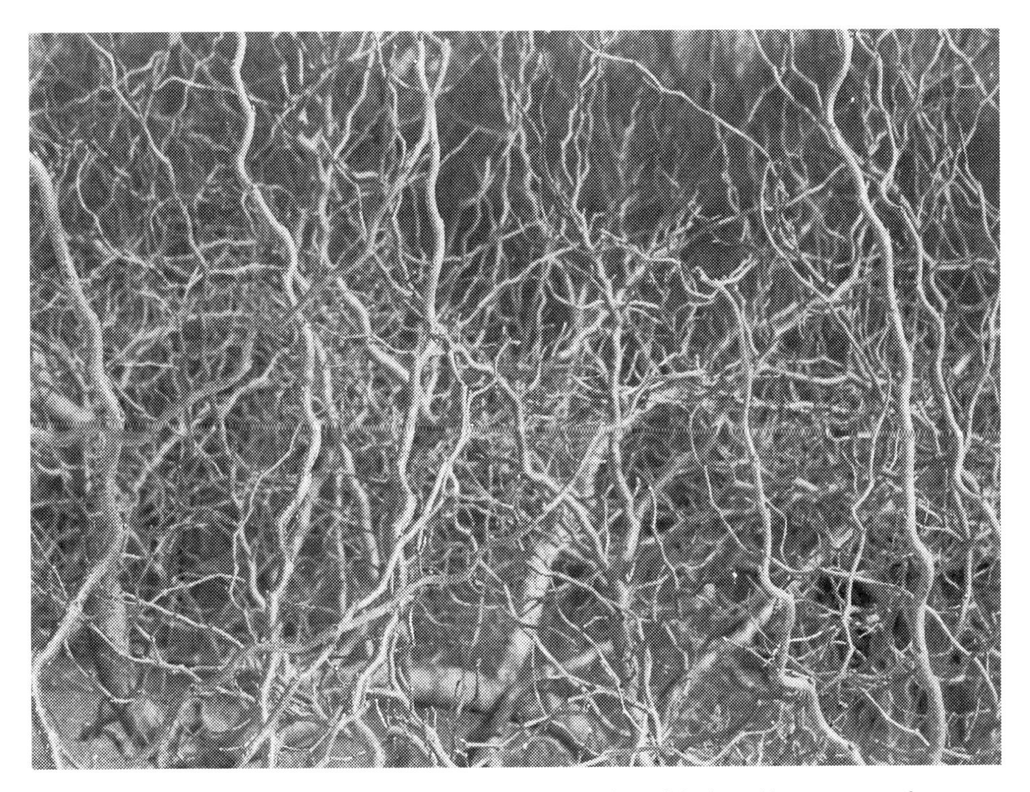

Plate 4.4. Dense branching structure. Closeup view of the branching structure of a mulberry tree. (Photographed by William D. Rich in Arnold Arboretum, Boston; Copyright © 2001 William D. Rich.)

drag, we must add the value given by Eq. 4.107 at the low-density end, and add nothing at the high-density end. We will assume the variation of the substrate component to be linear between these two extremes.

The field observations for Sitka spruce and for beans (Thom, 1971) validate the above reasoning. From these observations we note also that the proportionality coefficient,

$$\Psi \equiv \frac{\dfrac{u_*}{ku_0}}{\gamma L_t\left(\dfrac{h-d_0}{h-h_s}\right)}, \tag{4.111}$$

apparently will be about unity (dashed line in Fig. 4.26b) over the practical range of crown density.

I Eddy viscosity for open canopies: M ≪1

For an open canopy with $M \ll 1$, we assume that the scattered trees are exposed to a free stream velocity distribution determined solely by the frictional properties of the substrate. In the limiting case, $M = 0$, the spatial average eddy viscosity within the "canopy" is given by Eq. 4.38 averaged over the height of the trees present when $M \ll 1$. That is, using Eqs. 4.38 and 4.07 in Eq. 4.105

$$\widehat{K_m^o} = \frac{1}{h-h_s}\int_{h_s}^{h}\frac{ku_*(z-d_{os})}{ku_*(h-d_{os})}dz, \tag{4.112}$$

in which d_{os} is the zero plane displacement height for the substrate. Integrating

$$\widehat{K_m^o} = \frac{1}{2}\left(1+\frac{h_s}{h}\right), \quad M = 0. \tag{4.113}$$

We will use Eq. 4.113 later, along with the multilayer and monolayer values of $\widehat{K_m^o}$ just derived for $M = 1$, to estimate the spatial average eddy viscosity within the crowns of canopies having $0 < M < 1$.

J Flux of mass and of sensible heat

The flux–gradient equations

In the preceeding sections of this chapter we have seen that the vertical flux density, τ, of horizontal momentum is given by

$$\tau = \rho K_m\frac{du}{dz}, \tag{4.114}$$

in which ρ is the mass density of the air; the driving gradient, $\frac{du}{dz}$, and eddy viscosity, K_m, can be calculated theoretically from first principles given the crown shape and density. However, for the purposes of this book, which concern optimal forest structures in a given resource environment, we need to estimate the vertical fluxes of mass (water vapor and carbon dioxide) and of sensible heat. These are given by the following analogous variants of the general flux–gradient relation derived in Eq. 4.2:

water vapor

$$E = \rho K_v \frac{dq_v}{dz}, \tag{4.115}$$

carbon dioxide

$$Q = \rho K_c \frac{dc}{dz}, \tag{4.116}$$

in which E and Q are the vertical flux densities (mass · area^{-1}· time^{-1}) of water vapor (i.e., evaporation) and carbon dioxide respectively; ρ is the moist air mass per unit volume; K_v and K_c are the respective eddy diffusivities (cm^2 s^{-1}), and q_v and c are the masses of water and carbon dioxide per unit mass of moist air respectively. The quantity q_v is commonly called the *specific humidity*.

sensible heat

$$H = \rho c_p K_h \frac{dT}{dz}, \tag{4.117}$$

in which H is the vertical flux density (cal cm^{-2} s^{-1}) of heat, c_p is the specific heat of air at constant pressure (cal g^{-1} K^{-1}), T is the temperature (K), and K_h is the eddy diffusivity for heat (cm^2 s^{-1}).

It has been shown by Monin and Obukhov (1954), and verified since by many others (see Brutsaert, 1982, for a review), that

$$\frac{K_v}{K_m} = a_1(1 \pm b_1 \boldsymbol{R}_i), \tag{4.118}$$

and

$$\frac{K_h}{K_m} = a_2(1 \pm b_2 \boldsymbol{R}_i), \tag{4.119}$$

in which a_1 and a_2 are approximately unity, b_1 and b_2 are of order 10^2, $\boldsymbol{R}_i$ is the Richardson number representing the relative importance of buoyancy forces, and the + or − sign designates the direction of the buoyant instability. The Richardson number is defined by

$$\boldsymbol{R}_i \equiv \frac{g \dfrac{dT}{dz}}{T \left(\dfrac{du}{dz} \right)^2}. \tag{4.120}$$

in which "g" is the gravitational constant (m s^{-2}).

It is consistent with the accuracy of this work to endow our climates with atmospheres which are, on the average, neutrally stable with $dT/dz \approx 0$. This eliminates consideration of buoyancy (i.e., thermal convection) and justifies our invoking the *Reynolds analogy* which states

$$K_h = K_m = K_v = K_c. \tag{4.121}$$

Among the many observational confirmations of Eq. 4.121, Wright and Brown (1967) show that $K_h = K_m$ in maize, Denmead and McIlroy (1970) demonstrate that $K_h = K_v$ in wheat, and Denmead (1976) shows that $K_h = K_v = K_m$ in temperate cereals under "near-neutral" stability conditions. Of course there are ample demonstrations that Eq. 4.121 breaks down in the presence of instability (e.g. Daily and Harleman, 1966; Businger *et al.*, 1971; Pruitt *et al.* 1973).

Using Eq. 4.121, flux estimation now depends upon our ability to estimate the relevant concentration gradient. In the case of evaporation, it is common practice to circumvent this difficult task through joint use of the flux–gradient and thermal energy conservation relations to derive the rate of evaporation from a saturated surface for

Plate 4.5. Oriental beech. (Photographed by William D. Rich in Arnold Arboretum, Boston; Copyright © 2001 William D. Rich.)

use as a reference maximum (i.e., *potential*) rate. The details of this calculation are given here from two points of view: (1) In the next section we use the traditional flux–gradient approach introduced by Penman (1948), refined by Van Bavel (1966), and common to the literature of hydrology; (2) in Chapter 5 we revisit this question from the state–space (i.e., vapor pressure vs. temperature) viewpoint pioneered by Monteith (1965) and most familiar to plant scientists.

K Evaporation from a saturated surface

Penman (1948) used Eq. 4.115 as the starting-point for what has turned out to be the most widely used estimator of evaporation from large-scale natural surfaces. We will derive it here following Eagleson (1970, pp. 215 *et seq.*)

Equation 4.115 gives the evaporative mass flux density (g cm^{-2} s^{-1}). From thermodynamics (cf. Solot, 1939)

$$q_{\mathrm{v}} \approx 0.622\frac{e}{p}, \tag{4.122}$$

in which

e = partial pressure of water vapor (i.e., the *vapor pressure*), and
p = atmospheric (total) pressure.

Thus, Eq. 4.115 is

$$E = 0.622K_{\mathrm{v}}\frac{\rho}{p}\frac{\mathrm{d}e}{\mathrm{d}z}. \tag{4.123}$$

Using Eq. 4.27 we have, above the canopy,

$$\tau_0 = \rho u_*^2, \tag{4.124}$$

so that Eq. 4.114 becomes there

$$K_{\mathrm{m}} = \frac{u_*^2}{\mathrm{d}u/\mathrm{d}z}. \tag{4.125}$$

Multiplying and dividing Eq. 4.123 by Eq. 4.125 we obtain

$$E = 0.622\frac{K_{\mathrm{v}}}{K_{\mathrm{m}}}\frac{\rho u_*^2}{p}\frac{(e_1 - e_2)}{(u_2 - u_1)}, \tag{4.126}$$

where the subscript "1" refers to the elevation, $d_0 + z_0$, at which u vanishes, and "2" refers to a higher elevation. Using Eq. 4.26 in Eq. 4.126, we obtain

$$E = B_{\mathrm{o}}(e_1 - e_2), \tag{4.127}$$

in which

$$B_o = 0.622 \frac{K_v}{K_m} \frac{\rho k^2}{p} \frac{u_2}{\ln^2\left(\frac{z_2 - d_0}{z_0}\right)}. \tag{4.128}$$

The flux of latent heat of vaporization (i.e., the heat equivalent of the evaporation rate) is written

$$Q_v = \lambda E, \quad (\text{cal cm}^{-2}\ \text{s}^{-1}), \tag{4.129}$$

in which

$$\lambda = \text{latent heat of vaporization of water} = 597.3\ \text{cal g}^{-1}\ (\text{at } 0\ ^\circ\text{C}).$$

Equations 4.127 and 4.129 then give

$$Q_v = \lambda B_o(e_1 - e_2). \tag{4.130}$$

In a similar manner we use Eqs. 4.117, 4.123, and 4.129 to write the *Bowen Ratio*, $\boldsymbol{R}_b$, as

$$\boldsymbol{R}_b \equiv \frac{H}{Q_v} = \gamma_o \frac{T_s - T_2}{e_s - e_2}, \tag{4.131}$$

which assumes saturation conditions (subscript "s") at elevation "1" and defines

$$\gamma_o \equiv \frac{c_p}{0.622\lambda} \frac{K_h}{K_v} p \tag{4.132}$$

as the *psychrometric constant.* Equations 4.130 and 4.131 then give

$$H = \lambda B_o \gamma_o (T_s - T_2), \quad (\text{cal cm}^{-2}\ \text{s}^{-1}). \tag{4.133}$$

We now introduce the slope of the saturation vapor pressure vs. temperature curve

$$\Delta \equiv \frac{e_s - e_{s2}}{T_s - T_2}, \tag{4.134}$$

which allows us to eliminate the troublesome need to know the surface temperature, T_s. Equation 4.133 then becomes

$$H = \frac{\gamma_o}{\Delta} \lambda B_o(e_s - e_{s2}) = \frac{\gamma_o}{\Delta} \lambda B_o(e_s - e_2) - \frac{\gamma_o}{\Delta} \lambda B_o(e_{s2} - e_2), \tag{4.135}$$

or

$$H = \frac{\gamma_o}{\Delta} Q_v - \frac{\gamma_o}{\Delta} \lambda B_o(e_{s2} - e_2). \tag{4.136}$$

Neglecting horizontal flux divergences, storages within canopy and substrate, and biological metabolism,† conservation of energy gives

$$Q_v = R_n - H, \tag{4.137}$$

† These assumptions are discussed in Chapter 5.

in which Q_v = vertical flux of latent heat from canopy to atmosphere,
R_n = net influx of solar radiation, and
H = vertical flux of sensible heat from canopy to atmosphere.

Remember that the units of E are g cm^{-2} s^{-1} while the conventional measure of evapotranspiration rate, E_T, has the units cm s^{-1}. To make the conversion we use the standard unit mass density of fresh water, ρ_w (g cm^{-3}) = 1, to write

$$E_T\ (\text{cm s}^{-1}) = E(\text{g cm}^{-2}\ \text{s}^{-1}).$$

Plate 4.6. Eastern white pine. (Photographed by William D. Rich in Arnold Arboretum, Boston; Copyright © 2001 William D. Rich.)

Using Eqs. 4.129 and 4.137, Eq. 4.136 becomes the standard *combination form* (i.e., diffusion and energy balance) of the *Penman equation* for evaporation from a saturated surface

$$\lambda E_T = \frac{\Delta \cdot R_n + \gamma_o \lambda B_o (e_{s2} - e_2)}{\Delta + \gamma_o}. \tag{4.138}$$

L The equivalent atmospheric resistance

To define the *equivalent atmospheric resistance* we first note, after Thom (1975, p. 65) that a lumped linearization admits an analogy to Ohm's law in the form

$$\text{aerodynamic resistance} = \frac{\text{momentum concentration difference}}{\text{momentum flux}}. \tag{4.139}$$

The aerodynamic resistance, r_a, to the transfer of momentum from an elevation "2" at which the wind speed is u_2 to a lower elevation at which the wind speed is zero, is then

$$r_a = \frac{\rho(u_2 - 0)}{\tau_0}, \tag{4.140}$$

in which the momentum flux to the lower surface is given by the shear stress, τ_o, thereon.

Using Eqs. 4.27 and 4.28, Eq. 4.140 can be written

$$r_a = \frac{u_2}{u_*^2} = \frac{\ln\left(\frac{z_2 - d_0}{z_0}\right)}{k u_*} = \frac{\ln^2\left(\frac{z_2 - d_0}{z_0}\right)}{k^2 u_2}. \tag{4.141}$$

Invoking Reynolds similarity (i.e., $K_h = K_m$), B_o can be simplified and Eq. 4.138 finally takes the commonly used form of the Penman equation: for evaporation from a square centimeter of *saturated surface*:

$$\lambda E_T = \frac{\Delta \cdot R_n + \frac{\rho c_p}{r_a}[e_s(T_2) - e_2]}{\Delta + \gamma_o}. \tag{4.142}$$

in which the first term on the right-hand side is the so-called "available energy forcing", and the second term is the "air dryness forcing" (Kim and Entekhabi, 1997). Elevation "2" is usually taken at screen height (normally 2 m) above the evaporation surface, making $u_2 \equiv u_0$, the free-stream velocity, and $z_2 = h + 2$, in Eq. 4.141. Note that when $h \gg 2$ m, Eq. 4.62 reduces Eq. 4.141 to the estimator

$$r_a \approx (k^2 u_0)^{-1}. \tag{4.143}$$

M Transpiration from saturated stomatal surfaces

Resistances at stomatal scale

In Fig. 4.27a we sketch a cross-section of a typical leaf showing a single stomate as adapted from Larcher (1983, Fig. 3.11, p. 88). Light is incident upon the upper surface, while water vapor and carbon dioxide are exchanged (largely) through the stomatal opening on the lower surface. In this diagram we continue our use of the symbol r to represent the equivalent lumped resistance (s cm^{-1}) of a canopy element to flow through the element. Here the element is a single stomate. There are several such lumped resistances; beginning at the top of the diagram we identify, for transpiration:

r_r = *residual resistance*, governing flow across the cell walls into the stomatal cavity
r_{ic} = *intercellular resistance*, controlling flow within the stomatal cavity
r_s = *stomatal resistance*, (a physiological "valve") regulating plant water loss and carbon dioxide assimilation
r_{cu} = *cuticular diffusion resistance*, impeding diffusion through the leaf cuticle, and
r_{l_a} = *leaf boundary layer resistance*, restricting flow out to the ambient atmosphere.

These resistances combine as in Fig. 4.27b to form an equivalent *stomatal cavity resistance*, r_o, written

$$r_o = \frac{1}{\dfrac{1}{r_{cu}} + \dfrac{1}{r_r + r_{ic} + r_s}} + r_{la}. \tag{4.144}$$

In applying the Penman equation to transpiration, the saturated evaporating surface becomes the cell wall within the stomate. With this surface *always saturated*, we may assume that $r_r = 0$ (Monteith, 1965, p. 209). In addition, for trees:

(1) the cuticular diffusion resistance, r_{cu}, is usually very high with a value $r_{cu} = O(10^4 \text{ s m}^{-1})$ (Larcher, 1983, p. 228); and
(2) the intercellular resistance, r_{ic}, is "small" (Gaastra, 1963, p. 119) with a value $r_{ic} = O(10 \text{ s m}^{-1})$ (Gates, 1980, p. 327).

Thus, Eq. 4.144 can be simplified to yield

$$r_o \approx r_s + r_{la}. \tag{4.145}$$

In addition, for homogeneous multilayers at least, the leaf boundary layer resistance, r_{la}, is small when compared with the stomatal resistance (Gates, 1980, p. 329), at least when the wind speed $u_0 > 2$ m s^{-1} (Larcher, 1983, p. 224). Then Eq. 4.145 becomes

$$r_o \approx r_s, \tag{4.146}$$

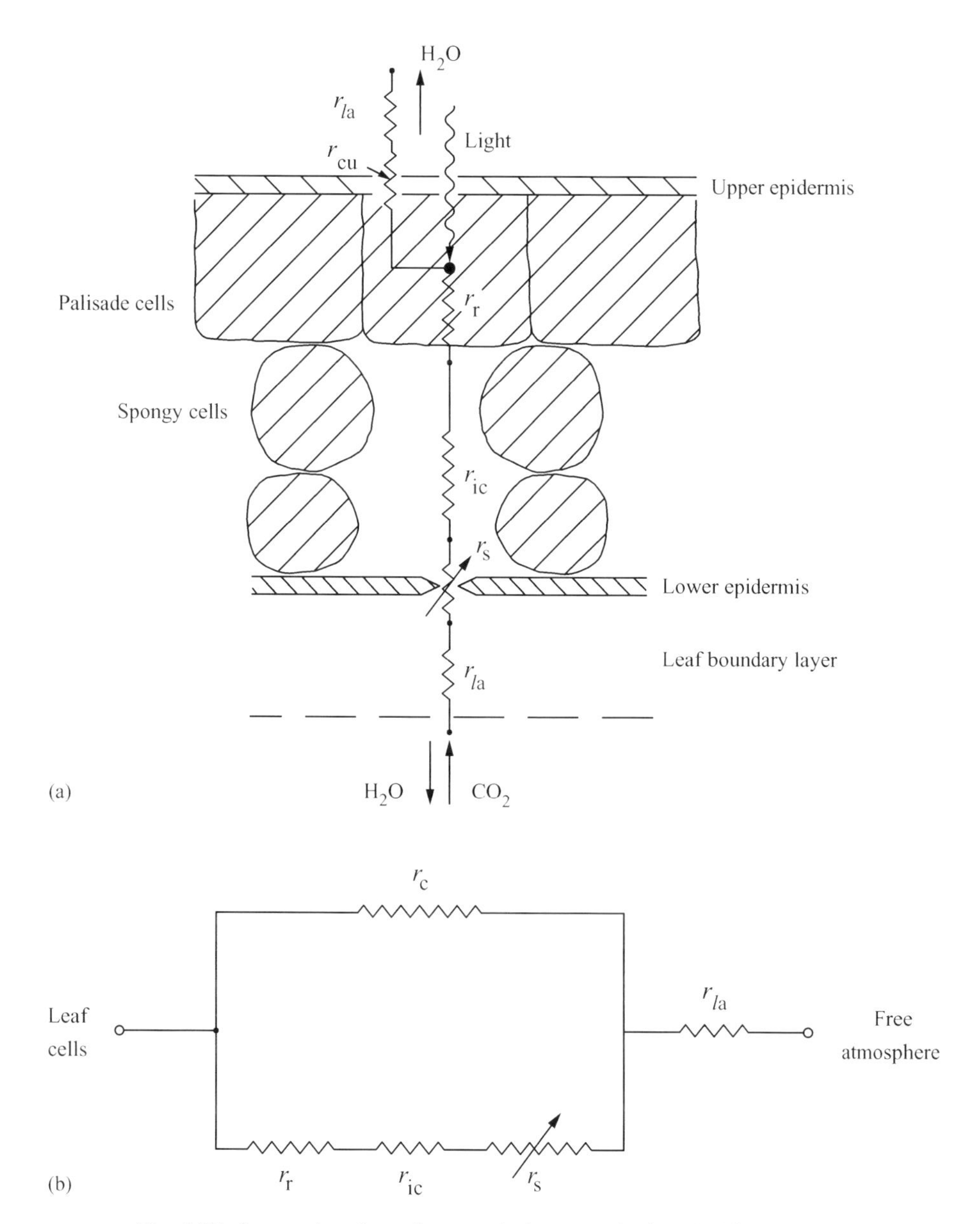

Fig. 4.27. Stomatal cavity resistances during transpiration. (a) The structure of a stomatal cavity. (From Larcher, W., *Physiological Plant Ecology*, corrected printing of 2nd Edition, Fig. 3.11, p. 88, Copyright © 1983 Springer-Verlag GmbH & Co. KG, with kind permission from both Springer-Verlag GmbH & Co. KG, Heidelberg, and W. Larcher. (b) The equivalent resistance of a stomatal cavity.

and we see that in this common case the total evaporative resistance of the stomatal cavity is approximated by the stomatal resistance alone.

To transfer moisture from the leaf boundary layer out to the ambient free stream requires the equivalent atmospheric resistance, r_a, to be placed in series with r_o.

Resistances at leaf scale

At the scale of the whole leaf, there are n_s stomatal resistances in parallel, where n_s is the number of stomata per leaf. However, for transpiration calculations we restrict our interest to a unit area (1 cm^2) of (horizontal) leaf surface from the leaf which has total (one-sided) area A_l. The *equivalent leaf stomatal resistance*, r_{ls}, of this unit leaf area is

$$r_{ls} = \frac{r_s}{n_s/A_l}, \tag{4.147}$$

while including the leaf boundary layer resistance gives the *equivalent leaf resistance*, r_l, as

$$r_l = r_{ls} + r_{la}. \tag{4.148}$$

Once again, to transfer moisture from the leaf boundary layer out to screen height in the ambient atmosphere requires that we add the equivalent atmospheric resistance, r_a, in series with r_{la} as is shown in Fig. 4.28.

Typical measured values of the important stomatal and leaf properties are given in Tables 4.2, and 4.3 for various vegetation classes as taken from the literature (Larcher, 1983; Gates, 1980; Kozlowski, 1968). For the moment consider Table 4.3 where we see that to within an average of 15% we may simplify Eq. 4.148 to

$$r_l \approx r_{ls}. \tag{4.149}$$

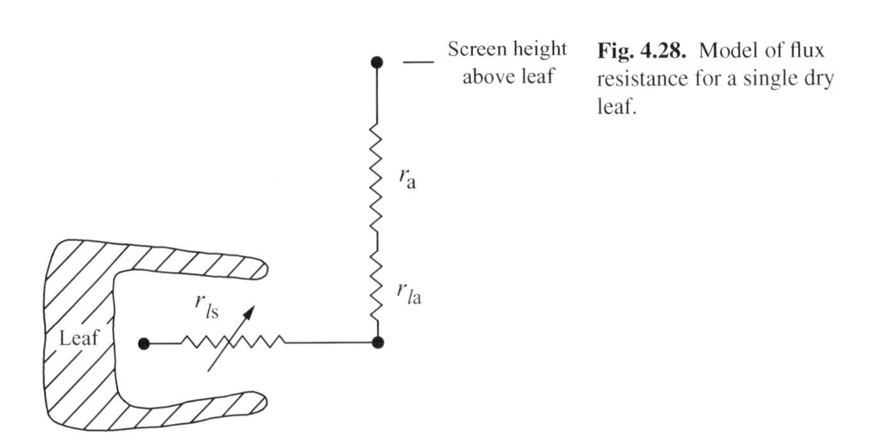

Fig. 4.28. Model of flux resistance for a single dry leaf.

N Transpiration from a dry leaf surface

To convert Eq. 4.138 into an estimator of transpiration, E_v, Monteith (1965) first assumed equilibrium between the rates of diffusion of water vapor in the external air and within the leaf. He argued as follows:

If the atmospheric resistance, r_a, is the time during which 1 cm^3 of air exchanges water vapor with 1 cm^2 of leaf surface, then r_a^{-1} is the rate of vapor exchange in $cm^3\ cm^{-2}\ s^{-1}$. By analogy with Ohm's law (i.e., potential difference = current · resistance), Monteith (1965, p. 209) wrote, in consistent units

$$\frac{\gamma_0 \lambda}{\rho c_p} \cdot E = \frac{\Delta e}{r} = \frac{e_0 - e_2}{r_a}, \quad (\mathrm{g\ cm^{-2}\ s^{-1}}), \tag{4.150}$$

in which e_0 = vapor pressure at the leaf surface. Similarly, from the saturated cell walls of the stomatal cavity to the leaf surface,

$$\frac{\gamma_0 \lambda}{\rho c_p} \cdot E = \frac{e_s(T_0) - e_0}{r_l} \approx \frac{e_s(T_0) - e_0}{r_{ls}}, \tag{4.151}$$

in which

e_s = saturation vapor pressure at the temperature of the stomatal cavity (assumed equal to the leaf surface temperature, T_0).

Equating Eqs. 4.150 and 4.151 at equilibrium gives

$$\frac{e_s(T_0) - e_0}{r_{ls}} = \frac{e_0 - e_2}{r_a}. \tag{4.152}$$

Table 4.2. *Typical stomatal properties*

Plant type	Number of stomata per mm^2	r_{ls} ($s\ m^{-1}$)[a]
Winter deciduous trees	100–300[b]	170–650
Evergreen conifers	40–120	200–650
Tropical forest trees	200–600	
Palms	150–180	600
Desert shrubs	150–300	160–780
Grasses	50–100	90–520
Average:	O(10^4 per cm^2)	435

[a]Open stomata.
[b]As modified in Larcher (1995).
Source: Larcher, W., *Physiological Plant Ecology*, corrected printing of 2nd Edition, Table 3.2, p. 91; Copyright © 1983, Springer-Verlag GmbH & Co. KG, with kind permission from Springer-Verlag GmbH & Co. KG, Heidelberg, and from W. Larcher.

Table 4.3. *Typical tree leaf properties*

Species	$h - h_s$ [a] (m)	Leaves per crown,[a] n_l	Single leaf area,[a] A_l (cm^2)	10^4 Stomata per leaf[b]	Minimum r_{ls} [c] ($s\ m^{-1}$) (per unit area)	r_{cu} [d] ($s\ m^{-1}$)	r_{la} [d] ($s\ m^{-1}$)
Black alder	2.5	2 500	16	16			
European ash	3	500	70	70			
European aspen					230		60
Silver birch					120	7 000	80
Cottonwood					400		
Southern magnolia					600		
Norway maple	2.5	900	33	33	790	10 100	80
Sycamore maple	2.5	300	50	50			
Chestnut-oak					1 160	29 000	90
Cork oak					600		
Red pine					934[e]		
Scots pine					200		
					300[e]		
Yellow poplar					540		
Sweet gum					402		
White willow	3.5	25 000	2	2			
Averages	2.8	5 840	34	30	505	15 367	78

[a]Larcher (1983, Table 5.4, p. 227). [b]Using 10^4 stomata per cm^2 (Larcher, 1995, Table 2.2, p. 81).
[c]Rutter (1968, Table XIV, p. 65) (from ratio of vapor pressure gradient to transpiration rate except where noted), all values are per unit of leaf area.
[d]Holmgren *et al.* (1965), as given by Gates (1980, Table 10.2, p. 329).
[e]From geometric measurement (cf. Gates, 1980, pp. 333–335).

Adding and subtracting e_2/r_{ls} gives

$$e_0 - e_2 = \frac{e_s(T_0) - e_2}{1 + \dfrac{r_{ls}}{r_a}}, \tag{4.153}$$

which shows that we can represent the vapor pressure difference $(e_0 - e_2)$ above the *dry* leaf surface by a fraction of that above a water surface. We can thus maintain the appropriate dry surface Bowen ratio while still obtaining the Penman equation from a *saturated* leaf surface provided we augment γ_o by the factor $(1 + \frac{r_{ls}}{r_a})$. Now

$$\frac{H}{Q_v} = \gamma_o \frac{T_0 - T_2}{e_0 - e_2} = \gamma_o \left(1 + \frac{r_{ls}}{r_a}\right) \frac{T_0 - T_2}{e_s(T_0) - e_2}, \tag{4.154}$$

which is identical to Eq. 4.131 since we assume that with thin leaves at equilibrium, $T_0 \approx T_s$. We therefore use the augmented γ_o in Eq. 4.142 to get the Penman–Monteith equation for transpiration from a square centimeter of *dry leaf surface*:

$$\lambda E_v = \frac{\Delta \cdot R_n + \dfrac{\rho c_p}{r_a}[e_s(T_2) - e_2]}{\Delta + \gamma_o\left(1 + \dfrac{r_{ls}}{r_a}\right)}, \tag{4.155}$$

in which r_{ls} will vary with the availability of soil moisture.†

O Transpiration from a dry forest canopy

Continuing to scale up, a whole canopy consists of a massive parallel array of individual leaves each responding to its own micro-environment through a separate Penman–Monteith equation. Accurate formulation of this composite is an unrealistic goal; however, complex approximations do exist (e.g., Sellers *et al.*, 1986; Dickinson *et al.*, 1991). Their simplified, generic form is illustrated by the so-called "big leaf" or "single source" canopy. This model represents the entire canopy by a single "equivalent" leaf as is shown in Fig. 4.29, and forms the basis for our work.

In Fig. 4.29 the left-hand side represents the pathway for water vapor in which the equivalent *canopy resistance*, r_c, is defined as the sum of an internal resistance and an external resistance. That is

$$r_c \equiv r_{cs} + r_{ci}, \tag{4.156}$$

where the internal resistance, r_{cs} = *equivalent canopy stomatal resistance*, represents the effect of variable plant stress in all the leaves of the crown, and the external resistance, r_{ci} = *equivalent interleaf resistance*, represents the effective labyrinthine atmospheric pathway for fluxes through the crown. The pathway on the right-hand

† McNaughton and Jarvis (1983) write $\lambda E_v = \Omega E_r + (1 - \Omega)E_a$ in which E_r depends mainly upon radiation and E_a upon the atmosphere. Their *decoupling factor*, Ω, is controlled by the resistance ratio.

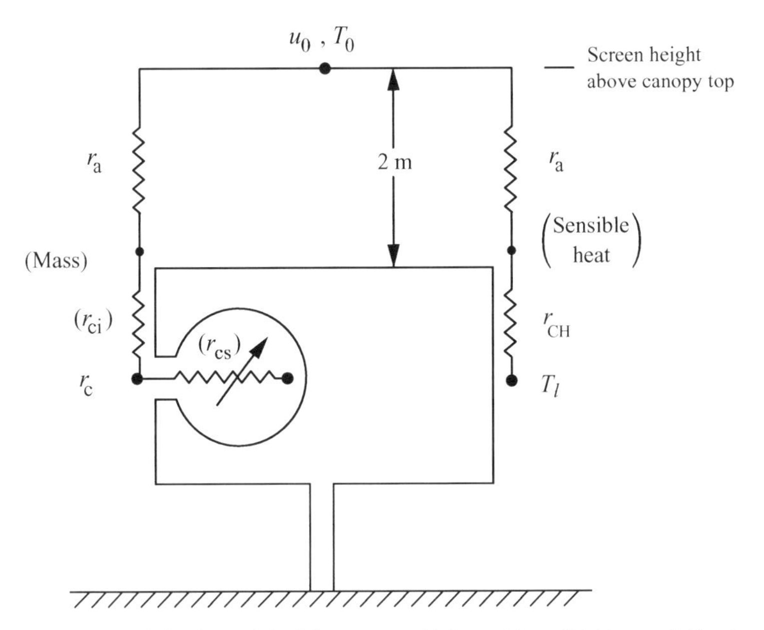

Fig. 4.29. "Big leaf" model of dry canopy. (Adapted from Dickinson, R.E., A. Henderson-Sellers, C. Rosenzweig and P.J. Sellers, Evapotranspiration models with canopy resistance for use in climate models: A review, Fig. 3, p. 377, *Agricultural and Forest Meteorology*, 54; Copyright © 1991 Elsevier Science, with kind permission from Elsevier Science.)

side of Fig. 4.29 is that for sensible heat in which the canopy resistance, r_{cH}, is fixed. Invoking the Reynolds analogy, the effective interleaf resistance will be the same for mass and heat giving

$$r_{cH} \equiv r_{ci}. \tag{4.157}$$

The atmospheric resistance, r_a, regulates the flux from canopy top out to screen height (above the canopy), and is given by Eq. 4.141 (or 4.143) independent of our canopy model.

By analogy with Eq. 4.155, the governing *Penman–Monteith equation for transpiration from a square centimeter of dry canopy* is then

$$\lambda E_v = \frac{\Delta \cdot R_n + \dfrac{\rho c_p}{r_a}[e_s(T_2) - e_2]}{\Delta + \gamma_o\left(1 + \dfrac{r_c}{r_a}\right)}, \tag{4.158}$$

where elevation "2" is now taken at "screen" height above the canopy top. The canopy resistance, r_c, is the primary flux-controlling plant–atmosphere property, and its estimation poses the greatest challenge to successful use of Eq. 4.158. Monteith (1963, discussion) considered r_c (r_s in his notation) to be strongly dependent upon stomatal resistance, whereas Philip (1966, p. 264) suggested that further work would show r_c (again r_s in his notation) "is subject to great variability and is relatively weakly linked with stomatal resistance". As we show below, our model of the big leaf canopy provides theoretical support for Philip's conjecture.

As a canopy grows from a single leaf where $r_c \equiv r_l = r_{ls} + r_{la}$ (Eq. 4.148), to many, many leaves where $r_c = r_{cs} + r_{ci}$, the relative importance of the internal and external resistances reverses.

The conceptual crown

In transferring from leaf scale to big leaf canopy scale we will use an intermediate lumped conceptual state called the *leaf layer*. A leaf layer is defined as a horizontal crown layer containing leaves whose one-sided surface area equals the plan area of the crown. In other words, there are $L \approx L_t$ leaf layers in a crown. In our lumped conception, sketched here in Fig. 4.30, there are only $L_t - 1$ actively transpiring leaf layers, because the bottommost leaf at every radius (i.e., the L_tth layer) uses water at a negligible rate associated solely with maintenance respiration (see Chapter 3). Each of these photosynthetically active $L_t - 1$ leaf layers is located at the center of an interleaf atmospheric layer having thickness Δz_L representing the effective length of the diffusion path between adjacent leaf layers. For homogeneous canopies composed of cylindrical crowns having uniform foliage area density

$$\Delta z_L \approx \frac{h - h_s}{L_t - 1}. \tag{4.159}$$

For non-homogeneity this definition will change as we will see later in this chapter.

The canopy can now be modeled as a much simpler network of equivalent *leaf layer stomatal resistances*, r_{lls}, and *interleaf layer resistances*, r_i. We assume that all leaf surfaces are at a common potential level, enabling us to draw the canopy resistance

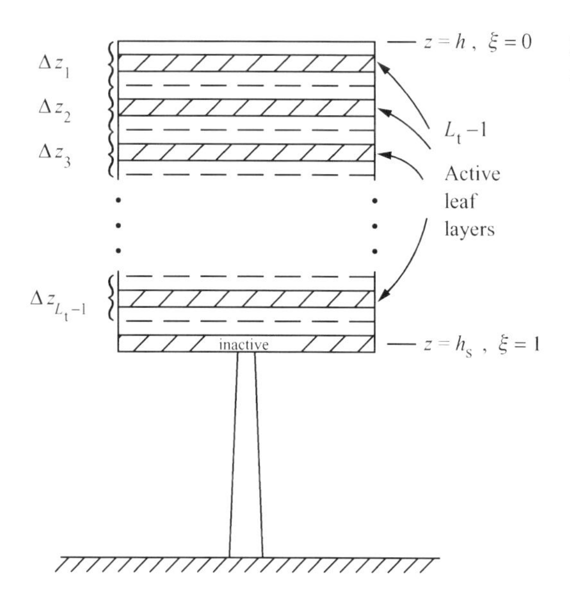

Fig. 4.30. Conceptual model of crown.

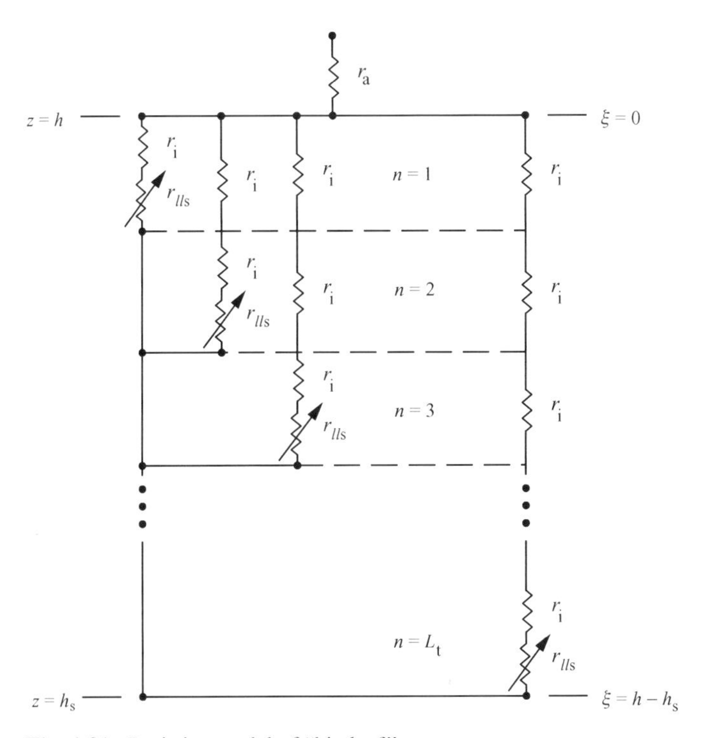

Fig. 4.31. Resistive model of "big leaf" canopy.

diagram shown in Fig. 4.31. In this model, the water vapor emanating from the nth leaf layer down from the canopy top must diffuse serially through n interleaf atmospheric resistances, while all of these n separate diffusion paths are in parallel. From this series–parallel circuit we have

$$\frac{1}{r_c} = \frac{1}{r_{lls} + r_i} + \frac{1}{r_{lls} + 2r_i} + \cdots + \frac{1}{r_{lls} + (L_t - 1)r_i}. \tag{4.160}$$

The leaf layer stomatal resistance

For a crown having n_l leaves and L_t leaf layers, each leaf layer will have n_l/L_t single leaf stomatal resistances, r_{ls}, connected in parallel. The equivalent leaf layer stomatal resistance, r_{lls}, is then

$$\frac{1}{r_{lls}} = \frac{n_l/L_t}{r_{ls}}. \tag{4.161}$$

Incidentally, when inferring canopy resistance from observations of single-leaf resistance, this step seems to have been overlooked by previous investigators, e.g., Monteith (1965, p. 217).

The single-leaf stomatal resistance may vary from infinity, when the stomata close under stress, to the minimum value given in Table 4.3 when the stomata are fully open under no stress. A basic assumption of this work is that natural selection will set the average operating state as close to the stressless state as possible. Therefore, using the *minimum* single leaf resistances (*per unit of leaf area*), r_{ls}, of Table 4.3 in Eq. 4.161 yields the *mimimum* equivalent leaf layer resistance (*per unit of leaf layer area*),

$$r_{lls} = \frac{L_t r_{ls}}{n_l}. \tag{4.162}$$

From Tables 2.2, 2.3, and 4.3 we estimate

$$r_{lls} = \frac{L_t r_{ls}}{n_l} \approx \frac{5 \times 500}{6000} = 0.42, \quad (\text{s m}^{-1}). \tag{4.163}$$

Using Eq. 4.143 with Eq. 4.163 an estimate of the dimensionless resistance ratio is

$$r_{lls}/r_a \approx 0.07 u_0 \, (\text{m s}^{-1}). \tag{4.164}$$

The interleaf layer resistance

Combining Eqs. 4.37 and 4.139 in the manner of Eq. 4.140, and using Eq. 4.114, the interleaf layer resistance is written

$$r_i = \rho \frac{\Delta u}{\tau} = \frac{(h - h_s)\Delta\xi}{K_m(\xi)}, \tag{4.165}$$

in which K_m has been lumped at the scale of Δz_L. Accuracy of Eq. 4.165 will be inversely related to this lumping scale. Although the r_i will actually vary with ξ due to the dependence of r_i upon $K_m(\xi)$, we will neglect this variation here, replacing $K_m(\xi)$ by $\widehat{K_m}(\xi)$. With this "first order" approximation, the r_i become identical, and we use Eqs. 4.38 and 4.159 to write Eq. 4.165 as

$$r_i = \frac{1}{ku_*} \cdot \left(\frac{h - h_s}{h - d_0}\right) \cdot \frac{1}{(L_t - 1)\widehat{K^o_m}}. \tag{4.166}$$

For homogeneous cylindrical multilayers, we use Eqs. 4.35, 4.59, and 4.41 in Eq. 4.166, to get the first order estimate of r_i

$$r_i = \left(\frac{1}{ku_*}\right) \frac{\frac{1-m}{m}(\gamma L_t)^2}{(L_t - 1)\left[1 - \exp\left(-\frac{1-m}{m}\gamma L_t\right)\right]}, \tag{4.167}$$

and with Eqs. 4.59 and 4.143, the associated resistance ratio is

$$\frac{r_i}{r_a} = \frac{[(1-m)/m](\gamma L_t)^2/(L_t - 1)}{1 - \exp\{-[(1-m)/m]\gamma L_t\}}, \tag{4.168}$$

while Eqs. 4.59, 4.162, and 4.167 give, finally

$$\frac{r_{lls}}{r_i} = k^2 u_0 \left(\frac{r_{ls}}{n_l}\right) \left\{ \frac{1 - \exp\left[-\left(\frac{1-m}{m}\gamma L_t\right)\right]}{\gamma^2 L_t(1-m)/[m(L_t-1)]} \right\}. \tag{4.169}$$

For the common case: $m = 0.5, n = 2, \beta = 0.5, L_t = 5$, Eq. 4.168 gives the estimate

$$\frac{r_i}{r_a} \approx 1.7, \tag{4.170}$$

and Eq. 4.169 determines

$$\frac{r_{lls}}{r_i} \approx 0.04\, u_0 \text{ (m s}^{-1}\text{)}. \tag{4.171}$$

Therefore, with u_0 in the normal range 2–3 m s^{-1}

$$\frac{r_{lls}}{r_i} = O(10^{-1}). \tag{4.172}$$

Plate 4.7. Moreton Bay fig. (Photographed by Peter S. Eagleson in the Royal Botanic Gardens, Sydney. Identification kindly supplied by Eric and Daphne Laurenson.)

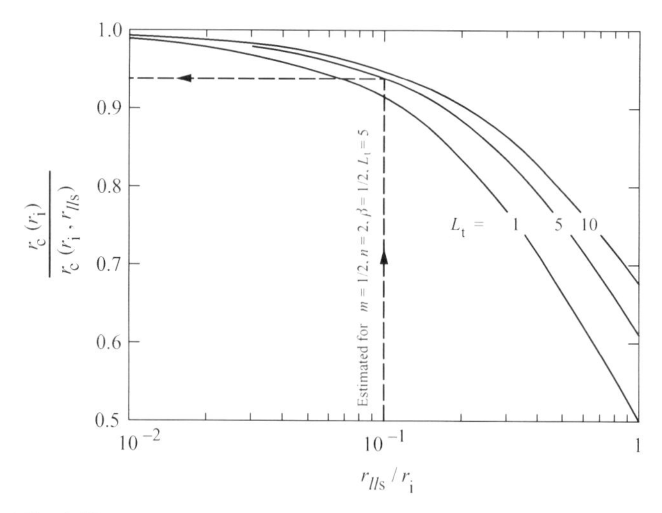

Fig. 4.32. Dominance of interleaf layer resistive component of "big leaf" canopy model.

The canopy resistance

Equation 4.172 encourages us to neglect the leaf layer stomatal resistance with respect to the interleaf layer resistance and approximate the conceptual big leaf canopy resistance of Eq. 4.160 by

$$\frac{1}{r_c} \approx \frac{1}{r_i} + \frac{1}{2r_i} + \cdots + \frac{1}{(L_t - 1)r_i}. \tag{4.173}$$

We can find the effect on r_c of neglecting $r_{l/s}$ by dividing Eq. 4.160 by Eq. 4.173

$$r_c(r_i)/r_c(r_i, r_{l/s}) = \frac{\frac{1}{1 + r_{l/s}/r_i} + \frac{1}{2 + r_{l/s}/r_i} + \cdots + \frac{1}{(L_t - 1) + r_{l/s}/r_i}}{1 + 1/2 + 1/3 + \cdots + 1/(L_t - 1)}. \tag{4.174}$$

Equation 4.174 is plotted in Fig. 4.32 for various values of L_t. We see there that the estimate, $\frac{r_{l/s}}{r_i} = O(10^{-1})$, leads to an error of only 6% in r_c when we estimate the latter using Eq. 4.173 in which the stomatal resistance is neglected. This is contrary to "conventional wisdom" but provides a theoretical basis for the Philip (1966, p. 264) conjecture mentioned earlier.

We will use the approximation $r_c = r_c(r_i)$ to evaluate the resistance ratio, r_c/r_a, in Chapter 7.

5

Thermal energy balance

The long-term average thermal energy balance of an infinite closed canopy is formulated under idealized conditions, the thermodynamics of evaporation are considered in state space, and the special case of potential evaporation is derived. Estimates of the Bowen ratio from climatic observations of wet surfaces at catchment scale indicate small positive values implying that the average "big leaf" canopy temperature approximates the average ambient atmospheric temperature at screen height yielding near-isothermal transpiration from such surfaces on the seasonal average.

A Definitions and assumptions

Figure 5.1 defines the control volume that we will use to study the thermal energy balance for the special case $M = 1$.[†] The volume is bounded on the top by a plane located at screen height above the canopy and is bounded on the bottom by the land surface. The arrows at the control surfaces represent the energy flux densities (W m^{-2}) with the arrow heads showing the sign convention:

R_n = net radiation from the Sun,
H = vertical transfer of sensible heat from the canopy to the atmosphere,
Q_v = vertical transfer of latent heat from the canopy to the atmosphere,
G = sensible heat transfer into the substrate,
D_H = horizontal flux divergence of sensible heat, and
D_{Q_v} = horizontal flux divergence of latent heat.

Within the control volume are the rates of change of stored energy density (W m^{-2}):

S = rate of heat storage within the canopy, and
B = rate at which plant metabolism uses energy.

[†] We restrict this analysis to surfaces which are closely homogeneous laterally in support of our neglect of lateral energy fluxes.

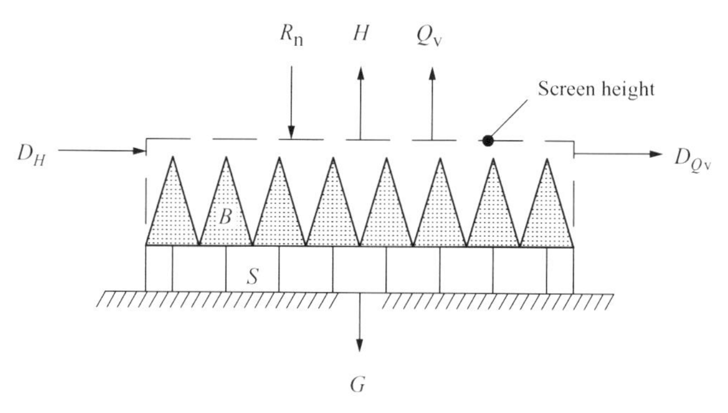

Fig. 5.1. Control volume for thermal energy balance.

With these definitions the thermal energy balance for the control volume is

$$R_n + D_H = Q_v + H + G + B + S + D_{Qv}. \tag{5.1}$$

Let us examine the relative importance of the individual terms of Eq. 5.1:

(1) There is general agreement in the literature (e.g., Gates, 1980, p. 26; Larcher, 1983, p. 20) that the biological rate of energy consumption, B, is negligible in comparison with the other terms of this energy balance.

(2) The substrate heat storage, G, will have a high diurnal and seasonal fluctuation, particularly in the presence of high soil moisture, but its long-term average rate is zero. Similarly, if we confine our interest to the long-term average, we can omit S as being identically zero.

(3) For a "typical" energy budget over vegetation, Thom (1975, Table I, p. 98) shows that D_H and D_{Qv} are small enough to be neglected for the control volume chosen. However, for non-zero H, there *must* be flux divergence external to this control volume and fixed by conditions in the planetary boundary layer.† Such geographically and seasonally variable fluxes may drastically alter the findings of this chapter. To promote generalization, we make the simplifying assumption that these active connections to the external system are severed and replaced by prescribed climatic boundary conditions, T_0, S_r, u_2, etc.

With the above assumptions

$$D_{Qv} = D_H = G = S = B = 0, \tag{5.2}$$

and, for an evaporating (as opposed to a condensing) surface, the *long-term average* climatic thermal energy balance reduces to

$$R_n = Q_v \pm H, \tag{5.3}$$

† For example, does the boundary layer entrain dry air (cf. de Bruin, 1983)?

in which we have introduced the possibility that sensible heat may be *added* to (i.e., $-H$) the control volume of Fig. 5.1 rather than being *rejected* by it (i.e., $+H$) as the figure shows. Using Eq. 4.129, we introduce the evapotranspiration, E_T, to write Eq. 5.3

$$R_{\mathrm{n}} = \lambda E_T \pm H, \tag{5.4}$$

where λ is the latent heat of vaporization of water (i.e., 585 cal $\mathrm{g_w^{-1}}$ at 20 °C). With $D_{Q\mathrm{v}} = D_H = 0$, Eq. 5.4 applies *separately* to both the bare soil and the vegetated fractions of the surface.

Plate 5.1. Royal palm. Elevation of leaf surface is one mechanism for temperature control as employed by this palm in India. (Photograph by Peter S. Eagleson.)

Dividing Eq. 5.4 by λE_T we have

$$\frac{R_n}{\lambda E_T} = 1 \pm \frac{H}{\lambda E_T} = 1 \pm \boldsymbol{R}_b, \tag{5.5}$$

in which $\boldsymbol{R}_b$ is the long-term average Bowen ratio introduced earlier in Eq. 4.131.

B Evaporation in state space

It is illuminating to consider the thermodynamic relationship among the three terms of Eq. 5.4 using the state space approach described in the classic paper by Monteith (1965).

When integrated, the Clausius–Clapeyron differential equation for liquid–vapor equilibrium gives the familiar vapor pressure–temperature relationship of Fig. 5.2. In this illustration, three curves of constant relative humidity are indicated. The upper, solid curve is for a surface separating the liquid and vapor phases, and along which saturation conditions, $e_s(T)$, exist. The lower, dashed curves are for progressively smaller constant relative humidities within the unsaturated vapor phase.

Change of atmospheric state at a saturated surface

In the manner of Monteith (1965), we use Fig. 5.2 to consider the transient change in state of a parcel of unsaturated, moist air placed *adjacent to a saturated surface.* In doing so we follow a path indicated by the lines connecting the numbered points, and to insure the unlimited moisture supply that the saturated surface offers we assume the parcel always contains that surface. The path that we take from one point to another through the state space of Fig. 5.2 is arbitrary (cf. First Law of Thermodynamics); that is, only the end states matter. It is therefore possible as well as instructional to deal only with adiabats (i.e., paths of constant heat) such as 1–2 (or 2–1) and 3–4 (or 4–3) and with the saturated state, $e_s(T)$.

Assume point no. 1 of Fig. 5.2 to represent the *initial* state of the parcel *before* any evaporation occurs, and point no. 4 to represent its *final* state. We move first adiabatically from an unsaturated state at point no. 1 to saturation at point no. 2 which is located on the line representing $e_s(T)$. In so doing, the parcel redistributes its energy; it undergoes an increase in relative humidity and hence in latent heat, but to satisfy the adiabatic requirement of constant heat, it simultaneously undergoes an equal decrease in sensible heat. The Bowen ratio of this step is thus $R_b = 1$, giving (cf. Eq. 4.131)

$$e_s(T_2) - e_1 = \gamma_o(T_1 - T_2) = \gamma_o D_{12}, \tag{5.6}$$

in which D_{12} is called the *wet-bulb depression*, that is the difference between the actual temperature and the saturation temperature at constant heat. With φ = angle of inclination of the adiabatic line 1–2 in Fig. 5.2, we thus have

$$\tan\varphi = \gamma_o. \tag{5.7}$$

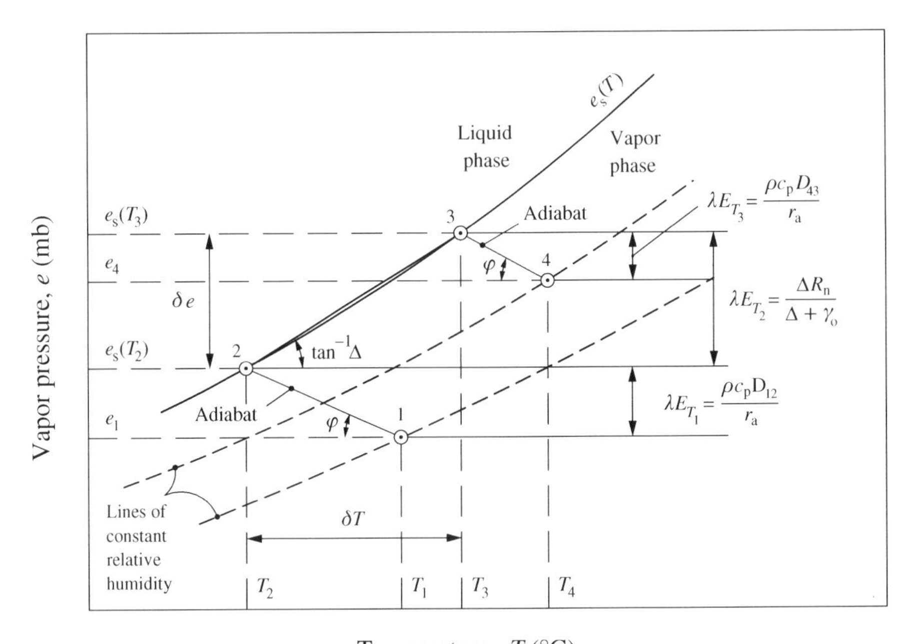

Fig. 5.2. The change of moisture state. (After Monteith, J.L., Evaporation and environment, in *The State and Movement of Water in Living Organisms*, edited by G.E. Fogg, Part II, *Water in the Plant*, Fig. 1, p. 207; Copyright © 1965 The Society for Experimental Biology; with kind permission from J.L. Monteith.)

To move along the *saturated* path from point no. 2 to point no. 3, we must increase both the latent heat (higher vapor pressure) and the sensible heat (higher temperature) of the parcel. This requires forcing from outside the parcel through the *addition* of heat at the long term average rate, Q. In our simplified case of negligible advection and storages this added energy is the net solar radiation, and hence $Q \approx R_n$. The flux of this driving energy is partitioned between increases in latent heat (δQ_v) and sensible heat (δH) along the saturated path 2–3. If the changes δe and δT are small we can replace the curve $e_s(T)$ by its chord having slope, Δ, as given by Eq. 4.134. The Bowen ratio definition (cf. Eq. 4.131) then gives

$$\frac{\delta H}{\delta Q_v} = \frac{\gamma_o}{\Delta}. \tag{5.8}$$

Equations 5.8 and 5.3 yield

$$\delta H = \frac{\gamma_o R_n}{\Delta \pm \gamma_o}, \tag{5.9}$$

and, for either direction of H,

$$\delta Q_v \equiv \lambda E_{T_2} = \frac{\Delta R_n}{\Delta + \gamma_o}. \tag{5.10}$$

The saturated energy partition parameter, Δ/γ_o, is a function (through Δ) of the atmospheric temperature, T_0, as is given by VanBavel (1966) and reproduced here as Table 5.1.

We conclude the parcel's transformation with an adiabatic return to an unsaturated state through condensation (i.e., negative evaporation) and release of latent heat along path 3-4 where, as before

$$e_s(T_3) - e_4 = \gamma_o(T_4 - T_3) = \gamma_o D_{43}, \tag{5.11}$$

and Eq. 5.7 again applies. In this last condensation, 1 cm^3 of saturated air will give up $\rho c_p\,(T_4 - T_3)$ calories of latent heat which will be absorbed by the high sensible heat capacity of the liquid water.

With the parcel being adjacent to a saturated surface, the source and depository of the parcel's liquid water is this surface. The time required for the exchange of the released sensible heat with 1 cm^2 of the saturated surface is called the *atmospheric resistance*, r_a (s cm^{-1}), and the rate at which the latent heat is *removed* from the parcel in this step becomes

$$\lambda E_{T_3} = \frac{\rho c_p D_{43}}{r_a}. \tag{5.12}$$

Table 5.1. Δ/γ_o *vs. temperature*, T, °C

T	Δ/γ_o	T	Δ/γ_o	T	Δ/γ_o	T	Δ/γ_o	T	Δ/γ_o	T	Δ/γ_o
0.0	0.67	10.0	1.23	20.0	2.14	30.0	3.57	40.0	5.70	50.0	8.77
0.5	0.69	10.5	1.27	20.5	2.20	30.5	3.66	40.5	5.83	50.5	8.96
1.0	0.72	11.0	1.30	21.0	2.26	31.0	3.75	41.0	5.96	51.0	9.14
1.5	0.74	11.5	1.34	21.5	2.32	31.5	3.84	41.5	6.09	51.5	9.33
2.0	0.76	12.0	1.38	22.0	2.38	32.0	3.93	42.0	6.23	52.0	9.52
2.5	0.79	12.5	1.42	22.5	2.45	32.5	4.03	42.5	6.37	52.5	9.72
3.0	0.81	13.0	1.46	23.0	2.51	33.0	4.12	43.0	6.51	53.0	9.92
3.5	0.84	13.5	1.50	23.5	2.58	33.5	4.22	43.5	6.65	53.5	10.1
4.0	0.86	14.0	1.55	24.0	2.64	34.0	4.32	44.0	6.80	54.0	10.3
4.5	0.89	14.5	1.59	24.5	2.71	34.5	4.43	44.5	6.95	54.5	10.5
5.0	0.92	15.0	1.64	25.0	2.78	35.0	4.53	45.0	7.10	55.0	10.8
5.5	0.94	15.5	1.68	25.5	2.85	35.5	4.64	45.5	7.26	55.5	11.0
6.0	0.97	16.0	1.73	26.0	2.92	36.0	4.75	46.0	7.41	56.0	11.2
6.5	1.00	16.5	1.78	26.5	3.00	36.5	4.86	46.5	7.57	56.5	11.4
7.0	1.03	17.0	1.82	27.0	3.08	37.0	4.97	47.0	7.73	57.0	11.6
7.5	1.06	17.5	1.88	27.5	3.15	37.5	5.09	47.5	7.90	57.5	11.9
8.0	1.10	18.0	1.93	28.0	3.23	38.0	5.20	48.0	8.07	58.0	12.1
8.5	1.13	18.5	1.98	28.5	3.31	38.5	5.32	48.5	8.24	58.5	12.3
9.0	1.16	19.0	2.03	29.0	3.40	39.0	5.45	49.0	8.42	59.0	12.6
9.5	1.2	19.5	2.09	29.5	3.48	39.5	5.57	49.5	8.60	59.5	12.8
10.0	1.23	20.0	2.14	30.0	3.57	40.0	5.70	50.0	8.77	60.0	13.1

Source: Van Bavel (1966), with permission of the American Geophysical Union.

Similarly, for the first adiabatic step, latent heat was *added* to the parcel at the rate

$$\lambda E_{T_1} = \frac{\rho c_p D_{12}}{r_a}. \tag{5.13}$$

The net evaporation rate, in heat units, is then

$$\lambda E_T = \lambda\left(E_{T_1} + E_{T_2} - E_{T_3}\right), \tag{5.14}$$

or, using Eqs. 5.10, 5.12, and 5.13,

$$\lambda E_T = \frac{\Delta R_n}{\Delta + \gamma_o} + \frac{\rho c_p (D_{12} - D_{43})}{r_a}. \tag{5.15}$$

The first term of Eq. 5.15 is commonly referred to as the "radiational forcing", and the second term as the (net) "drying power" of the air, that is the capacity of the ambient atmosphere to augment (or diminish) the radiational forcing through adiabatic transformation of excess sensible heat into additional latent heat.

It is important to remember that Eq. 5.15 is a *rate* equation which we have examined over a finite interval of time, Δt. The thermodynamic state of a particular air mass is therefore changing with time, and without continuous replacement of the moistened air mass at location x_{ijk} with a dry one at the same location, the evaporation rate there will decline steadily. For homogeneous surfaces of infinite extent, horizontal air movement alone will not replenish the evaporative capacity, and vertical transport of heat and mass, primarily due to turbulence in the surface atmospheric boundary layer, is required. We have discussed this transport at length in the preceeding chapter.

Potential evaporation from a free water surface

We now apply the ideas of the previous section to evaporation from a free water surface. Free water surface evaporation is that which occurs from a thin film of water having insignificant heat storage and thus it closely represents the potential evaporation from adequately watered "simple" natural surfaces such as individual wet leaves and saturated bare soil. In this case E_T in Eq. 5.15 is called the "potential" bare soil evaporation, E_{ps}, because it is the maximum evaporation that the surface can produce under fixed atmospheric conditions. Although this is a special case of the atmospheric moisture state change considered in Fig. 5.2, we redraw it in Fig. 5.3 for later convenience using the same notation but adding an energy flux density scale on the right-hand ordinate.

Once again, in Fig. 5.3 a moist parcel of air lying in contact with a saturated soil surface has the initial state indicated either by point no. 1a or no. 1b, but in this case the final state is the saturated one shown as point no. 3. As before, part of the radiant energy produces evaporation

$$\delta Q_v \equiv \lambda E_1 = \frac{\Delta R_n}{\Delta + \gamma_o}, \tag{5.16}$$

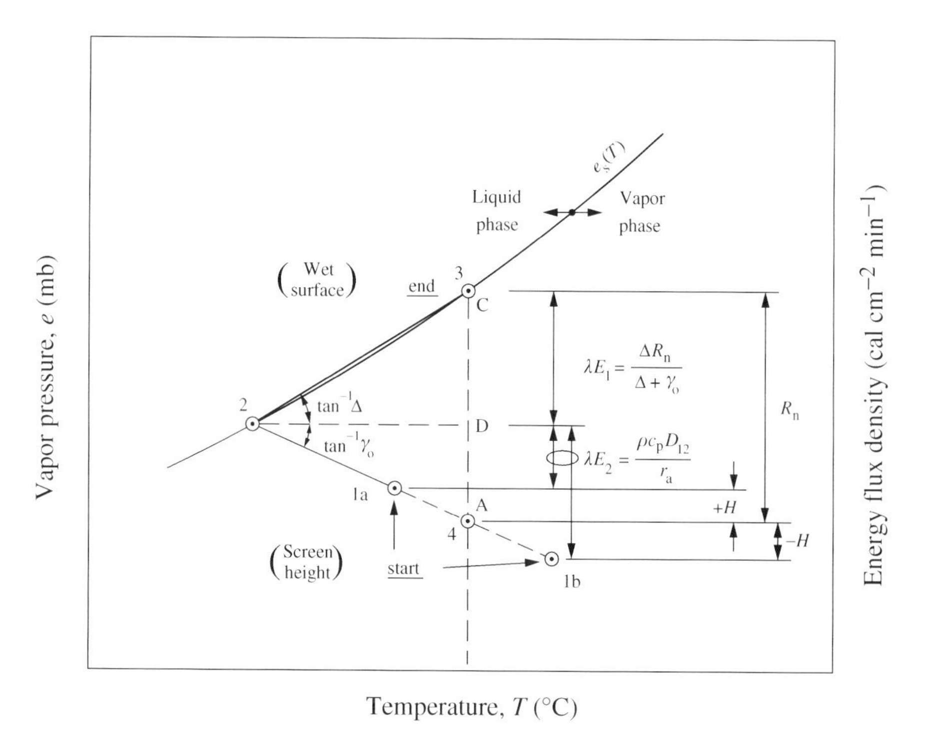

Fig. 5.3. Evaporation from a warm, wet surface.

while the drying power of the ambient atmosphere leads to the additional evaporation

$$\lambda E_2 = \frac{\rho c_p D_{12}}{r_a}, \tag{5.17}$$

which has different numerical values for the different starting-points no. 1a and no. 1b. The total potential evaporation is then

$$\lambda E_{ps} = \lambda(E_1 + E_2) = \frac{\Delta R_n}{\Delta + \gamma_o} + \frac{\rho c_p D_{12}}{r_a}, \tag{5.18}$$

as was originally presented by Slatyer and McIlroy (1961).

The wet bulb depression, D_{12}, is related to the *saturation vapor pressure deficit*, $e_s(T) - e$, that is the difference between saturated and actual vapor pressure at the actual temperature, by

$$\frac{e_s(T) - e}{D_{12}} \equiv \frac{e_s(T_1) - e_1}{T_1 - T_2} = \frac{e_s(T_1) - e_s(T_2)}{T_1 - T_2} + \frac{e_s(T_2) - e_1}{T_1 - T_2} \approx \Delta + \gamma_o, \tag{5.19}$$

whereupon Eq. 5.18 is written

$$\lambda E_{ps} = \frac{\Delta R_n + \rho c_p [e_s(T) - e]/r_a}{\Delta + \gamma_o}, \tag{5.20}$$

where E_{ps} = *bare soil potential evaporation rate*, and in which for practical applications all atmospheric variables are measured at screen height above the surface.

Plate 5.2. Asoka tree. Leaf angle is another mechanism for temperature control as employed by this Asoka tree in Jodhpur, India. (Photograph by Peter S. Eagleson.)

Equation 5.20 is recognized as the classic Penman equation (Penman, 1948) for potential evaporation from a saturated surface and was derived here in another fashion in Chapter 4 (cf. Eq. 4.142). For later convenience, Eq. 5.20 is also written

$$\lambda E_{ps} = \frac{1}{1 + \gamma_o/\Delta}(R_n + D_p), \tag{5.21}$$

in which the variable, D_p, is defined as

$$D_p \equiv \frac{\rho c_p}{r_a}\frac{e_s}{\Delta}(1 - S_r), \tag{5.22}$$

and is called the *drying power*, while S_r is the (constant temperature) *saturation ratio*, $S_r \equiv e/e_s$, i.e., the (fractional) relative humidity.

Understanding the state space diagram

In understanding the significance of Eq. 5.20 and Fig. 5.3 it is helpful to introduce the vertical coordinate, z. We let the parcel of air in Fig. 5.3 span the vertical distance between the surface ($z = 0$) and the screen height ($z = 2$ m). The parcel's initial state no. 1 ("a" or "b") is representative of the ambient atmosphere as measured at screen height, while the final state no. 3 is representative of the saturated underlying surface.

Consider this system to obey our now-familiar simplifying assumptions of no net lateral advection of energy, no long-term average energy storage, and a stable atmosphere. Imagine that we turn on the flow of solar energy at the rate R_n to the free water surface where a fraction (cf. Eq. 5.10) is converted immediately to latent heat. In response to this transient, the surface (saturation) vapor pressure, e_3, and along with it the surface temperature, T_3, rise until the vertical gradients of vapor pressure and temperature between the surface and screen height become such that the *net* vertical energy flux vanishes and equilibrium is established. It is perhaps easiest to understand the range of resulting equilibria by considering the surface state to be fixed and varying the ambient atmospheric vapor pressure over a range of values:

Consider the surface state to be fixed at point C in Fig. 5.3 while we increase the ambient (i.e., point no. 1) saturation vapor pressure deficit. Starting with a saturated atmosphere, $e_s(T) - e = 0$, the ambient state is identical with state no. 2, the drying power, D_p, is zero, and the fraction of solar radiation not used in evaporation must be rejected to the ambient atmosphere in a positive flux of sensible heat. Note that this state minimizes the evaporation and thus (cf. Eq. 5.4) maximizes the sensible heat flux. Let us now increase the saturation vapor pressure deficit by moving point no. 1 along the adiabat 2–A in Figure 5.3 to an ever-drier ambient state. The vapor pressure deficit at any T along this adiabat is simply the constant temperature base of the triangle it forms with opposite angle 4–2–D. As the deficit increases, the drying power increases and

Plate 5.3. Branched palm. Note the symmetry of the branching structure in this unusual Kenyan species. (Photograph by Peter S. Eagleson.)

the sensible heat rejection decreases. When point no. 1 coincides with A, the system is isothermal, thus $H \equiv 0$ and $\lambda E_{ps} = R_n$ (cf. Eq. 5.4). Moving still further along this adiabat to an even drier ambient atmosphere at point no. 1b, the evaporation potential exceeds the available solar radiation and is met by the importation of additional energy through a negative flux of sensible heat. In all these cases, the atmospheric resistance, r_a, plays a key role in determining H for a given vapor pressure deficit. As we will see later, the role of this resistance is markedly increased when the surface layer contains a high-impedance vegetal canopy.

C Bounds to the Bowen ratio

The limiting evaporative conditions just discussed place corresponding limits upon the Bowen ratio. Referring again to Fig. 5.3, the isothermal condition $T_1 = T_4 = T_3$ means $H \equiv 0$, where the Bowen ratio (cf. Eq. 5.5) takes the limiting value

$$(\boldsymbol{R}_b)_{min} = 0. \tag{5.23}$$

This condition might be approached, but not equaled, over deeper water bodies where mixing in both water and air prevents the maintainance of a vertical atmospheric temperature gradient near the surface. In such cases Eq. 5.4 gives

$$\lambda E_{ps} = R_n. \tag{5.24}$$

When points no. 1 and no. 2 are coincident, the atmosphere is saturated to screen height and the vertical humidity gradient is zero. In this case, $D_{12} \equiv 0$ and, using Eqs. 5.16 and 5.5, the Bowen ratio takes on the limiting value for positive H

$$(\boldsymbol{R}_b)_{max} = \frac{\gamma_o}{\Delta}, \tag{5.25}$$

which is less than unity for temperatures above 6.5 °C (cf. Table 5.1).† This condition might be expected over smooth, wet surfaces such as mud flats where the near-surface mixing may be low enough and thus the resistance high enough to make the second term of Eq. 5.20 negligible with respect to the first term. The same condition may occur over *shallow*, wave-free water where the temperature difference between water and atmosphere (i.e., D_{12} in Eq. 5.20) is small, that is whenever water is evaporating into saturated air. In these cases, Eq. 5.20 leads to

$$\lambda E_{ps} = \frac{\Delta R_n}{\Delta + \gamma_o}. \tag{5.26}$$

It is important to note that for negative H there is no obvious upper bound to $\boldsymbol{R}_b$.

Short-term observations of the resistances of a pine canopy reported by Monteith (1965, Table 2, p. 218) give $\boldsymbol{R}_b = 0.42$, while short-term flux observations over Sitka spruce, Scots pine, and Douglas fir (see Jarvis *et al.*, 1976, Fig. 11, p. 210),

† Philip (1987) found a higher bound above very cold surfaces at the onset of supersaturation.

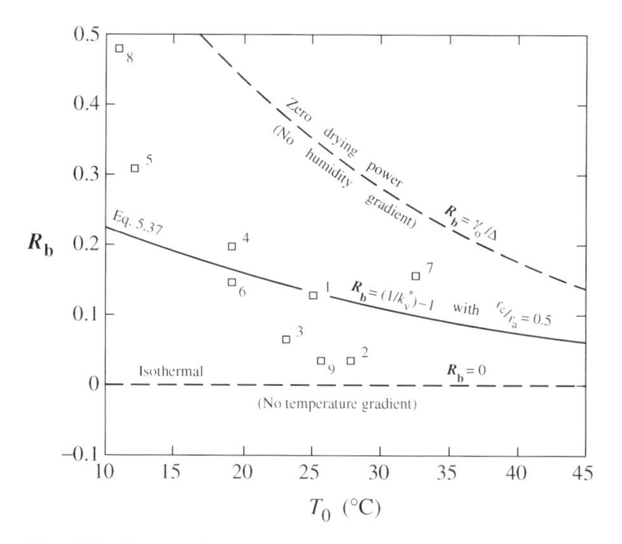

Fig. 5.4. Observed Bowen ratios under wet surface conditions. No energy advection, no energy storage, and stable atmosphere.

Observation	Location	Surface
1	Indian Ocean[a]	Water
2	Indian Ocean[a]	Water
3	Indian Ocean[a]	Water
4	Lake Eucumbene, Australia[a]	Water
5	Australia	Lysimeter, pasture crop
6	Wisconsin	Lysimeter, snap beans
7	Gurley, Australia	Fluxatron, plowed bare soil
8	Hay, Australia[b]	Unknown
9	Equatorial Atlantic Ocean[a]	Water

[a]Conditions selected to be "advection-free".

[b]Dormant season.

(After Priestley and Taylor, 1972, Fig. 2, p. 87, courtesy of the American Meteorological Society.)

lead to $\boldsymbol{R}_b$ values of 1.32, 2.31, and 0.40 respectively. Being greater than unity, the first two of these are undoubtedly for surfaces having negative sensible heat flux. Day-long observations over a variety of wet surfaces reported by Priestley and Taylor (1972) and selected for the apparent absence of energy advection (1972, Fig. 2, p. 87), yield the much smaller measured Bowen ratios plotted here in Fig. 5.4 where they are seen to fall within the limits provided by Eqs. 5.23 and 5.25.

D Transpiration

The Penman–Monteith equation

It was shown in Chapter 4 that by using the so-called "big leaf" approximation for a complex canopy of simple surfaces having aggregate flux resistance r_c, Eq. 5.20

Plate 5.4. Blue spruce. (Photographed in Arnold Arboretum, Boston, by Beverly G. Eagleson.)

can be used to approximate the transpiration from the entire *dry* vegetated fraction. Monteith (1965) pointed out that the effect of the canopy resistance is as though the simple surface of Eq. 5.20 was immersed in a different fluid having the augmented psychrometric constant

$$\gamma_o^* \equiv \gamma_o(1 + r_c/r_a). \tag{5.27}$$

That is, for the dry vegetated fraction we have the *Penman–Monteith equation*

$$\lambda E_v = \frac{\Delta R_n + \rho c_p[e_s(T) - e]/r_a}{\Delta + \gamma_o(1 + r_c/r_a)}, \tag{5.28}$$

in which $E_v =$ rate of *canopy transpiration*. Remember that the horizontal convergence of atmospheric sensible heat over the vegetated fraction was neglected in obtaining Eq. 5.28 when we assumed $Q \approx R_n$. This approximation deteriorates for $M \ll 1$ as heat is advected from the hotter bare soil to the cooler vegetation.

The canopy conductance

When deriving Eq. 5.28 (i.e., Eq. 4.158) we assumed the evaporation surface within the stomates to be saturated *always*, independent of the state of plant stress caused by soil moisture shortage. We now follow the lead of Penman (1948) and introduce a "reference canopy" with which to compare the transpiration from the canopy of a particular species. Penman (1948) chose short, green grass as this reference, but here we choose bare soil which we will find useful when normalizing the climatic water balance equation in Chapter 6. Let us imagine the bare soil to be saturated under atmospheric conditions at its reference level, $z = 2$ m, that are *identical* to those defining E_v (see Eq. 5.28) at its reference level, $z = h + 2$ m. (Alternatively, we imagine the reference bare soil to be located at the effective transpiration elevation of the canopy with a common atmospheric reference level for both.) Under such hypothetical circumstances the numerator of the bare soil Eq. 5.20 will approximate that of the plant canopy Eq. 5.28.[†] Equations 5.20 and 5.28 may then be combined to define the *canopy conductance*[‡]

$$k_v \equiv \frac{E_v}{E_{ps}} = \frac{\Delta + \gamma_o}{\Delta + \gamma_o\left(1 + \frac{r_c}{r_a}\right)} = \frac{1}{1 + \frac{r_c/r_a}{1 + \Delta/\gamma_o}}, \tag{5.29}$$

which is less than unity.

In this general case, the resistance ratio, r_c/r_a, will depend upon the stomatal resistance, the crown structure, and the canopy density, M, as we will see in Chapter 8. Eq. 5.29 is attributed to Monteith (1965, p. 220) and has also been cited by Jarvis *et al.* (1976, p. 212) and Shuttleworth (1979, p. 55) among others. The canopy conductance, k_v, and the canopy cover, M, are the two state variables we will use to define the canopy of a given species in the water balance equations of Chapter 6. In Chapter 7 we will add a third state variable, the long-term spatial average canopy temperature, T_l.

Empirical estimates of k_v abound in the literature for various tree and agricultural plant species (e.g., Doorenbos and Pruitt, 1977). It takes on values $0 \le k_v < 1$, and may vary over its full range in a single day according to the many environmental influences on r_a, r_c, Δ, and γ_o. Since some of these influences are poorly understood, and many are difficult to quantify, agriculturalists often use a single value of k_v, time-averaged over the growing season, to reflect the relative water use of a given species at maturity. We follow that simplification here.

[†] Differences in surface albedo, emissivity, and shear stress (see Eqs. B.3, B.4, and B.11) will weaken this approximation.

[‡] This is also called the *transpiration efficiency* (Eagleson, 1978d), or *crop coefficient* (Doorenbos and Pruitt, 1977; Shuttleworth, 1979), and others have used the term *plant factor*.

For the special case of open stomates, E_v will take on its maximum value, $E_v = E_{pv}$, where E_{pv} = *potential canopy transpiration*, whereupon k_v has its maximum value ($k_v < 1$), called the *potential canopy conductance*

$$k_v^* \equiv \frac{E_{pv}}{E_{ps}}. \tag{5.30}$$

It is important to recognize the difference between Eq. 5.29 as derived here and the same equation as it commonly appears in the literature:

1. Here, with the stomates assumed always fully open, r_c is dominated by the *interleaf layer atmospheric resistance*, r_i (see Chapter 4), which doesn't vary with canopy water supply. In particular, r_c doesn't vanish identically when the canopy water supply is unlimited. Therefore, in this work the canopy conductance takes on the maximum, open stomates value, $k_v \equiv k_v^*$, for the given species. Note in particular that with r_c dominated by the interleaf atmospheric resistance, the rate of canopy evapotranspiration is approximately the same when the exterior leaf surfaces are wet (such as after rain) as it is when the wet evaporating surface is the interior lining of the stomatal cavity (dry leaf stressless transpiration). The condition $k_v^* < 1$ reflects that canopy–atmosphere control of the energy exchange discussed in the previous section.
2. In the literature r_c is assumed to be dominated by the effective canopy *stomatal* resistance. Equation 5.30 is obtained in such cases by normalizing the E_v of Eq. 5.28 with a rate given by Eq. 5.28 in which the plant canopy has saturated leaf surfaces so that this stomatal resistance and hence r_c vanish. In such cases $k_v \to 1$ as the water supply increases and the canopy (i.e., stomatal) resistance vanishes.

Isothermal evaporation in *e*–T space: (the Priestley–Taylor equation)

As seen above, the isothermal condition, $\boldsymbol{R}_b = 0$, provides a baseline for the estimation of evapotranspiration from real surfaces, and we can benefit from its analysis in *e*–*T* space as we did earlier (see Fig. 5.2) for the general case. When $T_1 = T_A$, Fig. 5.3 shows this schematic of isothermal evaporation from a wet surface under the continued assumption that the net radiation is the only heat available (i.e., $Q \approx R_n$). Note that in this special case the triangles ABD and CBD share a common side. That is

$$\frac{\rho c_p D_{12}/r_a}{\text{BD}} = \gamma_o, \tag{5.31}$$

and

$$\frac{\Delta R_n}{\Delta + \gamma_o} \Big/ \text{BD} = \Delta. \tag{5.32}$$

Plate 5.5. Jeffrey pine. (Photographed by William D. Rich in Arnold Arboretum, Boston; Copyright © 2001 William D. Rich.)

Eliminating BD between Eqs. 5.31 and 5.32 gives

$$\frac{\rho c_p D_{12}}{r_a} = \frac{\gamma_o}{\Delta} \cdot \frac{\Delta R_n}{\Delta + \gamma_o}, \tag{5.33}$$

which is substituted into Eq. 5.18 to regain Eq. 5.4 for isothermal evaporation from bare wet soil. That is, $E_T \equiv E_{ps}$, giving

$$\lambda E_{ps} = \left[1 + \frac{\gamma_o}{\Delta}\right] \frac{\Delta R_n}{\Delta + \gamma_o} \equiv R_n, \tag{5.34}$$

which demonstrates that without the addition of sensible heat transferred from the atmosphere, that is as long as $T_1 < T_3$, only a fraction of the available energy, R_n, can be converted into latent heat. This fraction is unity when the atmosphere and the surface are at the same temperature or when the atmosphere is warmer than the surface. Equation 5.34 may be written, albeit less revealingly, as

$$\lambda E_{ps} = \alpha \frac{\Delta R_n}{\Delta + \gamma_o}, \tag{5.35}$$

where, in the isothermal case,

$$\alpha = 1 + \frac{\gamma_o}{\Delta}. \tag{5.36}$$

Briscoe (1969) was one of the first to demonstrate Eq. 5.35 empirically by showing that *transpiration* is also closely proportional to the available energy in studies of barley. Without noting its implication of isothermality, Priestley and Taylor (1972) demonstrated empirically that for open water and saturated soil surfaces, the potential evaporation rate is proportional to available energy through Eq. 5.35 in which the so-called Priestley–Taylor coefficient, α, was presented as an empirical coefficient having the value $\alpha = 1.26$ for these surfaces. A collection of observations and theoretical estimates of α from the literature is summarized here in Table 5.2.

The relationship between Bowen ratio and canopy conductance

We use the isothermal state, $\boldsymbol{R}_b = 0$, as the base upon which to build a second approximation transpiration model which incorporates the true, $\frac{\gamma_o}{\Delta} \geq \boldsymbol{R}_b > 0$, condition. To do this we substitute Eq. 5.35 into Eq. 5.5 obtaining, for positive H,

$$\alpha = \left(1 + \frac{\gamma_o}{\Delta}\right)(1 + \boldsymbol{R}_b)^{-1}, \tag{5.37}$$

which was first proposed by Kim and Entekhabi (1997). But how do we estimate the Bowen ratio a priori?

Returning to Fig. 5.3 we note again that whenever $T_1 < T_3$ some of the net solar radiation is not used in evaporation. In the absence of energy storage and lateral advection, the excess energy is shed back to the atmosphere through sensible heat transfer. Given that there is unlimited liquid water at the surface, this departure from isothermality must be due to flux impedance in the water vapor path between surface and atmosphere (i.e., screen height). For bare soil surfaces this path involves only the atmospheric boundary layer over the 2-meter screen height so the impedance is small. However, for vegetal canopies the vapor flux impedance can be high as we have seen by the departure of the maximum canopy conductance, k_v^*, from unity (Eq. 5.30). Invoking Reynolds similarity between the fluxes of mass and heat, we expect the Bowen ratio to be greater than zero to a degree that is proportional to the difference of k_v^* from unity (see the discussion following Eq. 5.22). In a successive approximation we use Eqs. 5.30 and 5.34 in Eq. 5.5 to get the estimators

$$\boldsymbol{R}_\mathrm{b} \equiv \frac{H}{\lambda E_\mathrm{v}} = \frac{R_\mathrm{n}}{\lambda E_\mathrm{v}} - 1 \approx \frac{\lambda E_\mathrm{ps}}{\lambda E_\mathrm{pv}} - 1 = \frac{1}{k_\mathrm{v}^*} - 1, \quad +H \tag{5.38}$$

and

$$\boldsymbol{R}_\mathrm{b} \equiv \frac{H}{\lambda E_\mathrm{v}} = 1 - \frac{R_\mathrm{n}}{\lambda E_\mathrm{v}} \approx 1 - \frac{\lambda E_\mathrm{ps}}{\lambda E_\mathrm{pv}} = 1 - \frac{1}{k_\mathrm{v}^*}, \quad -H. \tag{5.39}$$

Equation 5.38 is compared with observations in a later section of this chapter.

Table 5.2. *Values of the Priestley–Taylor coefficient*

Surface conditions	α	Reference
Observations		
Pine forest		
Dry (Norfolk, England)	0.74	Shuttleworth and Calder (1979)
Wet (Norfolk, England)	9.69	Shuttleworth and Calder (1979)
Wet	1.50	Gash and Stewart (1975)
Sitka and Norway spruce		
Dry (Central Wales)	0.64	Shuttleworth and Calder (1979)
Wet (Central Wales)	1.50	Shuttleworth and Calder (1979)
Bush (wet soil)	1.33	Priestley and Taylor (1972)
Pasture crop (wet soil)	1.34	Priestley and Taylor (1972)
Snap bean crop (wet soil)	1.30	Priestley and Taylor (1972)
Unstressed agricultural crops and grasses	1.26	Davies and Allen (1973)
Plowed bare wet soil	1.08	Priestley and Taylor (1972)
Indian Ocean (CSIRO)	1.26	Priestley and Taylor (1972)
No advection	1.25	Priestley and Taylor (1972)
Indian Ocean (CSIRO)	1.30	Priestley and Taylor (1972)
No advection	1.31	Priestley and Taylor (1972)
Indian Ocean (University of Washington)	1.20	Priestley and Taylor (1972)
Lake Eucumbene	1.25	Priestley and Taylor (1972)
Equatorial Atlantic Ocean	1.30	Priestley and Taylor (1972)
Theory		
Forest	3.62	Lhomme (1997)
Grass	1.26	Lhomme (1997)
Unsaturated surface	1	McNaughton (1976), Perrier (1982), Brutsaert (1982)
Moderately dry surface	<1	
Realistic atmospheric conditions, growing convective boundary layer	<1.46; >1.08	Culf (1994)
Moderate weather in the Netherlands	slightly >1.26	McNaughton and Spriggs (1989)

The Priestley–Taylor coefficient

The observed values of α associated with the $\boldsymbol{R}_b$ observations of Priestley and Taylor (1972) (cf. Fig. 5.4) are presented as the circular plotted points in Fig. 5.5 where they are bounded by the limiting Bowen ratios: $\boldsymbol{R}_b = 0$ corresponding to $\alpha = 1 + \gamma_o/\Delta$, and $\boldsymbol{R}_b = \gamma_o/\Delta$, corresponding to $\alpha = 1$. The solid line is the theoretical value for unstressed canopies as given by Eqs. 5.29 and 5.38 for the typical value $r_c/r_a = 0.5$ (cf. Chapter 8), and is seen to agree fairly well with the only vegetal observations (points no. 5 and no. 6).

Shuttleworth and Calder (1979) and Gash and Stewart (1975) present observations of α (cf. Table 5.2) for wet pine and wet spruce forests in Britain. The average July temperature in Britain is estimated (Trewartha, 1954, Fig. 1.25, p. 35) as 60 °F (15.5 °C) which allows adding these points to Fig. 5.5. Two of the three tall, wet canopies are seen to approach the isothermal condition as was reasoned earlier (Eq. 5.23) for elevated saturated surfaces, while the third observation is well out of the plotted range.

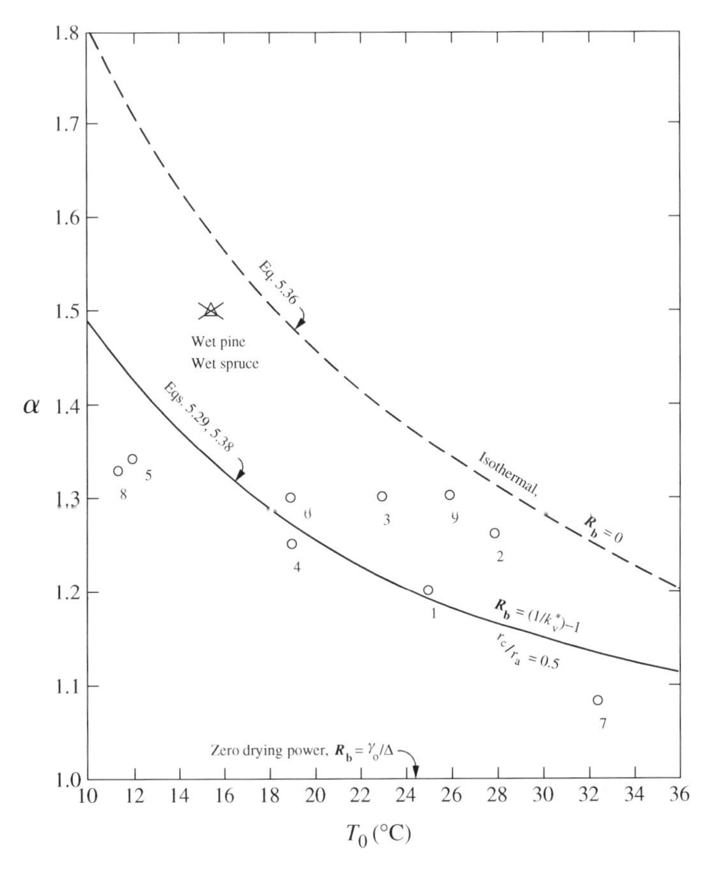

Fig. 5.5. Priestley–Taylor coefficients. Numbers refer to caption of Fig. 5.4. ○, Priestley and Taylor (1972); △, Shuttleworth and Calder (1979); ×, Gash and Stewart (1975).

Observed Bowen ratios in canopy transpiration

From water balance studies, Hamon (1961) determined the total annual evapotranspiration, E_T, for a variety of natural watersheds. In the interest of estimating transpiration, E_v, we disregard those of his watersheds experiencing ice and/or snow to obtain the reduced list summarized here in Table 5.3. For these six watersheds we assume $E_T \approx E_v = E_{pv}$. Also listed in Table 5.3 for each watershed are assumed surface albedos, along with the *mean annual* values of screen height atmospheric temperature, T_0; fractional cloud cover, N; and saturation ratio, S_r; all as given by the National Weather Service Climatological Summary at the nearest station. Using the material of Appendix B, these parameters permit estimation of the *mean annual* net radiation flux, R_n, which is also tabulated. Comparing paired values of E_v and R_n/λ for each of the six watersheds gives an estimate of the Bowen ratio, $\boldsymbol{R}_b$, using Eq. 5.5.† As is shown in Table 5.3, these values are all positive and range from $\boldsymbol{R}_b = 0.22$ to 0.33 for the desert shrubs of New Mexico and Texas, to $\boldsymbol{R}_b = 0.10$ to 0.17 in humid Georgia, Kentucky, and Missouri.

Using the method of Appendix B, we can estimate the free water surface evaporation, E_{ps}, for the six selected catchments of Hamon (1961). These values are also listed in Table 5.3, and in conjunction with $E_{pv} \approx E_T$, they provide an estimate of $k_v = k_v^*$ through Eq. 5.30. The pairs of k_v^* and $\boldsymbol{R}_b$ so calculated are very sensitive to the shortwave albedo of the surface. Here we have assumed the surfaces to be half grass and half forest and have used average albedo values for each as given by Kondratyev (1969, Table 7.5, p. 425). The resulting estimates are listed in the last two columns of Table 5.3 and are plotted in comparison with Eq. 5.38 (positive H) in Fig. 5.6.

E The canopy–atmosphere temperature difference

Using the flux–gradient relationships Eqs. 4.115 and 4.117, and invoking the Reynolds analogy, the definition of $\boldsymbol{R}_b$ gives

$$\boldsymbol{R}_b \equiv \frac{H}{\lambda E_v} = \frac{\rho c_p K_h \Delta T/\Delta z}{\lambda \rho K_v \Delta q_v/\Delta z} \approx \frac{c_p}{\lambda} \cdot \frac{\Delta T}{\Delta q_v}, \tag{5.40}$$

where q_v is the specific humidity given by the thermodynamic approximation (Solot, 1939)

$$q_v \approx 0.622\frac{e}{p}. \tag{5.41}$$

Multiplying and dividing by the saturation vapor pressure, e_s, allows us to use Eq. 5.38 with Eq. 5.40 to obtain the temperature difference between canopy and

† Note that $\boldsymbol{R}_b$ is a ratio of flux rates. Therefore in this calculation the observed gross annual evaporation is doubled to obtain an estimate of the evaporation rate during daylight hours for comparison with R_n.

Table 5.3. *Estimation of potential transpiration for vegetated surfaces (negligible snow and ice)*

Site locations[a]	Φ(°N)	I_0[b] (ly min^{-1})	ρ_T[c]	T_0[d](°C)	N[d]	S_r[d]	E_T[a](Obs.) (m y^{-1})	q_i[e] (ly min^{-1})	q_b[f] (ly min^{-1})	R_n[g] (ly min^{-1})	D_p[h] (ly min^{-1})	$1 + \frac{\gamma_0}{\Delta}$[i]	E_{ps}[j] (calc.) (m y^{-1})	k_v^*[k]	$\boldsymbol{R}_b$[l]
10 Mesilla, NM	32.3	0.329	0.18	14.6	0.18	0.53	0.86	0.270	0.125	0.145	0.069	1.63	1.15	0.75	0.23
11 Pecos, NM	35.6	0.316	0.18	13.8	0.18	0.45	0.90	0.259	0.126	0.133	0.081	1.65	1.14	0.79	0.33
12 Green River, KY	37.9	0.307	0.18	13.1	0.46	0.73	0.80	0.252	0.093	0.159	0.029	1.68	0.98	0.82	0.10
13 Tallapoosa River, GA	32.5	0.328	0.18	15.4	0.36	0.77	0.84	0.269	0.103	0.166	0.028	1.60	1.06	0.79	0.10
14 West Fork of the White River, MO	37.0	0.309	0.18	13.4	0.33	0.72	0.79	0.253	0.109	0.144	0.037	1.67	0.95	0.83	0.17
15 Cyprus Creek, TX	32.0	0.329	0.18	17.9	0.20	0.75	0.92	0.270	0.113	0.157	0.033	1.52	1.09	0.84	0.22

[a] Hamon (1961).

[b] Average of observations, Fig. B.2.

[c] Kondratyev (1969, Table 7.5, p. 425); no attempt made to distinguish variations: assumed for all sites: 50% grass @ $\rho_T = 0.22$ and 50% forest @ $\rho_T = 0.14$.

[d] U.S. Weather Service Annual Climatological Summary, 1941–70 (at nearest station).

[e] $q_i = (1 - \rho_T)I_0$.

[f] Eq. B.12.

[g] $R_n = q_i - q_b$.

[h] Eq. B.18.

[i] Table 5.1.

[j] Eq. B.1.

[k] Eq. 5.30 with the assumption that $E_{pv} \approx E_T$.

[l] Eq. 5.5, using twice the observed annual E_T as an estimate of the daylight-hour rate.

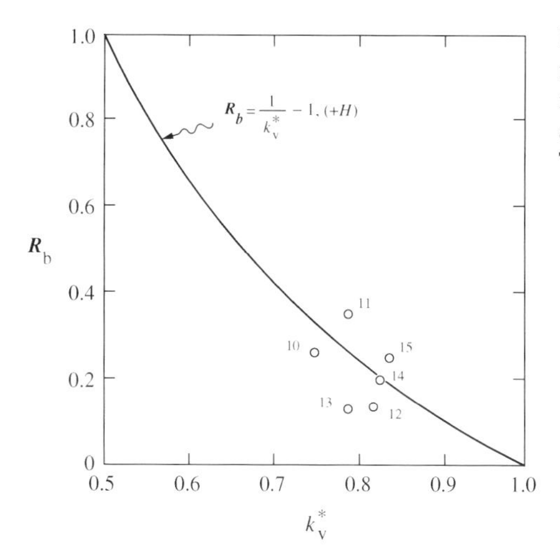

Fig. 5.6. The Bowen ratio–canopy conductance relation. Daylight-hour average; numbers refer to Table 5.3.

atmosphere

$$\Delta T \approx 0.622 \frac{\lambda}{c_p} \cdot \frac{e_s}{p} \cdot \Delta S_r \cdot \boldsymbol{R}_b \tag{5.42}$$

with

$$\lambda \approx 597.3 \text{ cal g}^{-1},$$
$$c_p \approx 0.24 \text{ cal g}^{-1}\text{K}^{-1},$$
$$p \approx O(1000 \text{ mb}), \text{ and}$$
$$e_s \approx O(10 \text{ mb}),$$

Eq. 5.42 gives the estimator

$$\Delta T \approx 15 \Delta S_r \boldsymbol{R}_b, \quad (\text{K}). \tag{5.43}$$

Under normal conditions, $\Delta S_r = O(10^{-1})$, so with $\boldsymbol{R}_b = O(10^{-1})$ as was found in the previous section, Eq. 5.43 gives $\Delta T = O(10^{-1})$, and the system is closely isothermal. In this work we will make this approximation, assuming the long-term average "big leaf" canopy temperature, T_l, equals the ambient (i.e., screen height) atmospheric temperature, T_0. That is

$$T_l \approx T_0. \tag{5.44}$$

Gates *et al.* (1968) show that $T_l - T_0 \leq 2$–3 °C for many small leaves. Monteith (1973, p. 183) notes that "in bright sunshine and in a breeze, small leaves are expected to be only 1–2 °C hotter than the surrounding air. Greater excess temperatures are observed on very large leaves in a light wind. . .". Measurements within a spruce canopy bear this out (Larcher, 1983, Fig. 2.16, p. 27). They show a temperature difference, $T_l - T_0 \approx -1$ °C with averages taken over a typical midsummer day, while finding for a single

Canna leaf that this difference is about +1 °C with high wind but rises to about +20 °C for low wind. (Note that compared to a single leaf, the long-term canopy average T_l is much less sensitive to wind, and the canopy T_l is much smaller than for a single leaf.)

Looking at Eq. 5.42 from another direction, we can use the observed small temperature gradients over forests (cf. Jarvis *et al.*, 1976, Table IX, p. 204) to obtain $\Delta T = O(10^{-1})$, and with $\Delta S_r = O(10^{-1})$ as before, we again obtain $\boldsymbol{R}_b = O(10^{-1})$. This is in contrast to the much higher values, $\boldsymbol{R}_b > 1$, calculated by Jarvis *et al.* (1976, Fig. 10, p. 209) using much larger values of r_c/r_a than those derived here in Chapter 7.

Plate 5.6. Siberian larch. Author alongside larch in Arnold Arboretum, Boston. (Photograph by Beverly G. Eagleson.)

F Relevance to nutrient and carbon flux

Mineral nutrients, particularly nitrogen, calcium, potassium, and phosphorus (cf. Bormann and Likens, 1979, Table 2-7, p. 69), enter the plant dissolved in the soil water. If the soil imposes no limit on their quantity or on their solution rate, the nutrient flux is proportional to the through-flow of water in the average transpiration stream, E_v. Nutrient availability may be a growth-limiting factor as has been shown for Swedish forests (Aronsson and Elowson, 1980; Jensén and Pettersson, 1980). Therefore, maximization of E_v (through maximization of k_v^*) in a nutrient-poor situation will insure that this productivity is the constrained maximum.

The canopy resistance to outgoing moisture flux represented in the control of E_v by k_v^* operates identically to control the influx of carbon dioxide. Thus, realization of maximum canopy dry matter productivity when the latter is limited by carbon dioxide supply also requires maximum k_v^* and hence minimum canopy impedance, r_c/r_a. We return to this issue in later chapters, particularly in Chapter 10.

6

Water balance

A statistical–dynamic model of the average canopy water balance during the vegetation growing season is presented in terms of the three state variables: average root-zone soil moisture, canopy cover, and canopy conductance. Approximate limiting solutions are offered.

An adopted *stress tolerance criterion* puts the plant in incipient stress at the end of the average growing season, and the associated *critical moisture state* allows expressing the maximum canopy moisture flux in terms of climate and vegetation properties. The critical moisture state is one of maximum productivity for the given species and climate since the stomates are fully open during the entire growing season, and since the flux of water-borne nutrients is then a climatic maximum. Detailed observations of three semi-arid forests confirm that the trees use all the water available during the growing season.

A Introduction[†]

At land surfaces the soil column responds dynamically to the climatic sequence of precipitation and evapotranspiration events. It accepts part of the moisture during periods of precipitation, pumps some of this back to the surface during evaporative periods, and rejects the remainder to the water table more or less continuously. This surface moisture exchange and the accompanying surface heat exchange, to a large degree, depend critically upon the physical properties of the soil and vegetation as well as upon the weather conditions during the alternate periods of precipitation and evapotranspiration. The quantitative relation among the long-term averages of this partition of precipitation is called the *water balance*.

We seek here to express this climate–land surface coupling in a simplified form which provides insight into the physical basis for the role played by water in the growth

[†] Much of the material of this chapter is taken from Eagleson (1978a through 1978g) incorporating the corrections and extensions provided in the MIT Master of Science thesis of Salvucci (1992). The latter work is also available in an MIT Laboratory Report by Salvucci and Eagleson (1992). This summary does not contain all the details of the assumptions and approximations which are available in the original works (Eagleson, 1978a–g; Salvucci and Eagleson, 1992).

of vegetation communities. Such insight can come only through retention in our model of the underlying physical determinism. At the same time however, uncertainty plays such a large role in the climatic forcing of the water balance at its critical time-scales that our approach must deal also with probability distributions. This is in contrast to the approach taken in describing the radiation forcing in Chapter 3 where variability at the time-scales of vegetation growth is dominated by deterministic planetary motions.

Analytical solutions, even approximate ones, are preferred to precise numerical solutions whenever the objective is behavioral insight, and this model is structured accordingly. Only an outline is presented here; details are available elsewhere (Eagleson, 1978b–g; Salvucci, 1992; Salvucci and Eagleson, 1992).

Plate 6.1. Giant *Ficus* tree exploits water source. At Ta Prom, Angkor, Cambodia, spong tree (*Ficus tetrameles nudiflora*) sends spreading roots down wall to exploit soil moisture in unvegetated courtyard. (Photograph by Peter S. Eagleson.)

B Model framework

The water balance model (Eagleson, 1978a) is a one-dimensional model of soil moisture dynamics forced by a stochastic climate. The model solves the equilibrium moist-season water balance in terms of a soil moisture which is averaged temporally over climate scale (so that storage changes may be neglected), and is averaged vertically over the vegetation rooting depth, z_r, (on the order of 1 meter). The resulting "equilibrium" space–time average soil moisture concentration, s_o, is the state variable determining the average flux rates of moisture into and out of the near-surface soil. Multiplied by a common time, season or year, these average flux rates become depths (per unit of suface area) and must sum to zero. This sum is the *water balance equation* and is commonly written on an average annual basis as

$$P_A = E_{TA} + R_{sA} + R_{gA}, \tag{6.1}$$

where

P_A = average annual precipitation (cm),
E_{TA} = average annual evapotranspiration (cm),
R_{sA} = average annual surface runoff (cm), and
R_{gA} = average annual groundwater runoff (cm).

The right-hand side of Eq. 6.1 is controlled by a complex interaction of climate, soil, and vegetation. The dynamics of soil moisture in the near-surface soil column are governed by the Richards (1931) concentration-dependent diffusion equation. Eagleson (1978c) modified Philip's (1957, 1960, 1969) approximate analytical solution to the Richards equation to incorporate a distributed vegetal root sink and introduced the Brooks and Corey (1966) model of unsaturated soil properties. He then averaged the solutions over the ensemble of surface boundary conditions arising from a Poisson arrival of rainstorms in the form of rectangular pulses as is shown in the upper part of Fig. 6.1. During the interstorm periods, the bare soil surface is subject to a constant potential evaporation rate, E_{ps}, and the vegetation to a canopy-referenced constant potential evaporation rate, E_{pv}, causing transpiration at the species-dependent rate, $k_v E_{ps}$ (cf. Eq. 5.29). With this forcing, the expected value of the stormscale surface fluxes: storm infiltration, $E[I_j]$; interstorm exfiltration, $E[E_{sj}]$; and interstorm

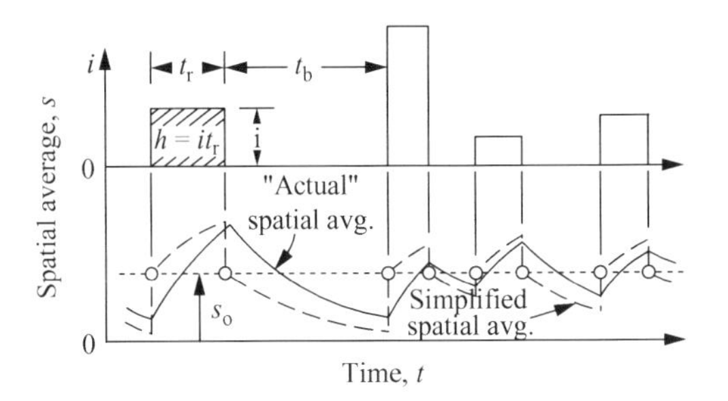

Fig. 6.1. Precipitation and soil moisture representation. (Eagleson, 1978a, Fig. 7, p. 708; with permission of the American Geophysical Union.)

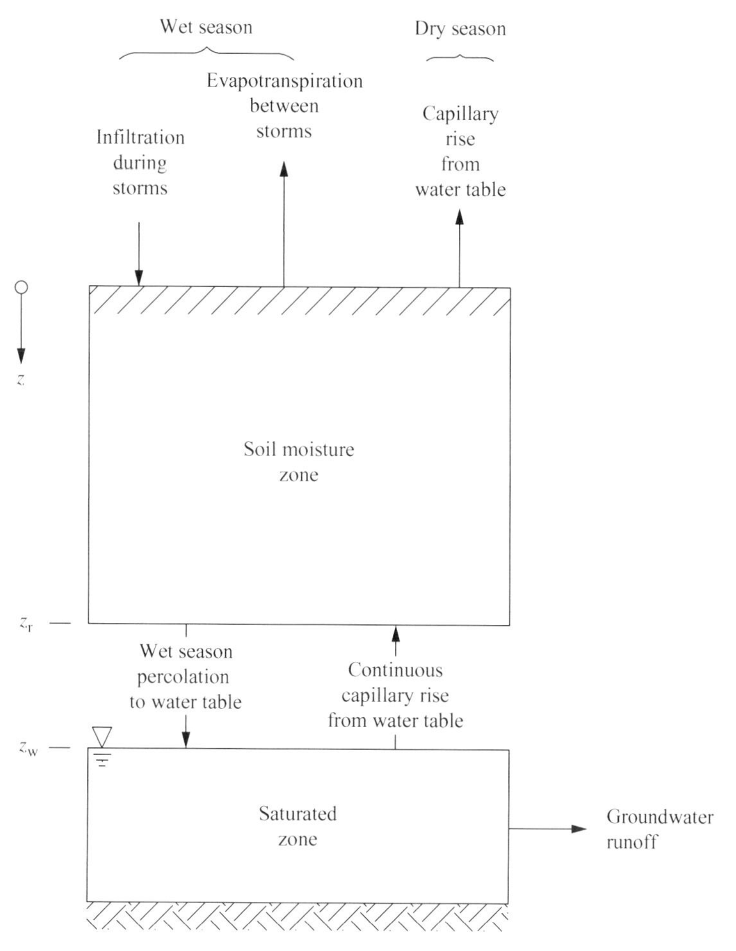

Fig. 6.2. Moisture fluxes in soil column. (Eagleson, 1978a, Fig. 6, p. 708; with permission of the American Geophysical Union.)

transpiration, $E[E_{vj}]$, are found using the simplifying assumption that the spatial average soil moisture concentration, s, is at its equilibrium value, s_0, at the beginning of each storm and interstorm period. A qualitative comparison between the simplified and actual variability of s is sketched in the lower part of Fig. 6.1.

The soil moisture dynamics at the bottom boundary of the near-surface soil column are predicated upon the restrictive assumption of a deep water table that does not influence the transient moisture dynamics of the upper soil layer. With this assumption, the vertical diffusive processes throughout the soil profile are assumed to attenuate the random near-surface variations so that at the lower boundary of the soil moisture zone the soil moisture is at steady state yielding a constant percolation, v, down to the water table during the wet season and a constant capillary rise, w, from the water table to the surface year around. Long-term mass fluxes into and out of the saturated zone are balanced by lateral groundwater runoff. These fluxes are illustrated in the sketch of the soil column shown as Fig. 6.2.

Following is a summary of the construction of the long-term average water balance from these assumed short-term behaviors.

C Precipitation

Point precipitation is represented by Eagleson (1978b) as Poisson arrivals of rectangular intensity pulses that have random depth and duration as is illustrated in the upper portion of Fig. 6.1. The probability density functions (i.e., pdfs) of these variables are incomplete gamma, $\gamma^*(\kappa_o, \lambda_o)$, for the storm depth, h, giving

$$f_h(h) = \gamma^*(\kappa_o, \lambda_o) \equiv \frac{\lambda_o(\lambda_o h)^{\kappa_o - 1} e^{-\lambda_o h}}{\Gamma(\kappa_o)}, \tag{6.2}$$

in which κ_o = shape parameter, or index of the distribution (dimensionless),
λ_o = scale parameter of the distribution (cm^{-1}),
$\Gamma(\cdot)$ = factorial gamma function (see Table 6.1a),

and the mean and variance are, repectively

$$m_h = \kappa_o/\lambda_o, \tag{6.3}$$

and

$$\sigma_h^2 = \kappa_o/\lambda_o^2; \tag{6.4}$$

and exponential for the storm duration, t_r,

$$f_{t_r}(t_r) = \delta e^{-\delta t_r}, \tag{6.5}$$

for which the mean and variance are, respectively

$$m_{t_r} = 1/\delta, \tag{6.6}$$

and

$$\sigma_{t_r}^2 = 1/\delta^2. \tag{6.7}$$

The potential bare soil evaporation rate is assumed to have the same constant value throughout all of the interstorm periods whose duration, t_b, is also exponentially distributed,

$$f_{t_b}(t_b) = \alpha e^{-\alpha t_b}, \tag{6.8}$$

with mean

$$m_{t_b} = 1/\alpha, \tag{6.9}$$

and variance

$$\sigma_{t_b}^2 = 1/\alpha^2. \tag{6.10}$$

The storm intensiy, i, is assumed to be exponentially distributed as well,

$$f_i(i) = \omega\, e^{-\omega i}, \tag{6.11}$$

with mean

$$m_i = 1/\omega, \tag{6.12}$$

and variance

$$\sigma_i^2 = 1/\omega^2. \tag{6.13}$$

The storm arrival rate is m_ν, storms per year (or season), so that the mean of the annual (or seasonal) precipitation, P, is simply

$$P_A \equiv m_P = m_\nu m_h, \tag{6.14}$$

and finally, the variance of the annual (or seasonal) precipitation, P, is (Eagleson, 1978b)

$$\sigma_P^2 = m_P{}^2\left(1 + \frac{1}{\kappa_o}\right)/m_\nu. \tag{6.15}$$

See Appendix F for the monthly climatology of these and other storm parameters at 74 first-order weather stations in the continental United States as determined by Hawk and Eagleson (1992).

Plate 6.2. Weeping beech. Author alongside beech in Arnold Arboretum, Boston. (Photograph by Beverly G. Eagleson.)

D Storm infiltration and surface runoff

In general, surface runoff can be caused by one of two mechanisms: (1) the rate of precipitation supply at the surface exceeds the soil moisture-dependent *infiltration capacity* of the soil column, or (2) the moisture storage capacity of the soil column has been filled to the soil surface. The first of these processes is called "Hortonian" runoff (Horton, 1933) and is found for high rainfall intensities with relatively deep water tables. Its calculation is independent of the water table elevation as long as the latter is "deep enough". The second process is often called "Dunne" runoff (Dunne and Black, 1970) and is found with shallow water tables and low rainfall intensity. Because the latter process requires knowledge of the water table elevation (cf. Salvucci, 1993), we omit its consideration here and assume all surface runoff to be Hortonian. This will lead to our underestimation of surface runoff in wet climates.

When the bare soil surface is saturated, the soil can infiltrate moisture at the maximum rate. We call this rate the *infiltration capacity*, f_i^*, and it is written by Philip (1957, 1960, 1969) as

$$f_i^* = \frac{1}{2} S_i t^{-1/2} + A_0, \tag{6.16}$$

in which Eagleson (1978e) sets the modified gravitational term to be

$$A_0 = \frac{1}{2} K(1) \left(1 + s_o^c\right) - w, \tag{6.17}$$

and gives the *infiltration sorptivity*, S_i, as

$$S_i = 2(1 - s_o)\{[5 n_e K(1)\, \psi(1)\, \phi_i(d, s_o)]/3\,\mathrm{m}\,\pi\}^{1/2}, \tag{6.18}$$

in which the dimensionless *sorption diffusivity*, $\phi_i(d, s_o)$, is defined as

$$\phi_i(d, s_o) = (1 - s_o)^d \left\{ \frac{1}{d + 5/3} + \sum_{n=1}^{d} \frac{1}{d + [(5/3) - n]} \binom{d}{n} \left(\frac{s_o}{1 - s_o}\right)^n \right\}, \tag{6.19}$$

and where

n_e = effective porosity of soil,
$K(1)$ = saturated effective hydraulic conductivity of soil, cm s^{-1},
$\psi(1)$ = saturated matrix potential of soil, cm (suction),
c = permeability index of soil,
d = diffusivity index of soil, and
m = pore size distribution index of soil.

Equation 6.19 is plotted in Fig. 6.3.

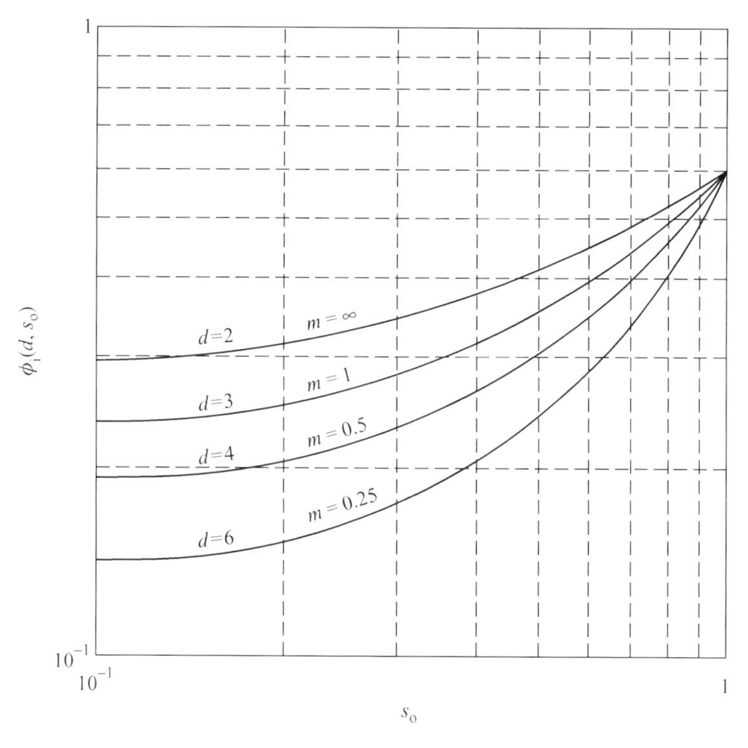

Fig. 6.3. Dimensionless sorption diffusivity; see Eq. 6.19. (Eagleson, 1978c, Fig. 9, p. 727; with permission of the American Geophysical Union.)

Remembering our restrictive assumption of a deep water table which prevents soil saturation from below, the process physics responsible for generation of storm runoff is as shown in Fig. 6.4 for storm durations, t_r, large enough such that all three distinct stages of infiltration are operative:

(1) in the first stage the *surface retention capacity*, h_o, is filled at rate i. This storage is isolated from the porous soil and so no infiltration takes place in time h_o/i;

(2) in the second stage infiltration takes place at rate i because the infiltration capacity, f_i^*, exceeds the rainfall intensity, i, and this continues until $f_i^* = i$ at time $t_o + h_o/i$; and

(3) in the third stage infiltration occurs at the declining rate f_i^* due to this rate being less than the rainfall intensity, i, and it continues, producing runoff at the rate $i - f_i^*$ until the storm ends at time t_r with a total *storm runoff* R_{sj}.

The average value of the storm surface runoff is obtained by taking the expected value of R_{sj} over the joint probability distribution of storm intensity, i, and duration, t_r. For mathematical convenience, Eagleson (1978e) made the simplifying assumption that i and t_r are independent random variables[†] so that with Eqs. 6.5 and 6.11, the

[†] Appendix F shows that in the continental United States $|\rho[i, t_r]| \leq O(10^{-1})$

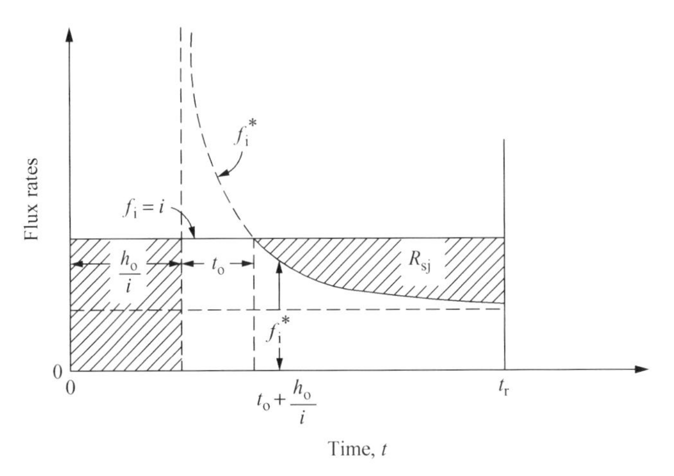

Fig. 6.4. Surface runoff generation during typical storm. (Eagleson, 1978e, Fig. 1, p. 742; with permission of the American Geophysical Union.)

desired joint pdf is simply

$$f_{i,t_r}(i, t_r) = f_i(i)\, f_{t_r}(t_r) = \omega\, \delta\, e^{-\omega i - \delta t_r}, \tag{6.20}$$

in which the mean storm depth appears in its alternate form

$$m_h = (\omega \delta)^{-1}. \tag{6.21}$$

With these definitions and assumptions, the expectation of *storm rainfall excess*, R_{sj}^*, is given by the sum of the two shaded areas of Fig. 6.4 (see Eagleson, 1978e, p. 745)

$$\frac{E[R_{sj}^*]}{m_h} = e^{-G-2\sigma}\Gamma(\sigma + 1)\sigma^{-\sigma} + \frac{E[E_r]}{m_h} \tag{6.22}$$

in which E_r is the *storm surface retention* depth,

$$G \equiv \omega K(1)\left[\frac{1 + s_o^c}{2} - \frac{w}{K(1)}\right], \tag{6.23}$$

$$\sigma \equiv \left[\frac{5 n_e \lambda_o^2 K(1)\psi(1)(1 - s_o)^2 \phi_i(d, s_o)}{6\pi\, \delta m \kappa_o^2}\right]^{1/3}, \tag{6.24}$$

and by definition of the gamma function (see Abramowitz and Stegun, 1964)

$$\Gamma(n + 1) \equiv n\Gamma(n) \equiv n!, \tag{6.25}$$

which is tabulated for $1 \leq n \leq 2$ in Table 6.1.

The *storm surface retention*, E_r, is supplied from rainfall excess, and we assume it to be evaporated during the following interstorm period, thus the expected value of

Table 6.1. *The gamma functions*

(a) Values of the gamma (factorial) function $\Gamma(n+1) = n\Gamma(n) = n!$

x	$\Gamma(x)$	x	$\Gamma(x)$	x	$\Gamma(x)$
1.00	1.000	1.34	0.892	1.68	0.905
1.01	0.994	1.35	0.891	1.69	0.907
1.02	0.989	1.36	0.890	1.70	0.909
1.03	0.984	1.37	0.889	1.71	0.911
1.04	0.978	1.38	0.889	1.72	0.913
1.05	0.974	1.39	0.888	1.73	0.915
1.06	0.969	1.40	0.887	1.74	0.917
1.07	0.964	1.41	0.887	1.75	0.919
1.08	0.960	1.42	0.886	1.76	0.921
1.09	0.955	1.43	0.886	1.77	0.924
1.10	0.951	1.44	0.886	1.78	0.926
1.11	0.947	1.45	0.886	1.79	0.929
1.12	0.944	1.46	0.886	1.80	0.931
1.13	0.940	1.47	0.886	1.81	0.934
1.14	0.936	1.48	0.886	1.82	0.937
1.15	0.933	1.49	0.886	1.83	0.940
1.16	0.930	1.50	0.886	1.84	0.943
1.17	0.927	1.51	0.887	1.85	0.946
1.18	0.924	1.52	0.887	1.86	0.949
1.19	0.921	1.53	0.888	1.87	0.952
1.20	0.918	1.54	0.888	1.88	0.955
1.21	0.916	1.55	0.889	1.89	0.958
1.22	0.913	1.56	0.890	1.90	0.962
1.23	0.911	1.57	0.890	1.91	0.965
1.24	0.909	1.58	0.891	1.92	0.969
1.25	0.906	1.59	0.892	1.93	0.972
1.26	0.904	1.60	0.894	1.94	0.976
1.27	0.903	1.61	0.895	1.95	0.980
1.28	0.901	1.62	0.896	1.96	0.984
1.29	0.899	1.63	0.897	1.97	0.988
1.30	0.897	1.64	0.899	1.98	0.992
1.31	0.896	1.65	0.900	1.99	0.996
1.32	0.895	1.66	0.902	2.00	1.000
1.33	0.893	1.67	0.903		

(b) The incomplete gamma function

$$\gamma(a, x) \equiv x^a e^{-x} \Gamma(a) \sum_{n=0}^{\infty} \frac{x^n}{\Gamma(a+n+1)} = x^a e^{-x} \sum_{n=0}^{\infty} \frac{x^n}{a(a+1)(a+2)\cdots(a+n)}$$

Source: Abramowitz and Stegun (1964, pp. 260–261).

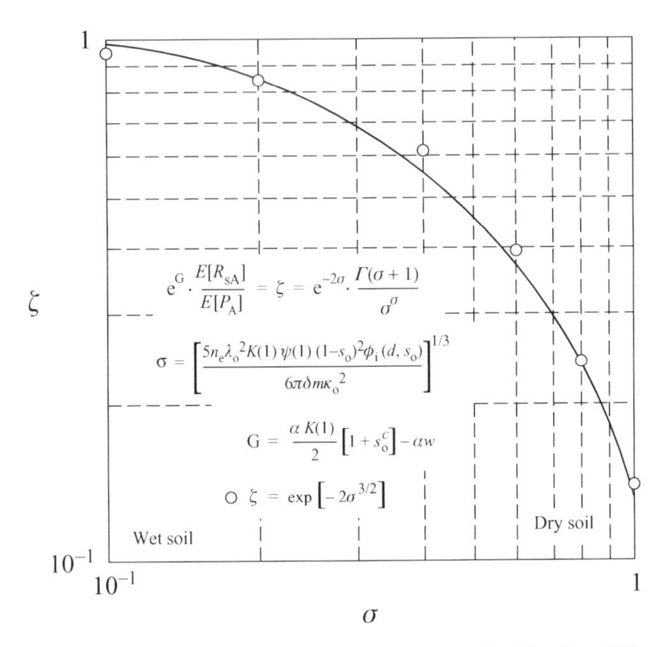

Fig. 6.5. Surface runoff function. (Eagleson, 1978e, Fig. 6, p. 746; with permission of the American Geophysical Union.)

the surface runoff, R_{sj}, is

$$E[R_{sj}] = E[R^*_{sj}] - E[E_r] = \omega^{-1}\delta^{-1}e^{-G-2\sigma}\Gamma(\sigma + 1)\sigma^{-\sigma}, \tag{6.26}$$

Equation 6.26 is presented graphically as the solid line in Fig. 6.5 where the plotted points represent the convenient approximation we call the *surface runoff function*,

$$\zeta(\sigma) \equiv \frac{\Gamma(\sigma + 1)}{\sigma^{\sigma}}e^{-2\sigma} \approx e^{-2\sigma^{3/2}}. \tag{6.27}$$

The *storm infiltration*, I_{sj}, is simply the storm depth minus the surface runoff and minus the surface retention. Averaging over all storms this becomes

$$E[I_{sj}] = E[h_j] - E[R^*_{sj}]. \tag{6.28}$$

E Potential (unstressed) transpiration

Immediately following a storm the soil is moist, the leaf stomates are fully open, and transpiration, E_v, begins at the maximum rate, E_{pv}, given by (cf. Eqs. 5.29 and 5.30)

$$E_v = E_{pv} = k^*_v E_{ps}, \tag{6.29}$$

where

E_v = long-term annual or seasonal average transpiration rate,
E_{pv} = long-term annual or seasonal average potential transpiration rate,
k_v^* = unstressed (i.e., maximum or potential) canopy conductance, and
E_{ps} = long-term annual or seasonal average potential rate of evaporation from the bare soil.

As deep percolation, interstorm evaporation, and transpiration reduce the transient soil moisture concentration, s, the soil moisture potential, $\psi_s(s)$, and the leaf moisture potential, $\psi_l(s)$, both fall in value (see Chapter 7). At the critical leaf moisture potential, $\psi_{lc}(s) \approx 10$ bar (Larcher, 1983, Fig. 3.46, p. 123), the leaf stomates begin to close, the resistance to moisture flux from the stomates rises, and transpiration begins to decline from the potential value. The critical soil moisture potential at which the decline in transpiration begins varies with the type of soil and with the structure of the tree connecting leaf and soil. According to Larcher (1983, p. 237), trees transpire to soil moisture potentials of approximately 15 bars negative pressure, while data from Brix (1962), Hinckley *et al.* (1975, 1978), Bunce *et al.* (1977), Havranek and Benecke (1978), and others, indicate that the transpiration rate of trees is variable only between soil moisture potentials of 5 and 15 bars negative pressure. We set the critical soil moisture potential marking the limit of potential transpiration at $\psi_{sc}(s) = 5$ bar (suction).

The equations of soil physics (e.g., Eagleson, 1978c) give

$$\psi_s = \psi_s(1)s^{-1/m}, \tag{6.30}$$

which, with the typical soil properties of Table 6.2, shows the critical soil moisture, $s_c \equiv s_5$, corresponding to $\psi_s = \psi_{sc} = 5$ bar to be: $s_5 = 0.16$ for clay; $s_5 < 0.003$

Table 6.2. *Hydraulic properties of soils*

	$k(1)$ (cm^2)	$K(1)$ ($cm\ d^{-1}$)	$\Psi_s(1)$ (cm suction)	n_e	m	c	d
(a) Typical soils[a]							
Clay	4×10^{-10}	2.94	90	0.45	0.44	7.5	4.3
Silty loam	4×10^{-9}	2.94	45	0.35	1.2	4.7	2.9
Sandy loam	4×10^{-8}	294	25	0.25	3.3	3.6	2.3
Sand	10^{-7}	743	15	0.20	5.4	3.4	2.2
(b) Beaver Creek soils[b]							
Springerville (watersheds 1–6)		17.2	4.5	0.46	0.48	7.1	4.1
Brolliar (watersheds 7–18)		15.1	92.8	0.37	0.71	5.8	3.4

[a]From Bras, R.L., *Hydrology*, Table 8.1, p. 352; Copyright © 1990 Addison-Wesley Publishing Co., Inc.; with kind permission from R.L. Bras.
[b]From Salvucci (1992).

for silty loam; $s_5 < 10^{-8}$ for sandy loam; and $s_5 < 10^{-14}$ for sand (Salvucci, 1992). Except perhaps for the clay soil, exhaustion of soil moisture *from* s_o to these small critical values at which stress appears will take a longer time, t_s, than all but the rarest of interstorm intervals, t_b. Considering the fact that *as modeled*, s_o lies at the midpoint of the interstorm soil moisture range, t_s is even larger and we conclude that transpiration will occur primarily at the stressless, potential rate. Such reasoning is fundamental to our basic assumption that natural selection fixes E_v, the expected value of interstorm transpiration, at the unstressed level for the given climate and soil. Using Eqs. 5.30 and 6.29, we have

$$E_v \equiv E[E_{vj}] = E_{pv}\, m_{t_b} = k_v^*\, E_{ps}\, \alpha^{-1}. \tag{6.31}$$

The time interval, m_{t_b}, is the mean evapotranspiration window over which the maximum expected values, $E[E_{sj}]$ and $E[E_{vj}]$ are calculated.

Salvucci (1992) pointed out that *in reality* the soil moisture may reach the critical value s_5 before the end of the interstorm period, whereupon the transpiration rate would be throttled by closure of the stomates, and $E[E_{vj}]$ could be much less than is given by Eq. 6.31. This can happen in arid and semi-arid climates for two reasons:

(1) large m_{t_b} will insure that during a fraction of the interstorm intervals the critical soil moisture is reached, and

(2) dry climates have large m_{t_b} and hence few storms (m_ν) annually or seasonally, but their storm depths have small distribution indices (κ_o), so that from Eq. 6.15 the variance of annual or seasonal precipitation is high, and thus the soil moisture at the start of an arbitrary interstorm period may be much less than the long-term average, s_o.

One of the more useful features of Eagleson's (1978f) *mean annual* water balance is its extension, for cases of small variance (σ_P^2), to the exploration of the *annual* water balance (Eagleson, 1978f). However, as might be expected, when this variance is not small the stress situation described above will occur and in a dry year Eagleson's (1978f) annual water budget cannot be balanced. Because the condition of incipient stress lies at the critical boundary between high productivity and disaster for the plant, it is important that our water balance formulation at least be robust there even if not terribly accurate. We will approach this as did Salvucci (1992) by incorporating the first of the above two possibilities (the simpler of the two and not necessarily the most important) in our analysis, but before doing this we must look at the bare soil evaporation.

F Bare soil evaporation

The process physics within a typical interstorm interval are illustrated in Fig. 6.6 for t_b large enough such that all three distinct stages of evaporation are operative:

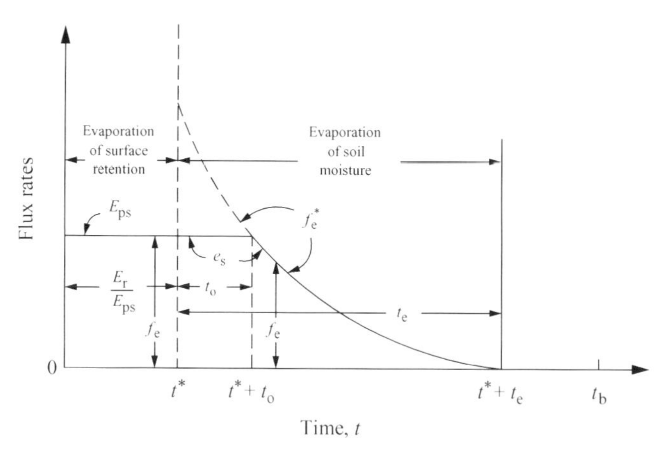

Fig. 6.6. Interstorm evaporation from bare soil. (Eagleson, 1978d, Fig. 2, p. 733; with permission of the American Geophysical Union.)

(1) the first stage is evaporation of the surface retention, E_r, which takes place at the surface-referenced bare soil potential rate, E_{ps}, until E_r is exhausted at time $t = t^*$;

(2) the second stage is evaporation of exfiltrated soil moisture at the potential rate, E_{ps}, due to this rate being less than the exfiltration capacity, f_e^* (defined below), and it continues until $E_{ps} = f_e^*$ at time $t = t^* + t_o$; and

(3) the third stage is evaporation of exfiltrated soil moisture at the declining rate f_e^* due to this rate being less than the potential rate, E_{ps}, and it continues until $f_e^* = 0$ and evaporation ceases at time $t = t^* + t_e$.

Eagleson (1978c) extended the infiltration equation of Philip (1969) to represent the rate at which the soil can return moisture to the surface. We call this rate the *exfiltration capacity*, f_e^*, of the bare soil, and it is given by

$$f_e^* = \frac{1}{2} S_e t^{-1/2} - M k_v^* E_{ps} + w, \tag{6.32}$$

in which, with f_e^*, E_{ps}, and w in cm s^{-1}, and the time, t, in seconds, the *exfiltration sorptivity*, S_e, is (Eagleson, 1978d)

$$S_e = 2 s_o^{1+d/2} \left[\frac{n_e K(1) \psi(1) \phi_e(d)}{\pi m} \right]^{1/2}, \tag{6.33}$$

where

$\phi_e(d) =$ *dimensionless desorption diffusivity* of soil.

Eagleson (1978c) reduced the number of independent soil parameters by using the results of Burdine (1958) and Brooks and Corey (1966) to write

$$d = \frac{c+1}{2} = c - (1/m) - 1 = 2 + 1/m, \tag{6.34}$$

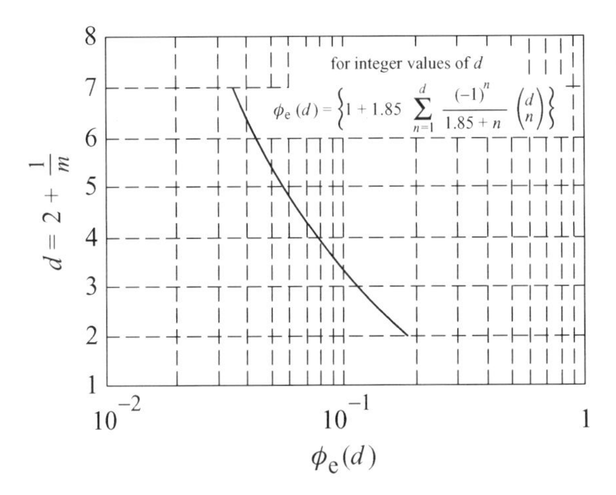

Fig. 6.7. Dimensionless desorption diffusivity. (Eagleson, 1978c, Fig. 10, p. 727; with permission of the American Geophysical Union.)

and that of Carslaw and Jaeger (1959, p. 80) and of Crank (1956, p. 256) to find

$$\phi_e(d) = \left[1 + 1.85 \sum_{n=1}^{d} (-1)^n \binom{d}{n} \frac{1}{1.85 + n}\right], \tag{6.35}$$

which is plotted in Fig. 6.7.

The above formulation of the bare soil evaporation is valid as long as the time to stress, t_s, is equal to or greater than the time, $t^* + t_e$, at which exfiltration stops. When $t_s < t^* + t_e$, the transpiration rate falls and the bare soil evaporation rate rises (see Eq. 6.32). In Appendix E we show that biological productivity is proportional to transpiration, thus we assume the circumstances insuring $t_s \geq t^* + t_e$ will be selected naturally. This is insured (Salvucci, 1992) as long as the root depth, z_r, is equal to or greater than z_e, the average penetration depth of exfiltration. Neglecting gravitational percolation, we approximate z_e from simple diffusion (cf. Eagleson, 1978c, p. 727). Using the results of Appendix C, the *characteristic diffusive penetration depth* over the average interstorm period is

$$z_e \approx 1.77 (D_e/\alpha)^{1/2}, \tag{6.36}$$

in which D_e, the *effective desorption diffusivity*, is

$$D_e \equiv \frac{\pi E_{ps}^2 E}{2\alpha n_e^2 s_o^2}, \tag{6.37}$$

while the dimensionless *bare soil evaporation effectiveness*, E, is

$$E \equiv \frac{2\alpha n_e K(1) \psi(1)}{\pi m E_{ps}^2} \phi_e s_o^{d+2}. \tag{6.38}$$

The average interstorm bare soil exfiltration is obtained by taking the expected value of the integral, E_{sj}, of the instantaneous bare soil evaporation rate, e_s (the smaller of E_{ps} and f_e^* in Fig. 6.6), over *all* the possible (exponentially distributed) times between storms. This is tedious and the somewhat unwieldy result is given in Appendix C as Eqs. C.1–C.3.

We define the *bare soil evaporation efficiency*, β_s, after Eagleson (1978d), as

$$\beta_s = \frac{E[E_{sJ}]}{\alpha^{-1} E_{ps}}, \tag{6.39}$$

and its analytical expression is developed in Appendix C as Eq. C.5. For the simplified case of zero surface retention ($h_o = 0$), bare soil, and deep water table ($w/E_{ps} \ll 1$), Eq. C.5 reduces to Eagleson's (1978d) important function

$$\beta_s = 1 - [1 + \sqrt{2}E]e^{-E} + (2E)^{1/2}\Gamma\left[\frac{3}{2}, E\right], \tag{6.40}$$

which is plotted here, along with its asymptotes, in Fig. 6.8. Approximating Eq. 6.40 by its asymptotes, we see that their intersection, at $E = 2/\pi$, divides the coupled soil–atmosphere behavior into two distinct zones:

(1) a soil-controlled zone ($E < 2/\pi$) where evaporation is limited because soil moisture cannot be raised to the soil surface fast enough to satisfy the atmospheric evaporative demand. This may occur because the soil properties either inhibit infiltration or adversely constrain the desorption. In either case, the system is "water-limited" from the standpoint of evaporation, and

(2) a climate-controlled zone ($E \geq 2/\pi$) where evaporation is limited only by the atmospheric evaporative demand. Because this demand can be expressed in terms of the net radiation (see Chapter 7), climate-controlled evaporation is "light-limited".

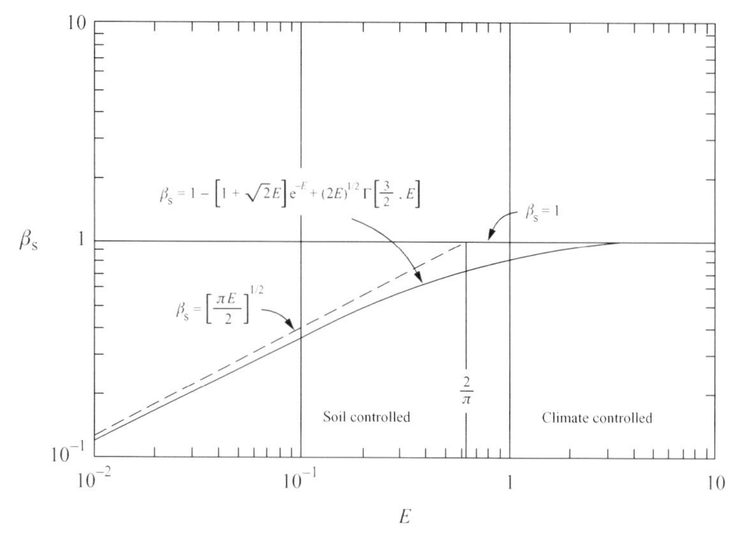

Fig. 6.8. The bare soil evaporation efficiency. (Eagleson, 1978d, Fig. 5, p. 737; with permission of the American Geophysical Union.)

Equation 6.40 gives a rational physical basis for the natural surface evaporation function developed empirically by earlier investigators such as Pike (1964) and Budyko (1974). Furthermore, the use of the asymptote intersection of such a "saturating" function to separate regimes of distinctly different behavior, each governed by different limiting processes, sets the stage for defining the limits to biological behavior in the remainder of this monograph.

G Moisture-constrained transpiration

Transpiration can continue only as long as there is soil moisture available to the plant roots. In Fig. 6.9 we sketch the average time variation of the root-zone spatial average soil moisture, s, from its maximum value, $s(0)$, at the end of the average recharging interval of storm precipitation, m_{t_r}, to its critical minimum value, $s_c = s_5$, at the end of the average drying interstorm interval, m_{t_b}. As a first approximation, we assume the decay of s to be linear with time as it would be if soil moisture was exhausted only by transpiration occurring at the constant unstressed (i.e., maximum) rate, $E_v = k_v^* E_{ps}$. The space–time average (i.e., "equilibrium") root-zone soil moisture is then $s_o = s_{oc} = [s(0) + s_5]/2$, and the volume of soil moisture, V_e, available for exchange with the atmosphere (per unit of surface area) during the average interstorm period is

$$V_e = 2n_e(s_o - s_5)z_r, \tag{6.41}$$

in which n_e is the effective soil porosity, and z_r is again the rooting depth.

Following the development of Salvucci (1992), we continue the assumption that after the surface retention has been evaporated at time t^*, evapotranspiration (only) draws down the soil moisture storage, V_e, until that too is exhausted at the "time to stress", t_s (cf. Fig. 6.10). That is

$$V_e = \int_{t^*}^{t_s} [(1 - M)e_s + ME_v]\,dt, \tag{6.42}$$

in which the vegetated fraction of the surface is denoted by the canopy cover, M, transpiring at rate E_v, and the remaining fraction, $(1 - M)$, is bare soil evaporating

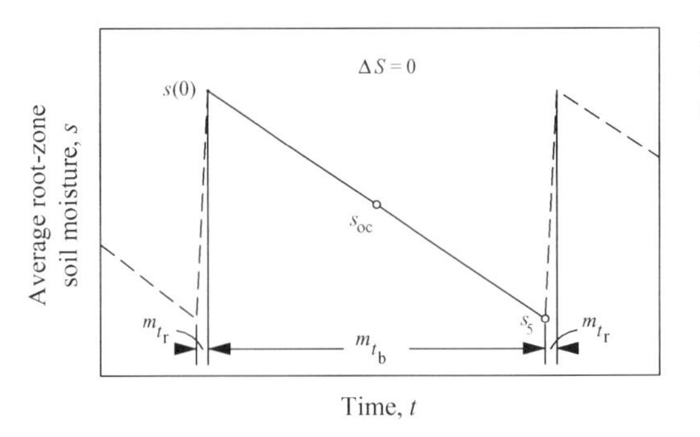

Fig. 6.9. Hypothetical time variability of the average root-zone soil moisture.

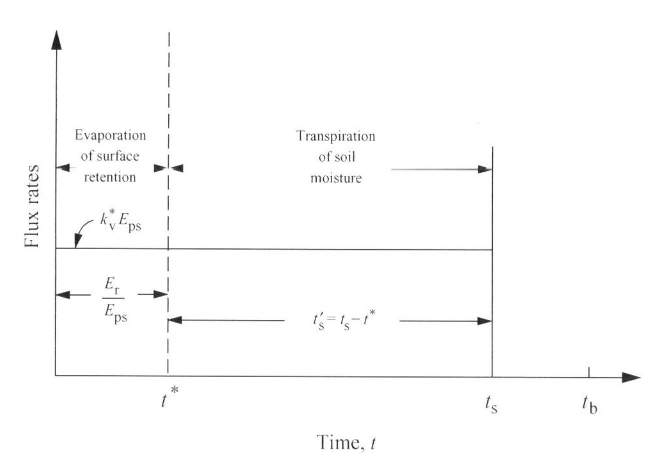

Fig. 6.10. Rate of vegetal transpiration. (Salvucci and Eagleson, 1992, Fig. 3.2, p. 59.)

at rate e_s. Using Eqs. 6.41, 6.29, and 6.32, with the guidance of Fig 6.4, Salvucci (1992) solved Eq. 6.42 for the time to stress, t_s, with the result given in Appendix C as Eq. C.6. The time, t_s', during which the vegetation transpires, is that between the evaporative exhaustion of surface retention at time t^*, and the time, t_s, when stress sets in. That is

$$t_s' \equiv t_s - t^*. \tag{6.43}$$

In the manner introduced by Salvucci (1992), we are now prepared to take the expected value of the integral of the rate, E_{vj}, of *stressed* interstorm transpiration over all the initial surface retention conditions (resulting from the gamma-distributed storm depths), and over all the possible (immediately following) exponentially distributed interstorm periods, i.e.,

$$E[E_{vj}] = \int \mathrm{d}t_b \int E_{vj}(h, t_b)\, \mathrm{f}_{h.t_b}(h, t_b)\, \mathrm{d}h, \tag{6.44}$$

in which $\mathrm{f}_{h.t_b}(h, t_b)$ is the joint distribution of storm depth and the immediately following interstorm interval. Assuming these to be independent random variables, we use Eqs. 6.2 and 6.8 to write

$$\mathrm{f}_{h.t_b}(h, t_b) = \mathrm{f}_h(h)\, \mathrm{f}_{t_b}(t_b) = \frac{\alpha\lambda_o (\lambda_o h)^{\kappa_o - 1} e^{-\lambda_o h - \alpha t_b}}{\Gamma(\kappa_o)}, \tag{6.45}$$

which facilitates integration of Eq. 6.44. The result is presented in Appendix C as Eq. C.7.

Using Eq. 5.30, we define the *canopy transpiration efficiency*, β_v, as

$$\beta_v \equiv \frac{E_v}{E_{pv}} = \frac{E_v}{k_v^* E_{ps}} = \frac{k_v}{k_v^*} = \frac{E[E_{vj}]}{\alpha^{-1} k_v^* E_{ps}}, \tag{6.46}$$

whereupon, Eqs. 6.31 and C.7 give

$$\beta_v \approx 1 - e^{-\alpha t_s'} \left\{ \left[1 + \frac{\alpha}{\lambda_o E_{pv}} \right]^{-\kappa_o} \frac{\gamma[\kappa_o, (\lambda_o + \alpha/E_{pv})]}{\Gamma(\kappa_o)} + e^{-\alpha h_o/E_{pv}} \left[1 - \frac{\gamma(\kappa_o, \lambda_o h_o)}{\Gamma(\kappa_o)} \right] \right\}, \tag{6.47}$$

in which t_s' is given by Eqs. C.6 and 6.43, and $\gamma[\cdot, \cdot]$ or $\gamma(\cdot, \cdot)$ is the incomplete gamma function (see Table 6.1b). Note from Eq. C.6 that as $t_s' \to \infty$, such as would occur for high soil moisture (s_o) and large root depth (z_r), $\beta_v \to 1$, and the transpiration is at the unstressed potential rate. Note also that for the simplified case of negligible surface retention (i.e., $h_o \to 0$)

$$\beta_v = 1 - e^{-\alpha t_s'} = 1 - e^{-t_s'/m_{t_b}}. \tag{6.48}$$

Salvucci (1992, Table 3.2, p. 62) computed t_s/m_{t_b} for a range of soil types and plant coefficients at Beaver Creek, Arizona, and found $t_s/m_{t_b} \geq 1$ for all but the clay soil (with $k_v^* = 1$). However, because of the exponential distribution of t_b, there will always be some interstorm periods during which $t_s' = t_s < t_b$, and β_v will begin to fall below unity before $t_s' = t_s = m_{t_b}$. As mentioned above in Section E, the primary cause of $\beta_v < 1$ is the variance of the initial soil moisture, s_o. Note that with a pdf of s_o, we could derive the distribution of t_s' from Eqs. 6.43 and C.6, and thus incorporate the variable moisture storage at the beginning of the interstorm interval as the important origin of transient stress. We will not attempt that complicated refinement in this work.

This completes the transpiration analysis. Using Eqs. 6.39 and 6.47, the integrand of Eq. 6.42 can be expanded to write the expected value of the total annual evapotranspiration, E_{TA}, (bare soil evaporation plus vegetal transpiration)† as

$$E_{TA} = \frac{m_\nu E_{ps}}{\alpha} [(1 - M)\beta_s + M k_v^* \beta_v]. \tag{6.49}$$

H Percolation and capillary rise

As was discussed earlier, a percolation component and, where a water table exists, a capillary rise component are active at the lower boundary of the soil column. Both are assumed to be steady fluxes (Eagleson, 1978c).

The percolation rate is

$$v(s_o) = K(1) s_o^c, \tag{6.50}$$

† We recall that E_{ps} was defined in Appendix B as an instantaneous rate keyed to the daylight hour average insolation. Hence, in using E_{ps} to calculate interstorm evapotranspiration volume we must use only the daylight hour fraction of the interstorm period, α^{-1}.

and occurs only during the rainy season, τ, having mean value, m_τ. The capillary rise rate is

$$w = K(1)\left[1 + \frac{3/2}{mc - 1}\right]\left[\frac{\psi(1)}{z_w}\right]^{mc}, \tag{6.51}$$

and occurs throughout the entire year, T_A. In the above, z_w is the depth to the water table.

I Evaporation from surface retention

Eagleson (1978e) gives analytical expressions for both the bare soil and the vegetated components of the temporal average evaporation of surface-retained precipitation, $E[E_r]$. However, these are more complicated than is warranted by the underlying assumptions so we will choose a simpler alternative here. There are four basic assumptions:

(1) The fraction of *total* foliage area retaining intercepted precipitation. We express this fraction by the product, $\eta_o L_t$, where η_o is the ratio of wetted leaf perimeter to leaf chord (cf. Chapter 2) For *broad leaves* only one side is wetted, while for *needle leaves* surface tension wets the entire surface and the curvature holds the moisture against gravity drainage. Thus $\eta_o = 1$ for broad leaves, $\eta_o \approx 2$ for the relatively flat needles of spruce, and $\eta_o \approx 2.36$ to 2.50 for the fatter cross-sections of pine needles.

(2) Surface retention capacity, βh_o. This is a complex function of surface properties, temperature (i.e., surface tension), leaf angle, wind speed, etc. We assume gravity drainage proportional to the leaf angle and use a constant nominal value, $h_o = 0.1$ cm on all surfaces retaining intercepted precipitation.

(3) The fraction of total (as opposed to "bare") soil surface retaining precipitation. The "bare" soil fraction, $1 - M$, always retains depth h_o (assuming $m_h \geq h_o$), while the vegetated fraction retains h_o only as long as the *interception capacity*, $\eta_o \beta L_t h_o < m_h$. Otherwise, no precipitation reaches the soil on the surface fraction M.

(4) The fraction of surface retention evaporated during the subsequent interstorm period. We assume that all is evaporated including nightly dew.

With these simplifying assumptions, we estimate the surface-retained precipitation as:

$$E[E_r] \equiv \overline{h_o} = (1 - M)h_o + M(1 + \eta_o \beta L_t)h_o = (1 + M\eta_o \beta L_t)h_o, \quad \frac{\eta_o \beta L_t h_o}{m_h} < 1$$

$$E[E_r] \equiv \overline{h_o} = (1 - M)h_o + M m_h, \quad \frac{\eta_o \beta L_t h_o}{m_h} \geq 1. \tag{6.52}$$

J The mean annual (or seasonal) water balance

Guided by Fig. 6.2, we equate the long-term average inflows and outflows of the soil column during the *wet season*, τ, to obtain the mean *growing season* (assumed to be the same τ) water balance

$$P_\tau - m_\nu E[E_r] + \Delta S = m_\nu E[R_{sj}] + E[E_{T\tau}] + m_\tau v - m_\tau w. \tag{6.53}$$

In the above, the notation (i.e., τ subscripts) was chosen to call attention to the fact that the long-term averages are taken for only a fraction of the year, the growing season of mean length, m_τ. There will normally be some difference, ΔS (cm), in the stored soil moisture at the two ends of a partial-year season. The major difficulty in this use of the long-term water balance is estimating the magnitude and sign of ΔS. The storage change is placed on the left-hand side of the equation as in Eq. 6.53 where it will be positive when there is carryover storage at the start of the growing season and negative when there is carryover storage at the end of the growing season. *All* quantities in Eq. 6.53, subscripted or not, are determined for the selected season.

Using Eqs. 6.26, 6.27, 6.49, 6.50, 6.51, and 6.52, and dividing through by the mean seasonal precipitation, Eq. 6.53 gives the normalized seasonal water balance for the common case $\frac{\eta_o \beta L_t h_o}{m_h} < 1$ to be

$$1 - e^{-G-2\sigma^{3/2}} - \frac{\overline{h_o}}{m_h} + \frac{\Delta S}{m_\nu m_h} = \frac{m_{t_b} E_{ps}}{m_h}[(1-M)\beta_s + Mk_v^* \beta_v] + \frac{m_\tau K(1)}{P_\tau} s_o^c - \frac{m_\tau K(1)}{P_\tau}\left[1 + \frac{3/2}{mc-1}\right]\left[\frac{\psi(1)}{z_w}\right]^{mc}, \tag{6.54}$$

which expresses the equilibrium relationship among the climate, soil, and vegetation for a deep water table.[†]

For the rarer case, $\frac{\eta_o \beta L_t h_o}{m_h} \geq 1$, in which interception capacity equals or exceeds rainfall, there will be no rainfall input to the substrate of the vegetated fraction, and the second term on the left-hand side of Eq. 6.54 must be multiplied by the factor $1 - M$.

Equation 6.54 is a complex function of the equilibrium soil moisture state variable, s_o, in which different physical processes (i.e., different terms in the equation) dominate in different ranges of s_o. A general analytical solution for s_o is unavailable. However, in keeping with our treatment in this work of other complex "saturating" functions such as bare soil evaporation (Fig. 6.8) and photosynthetic capacity (Fig. 8.7), we can solve simpler limiting cases as is detailed in Appendix C. For the simplest case of a very deep water table, surface retention that is small with respect to the rainstorms, and negligible carryover storage, Eq. 6.54 gives

$$1 - e^{-G-2\sigma^{3/2}} = \frac{m_{t_b} E_{ps}}{m_h}[(1-M)\beta_s + Mk_v^* \beta_v] + \frac{m_\tau K(1)}{P_\tau} s_o^c. \tag{6.55}$$

[†] Salvucci and Entekhabi (1995, corrected 1997) modify Eq. 6.54 to account for water table influence.

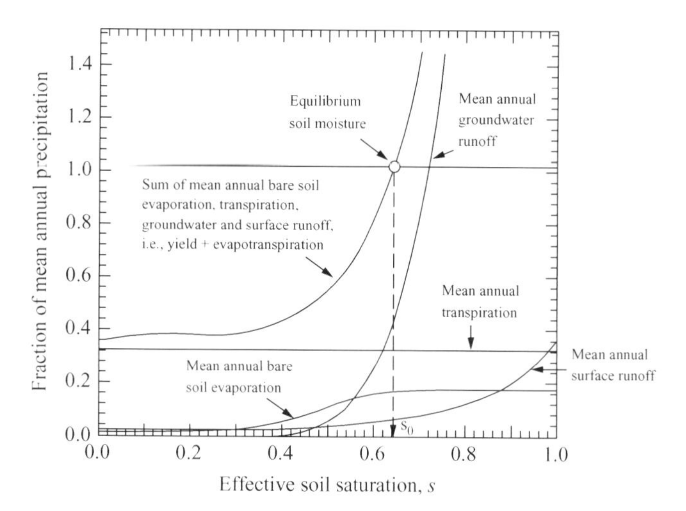

Fig. 6.11. Components of the annual water balance. Fixed climate, soil, and vegetation. (Salvucci and Eagleson, 1992, Fig. 2.3, p. 38.)

It is common to speak of the long-term average water *yield*, Y_τ, defined as the sum of the surface and groundwater components of runoff. From Eq. 6.55 this is

$$Y_\tau = P_\tau - \frac{m_{t_b} E_{ps}}{m_h}[(1 - M)\beta_s + M k_v^* \beta_v], \tag{6.56}$$

while the normalized evapotranspiration is

$$\Theta = \frac{E[E_T]}{E[E_{ps}]} = \frac{E[E_T]}{m_\nu m_{t_b} E_{ps}}. \tag{6.57}$$

The separate terms of Eq. 6.55 are plotted in Fig. 6.11 for a typical humid climate with clay soil and representative values of M and k_v^*. Summing all the normalized terms to unity locates the equilibrium soil moisture, s_o.

K The vegetation state–space

The role of vegetation in the water balance is expressed (Eq. 6.54) in terms of the two vegetation "state variables", M, and k_v^*. Their specification a priori, in terms of climate and soil, is at the core of this entire work, so we will examine their "state–space".

Eagleson (1978f) investigated the M-dimension of this space for several characteristic soils and found (see Fig. 6.12), for a fixed k_v^*, that the equilibrium soil moisture, s_o, has a maximum at an intermediate value of the canopy cover M which coincides with the minimum in evapotranspiration. In subsequent work (Eagleson, 1982) he demonstrated this coincidence analytically, and through decomposition of the evapotranspiration into its bare soil and vegetal components he explained the cause of the soil moisture maximum as follows.

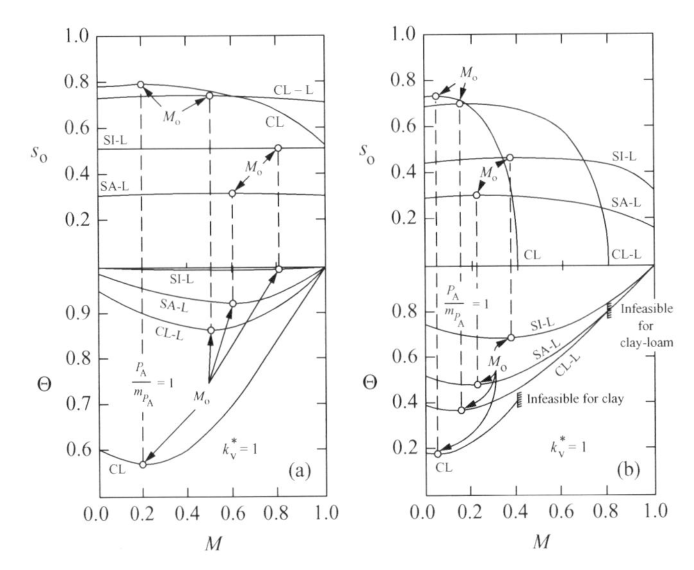

Fig. 6.12. The maximum soil moisture condition under average annual precipitation. (a) Clinton, MA: climate controlled; (b) Santa Paula, CA: soil controlled. CL, clay; CL-L, clay-loam; SA-L, sandy-loam; SI-L, silty-loam. (Eagleson, 1978f, Figs. 3 and 4, p. 755; with permission of the American Geophysical Union.)

The two components of Eq. 6.49 are each plotted dimensionlessly in Fig. 6.13 for two contrasting climate–soil systems under the condition of unstressed transpiration ($\beta_v = 1$) and for two values of k_v^*. Note first that the unstressed transpiration is linear in M regardless of climate and soil. However, the bare soil evaporation will be linear in M only when the soil properties do not influence the evaporation rate, that is in climate-controlled situations where bare soil evaporation is limited only by the atmospheric vapor transport capacity. Figure 6.13 clearly shows the Clinton climate–soil system to be climate-controlled by this criterion. As we move to drier, soil-controlled climates such as Santa Paula, the bare soil evaporation becomes progressively more non-linear. This non-linearity is due to the dependence of the desorption velocity upon the soil moisture which is in turn influenced by the M-dependent extraction of soil moisture by the plant roots. The non-linearity produces the minimum in total evapotranspiration which is marked in Fig. 6.12 by the plotted circle at the canopy cover labelled $M = M_o$. Looking now at the soil moisture in the upper part of Fig. 6.12, at small but non-zero M, where bare soil desorption is the dominant mode of soil moisture depletion, the composite evapotranspiration rate will be reduced over that for barren soil yielding higher equilibrium soil moisture. As M increases, the vegetated fraction, transpiring at the potential rate, becomes dominant and the composite evapotranspiration rate begins to increase, thereby reducing the equilibrium soil moisture. Eagleson (1982) showed analytically that the maximum soil moisture and the minimum evapotranspiration occur at the same $M = M_o$ as long as M and k_v^* are independent.

Eagleson (1978f) hypothesized that in natural systems M and k_v^* are subject to different mechanisms of change, M adapting to short-term variability in climate, and k_v^* changing on an evolutionary time-scale. Accordingly, he proposed separate equilibrium states: (1) a short-term or "growth" equilibrium reached by the canopy cover of a given species (i.e., of a given k_v^*) through adaptation to fluctuations in soil moisture at the time-scale of the life span of an individual tree and (2) a long-term or "evolutionary" equilibrium reached by natural selection of individual species to be optimally compatible with the given climate and soil. He further reasoned that the governing physical condition for the short-term is minimum stress and therefore that at this time-scale the equilibrium canopy cover is $M = M_o$, the value at maximum soil moisture. At the longer time scale he reasoned, after Odum (1959, p. 252), that maximum biomass productivity (with its attendant maximum reproductive potential) would guide evolution, and since plant productivity is directly proportional to its rate of transpiration (e.g., Rosenzweig, 1968, and Appendix E), the evolutionary equilibrium is governed by maximum transpiration. A later study by Eagleson and Tellers (1982) of

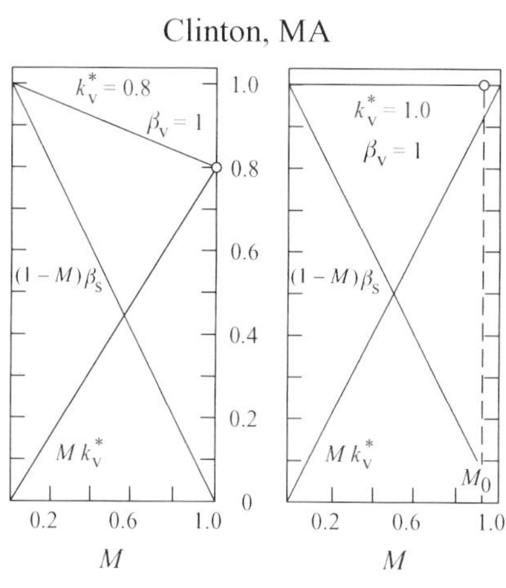

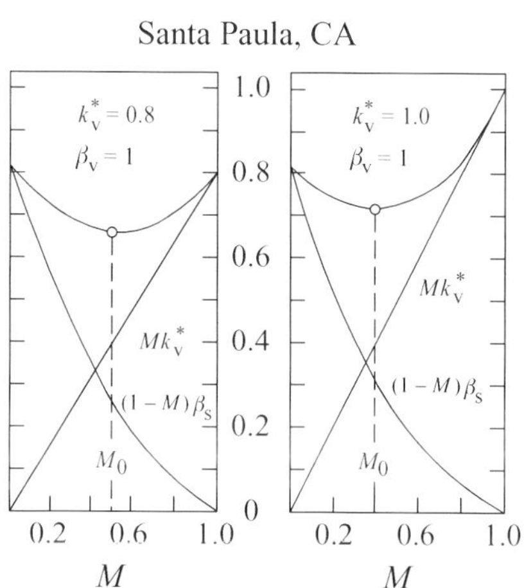

Fig. 6.13. Evapotranspiration composition in contrasting climates under average annual precipitation. (Eagleson, 1982, Fig. 10, p. 332; with permission of the American Geophysical Union.)

Plate 6.3. Tree–shrub–grass savannas. (a) Shrub–grass; (b) park-like (i.e., tree–shrub–grass): (c) tree–grass. All are in a common climate with soil differences determining the savanna type. (Unidentified species photographed in equatorial Kenya by Peter S. Eagleson.)

several forested watersheds, in which the soil properties and the plant coefficient were *estimated* rather than observed, gave reasonable support to the short-term equilibrium hypothesis. However, the work of Salvucci and Eagleson (1992) suggests a different basis for M as we will see later in this chapter.

Summary

Equation 6.54 or its three approximate solutions given in Appendix C describe all the possible soil moisture states in a given climate and soil for a canopy of given M and k_v^*. However, not all of these states are feasible, for we have seen in Section E of this chapter that there is a limiting soil moisture potential below which the canopy conductance, k_v, declines from its open stomate value, k_v^*, due to stomatal closure. As we will now see, incorporating the idea of this *critical moisture state* (cf. Salvucci, 1992) leads to an additional, independent expression for s_o.

L The critical moisture state

Introduction

For the moment let us consider as our moisture state variable the growing season soil moisture, spatially averaged over the root zone but varying with time. It is a random variable which we denote as $\langle s \rangle$, and its minimum value (also random) is a complex function of:

(1) the joint distribution of interstorm duration and prior storm depth,[†]
(2) interception,[‡] and
(3) initial soil moisture at the beginning of the growing season.[§]

The relation of $\langle s \rangle$ to the critical soil moisture s_c controls in some way which species thrive in a given climate–soil wherever the supply of water is constraining (cf. Chapter 8). How frequently and for how long can a given species tolerate stomatal shut-down due to $\langle s \rangle < s_c$? The answer depends (among other things) upon the innate internal water storage capacity of the species and is difficult to quantify. For grasses the tolerance is small while for some evergreens it is high enough to carry the tree through an entire dry season.

To use the concept of stomatal shut-down quantitatively, we must select a *tolerance criterion*. We will assume that plants have evolved to survive random stress of some unknown, species-dependent, intensity, duration, and frequency, but are vulnerable to any degree of stress in the mean. It is consistent with selection pressure for maximum biomass production (cf. Appendix E) that the plant transpire at its unstressed

[†] Rodríguez-Iturbe *et al.* (1987, 1988) infer that these two variables are essentially independent.
[‡] Evaporation from interception truncates the distribution of rainstorms incident upon the soil.
[§] This carryover moisture storage becomes important in seasonal climates.

Table 6.3. *Estimated and observed water balance parameters: Beaver Creek, Arizona*

Watershed number	Tree type	M	m_τ (day)[a]	E_{ps} (cm day^{-1})[b]	P_τ (cm)	Y_τ (cm)	m_{t_b} (day)	m_{t_r} (day)	λ_o (cm^{-1})	κ_o	h_o (cm)	T_0 (°C)	m	c	$\psi(1)$ (cm)	n_e	$K(1)$ (cm day^{-1})
1	Pinyon–Utah juniper woodland	0.40	205	0.35	23.7	0.17	5.33	0.12	0.96	0.61	0.1	18.5	0.48	7.13	4.51	0.46	17.19
2	Pinyon–Utah juniper woodland	0.42	202	0.35	20.2	0.51	6.20	0.12	0.72	0.45	0.1	19.3	0.48	7.13	4.51	0.46	17.19
3	Pinyon–Utah juniper woodland	0.40	238	0.34	25.9	0.76	6.06	0.14	0.80	0.53	0.1	17.7	0.48	7.13	4.51	0.46	17.19
4	Pinyon–alligator juniper woodland	0.52	181	0.33	24.6	1.02	5.69	0.12	0.62	0.49	0.1	16.4	0.48	7.13	4.51	0.46	17.19
5	Pinyon–alligator juniper woodland	0.53	181	0.32	24.6	0.95	5.69	0.12	0.62	0.49	0.1	15.8	0.48	7.13	4.51	0.46	17.19
6	Pinyon–alligator juniper woodland	0.52	184	0.32	22.6	0.11	5.85	0.12	1.0	0.73	0.1	15.5	0.48	7.13	4.51	0.46	17.19
7–18	Ponderosa pine	0.85	169	0.31	22.3	0.67	5.32	0.12	0.56	0.41	0.1	14.2	0.71	5.8	92.8	0.37	15.12

Watershed number	Tree type	Years of record	m_d (day)	E_{psd} (cm day^{-1})	P_d (cm)	Y_d (cm)	ΔS (cm)	E_o	m_ν[c]	m_h[d] (cm)	βL_t[e]	$G(0)$	$\sigma(0)$	$\overline{h_o}$[f] (cm)	$\frac{\overline{h_o}}{m_h}$	$\frac{\Delta S}{m_\nu m_h}$	$e^{-G(0)-2\sigma(0)^{3/2}}$
1	Pinyon–Utah juniper woodland	1958–62	160	0.14	22.0	0.78	7.13	3.02	38.5	0.62	2.27	1.63	1.05	0.33	0.53	0.30	0.04
2	Pinyon–Utah juniper woodland	1958–80	163	0.13	23.6	2.96	6.11	2.80	32.6	0.62	2.27	1.63	1.07	0.34	0.55	0.30	0.02
3	Pinyon–Utah juniper woodland	1958–67	127	0.13	18.1	1.55	5.14	2.91	39.3	0.66	2.27	1.81	1.08	0.33	0.50	0.20	0.02
4	Pinyon–alligator juniper woodland	1958–72	184	0.14	28.4	11.3	3.16	3.10	31.8	0.77	2.58	1.38	0.91	0.44	0.57	0.13	0.04
5	Pinyon–alligator juniper woodland	1958–72	184	0.14	28.4	11.0	3.97	3.20	31.8	0.77	2.58	1.38	0.91	0.44	0.57	0.16	0.04
6	Pinyon–alligator juniper woodland	1959–64	181	0.14	23.8	3.84	6.96	3.15	31.5	0.72	2.58	1.46	0.96	0.44	0.61	0.31	0.04
7–18	Ponderosa pine	1958–80	196	0.08	39.1	12.8	21.0	11.9	31.8	0.70	1.77	1.29	2.12	0.48	0.69	0.94	0.00

[a] Williams and Anderson (1967). [b] Appendix B, Table B.2. [c] $m_\nu = m_\tau / m_{t_b}$. [d] $m_h = P_\tau / m_\nu$.

[e] From Baker (1950) as given here in Table 3.9 using Baker's Northern white cedar for pinyon–Utah juniper and Baker's lodgepole pine for pinyon–alligator juniper as suggested by similarities of needle structure as sketched in Zim and Martin (1987, pp. 25, 36, 37, and 39).

[f] From Eq. 6.52 using: $h_o = 0.1$ cm; $\eta_o = 2.50$ (Waring, 1983).

Source: Modified from Salvucci (1992).

(i.e., open-stomate) rate for as long as possible; thus we estimate the minimum (i.e., limiting) soil moisture volume as that necessary to maintain open-stomate transpiration throughout the *mean* interstorm period. Furthermore, we will show that this limiting soil moisture volume is insensitive to the value of s_c and hence is species-dependent only through $\overline{h_o}$ and k_v^*.

Data base

It now becomes helpful to have a data base with which to evaluate comparatively terms of the water balance and against which to test our hypotheses. Jasinski and Eagleson (1990) located a group of forested watersheds at Beaver Creek (near Flagstaff), Arizona having 5 to 25 year hydrologic records (Baker, 1982, 1986) as well as a soil survey (Williams and Anderson, 1967), and canopy cover observations (Clary *et al.*, 1974; Baker, 1986).† The composite data (after reduction by Salvucci, 1992) are summarized here in Table 6.3, and the soil data are also compared with the properties of typical soils in Table 6.2b.

Definition

Using the climate and soil parameters of a typical pinyon–alligator juniper watershed at Beaver Creek, Salvucci (1992) solved the simplified water balance equation (Eq. 6.55) repetitively by trial for the equilibrium soil moisture, s_o, using a different assumed pair of the vegetation state variables, k_v^* and M, for each solution. From this field of feasible solutions he interpolated the isolines of s_o shown in Fig. 6.14a. This is an expansion into the full, two-dimensional, state–space of the marginal, one-dimensional explorations made earlier by Eagleson (1978f) and sampled here in the upper portion of Fig. 6.12. Two important features of Fig. 6.14a should be noted.

(1) The locus of $\frac{\partial s_o}{\partial M}|_{k_v^*} = 0$, i.e., the soil moisture "ridge" hypothesized by Eagleson (1978f) as the optimal operating point, is indicated by the dashed quasi-hyperbola, $s_{o\,max}$.

(2) Somewhere between the $s_o = 0.34$ and $s_o = 0.14$ contours, the soil moisture declines abruptly as the unstressed transpiration of this model dominates the extraction of soil moisture. The locus of this soil moisture decline is so topographically abrupt, particularly in arid and semi-arid climates, that Salvucci (1992) dubbed it the "cliff". This cliff is illustrated for the clay and clay-loam soils of Fig. 6.12b, and for a typical Beaver Creek watershed in Fig. 6.14b. In the latter illustration we have indicated the critical soil moisture at which the abrupt decline begins as $s_o = s_{oc} = 0.15$ which will be estimated in the next section.

† The availability of this unusually comprehensive data set prompted Salvucci's (1992) test of Eagleson's (1982) optimal vegetation hypotheses.

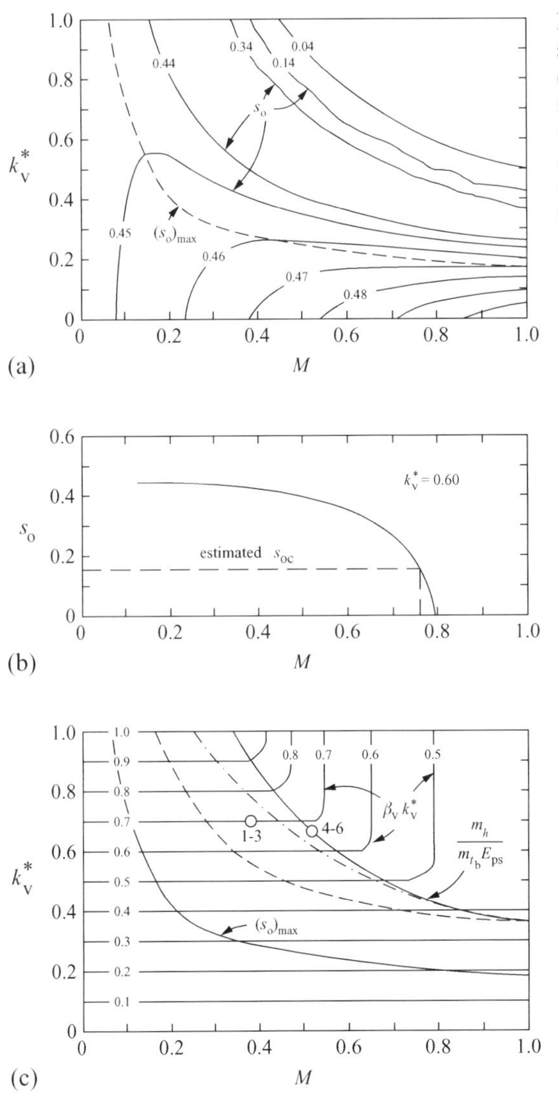

Fig. 6.14. Canopy state–space at Beaver Creek. (a) Contours of equilibrium soil moisture; (b) equilibrium soil moisture at $k_v^* = 0.60$; (c) contours of $\beta_v k_v^*$. (After Salvucci and Eagleson, 1992, Figs. 3.3 and 3.4.)

The cliff is illustrated more instructively by examining contours of the *effective canopy conductance*, $\beta_v k_v^*$, as is done in Fig. 6.14c for the same typical pinyon–juniper watershed. Let us conduct a thought experiment in which we start with a given bare soil, and a given climate producing no carryover soil moisture from the dormant season (i.e., $\Delta S \equiv 0$). We gradually add vegetation of given k_v^* to gain increasing M. As we saw in Fig. 6.12, the equilibrium soil moisture, s_o, rises initially with increasing M until the transpirational component of moisture loss becomes dominant; thereafter s_o falls as M continues to increase. The lower the space-time average s_o becomes, the closer to s_5 is the spatial average *transient* soil moisture, $\langle s \rangle$, at the end of the longest interstorm interval, and the closer the plants are to experiencing stress. As long as the plant is unstressed, $\beta_v = 1$ and $\beta_v k_v^* = k_v^*$. However, when M becomes so large that $\langle s \rangle$ reaches s_5 during a significant fraction of the t_b's, $\beta_v \to 0$ and the $\beta_v k_v^*$ contour curves upward. The true critical state lies at the beginning of this upturn in $\beta_v k_v^*$.

For a non-seasonal climate $\Delta S = 0$, and we assume the critical state to occur when s_o is such that $\langle s \rangle = s_5$ at the end of an interstorm interval of average length, m_{t_b}. For a seasonal climate $\Delta S \neq 0$, and we assume the critical state to occur at the end of the average growing season, $m_\tau \approx m_\nu m_{t_b}$. For the given species (i.e., k_v^*) and climate, *canopy* productivity will be a maximum at this moisture state because: (1) the transpiration rate, Mk_v^*, is a maximum there and thus the nutrient flux is a maximum, (2) as modeled here the stomates remain fully open throughout the growing season thereby maximizing the carbon flux, and (3) the canopy density is as large as the water supply will allow.[†] By definition, the critical state is not reached in the climate-controlled (i.e., light-limited) system.

Estimation of critical equilibrium soil moisture

From Fig. 6.9 and Eq. 6.41 the critical space–time average root-zone soil moisture is written

$$s_o = s_{oc} = s_5 + \frac{V_e}{2n_e z_r}, \tag{6.58}$$

in which z_r is the vegetation root depth, and V_e is the volume of soil moisture (per unit of surface area) available for exchange with the atmosphere during the average interstorm period. Using Eq. 6.58 we can eliminate s_o from Eq. 6.54 and obtain order-of-magnitude estimates for the individual terms of the water balance. Equation 6.58 is independent of the water balance equation (Eq. 6.54) enabling us to solve the two equations simultaneously to obtain the *maximum canopy moisture flux*, Mk_v^*. To do this we must first estimate V_e and z_r.

As a first approximation we let the available soil moisture (per unit of surface area) be the infiltration which will be larger for the bare soil than under the canopy due to the large interception loss in the latter case. We assume that the plant roots exploit the moisture under the entire surface and thus the critical condition is not reached until $s = s_5$ under the bare soil fraction where the interception is h_o. The limiting volume of soil moisture is then

$$V_e \approx m_h \left(1 - e^{-G(0)-\sigma(0)^{3/2}}\right) - \overline{h_o}. \tag{6.59}$$

Eagleson (1978c, p.727) assumed the rooting depth, z_r, to equal the so-called "penetration" depth, z_i, of the soil wetting processes, which he estimated to be the sum of gravitational percolation and diffusive sorption. Using the results of Appendix C,

[†] As we will see later, this is confirmed for carbon demand-limited productivity by Eq. 10.18.

Table 6.4. *Water balance summary for Beaver Creek Watersheds*

Watershed number	M (1)	k_v^* (2)	T_0 (°C) (3)	E_{ps} (cm day^{-1}) (4)	m_h (cm) (5)	E_o (6)	d (7)	m_{t_b} (day) (8)	$m_{t_b'}$ (day) (9)	$m_{t_b''}$ (day) (10)	$\frac{\eta_o \beta L_t}{m_h/h_o}$ (11)	$\frac{\overline{h_o}}{m_h}$ (12)	s_5 (13)	s_{oc} (14)	Term 4 ($s_o = s_{oc}$) (15)	Term 5 ($s_o = s_{oc}$) (16)	Term 6 ($s_o = s_{oc}$) (17)	Term 7[a] (18)
1–3	0.41	0.69[b]	18.5	0.35	0.63	2.88	4.1	5.86	5.06	5.57	0.90	0.53	0.04	0.15	0.02	0.03	0.0002	[a]
4–6	0.52	0.67[b]	15.9	0.32	0.75	3.23	4.1	5.74	4.74	5.43	0.86	0.58	0.04	0.15	0.01	0.05	0.0002	[a]
7–18	0.85	0.84[b]	14.2	0.31	0.70	11.9	3.4	5.32	4.61	5.00	0.63	0.69	0.06	0.10	0.01	0.001	0.0002	[a]

[a]Bedrock is at a nominal depth of 1 meter and thus the soil moisture profile does not correspond with that for a water table at $z_w = \infty$, the condition for which terms 6 and 7 (cf. Eq. 6.65) were derived. We assume rather arbitrarily here (after Salvucci, 1992) that term 7 is equal (and opposite) to term 6.

[b]Estimated using Eqs. 7.31 and 7.14. Columns (1)–(8) Averaged from Table 6.03; (9) Eq. 6.63; (10) Eq. 6.64; (11) and (12) $h_o = 0.1$ cm; $\eta_o = 2.50$ (Waring, 1983); $L_t = 4.0$ (average of all pines, Table 2.02); (13) Eq. 6.30 using $\psi_{sc} = 5$ bar; (14) Eq. 6.62; (15)–(18) Eq. 6.65.

Section E, for the diffusive component and Eq. 6.50 for the percolation, we have

$$z_r = z_i = 1.77\left(m_{t_r} D_i\right)^{1/2} + \frac{m_{t_r} K(1)\, s_{oc}^c}{n_e}, \tag{6.60}$$

in which the *sorption diffusivity*, D_i, in $cm^2\ s^{-1}$, is given by (Eagleson, 1978c, p. 726, following Crank, 1956, p. 256) as

$$D_i = \frac{5K(1)\,\psi(1)\,\phi_i(d, s_{oc})}{3mn_e}, \tag{6.61}$$

where for small s_{oc} (cf. Fig. 6.3) we let $\phi_i(d, s_{oc}) \approx \phi_i(d, 0)$.

Over the realistic range of soil properties as given in Table 6.2 and for storm durations on the order of 1 hour, the second (i.e., percolation) term of Eq. 6.60 is small with respect to the first and will be neglected.† At the critical state we can then combine Eqs. 6.58, 6.60, and 6.61 to obtain

$$s_{oc} \approx s_5 + \frac{m_h\left(1 - e^{-G(0)-\sigma(0)^{3/2}}\right) - h_o}{3.54\, n_e\, m_{t_r}^{1/2} D_i^{1/2}}. \tag{6.62}$$

Using $\psi_{Sc} = 5$ bars (cf. Section E of this chapter), the critical soil moisture, s_5, is calculated for the Beaver Creek watersheds from Eq. 6.30 and is listed in Table 6.4 along with the resulting s_{oc}.

Estimation of maximum canopy moisture flux

To estimate the maximum canopy moisture flux, Mk_v^*, we will first solve the full water balance equation, Eq. 6.54, for the canopy moisture flux, $Mk_v^*\beta_v$. Before doing so, we recognize that the average interstorm time, $m_{t_b'}$, available for transpiration, and the average interstorm time, $m_{t_b''}$, available for bare soil evaporation, are both less (often significantly so) than the full interstorm period, m_{t_b}, due to the time consumed in evaporating the stored interception. We approximate these available times as‡

$$m_{t_b'} = m_{t_b} - \frac{\eta_o \beta L_t h_o/2}{E_{ps}}, \tag{6.63}$$

in which we assume the collective transpiration rate to increase linearly with time as leaf surface is exposed, and

$$m_{t_b''} = m_{t_b} - \frac{h_o}{E_{ps}}. \tag{6.64}$$

† This is of course a physically unrealistic artifact of replacing random variables by their time averages for use in a non-linear equation, and demonstrates that it is the positive excursions of this wetting from their mean that generate groundwater recharge.

‡ Note that these are gross times in that they include the hours of darkness during which evapotranspiration is negligible, but E_{ps} must be the instantaneous value derived in Appendix B. Because the leaves dry from the top of the canopy downward, we start canopy transpiration when one-half the stored moisture has evaporated.

Incorporating these times into the water balance, Eq. 6.54, we obtain

$$\underbrace{M\,k_v^*\,\beta_v}_{1} \approx \frac{m_h}{m_{t_b'}\,E_{ps}}\left\{1 - \underbrace{\frac{\overline{h_o}}{m_h}}_{2} + \underbrace{\frac{\Delta S}{m_\nu\,m_h}}_{3} - \underbrace{\frac{m_{t_b''}\,E_{ps}}{m_h}(1-M)\,\beta_s}_{4} - \underbrace{e^{-G-2\sigma^{3/2}}}_{5}\right.$$

$$\left. - \underbrace{\frac{m_\tau\,K(1)}{P_\tau}s_o^c}_{6} + \underbrace{\frac{m_\tau K(1)}{P_\tau}\left[1+\frac{3/2}{mc-1}\right]\left[\frac{\psi(1)}{z_w}\right]^{mc}}_{7}\right\}. \tag{6.65}$$

The numbered terms of Eq. 6.65, normalized where appropriate by the mean storm depth, m_h, are:

(1) the canopy moisture flux; we assume open stomates giving $\beta_v = 1$, and k_v^* for the ponderosa pine watersheds is estimated from the internal canopy optimization presented in Chapter 7 (cf. Figure 7.4),
(2) space–time average surface retention; estimate $\overline{h_o}$ using Eq. 6.52,
(3) seasonal climate carryover soil moisture storage; estimate from site-specific data,
(4) bare soil evaporation; estimate β_s using Eq. 6.40,
(5) surface runoff; estimate σ and G using Eqs. 6.23 and 6.24 respectively,
(6) deep percolation; estimate from site-specific data,
(7) capillary rise; estimate from site-specific data.

Terms 4 to 6 of Eq. 6.65 contain the equilibrium soil moisture, $s_o = s_{oc}$, which we estimate using the independent Eq. 6.62. The resulting magnitudes of these terms for the Beaver Creek catchments are given in Table 6.4. Note that the sum of terms 4, 6, and 7 is less than 3% of the normalized storm depth, $m_h/m_h = 1$, allowing us to write, for $\beta_v = 1$,

$$Mk_v^* = \frac{V_e}{m_{t_b'}\,E_{ps}}, \tag{6.66}$$

where for $\frac{\eta_o \beta L_t h_o}{m_h} < 1$,

$$V_e \approx m_h\left[1 - \frac{\overline{h_o}}{m_h} - e^{-G(0)-2\sigma(0)^{3/2}} + \frac{\Delta S}{m_\nu\,m_h}\right], \tag{6.67}$$

and (cf. Section J of this chapter) for $\frac{\eta_o \beta L_t h_o}{m_h} \geq 1$,

$$V_e \approx m_h\left[1 - \frac{\overline{h_o}}{m_h} - (1-M)\,e^{-G(0)-2\sigma(0)^{3/2}} + \frac{\Delta S}{m_\nu\,m_h}\right]. \tag{6.68}$$

Note that both terms in the denominator of Eq. 6.66 are evaluated on the basis of consistent times.

Equations 6.66–6.68 express the principal role of hydrology in the structuring of vegetation communities. The right-hand side of Eq. 6.66 uses the seasonal and storm-scale hydroclimatology along with the soil properties dimensionlessly to define the available rate of water supply, while the left-hand side expresses the canopy water demand. We will return to these equations in Chapter 7 to determine separately the canopy state variables M and k_v^*, and in Chapter 11 to discuss their implications with respect to climate change.

Plate 6.4. Slash pine. (Photographed by Peter S. Eagleson, Jr. in dry, sandy soil near St. Petersburg, Florida.)

Comparison with observations from moderately dry catchments

Williams and Anderson (1967) report the Beaver Creek watersheds to have shallow soils which give way to rock at about 1 meter depth. The resulting shallow water table brings the capillary rise term (term 7) of Eq. 6.54 back into consideration, and with $z_w = O(1)$ meter, the saturated soil matric potential, $\psi(1)$, becomes critical to estimating this term. Table 6.3 shows that in going from the pinyon–juniper watersheds to the ponderosa pine watersheds, the estimated pore size distribution index, m, increases twofold while the estimated matric potential increases by an order of magnitude. According to observations of a wide range of soils (see Eagleson, 1978c, Fig. 6, p. 725), the value of $\psi(1)$ is exquisitely sensitive to m over just the m range of interest here, thus any estimate of capillary rise made from Eq. 6.51 may be wildly inaccurate. As an expedient, we assume (after Salvucci, 1992) that with these shallow water tables the capillary rise returns all percolated soil moisture to the surface for evaporation and transpiration thereby eliminating both capillary rise and percolation from the water balance equation.

The carryover storage, ΔS, is highly dependent upon both the soil profile and the seasonality. Salvucci (1992) estimated ΔS for all the Beaver Creek watersheds from the water balance equation written for the complementary (i.e., dormant) season. In this case ΔS is positive, representing a recharge of soil moisture. Using the subscript d for this season

$$\Delta S = P_d - E_{Td} - Y_d = P_d - (1 - M)E_{psd}\, m_d - Y_d. \tag{6.69}$$

The quantities P_d and Y_d were observed (Baker, 1982, 1986) along with the climatic variables needed to estimate E_{psd} and the season length, m_d, and they are tabulated here in the lower half of Table 6.3.

An attractive approximation to Eq. 6.66, strictly applicable only in very dry, non-seasonal climates, but useful later for generalizations concerning habitat is

$$Mk_v^* \approx \frac{m_h}{m_{t_b} E_{ps}}, \tag{6.70}$$

which is plotted on Fig. 6.14c as the topmost hyperbola using climate and soil parameters approximating an average pinyon–alligator juniper watershed at Beaver Creek. Equation 6.70 is seen there to give a reasonable representation of the critical moisture condition for such canopies at least in the semi-arid conditions for which the current approximations apply.[†] In the highly seasonal case of the ponderosa pine canopies, the carryover storage is large (see Table 6.3) and plays an important role in support of vegetation growth, but is difficult to estimate accurately.

The lowermost "hyperbola" of Fig. 6.14c is a reproduction of the maximum soil moisture "ridge" line of Fig. 6.14a. The two plotted points are the estimated canopy

[†] Walter (1962) states "In effect the density of the vegetational cover is proportional to the precipitation . . . ".

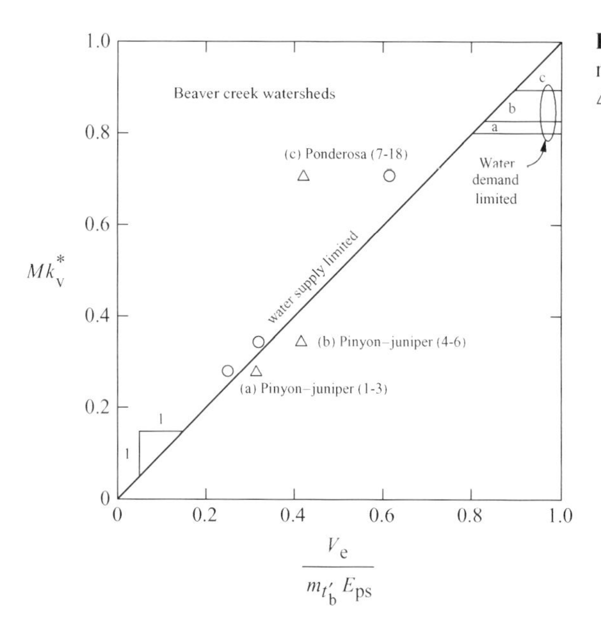

Fig. 6.15. Critical canopy moisture state. ○, Eq. 6.66; △, Eq. 6.70.

states for the two groups of pinyon–juniper watersheds as given by Salvucci (1992). Note that they lie much closer to the "cliff" than to the "ridge". The two dashed "hyperbolas" on Fig. 6.14c represent the locations of the ridge and the cliff following the above-mentioned modifications in the water balance introduced by Salvucci (1992) to accommodate the geological peculiarities of these particular watersheds.

Finally, the critical moisture state hyperbola, Eq. 6.66 and its approximation, Eq. 6.70, are compared, in Fig. 6.15, with observations from the three groups of Beaver Creek watersheds (two dry pinyon–juniper groups, watershed numbers 1–3 and 4–6, and one dry ponderosa pine group, watershed numbers 7–18). The relevant climate, soil, and vegetation parameters are summarized, with their sources, in Table 6.4 along with the estimated components of the available soil moisture V_e/m_h. Two conditions of this comparison should be noted:

(1) Observations of M (Salvucci, 1992 as listed in Table 6.3) and estimation of k_v^* (Eqs. 7.4, 7.14, and 7.31) allow comparison of Mk_v^* with Eq. 6.66 (plotted circles) and its approximation Eq. 6.70 (plotted triangles) for the two groups of pinyon–juniper watersheds as well as for the ponderosa pine watersheds. The product, Mk_v^*, of the two vegetation state variables fixes the ordinate for these observations while the abscissa is calculated from properties of the water environment using Eqs. 6.66 or 6.70.

(2) $\eta_o = 2.50$ for all pines as suggested for ponderosa by Waring (1983).

As we will see in Chapter 8, there is an upper limit to Mk_v^* for a given species at a given temperature beyond which the demand for water rather than the supply of water is constraining. This limit is $(k_v^*)_{M=1} \equiv k_{v1}^*$ and occurs, in a given soil, at the available moisture conditions, V_e, which produce a closed canopy. For larger V_e there is no change in Mk_v^*. This is indicated in Fig. 6.15 by the horizontal lines for the different species.

While the sample is small, the observations presented in Fig 6.15 strongly support Eq. 6.66 (plotted circles) as defining the operating state of these water-supply-limited canopies provided that we acknowledge an underrepresentation of the surface runoff (and hence a slight overestimation of V_e) in the wetter climates due to our expedient choice of exclusively Hortonian runoff. The approximation, Eq. 6.70, shown on Fig. 6.15 as the plotted triangles, is inadequate qualitatively as well as quantitatively in this highly seasonal climate due to its additional neglect of carryover storage.

In evaluating Fig. 6.15 we must remember that our analysis of a highly variable semi-arid climate has been based upon mean values producing the critical soil moisture. This idealization can distort reality in a significant way as is easily shown: Suppose the plant can adapt to a certain mild level of water stress. Under water-limiting circumstances the community of such plants can then develop a greater canopy cover than if no stress was tolerable and may generate higher production. This path to higher productivity seems certain to ensure that in the high variability of semi-arid climates some species will develop a measure of water stress tolerance.

The special case of riparian zones in semi-arid climates

For the small P_τ of semi-arid climates, where the soil moisture supply of their riparian zones comes primarily from groundwater recharged at a distance, term 7 may dominate the right-hand side of Eq. 6.65 giving, for $\beta_v = 1$

$$Mk_v^* \approx \frac{K(1)}{E_{ps}}\left[1 + \frac{3/2}{mc-1}\right]\left[\frac{\psi(1)}{z_w}\right]^{mc}, \tag{6.71}$$

in which case the soil properties are of paramount importance. Using Eq. 6.66, we have for this special case,

$$V_e \approx m_{t_b} K(1)\left[1 + \frac{3/2}{mc-1}\right]\left[\frac{\psi(1)}{z_w}\right]^{mc}. \tag{6.72}$$

M Conclusions

We have seen that the *concept* of critical soil moisture potential, ψ_c, is essential to understanding the limitations on stressless plant transpiration in a given climate, but we have also seen that its *magnitude* affects only the negligible terms of the water balance (i.e. terms 4–6 of Eq. 6.62) and hence is irrelevant to estimation of the canopy mass flux capacity, $M\, k_v^*\, E_{ps}$. *Assuming adequate nutrition*, we draw the important conclusion that species habitat is determined by parameters of the climate and not of the soil.

It is important for the work of later chapters to note that the remarkable validation of Eq. 6.66 presented in Fig. 6.15 may be interpreted as confirmation of our hypothesis that natural selection seeks stresslessly to maximize the productivity (cf. Appendix E) of biomass for a given species in a given climate. This is demonstrated by the adequacy of the canopy mass flux capacity to deliver the available moisture to the "free" atmosphere in the time available. Indeed, Fig. 6.15 shows the canopy mass flux capacity, $M\, k_v^*\, E_{ps}$, to be tuned to *exactly* match the rate, $V_e/m_{t_b'}$, required for moisture exhaustion to the point of incipient stress. Therefore, while *water resource-limited* by virtue of $M < 1$, *these systems are not limited by their moisture flux capacity*. It seems reasonable to conclude that this is no accident, and that natural selection in the given environment has fixed β and L_t at the appropriate values to bring about this productivity maximization. We return to this conjecture in Chapter 7.

This confirmation of the details of our water balance formulation gives reasonable confidence in its later use in testing the important habitat hypotheses of Chapter 9.

Part II

Darwinian ecology

7

Optimal canopy conductance

The canopy resistance ratio is calculated for a closed canopy of homogeneous multilayer crowns using a series–parallel network of interleaf-layer resistances having identical values obtained from the theoretical kinematic eddy viscosity.

For fixed leaf angle, cylindrical crowns show a minimum in the resistance ratio (corresponding to a maximum in the canopy water vapor conductance) at a leaf area index in the commonly observed range. At optical optimality, there is only one species having maximum canopy conductance at a given leaf angle. This is an *optimum foliage state* because it brings both maximum absorption of useful radiation and maximum flux of nutrients from soil to plant.

For fixed leaf angle, tapered multilayer crowns show a resistance ratio that decreases monotonically with increasing leaf area suiting these crowns for hot temperatures where high leaf areas will favor sensible heat rejection.

For fixed needle density, tapered monolayer crowns show a minimum in the resistance ratio at observed leaf areas due to a transition from foliage element to solid body aerodynamic resistance and thus are favored in cold climates due to their inhibition of heat transfer.

Conductance is approximated for open canopies by linear proportion to the estimated conductance of a stand-alone tree.

By jointly requiring the optimum foliage state and the critical moisture state, an *optimum canopy state* is defined that compares well with limited observations in open canopies.

A Compensation light intensity

In Chapter 3 we introduced the so-called *compensation* value of light intensity as that for which the light requirements of respiration are equal to those of the total photosynthesis. That is, at compensation value, the light incident upon a leaf has the intensity that is just sufficient to maintain life in the leaf but is not large enough to produce any leaf biomass. This is clearly an important limit in the plant's self-organization of its structure to maximize productivity. Horn (1971, p. 72) puts it clearly as follows:

> The optimal density of leaves is that which just reduces the light intensity below the tree to the compensation point; as long as there is more light, additional leaves would add more by photosynthesis than they would subtract by respiration, but once the light reaches the compensation point, additional leaves respire more than they photosynthesize.

Plate 7.1. London plane tree. (Photographed by William D. Rich in Arnold Arboretum, Boston; Copyright © 2001 William D. Rich.)

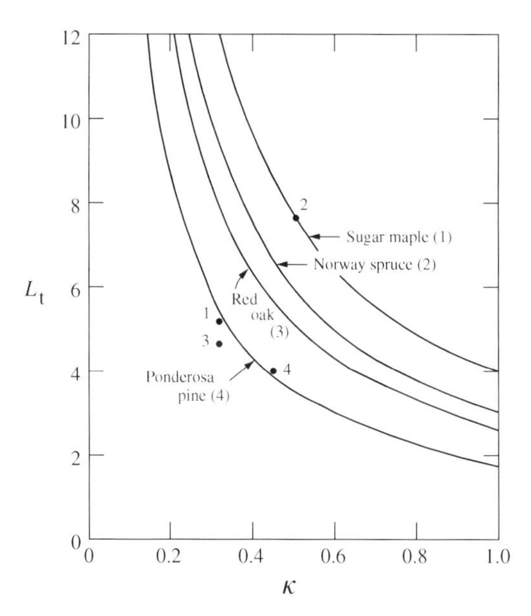

Fig. 7.1. Canopy parameters for optimal light transmission. Solid lines represent the minimum transmission of light to the lowest leaves as determined from experiments yielding so-called "compensation" light intensities (Baker, 1950, Table 12, p. 143) summarized here in Table 3.9. Plotted points represent the average of field observations assembled here in Tables 2.2 and 2.3 assuming $\kappa \approx \beta$, and are keyed to the curves by the associated numbers.

We saw in Chapter 3, that for homogeneous leafy canopies, the so-called Monsi–Saeki relationship for radiation extinction is

$$\frac{I(\xi)}{I_0} = \exp[-\kappa L_t \xi], \tag{7.1}$$

in which I_0 is the insolation at canopy top, $I(\xi)$ is the insolation transmitted to level ξ of the canopy, and κ is an empirical decay or extinction coefficient which is dependent primarily upon the quality of the light, the extent and arrangement of the foliage elements, and the solar elevation. At the base of the crown, $\xi = 1$, Horn's (1971) argument as quoted above puts

$$I(\xi = 1) = I_k, \tag{7.2}$$

where I_k is the compensation insolation, making the special case of Eq. 7.1

$$\frac{I_k}{I_0} \equiv I_k^o = \exp[-\kappa L_t]. \tag{7.3}$$

Values of I_k^o determined in laboratory experiments using artificial light with different tree species are listed in Table 3.9 as taken from Baker (1950, Table 12, p. 143). Using the condition of optical optimality from Chapter 3, i.e., that $\kappa = \beta$, these tabulated values define the desired species-specific *horizontal leaf area index*, βL_t,[†] as

$$\beta L_t = \kappa L_t = \ln\left[\frac{1}{I_k^o}\right]. \tag{7.4}$$

[†] The species specificity of βL_t is demonstrated in Appendix H.

which are plotted as the solid hyperbolae in Fig. 7.1. It should be noted that Baker (1950, p. 143) considers these values to be inaccurate (I_k^o too high) due to "conditions of the experiment".† Field observations of L_t, κ, and β have been presented in Tables 2.2 and 2.3 for various tree species. The average of these observed values is plotted in Fig. 7.1 in comparison with the experimental $\beta L_t = \kappa L_t$ curves for each of the species considered by Baker (1950) and is keyed to the Baker curves by the associated number in Table 3.9. Agreement of the values obtained by the two methods is obtained only for pine.

B The resistance ratio for closed canopies: M = 1

The homogeneous canopy of multilayer cylindrical crowns‡

In Chapter 4 we demonstrated that when the stomata are fully open, the canopy resistance, r_c, may be approximated by the composite interleaf atmospheric resistance, r_{ci}. That is

$$r_c \approx r_{ci}, \tag{7.5}$$

where $r_{ci} = r_c$ is assembled from a network of interleaf resistances, r_i, as is shown in the simplified canopy resistance diagram of Fig. 7.2a. Assuming the stomatal evaporating surfaces are all at a common potential§ *on one side of the leaf*, Fig. 7.2a reduces to Fig. 7.2b from which we write

$$\frac{1}{r_c} = \frac{1}{r_i} + \frac{1}{2r_i} + \frac{1}{3r_i} + \cdots + \frac{1}{(L_t - 1)r_i} = \frac{1}{r_i}\sum_{n=1}^{L_t - 1}\frac{1}{n}. \tag{7.6}$$

From Gradshteyn and Ryzhik (1980, p. 2, no. 0.131)

$$\sum_{k=1}^{n}\frac{1}{k} = \hbar + \ln(n) + \frac{1}{2n} - \sum_{k=2}^{\infty}\frac{A_k}{n(n+1)\cdots(n+k-1)}, \tag{7.7}$$

† In particular, it is not clear whether the "conditions of the experiment" included a different "full sunlight", I_0, representative of the habitat of each species, or used a common "full sunlight" such as the 107.6 klx defined by Gates (1980, p. 94). In addition, the compensation values may be sensitive to climate.

‡ Given the biological requirement for compensation light intensity at the lowest leaf (cf. Section A of this chapter), homogeneous canopy foliage area distribution is realistic only for cylindrical crowns. For tapered crowns the vertically averaged foliage density must increase with radius from the tree axis.

§ This feature distinguishes the formulation of canopy water vapor efflux from that of canopy CO_2 influx. The stomatal source of water vapor is at saturation vapor pressure which is constant at all elevations in isothermal crowns, while the CO_2 sink concentration within the stomata varies with depth in the crown in a manner that is dependent upon the foliage density and upon the concentration needed at the lowest leaf to support compensation-level photosynthesis there. The difficulty of estimating the gradients driving the canopy fluxes motivates and justifies the use of the "big leaf" approximation of the canopy intoduced for the derivation of canopy conductance in Chapter 5.

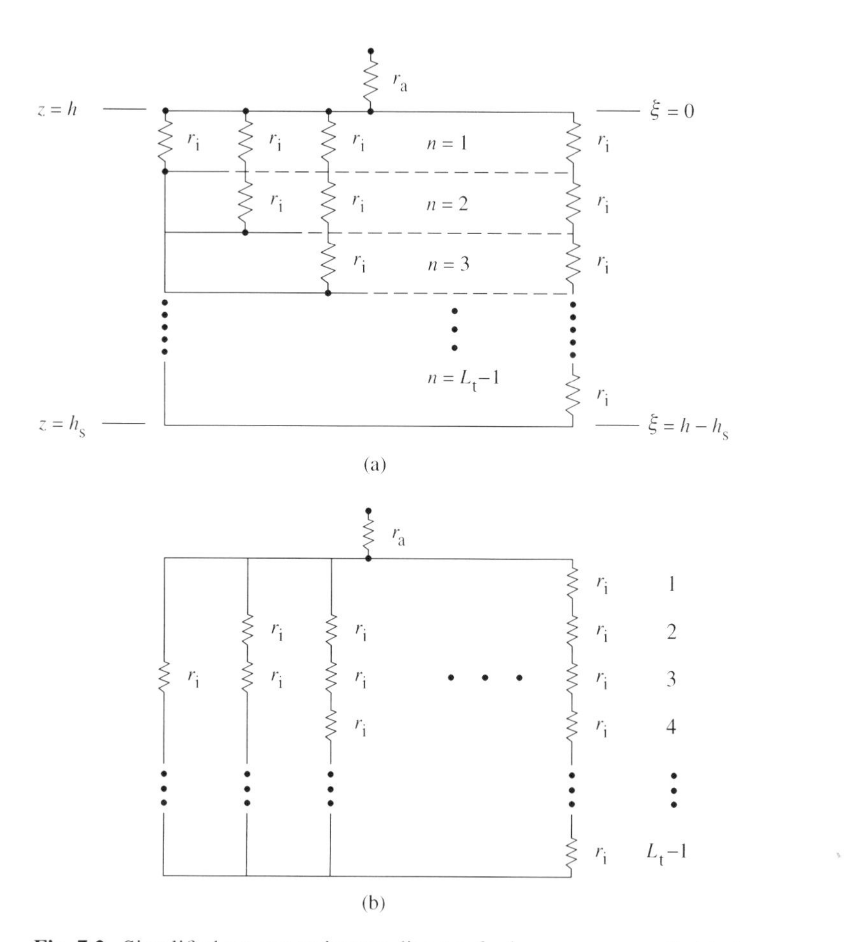

Fig. 7.2. Simplified canopy resistance diagram for homogeneous cylindrical multilayers. (a) Resistive analog to flow in crown; (b) equivalent crown resistive circuit.

where $\hbar$ = Euler's constant = 0.577 . . . , and

$$A_2 = \frac{1}{12}, \quad A_3 = \frac{1}{12}, \quad A_4 = \frac{19}{80}, \quad A_5 = \frac{9}{20}. \tag{7.8}$$

Thus, to within 0.2% at the common value $L_t = 5$, and with $L_t \geq 2$,

$$\sum_{k=1}^{L_t-1} \frac{1}{k} \approx 0.577 + \ln(L_t - 1) + \frac{1}{2(L_t - 1)} \equiv N(L_t). \tag{7.9}$$

The function $N(L_t)$ increases monotonically with L_t from $N(L_t) = 1.08$ at $L_t = 2$ to $N(L_t) = 2.83$ at $L_t = 10$. Introducing

$$n_s = \frac{\text{stomated leaf area}}{\text{projected leaf area}}, \tag{7.10}$$

to generalize for those species having stomates on more than one leaf surface, we have n_s parallel diffusion paths and we use Eqs. 7.6 and 7.9 to write

$$r_c = \frac{r_i}{n_s N(L_t)}. \tag{7.11}$$

For the homogeneous cylindrical multilayer, r_i is given by Eq. 4.167, and using Eq. 7.9, Eq. 7.11 becomes

$$r_c = \frac{\left[\dfrac{1}{ku_*}\right][(1-m)/m](\gamma L_t)^2}{n_s(L_t-1)\left[1-\exp\left(-\dfrac{1-m}{m}\gamma L_t\right)\right]\left[0.577+\ln(L_t-1)+\dfrac{1}{2}(L_t-1)^{-1}\right]}. \tag{7.12}$$

Dividing Eq. 7.12 by Eq. 4.143, and using Eq. 4.59, we finally have the important resistance ratio for the homogeneous cylindrical crown

$$n_s\frac{r_c}{r_a} = \frac{[(1-m)/m](\gamma L_t)^2}{(L_t-1)\left[1-\exp\left(-\dfrac{1-m}{m}\gamma L_t\right)\right]\left[0.577+\ln(L_t-1)+\dfrac{1}{2}(L_t-1)^{-1}\right]}. \tag{7.13}$$

Equation 7.13 is presented graphically by the solid lines in Fig. 7.3 for "leafy" plants (i.e., $m = 1/2$, $n = 2$) over the common range of β (see Fig. 3.18). We note a

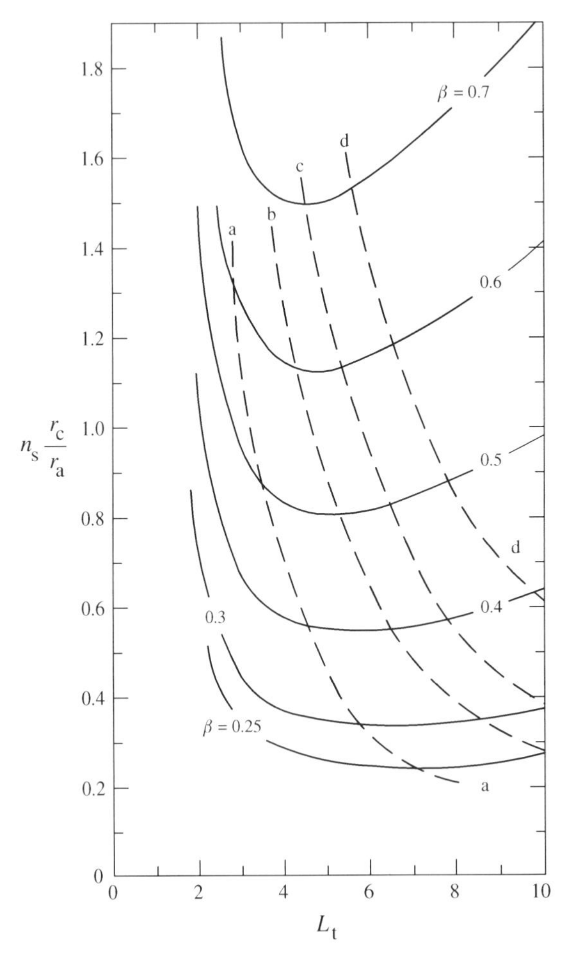

Fig. 7.3. Theoretical canopy resistance for homogeneous cylindrical multilayers ($\kappa = \beta$, $M = 1, m = 1/2, n = 2$). Dashed lines = optimal light transmission (i.e., compensation light intensity, Table 3.9): (a) ponderosa pine ($\beta L_t = 1.77$); (b) loblolly pine ($\beta L_t = 2.58$); (c) beech ($\beta L_t = 3.19$); (d) sugar maple ($\beta L_t = 3.91$).

minimum in the resistance ratio of $n_s r_c / r_a = O(1)$ at $4 \leq L_t \leq 6$. This minimum $n_s r_c / r_a$ translates, through Eq. 5.29, to a maximum in the evaporative flux of water vapor at the given temperature. The cause of this minimum in the resistance ratio can be seen from the behavior of Eq. 7.11. The denominator, $n_s N(L_t)$, increases monotonically with L_t, so the interesting behavior lies within the numerator, r_i, as given by Eqs. 4.165 and 4.167:

For low L_t, $\frac{dr_c}{dL_t}$ is controlled by the change in the dimension, Δz_L, the numerator of r_i, which declines rapidly with increasing L_t giving $\frac{dr_i}{dL_t} = -\frac{\beta}{(L_t - 1)^2}$.

For high L_t, $\frac{dr_c}{dL_t}$ is controlled by the change in the eddy viscosity, $K_m(z)$, the denominator of r_i, which declines rapidly with increasing L_t giving $\frac{dr_i}{dL_t} = \frac{\beta^2 L_t (L_t - 2)}{(L_t - 1)}$.

We will return to further discussion of the physical basis for this interesting behavior in Chapter 10.

In conjunction with Eqs. 5.29 and 5.30, we see that this minimum in $n_s r_c / r_a$ maximizes the canopy conductance at constant temperature.[†]

Optimum foliage state

We have seen in Eq. 7.4 that the compensation light requirement provides a constraint on the foliage density in the form of species-specific and optically optimal βL_t, (cf. Table 3.9). Accordingly, the horizontal leaf area index, βL_t, becomes established as the principal independent foliage parameter, and a few of these hyperbolas are superimposed on Fig. 7.3 as dashed lines covering the range of species tabulated. For a given β, determined by the light conditions as described in Chapter 3, there is one species for which the resistance ratio has its minimum value and thus the canopy water vapor flux, and with it the nutrient flux, is maximum (cf. Eq. 5.29). Note that at high β the shape of the resistance curve is such as to be highly selective of the species meeting the minimality condition, while as β gets smaller there is a broadening range of species lying within some arbitrary percentage of the minimum resistance ratio.

We now present the information of Fig. 7.3 in a more compact form. Differentiating Eq. 7.13 with respect to L_t while holding β, m, and n constant, we get the locus of β, L_t pairs producing minimum $n_s r_c / r_a$, and thus maximum canopy conductance, k_v, in a given climate:

$$\frac{\frac{1-m}{m} \gamma L_t \exp\left(-\frac{1-m}{m} \gamma L_t\right)}{1 - \exp\left(-\frac{1-m}{m} \gamma L_t\right)} = 2 - \frac{L_t}{L_t - 1} - \frac{\frac{L_t}{L_t - 1} - \frac{L_t}{2(L_t - 1)^2}}{0.577 + \ln(L_t - 1) + \frac{1}{2(L_t - 1)}}. \tag{7.14}$$

Equation 7.14 is plotted as the "maximum vapor flux" curve in Fig. 7.4 for the "leafy" plant case of $m = 1/2$, $n = 2$.

[†] In Chapter 10 we will see that the shape of this function also leads, for large βL_t, to a growth-limiting constraint on the flux of atmospheric carbon to the leaves.

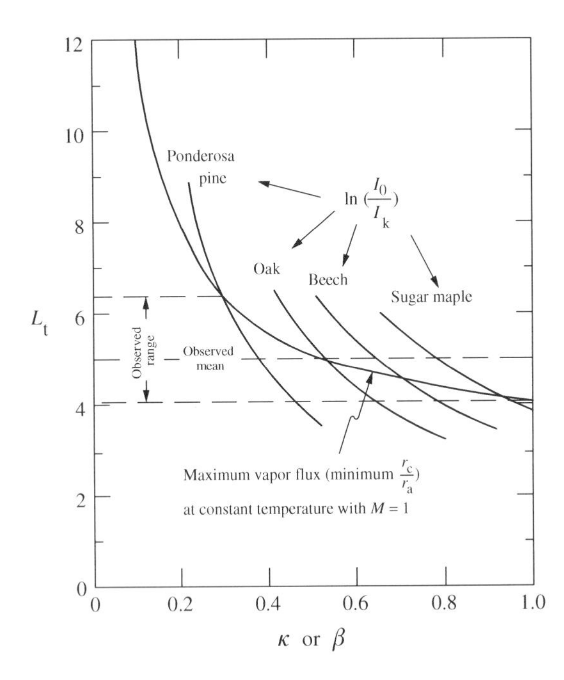

Fig. 7.4. Optimal foliage state (homogeneous cylindrical multilayers, $M = 1, m = 1/2, n = 2$).

Once again, Eqs. 7.4 and 7.14 together determine β and L_t for a given species in a given light environment. Their joint solution is illustrated graphically in Fig. 7.4 for four species covering the full range of those reported by Baker (1950) and repeated here in Table 3.9.[†] We refer to this foliage state as "optimal" because of the following accompanying flux conditions:

(1) The radiant energy absorbed is the maximum possible (i.e., the crown is optically optimal) – for larger βL_t the lower leaves will receive insufficient light, while for smaller βL_t unused energy will be transmitted through to the forest floor.

(2) The flux of transpirate from the soil to the atmosphere, and with it the flux of nutrients from the soil to the plant, will both be maximum for a given temperature because the minimization of r_c/r_a accompanying this joint solution assures maximization of the normalized canopy conductance, k_v^* (cf. Eqs. 5.29 and 5.30).[‡]

We wish to know if nature selects for this multiply optimum state. However, we are assuming natural selection to favor increased annual productivity which depends upon factors other than just the average canopy conductance. We return to this question in Chapter 10 and for now offer only the following observations relative to this theoretical optimum.

[†] There is evidence (Baker, 1950, pp. 141–142) that these values may be climate-dependent.
[‡] This statement is strictly true only for the limiting, $M = 1$ case. As we will see later in this Chapter, for $M < 1$ we must also specify the moisture supply and the geometry of the individual plant.

Plate 7.2. Cedar. (Photographed by Peter S. Eagleson in the Borghese Gardens, Rome.)

Observations of foliage state

(1) Note the relatively narrow range $4 \leq L_t \leq 6.4$ associated with the optimum foliage state for the species of Fig. 7.4 as indicated thereon by the horizontal dashed lines.† Tables 2.2 and 2.3 show the average observed L_t for deciduous trees to be $(L_t)_{avg} = 5.03$. Kestemont (1973), for the deciduous temperate forest, and Carbon *et al.* (1979, 1981) for eucalyptus, observed the average to be $4 \leq (L_t)_{avg} \leq 6$, while Mooney (1972) observed the optimum leaf area index for photosynthesis to be approximately 5. The agreement here between theory and observation is striking. The relatively narrow range of β called for

† Doley (1982) found the same range of optimum L_t from his model of the dependence of dry matter production on the light extinction coefficient, κ.

by growing-season light conditions over the latitudes supporting tree growth (cf. Fig. 3.18), along with the species-constant βL_t required for "compensation" level radiative flux (cf. Table 3.9), seriously constrain the maximum L_t that can be supported by photosynthesis for a given species in a given climate.

(2) Monteith (1965, p. 221) noted that field crops with adequate water have a "minimum" r_c/r_a between 1 and 2, but that r_c/r_a for a pine forest is 15 (not specified to be a minimum however). Similarly, for Sitka spruce and Scots pine Jarvis *et al.* (1976, pp. 212–213) give $20 < r_c/r_a < 40$ and $20 < r_c/r_a < 70$ respectively (both having $n_s \approx 1$), again with no mention of minimality in either case. These reported observations *for trees* are an order of magnitude larger than those just predicted theoretically. We will offer an explanation for this difference shortly.

(3) Further, but more indirect, confirmation of our small effective resistance ratios is obtained using the few estimates of k_v^* found in the literature and assembled here in Table 7.1. From this table we see that the average value of the canopy conductance for trees is about $k_v^* \approx 0.73$. Taking an average growing-season atmospheric temperature of $T_0 \approx 21\,^\circ\text{C}$, we can use Eq. 5.29 , with $k_v = k_v^* = 0.73$, to find the average resistance ratio for trees to be $r_c/r_a \approx 1.20$, which is consistent with the optimum values derived herein. In contrast, using $r_c/r_a \approx 30$, such as reported by Jarvis *et al.* (1976, pp. 212–213), at the same temperature, gives a plant coefficient, $k_v^* = 0.10$ which is far less than the estimates in Table 7.1.

Table 7.1. *Collected estimates of canopy conductance for trees*

Tree type	k_v^*	Reference
Citrus	0.70[a]	Doorenbos and Pruitt (1977, Table 25, p. 47)
Apple, cherry	0.85[a]	Doorenbos and Pruitt (1977, Table 26, p. 49)
Peach, apricot, pear, plum	0.75[a]	Doorenbos and Pruitt (1977, Table 26, p. 49)
Boreal forest	0.51[b,c]	Arris and Eagleson (1994, Table 1, p. 6)
Deciduous forest	0.79[b,c]	Arris and Eagleson (1994, Table 1, p. 6)
Southern pine forest	0.62[b,d]	Arris and Eagleson (1994, Table 1, p. 6)
Oak forest	0.88[e]	Rauner (1976, Fig. 13, p. 260)
Average	0.73	

[a]From transpiration measurement normalized by that from short, green grass.
[b]Calculated from observations of net primary productivity.
[c]Assuming $M = 0.84$ (maximum packing of circular discs).
[d]Assuming $M = 0.50$ (estimated from Eyre, 1968, Plate IIIA).
[e]From transpiration measurement normalized by that calculated from the heat balance of a wet surface.

We conclude from this analysis that at typical temperatures, the reported values of k_v^* for trees cannot be achieved *as modeled* with $r_c/r_a = O(10)$ but can be achieved with $r_c/r_a = O(1)$. The deciding element, missing from our formulation, seems to be advection and hence the value of the Bowen ratio, $\boldsymbol{R}_b$, at the time of the observations of either k_v^* or r_c/r_a. Equations 5.29 and 5.38 give the rough estimate

$$\frac{r_c}{r_a} = \left(\frac{1}{k_v^*} - 1\right)\left(1 + \frac{\Delta}{\gamma_o}\right) \approx |\boldsymbol{R}_b|\left(1 + \frac{\Delta}{\gamma_o}\right), \tag{7.15}$$

which shows that at typical temperatures, the Bowen ratio must be greater than unity and considerable advected energy is required for the resistance ratio to be of order ten.

The non-homogeneous canopy of multilayer hemispherical crowns

Allowing for multiple stomated leaf surfaces as in Eq. 7.13, the resistance ratio for a hemispherical multilayer crown having homogeneous leaf area density (cf. Chapter 4 for a discussion of this simplifying departure from reality) is given by Eq. A.74 as

$$n_s \frac{r_c}{r_a} = 0.97 \frac{\beta L_t}{\widehat{K_m^o}(L_t - 1)^{0.54}} \cdot \left[\sum_{n=1}^{L_t - 1} \frac{1}{n^{0.54}}\right]^{-1}, \tag{7.16}$$

in which, from Eq. A.25 for the full hemisphere

$$\widehat{K_m^o} = 0.55 + 0.052\,(\beta L_t) + 0.008\,(\beta L_t)^2. \tag{7.17}$$

Equation 7.16 (using Eq. 7.17) is plotted in Fig. 7.5 where we see that in contrast to the multilayer cylinder there is no minimum in the resistance ratio of the multilayer hemisphere. As was pointed out in Chapter 4, this is due to the monotonic increase in the eddy viscosity, $\widehat{K_m^o}$, with increasing βL_t, brought about by the "open" (i.e., rough) surface of tapered-crown canopies.

The non-homogeneous canopy of multilayer conical crowns

With homogeneous crown leaf area density, Eq. A.86 gives the resistance ratio for a canopy of conical multilayer crowns as

$$n_s \frac{r_c}{r_a} = n_s \left(\frac{r_c}{r_a}\right)_s = 1.23 \frac{\beta L_t}{\widehat{K_m^o}(L_t - 1)^{1/3}} \cdot \left[\sum_{n=1}^{L_t - 1} \frac{1}{n^{1/3}}\right]^{-1}, \tag{7.18}$$

where the subscript "s" is added to the resistance ratio to distinguish its density as "sparse" in contrast to that of the conical monolayer to be considered next, and in

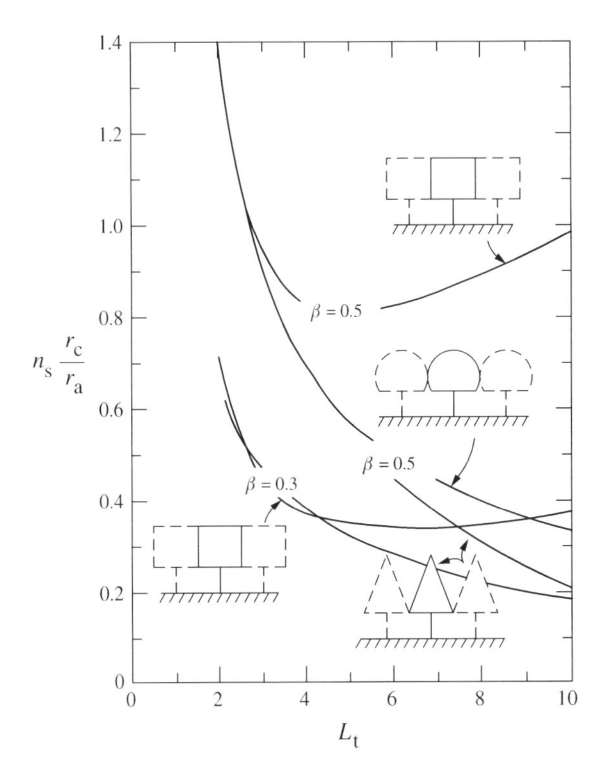

Fig. 7.5. Effect of crown shape on canopy resistance (homogeneous multilayers, $M = 1, m = 1/2, n = 2$; no consideration given to species-dependent light limitations on L_t).

which from Eq. A.27

$$\widehat{K_{\mathrm{m}}^{\mathrm{o}}} = 0.605 + 0.108\,(\beta L_t) + 0.042\,(\beta L_t)^2. \tag{7.19}$$

Equation 7.18 (using Eq. 7.19) is also plotted in Fig. 7.5 and shows the same monotonic behavior and for the same reasons explained above for the hemispheres.

The significance of crown shape

From our discussions thus far (including Fig. 7.5), it would appear that tapered multilayer crowns should be universally preferred whenever the availability of water is not limiting, since with all else being constant, lower flux resistance can always be obtained through an increase in its photosynthetically active foliage area.† Selection and/or adaptation should then drive the development of such crowns toward ever higher L_t for given β. Other factors come into play however. In the first place (considering only the conical crown), except when $2\frac{h}{d} \geq \tan(h_\otimes)$, one crown shadows another thereby reducing the productive potential. Furthermore, there must be an upper limit to L_t in a given climate for at least three reasons:

† For each species there is a maximum L_t for each β, imposed by the compensation light requirement of that species (cf. Table 3.9).

(1) Water is never truly unlimited and its vertical flux, which rises with L_t, is constrained by the presumed need to avoid water stress.

(2) As we will see in Chapter 10, biomass productivity is indifferent to crown shape at low βL_t since there productivity is water-limited. However at large βL_t where water is plentiful, CO_2 influx becomes limiting for cylindrical multilayers and the resistive advantages of tapered multilayer crowns do become important. Examples of the latter case are given by the conically crowned Japanese larch which is found in wet climates, and by the hemispherically crowned rainforest that displays a canopy-top looking like scoops of ice cream (cf. Odum, 1975, Fig. 7-10, p. 197).

(3) As the foliage area increases within a crown of given dimensions, the increasing crown foliage density causes a fundamental change in the mechanism of canopy resistance. As was argued in Chapter 4, increasing foliage resistance to the passage of air laterally *through* the tapered crowns causes a greater fraction of the air to flow *around* the crowns. This increases the role of crown shape in canopy resistance while decreasing the role of the foliage elements. A common instance of this occurs in cold climates where maintenance of biochemically functional leaf temperature becomes important. Structural tree forms having drooping branches that overlap closely in a "shingled" fashion at their extremities may be a selective response to this heat problem. In such structures, leaves, usually needles because of the singular ability of needle-leaved species to survive extremely low temperatures,[†] are confined to the extremities where their proximity forms an insulating layer restricting the loss of canopy heat (Gates, 1980, p. 364). The density of needles necessary to achieve the needed insulation is such that solid body resistance is generated. This high foliage density also causes the available light to decay to compensation intensity over a very short distance leaving a relatively thin "skin" of active vegetation at the crown surface referred to herein as a "monolayer". Of course, such limitation of heat efflux also operates to limit water vapor efflux as well as CO_2 influx, and thereby restricts the supply of both carbon and nutrients to the plant. We examine the effect of this limitation on plant productivity in Chapter 10, but for now we are only concerned with the production of solid body resistance by the monolayer crown, and how this resistance affects vertical mass flux in the canopy.

Accurate description of the transition, with increasing foliage density, from dynamic domination by foliage element resistance to that by solid body resistance in a canopy of tapered crowns is problematic. We base our estimate of this transition upon the limiting cases of low-density multilayers and high-density monolayers. Having just examined the lower limit provided by the homogeneous conical multilayer, we now look at the upper limit given by the high-density conical monolayer.

[†] See for example: Sakai and Weiser (1973), George *et al.* (1974), and Sakai (1978, 1979).

Plate 7.3. Fir foliage texture: Douglas fir. (Photographed by William D. Rich in Arnold Arboretum, Boston; Copyright © 2001 William D. Rich.)

The non-homogeneous canopy of high-density conical monolayers

In climates where the low temperatures are a result of high latitude and/or altitude, the relationship between leaf angle and solar altitude on the one hand (see Fig. 3.18), and that between leaf angle and crown surface inclination on the other hand (see Fig. 3.21), call for monolayers of conical shape. We have demonstrated in Fig. 3.18 that observations of evergreens at high latitude satisfy this relation for beam radiation and we restrict our consideration to these.

As developed in Appendix A (Eq. A.106) using the definitions of Fig. 7.6a, the limiting (i.e., solid body) reistance ratio is

$$n_s \frac{r_c}{r_a} = n_s \left(\frac{r_c}{r_a}\right)_d = \frac{2}{3} \left[\frac{n\beta L_t}{1 - \dfrac{n\beta L_t \exp(-n\beta L_t)}{1 - \exp(-n\beta L_t)}} \right] \frac{1}{\left(\widehat{K_m^o}\right)_d}, \tag{7.20}$$

in which the subscript "d" on the resistance ratio is to signify its "dense" character in contrast to that of the homogeneous multilayer discussed previously, and in which $(\widehat{K^o})_d$ is given by Eq. A.57

$$\left(\widehat{K_m^o}\right)_d = 0.454 + 0.009 L_t + 0.002 L_t^2. \tag{7.21}$$

Equation 7.20 (using Eq. 7.21) is plotted as the upper dashed line in Fig. 7.7 and Eq. 7.18 (using Eq. 7.19) is plotted as the lower dashed line in the same figure. Together they represent the upper and lower limits respectively of the resistance ratio for conical monolayers.

We now estimate a transition function between these limits in terms of the monolayer foliage density.

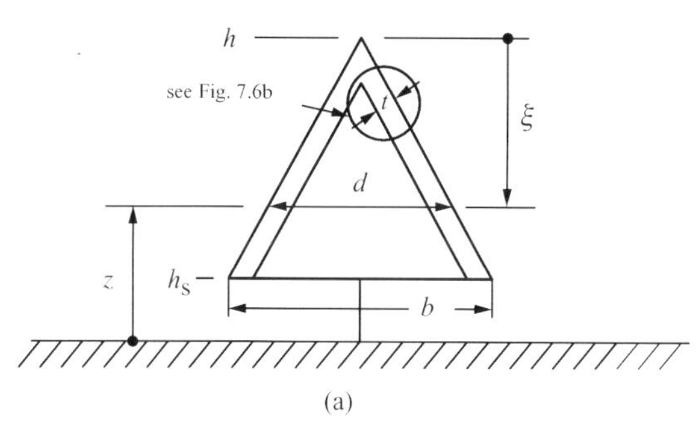

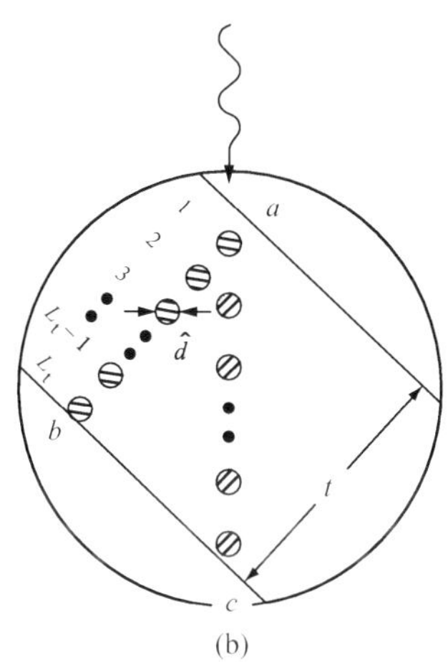

Fig. 7.6. Definition sketches for conical monolayer. (a) Cross-section of crown; (b) detail of idealized monolayer.

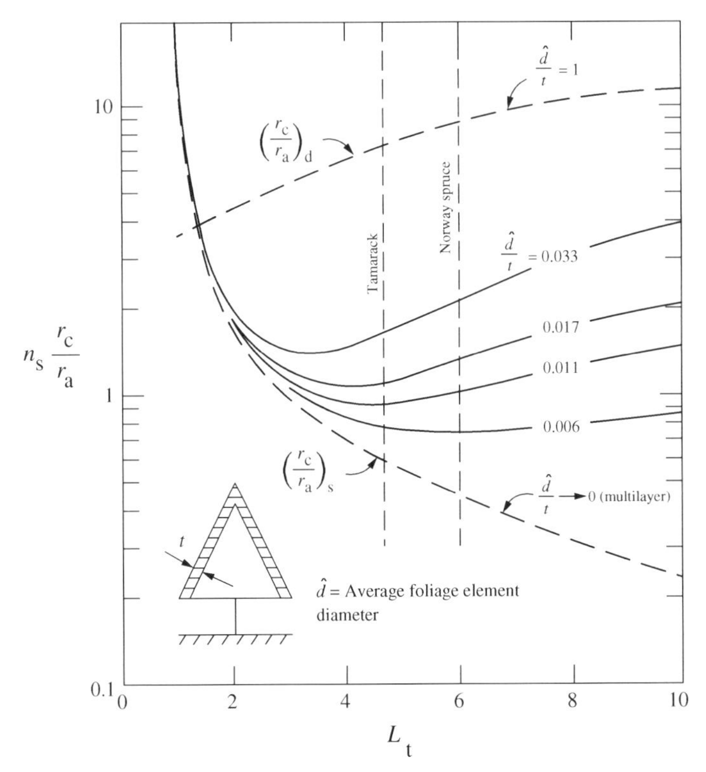

Fig. 7.7. Theoretical canopy resistance for conical monolayers ($M = 1$, $\beta = 0.50$).

Resistance ratio weighting function for conical monolayers

It is reasonable to expect the total resistance (ratio) to be composed of fractions of $\left(\frac{r_c}{r_a}\right)_s$ and $\left(\frac{r_c}{r_a}\right)_d$ proportioned, linearly we assume, according to the foliage volume density within the monolayer. In Fig. 7.6b we sketch an enlarged cross-section of the monolayer throughout which the foliage volume is assumed to be homogeneous. Because its thickness, t, is small with respect to the crown height, $h - h_s$, the thickness is measured perpendicular to the crown surface. When $t \rightarrow 0$, the crown surface is solid and contains all the foliage elements. (The foliage area index, L_t, is then given by the lateral crown surface divided by the crown basal area.) As $t > 0$, the crown surface must open to admit light for the leaves that are beneath the crown surface. The average diameter of the cylindrical foliage elements within this monolayer is indicated by $\hat{d}$, and a single $\hat{d}$ in Fig. 7.6b represents the average thickness of a single leaf layer. There will be an average of L_t such layers in any cross-section. From this one-dimensional approximation, the *monolayer foliage volume density*, δ_f, is written

$$\delta_f = \frac{\hat{d}}{t} L_t. \tag{7.22}$$

Landsberg and Thom (1971) give average diameters for the needles and twigs of spruce as 1 mm and 3 mm respectively, thus we take $\hat{d} \approx 2$ mm. Further, we estimate (cf. Landsberg and Thom, 1971, Fig. 1) the diameter of the spruce shoot at 3 cm. From these dimensions we consider four arbitrary but reasonable *monolayer densities*, $\hat{d}/t$; t = two shoots, $\frac{\hat{d}}{t} = 0.067$; t = four shoots, $\frac{\hat{d}}{t} = 0.017$; t = six shoots, $\frac{\hat{d}}{t} = 0.011$, and t = ten shoots, $\frac{\hat{d}}{t} = 0.006$.

With the definition of Eq. 7.22, the crown surface becomes solid as $\delta_f \rightarrow 1$ making

$$\frac{r_c}{r_a} = \left(\frac{r_c}{r_a}\right)_d, \tag{7.23}$$

and we can then write the general expression for the resistance ratio

$$\frac{r_c}{r_a} = \delta_f \left(\frac{r_c}{r_a}\right)_d + (1 - \delta_f)\left(\frac{r_c}{r_a}\right)_s. \tag{7.24}$$

Equation 7.24 is evaluated using Eqs. A.100 and A.106, and is plotted as the solid lines in Fig. 7.7 for the four chosen values of $\frac{\hat{d}}{t}$, and for the representative value for spruce crowns, $\beta = 0.50$ (cf. Fig. 3.18). Note that the resistance ratio now has a minimum indicating a preferred L_t in the range $3 \leq L_t \leq 6$ which is to be compared with the value $L_t = 3.3$ observed by Landsberg and Jarvis (1973) for Sitka spruce in northeast Scotland. Note also that over the fairly wide range of the parameters displayed, the minimum resistance ratio lies in a quite narrow range, $0.7 < r_c/r_a < 1.4$.

The vertical dashed lines on Fig. 7.7 represent the L_t for compensation light intensity at $\beta = 0.50$ for two densely foliated, conically crowned species, tamarack (i.e., larch)

Plate 7.4. Painted maple. The author alongside maple in Arnold Arboretum, Boston. (Photograph by William D. Rich; Copyright © 2001 William D. Rich.)

and Norway spruce as given by the values of Baker (1950, Table 12, p. 143) and reproduced here in Table 3.9. Note that they fall in the L_t range of minimum resistance for these reasonable monolayer densities.

C The resistance ratio for open canopies: M $\ll$ 1

The basic assumption

For closed canopies we saw that the vertical fluxes occur by turbulent diffusion through the leaf structure of the crowns. However, as the canopy cover, M, gets small, the wind enters the crown from the side and horizontally advects the water vapor and sensible heat from the leaf surfaces to the atmosphere over the downwind bare soil where it then diffuses vertically out into the "free" atmosphere. Of course CO_2 moves toward the leaf from the air under the same assumed processes. Irrespective of the crown shape, we will assume the velocity distribution that is effective in the vertical diffusion to be the logarithmic distribution (Eq. 4.26) characteristic of the upwind and downwind uniform, unvegetated substrate for which the kinematic eddy momentum viscosity is given by Eq. 4.38. Continuing our invocation of the Reynolds analogy, we will use the average of these two eddy momentum viscosities to make a *very crude* estimate of the vertical flux of transpirate.

An approximation for all crown shapes

Averaging over the vertical distance between the bottom of the crown ($z = h_s$) and the "free" atmosphere at canopy top ($z = h$), we approximate the average lumped "canopy" resistance, r_c, using Eqs. 4.38 and 4.165 as

$$r_c \approx \Delta z/\widehat{K_m} \approx \frac{\frac{1}{2}(h - h_s)}{u_* k\left(\frac{h + h_s}{2} - d_{os}\right)}, \tag{7.25}$$

in which d_{os} is the zero-plane displacement of the unvegetated surface. Equation 7.25 simplifies to

$$r_c \approx \frac{1}{ku_*}\left(\frac{1 - \frac{h_s}{h}}{1 + \frac{h_s}{h} - 2\frac{d_{os}}{h}}\right). \tag{7.26}$$

Equation 4.140 gives the lumped atmospheric resistance as

$$r_a = \frac{u_0}{u_*^2}, \tag{7.27}$$

where u_0 is the wind speed at $z = h$. Equations 7.26 and 7.27 give the important resistance ratio as

$$\left(\frac{r_c}{r_a}\right)_{M=0} = \frac{u_*}{ku_0} \cdot \left[\frac{1 - \dfrac{h_s}{h}}{1 + \dfrac{h_s}{h} - 2\dfrac{d_{os}}{h}}\right]. \tag{7.28}$$

The bulk drag parameter, $\frac{u_*}{ku_0}$, and the velocity distribution parameter d_{os}/h are dependent upon the substrate and the structure of both the crown and the canopy and hence are variable with the species, the climate, and the soil. For this case of sparse roughness elements (i.e., $M \ll 1$) their estimation is problematic. For $M \ll 1$, Raupach (1992, Fig. 5, p. 393) finds empirically that smooth substrates give $\frac{u_*}{ku_0} = 0.137$, while forest substrates maintain the value $\frac{u_*}{ku_0} \approx 1$ found theoretically in Chapter 4 for $M = 1$. For very sparse roughness elements, the displacement height, d_{os}, of a *fully developed* velocity profile must approach zero (Brutsaert, 1982, p. 116), while for homogeneous canopies of $M = 1$, the average of many observations on vegetated surfaces gives $d_o/h = 2/3$ (Brutsaert, 1982).

For our current purposes, we assume the bottom shear stress to remain $\frac{u_*}{ku_0} = 1$, and the displacement height to have the value $d_{os}/h = 1/3$, midway between that for the smooth substrate and that for the closed forest. With these values Eq. 7.28 becomes

$$\left(\frac{r_c}{r_a}\right)_{M=0} \approx \left[\frac{1 - \dfrac{h_s}{h}}{0.33 + \dfrac{h_s}{h}}\right], \tag{7.29}$$

which is plotted as the solid line in Fig. 7.8 along with a dashed line representing the limiting curve for $d_0/h = 0$ (i.e., smooth, bare substrate).

Note that our assumption of reduced effective eddy viscosity makes these estimated resistance ratios for $M \ll 1$ much larger than those for the $M = 1$ case presented in

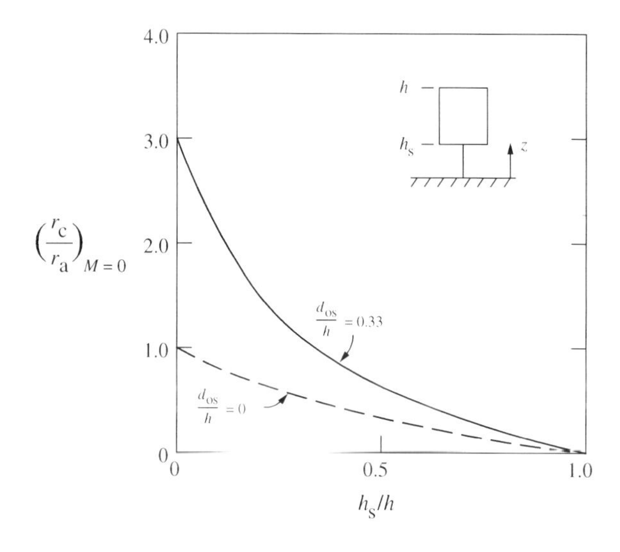

Fig. 7.8. Estimated resistance ratio for stand-alone trees.

Fig. 7.5 as long as h_s/h is small. We thus expect aridity to decrease land-surface transpiration by reduction in the canopy conductance (see Eq. 5.29) as well as by the reduction of canopy cover. Note also in Fig. 7.8 the rapid decline of the resistance ratio with increasing h_s/h due to the elevation of the average evaporating surface rising higher in the boundary layer. Keeping in mind that for a given temperature the canopy conductance increases with decreasing resistance ratio (Eq. 5.29 again), the behavior indicated in Fig. 7.8 provides ample evolutionary pressure for the development of plant stem height, h_s. Presumably, the stems will reach an equilibrium height at which a further growth increment causes an increase in the biological "overhead" for stem growth and maintenance that just balances the increase in CO_2 flux brought about by the added stem height. We will not consider this optimization here.

To make use of Eq. 7.29 we must estimate the stem fraction, h_s/h. MacMahon (1988, pp. 238–250) gives photographs of creosote bush deserts showing $h_s/h \approx 0$. Rauner (1976, Table 1, p. 242) lists observations of h_s/h for various deciduous forests, but with water plentiful (i.e., $M \to 1$), which give for aspen, $h_s/h = 0.52$; for linden, $h_s/h = 0.48$; for oak, $h_s/h = 0.46$; for maple, $h_s/h = 0.57$; and for birch, $h_s/h = 0.46$. The average of these deciduous observations is $h_s/h = 0.50$. For evergreen forests, Landsberg and Jarvis (1973) report $h_s/h = 0.67$ for a Sitka spruce canopy in the moist Fetteresso forest in Scotland. Photographs show $h_s/h \approx 0.57$ for ponderosa pine (Williams and Anderson, 1967), and $h_s/h \approx 0.15$ for Utah juniper (Clary *et al.*, 1974), both near Flagstaff, Arizona. In the same manner, we measure $h_s/h \approx 0.54$ for longleaf pine in central Florida (Christensen, 1988, Fig. 11.5, p. 328), and finally, we assume $h_s/h \to 0$ for a stand-alone spruce (cf. Plate 5.4).

D The resistance ratio for intermediate canopy cover

For intermediate canopy densities we will estimate the resistance ratio by linear interpolation between its values at $M = 0$ and at $M = 1$. That is

$$\left\langle \frac{r_c}{r_a} \right\rangle = (1 - M)\left(\frac{r_c}{r_a}\right)_{M \to 0} + M\left(\frac{r_c}{r_a}\right)_{M=1}. \tag{7.30}$$

Together with Eq. 5.29, Eq. 7.30 provides the important relation between the two canopy state variables, k_v^* and M:

$$k_v^* = \frac{1 + \Delta/\gamma_o}{1 + \Delta/\gamma_o + (1 - M)\left(\frac{r_c}{r_a}\right)_{M \to 0} + M\left(\frac{r_c}{r_a}\right)_{M=1}}. \tag{7.31}$$

This relation is a function of the crown structure through the resistance ratio, $\widehat{r_c/r_a}$; of the climate and soil through the canopy cover, M; and of the climatic growing season temperature through $\frac{\Delta}{\gamma_o}$. It serves to join the "internal optimization" of the canopy resistance to the "external optimization" of the security–productivity tradeoff as we will see in the next chapter.

Some typical examples of the M–k_v^* relation are given in Fig. 7.9 for broad vegetation classes using representative parameter values cited herein and which are listed in tabular form in the caption. In interpreting this figure, it cannot be overemphasized that the characteristic shape of the function, that is rising k_v^* with increasing M, is dominated both in kind and in degree by our crude model of the canopy resistance for $M \ll 1$ under which the relative values of h_s/h and d_{os}/h fix the $M = 0$ intercept and hence essentially determine the slope.

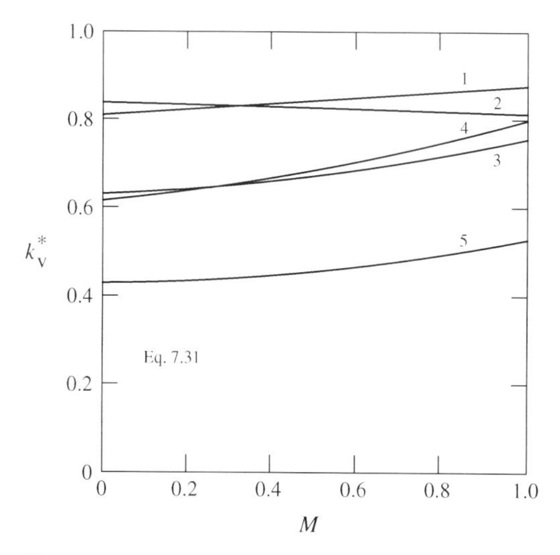

Fig. 7.9. Canopy conductance vs. canopy cover (multilayers). 1, deciduous; 2, loblolly pine; 3, ponderosa pine; 4, pinyon–juniper; 5, spruce.

	$M = 0$				$M = 1$		
Species	T_0 (°C)	d_{os}/h[a]	h_s/h[a]	r_c/r_a[b]	β[c]	βL_t[d]	r_c/r_a[e]
Deciduous[f]	15.6[i]	0.33	0.50	0.60	0.30	3.23[l]	0.38
Pine[f]							
Loblolly	21.0[i]	0.33	0.50[j]	0.60	0.45	2.58[l]	0.68
Ponderosa	14.2[h]	0.33	0.20[k]	1.51	0.50	1.77[l]	0.88
Spruce[g]	10.0[i]	0.33	0[i]	3.00	0.50	3.02[l]	0.82
Pinyon–juniper							
Nos. 1–3[f,h]	18.5[h]	0.33	0.10	1.77	0.42	2.27[m]	0.83
Nos. 4–6[f,h]	15.9[h]	0.33	0.10	1.77	0.52	2.58[n]	0.76
Creosote bush[f]	21.0[i]	0.33	0.50	0.75	—	—	—

[a] See text.
[b] Eq. 7.29.
[c] Fig. 3.18.
[d] $\beta = \kappa$ from Eq. 3.52.
[e] Fig. 7.03.
[f] Assumed cylindrical crown.
[g] Assumed conical crown.
[h] Table 3.03.
[i] Assumption.
[j] Christensen (1988, Figs. 11.4, 11.5).
[k] Williams and Anderson (1967, Fig. 5).
[l] Table 3.09.
[m] Table 3.09, Northern white cedar.
[n] Table 3.09, Lodgepole pine.

As we have seen in Chapter 6, the case of $M < 1$ is one in which available soil moisture, V_e, determines the transpiration rate. Therefore, an alternative to Eq. 7.31 for evaluating k_v^* in such cases is Eq. 6.66. Of course this doesn't remove the large uncertainty in estimating k_v^* but merely shifts it from $(r_c/r_a)_{M\to 0}$ in Eq. 7.31 to V_e and the open-canopy E_{ps} in Eq. 6.66.†

E The canopy state variables: cover and conductance

Optimum canopy state

In Chapter 6, we defined a critical soil moisture state at which canopy productivity is maximum and proposed this as the naturally selected canopy moisture state. From Eq. 6.66 this is

$$Mk_v^* = \frac{V_e}{m_{t_b'} E_{ps}}. \tag{7.32}$$

Equations 7.31 and 7.32 are sketched in Fig. 7.10. If the climate and soil are such that the soil moisture, V_e, available for stressless transpiration of a given species during the average interstorm period, produces an intersection with the $k_v^*(M)$ line (i.e., Eq. 7.31)

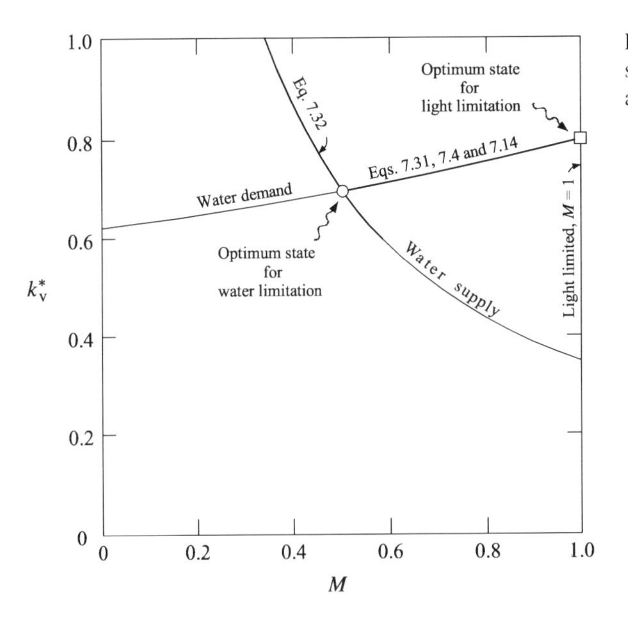

Fig. 7.10. Optimum canopy state (fixed species, climate and soil).

† Comparison of k_v^* as estimated by these two very different methods should be a useful and informative exercise. We carry it out implicitly in Fig. 7.11.

at $M < 1$, i.e., if

$$\frac{V_e}{m_{t_b'} E_{ps}} < \frac{1 + \Delta/\gamma_o}{1 + \Delta/\gamma_o \left(\frac{r_c}{r_a}\right)_{M=1}}, \tag{7.33}$$

the canopy will be *water-limited* and its optimum M–k_v^* state is as shown in Fig. 7.10 by the intersection of the two functions. On the other hand, should

$$\frac{V_e}{m_{t_b'} E_{ps}} \geq \frac{1 + \Delta/\gamma_o}{1 + \Delta/\gamma_o \left(\frac{r_c}{r_a}\right)_{M=1}}, \tag{7.34}$$

the canopy is *light-limited* and its optimum M–k_v^* state is $M = 1$ with

$$k_v^* \leq \frac{V_e}{m_{t_b'} E_{ps}}. \tag{7.35}$$

Some observed canopy states

Tables 6.3 and 6.4 contain a rare data set consisting of observations of climate, soil, and vegetation properties at the sites of several pinyon–juniper and ponderosa pine watersheds at Beaver Creek (near Flagstaff), Arizona, and a summary of their reduction into useful terms for water balance analysis. We now use these data to test the optimum canopy state estimation just discussed and illustrated in Fig. 7.10.

We begin by establishing the horizontal leaf area index, βL_t, for a particular species through optimal use of the available light using the compensation light intensity at the lowest leaves as given by Baker (1950) and reproduced here in Table 3.9,[†] along with our empirically confirmed condition of optical optimality, $\beta \approx \kappa$ (cf. Chapter 3).

We next obtain the L_t corresponding to the joint satisfaction of compensation light intensity and minimum r_c/r_a (for $M = 1$) through simultaneous solution of Eqs. 7.4 and 7.14 which can be carried out graphically using Fig. 7.4. Using this minimum r_c/r_a (for $M = 1$), along with the r_c/r_a (for $M = 0$) estimated earlier and listed in the caption of Fig. 7.9, we now calculate the optimal canopy conductance, $k_v^*(M)$ from Eq. 7.31 at the observed temperature for the catchment in question (cf. caption of Fig. 7.9). To summarize, this $k_v^*(M)$ is optimal because it maximizes both light use and water *demand* for the given species in the given climate, and it is plotted as curves (a) in Fig. 7.11.

[†] Of the Beaver Creek species, only ponderosa pine is tabulated by Baker (1950). We use the physical resemblance of needle structure (cf. Zim and Martin, 1987, pp. 36, 39) to represent the pinyon–alligator juniper as pinyon using Baker's (1950) κL_t for lodgepole pine, and to represent the pinyon–Utah juniper as juniper using Baker's (1950) κL_t for Northern white cedar.

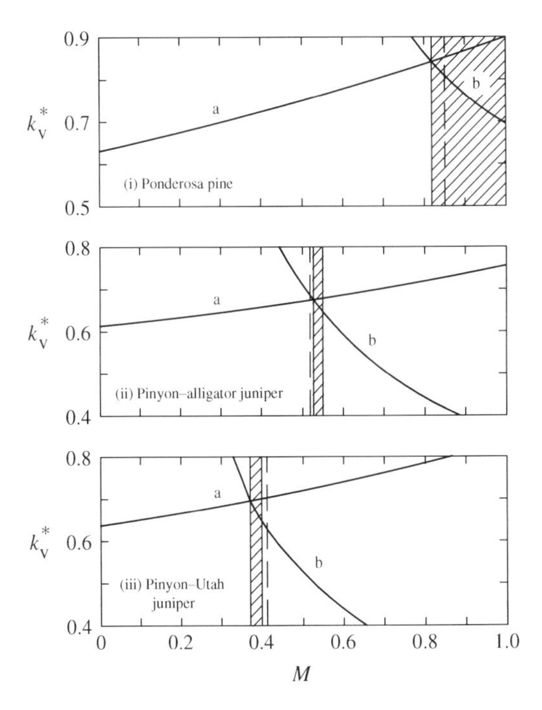

Fig. 7.11. Limited test of optimum canopy state (Beaver Creek watersheds). Solid lines: (a) Eq. 7.31 with Eqs. 7.4 and 7.14; (b) Eq. 7.32 using data of Tables 6.3 and 6.4. Dashed lines, observed canopy cover; shaded areas, sensitivity of theoretical M to non-optimal β.

The final step in resolving the optimal canopy state is to plot, for each of the three Beaver Creek species, the Mk_v^* hyperbola of Eq. 7.32 representing maximum water *supply*. From Chapter 6, this equation expands to

$$Mk_v^* = \frac{m_h}{m_{t_b'} E_{ps}} \left[1 - \frac{\overline{h_o}}{m_h} - (1 - M)\, e^{-G(0) - 2\sigma(0)^{3/2}} + \frac{\Delta S}{m_\nu m_h} \right], \tag{7.36}$$

the terms of which are quantified in Tables 6.3 and 6.4. These hyperbolae are presented as curves (b) on Fig. 7.11[†] and their intersections with curves (a) define the *optimal canopy state*.

We have observations only of M (cf. Table 6.3) and these are shown as the vertical dashed lines in Fig. 7.11. Their proximity to the theoretical optimal state is remarkable but must be viewed in relation to the sensitivity of the (a)–(b) intersection to use of other than the optimal value of β. To do this we return to Fig. 7.3 and for the βL_t of the given species (cf. Table 3.9) we pick off the resistance ratio corresponding to the range of β from $\beta = 0.25$ to $\beta = 0.60$. This allows recalculation of curve (a) for this range of $(r_c/r_a)_{M=1}$, and determines a corresponding range of non-optimal intersections with curve (b). This last range is highlighted on Fig. 7.11 by the shaded areas. The optimal state is that of mimimum resistance ratio (and hence highest k_v^*), and thus coincides with the lower limit of this range in M. We see that for dry climates such as the pinyon–Utah juniper, the range is quite narrow and non-optimality makes

[†] Note that Fig. 7.11 offers nothing not provided earlier by Fig. 6.15. The data are the same, but the coordinate axes have been changed for the particular purpose of this chapter.

Plate 7.5. Paper birch. (Photographed by William D. Rich in Arnold Arboretum, Boston; Copyright © 2001 William D. Rich.)

little difference in our forecast of M while making a $\pm 5\%$ difference in k_v^*. For wetter climates such as the ponderosa pine, sensitivity to the choice of β is significant and use of the postulated optimal state seems important.

We take these findings as moderate (albeit limited) support of the optimality hypothesis presented herein, but why should nature select for minimum flux resistance? We will discuss this further in Chapters 8 and 10.

8

Optimal bioclimate

The sensitivities of photosynthesis to leaf temperature and solar radiation are examined for various plants under conditions in which water supply is non-limiting. The photosynthetic capacity curve for C_3 leaves is fitted with a saturating exponential function rather than the classical Michaelis–Menton hyperbola for mathematical convenience and in acknowledgement of the dominant role of diffusive processes in CO_2 assimilation. The relative assimilation rate, here called the *leaf light characteristic*, is then described in terms of its asymptotes, with their intersection marking the minimum insolation, I_{SL}, for effectively fully open stomates. The rising asymptote is shown empirically to have the same slope for a wide range of C_3 plants thereby defining the *biochemical assimilation capacity* for C_3 leaves.

With water non-limiting, bioclimatic optimality is postulated to occur when: (1) the leaf temperature equals the climatic atmospheric temperature (i.e., negligible sensible heat transfer) which by selection is the leaf's photosynthetically optimum temperature, and (2) the climatic insolation, I_0, is just equal to the minimum insolation, I_{SL}, for fully open stomates.

With light non-limiting, the inverse relationship of root-zone soil moisture to solar radiation is demonstrated and leads to definition of the *leaf water characteristic* where relative assimilation falls from its peak when the insolation exceeds the maximum value, I_{SW}, for effectively fully open stomates.

It is shown that the relative values of the three defining insolations, I_0, I_{SL}, and I_{SW}, determine the feasibility, stability, and maximum productivity of a given climate–vegetation system, the soil being found only weakly influential for deep water tables. Reasoned consideration of the six permutations leads to the optimal bioclimatic state or climax state, $I_0 = I_{SL} \leq I_{SW}$, of the individual leaf. This leaf state is then scaled up to obtain an hypothesized *bioclimatic optimality* for the canopy having separate necessary propositions for heat, light, and water, and including a *climatic climax condition*.

A Photosynthesis

The concept

Photosynthesis is the primary process of life for green plants (and for some bacteria). It is the assimilative (i.e., *anabolic*) process through which these *primary producers* utilize sunlight to synthesize carbohydrates from CO_2 and water. In this process they must incorporate inorganic nutrients (principally nitrogen) obtained from the soil, thereby raising them to a higher energy level.

In biophysical and biochemical terms (Gaastra, 1963, p. 114; Gates, 1980, pp. 8, 47–48), photosynthesis consists of three sub-processes: (1) a gas-diffusion process through which the green leaf takes up CO_2 from the air and transports it to the reaction site in the chloroplasts of the palisade cells; (2) a photochemical reaction taking place within the leaf chloroplasts which converts absorbed light energy into chemical energy for the reduction of CO_2 to carbohydrates; and (3) biochemical enzymatic processes that both precede and follow the CO_2 reduction and are *highly temperature dependent.* The photochemical energy conversion can occur through one of several processes which each release gaseous oxygen while manufacturing amino acids, proteins, and other biologically useful compounds; they depend upon the availability of light (i.e., PAR) and of CO_2. The evaporation of plant water within the leaf stomatal cavities contributes to the regulation of leaf temperature, and the evaporate diffuses out from the leaf stomates at a rate which is closely related to that at which CO_2 diffuses in through those stomates.

Simultaneously, by both day and night, within the cells of *all* the plant organs, green and non-green, the destructive (i.e., *catabolic*) process of *respiration*, sometimes referred to as *dark respiration*, utilizes oxygen to break down substances in order to provide energy for cell metabolism. Respiration releases CO_2 which is diffused outward into the atmosphere. In addition, as a component of leaf photochemistry, there is release of CO_2 in a process called *photorespiration* or sometimes *light respiration.* Measurements of the light-induced CO_2 assimilation by leaves necessarily include all daylight respiration within their estimates of net photosynthesis.

During the day, the rate of CO_2 uptake per unit of plant mass required for photosynthesis is normally greater than the rate of release of CO_2 in respiration. The resulting net uptake of CO_2 during daylight hours is proportional to and defines the net rate of photosynthesis and hence plant productivity.

Plants may be distinguished in part by the predominant chemical process or "pathway" they use in the fixation of carbon during photosynthesis. These processes are several in number but the most common are (Osmond *et al.*, 1980):

(1) *Dicarboxylic acid pathway* ("C_4") – in which the uptake of CO_2 (and hence water loss) follows closely the changes in radiation intensity. These plants can utilize even the most intense solar radiation, and thus their family includes tropical grasses as well as agricultural plants such as millet, sorghum, and maize.

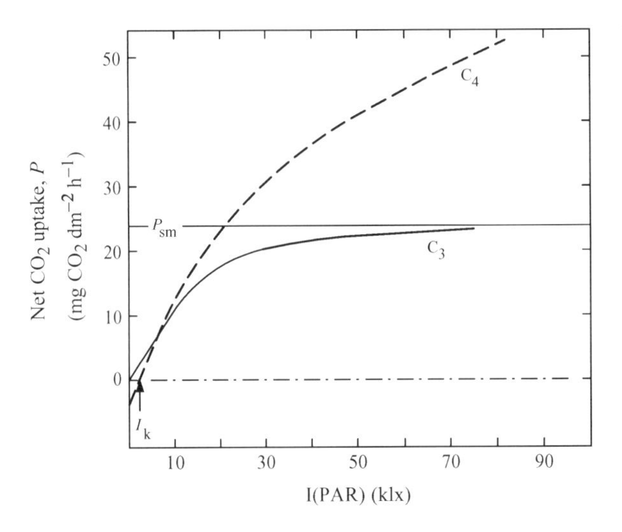

Fig. 8.1. Typical photosynthetic capacity of C_3 and C_4 plants (water non-limiting, optimal temperature, and ambient CO_2).

(2) *Pentose phosphate pathway* ("C_3") – in which the process becomes light-saturated at some radiation intensity so that strong light is not completely utilized. Most plants, including most agricultural crops and trees (both hardwoods and conifers) fall in this category.

(3) *Crassulacean acid metabolism pathway* ("CAM") – in which the stomata are open at night to take up and store CO_2 for processing during the day when the solar energy is available. This mechanism minimizes water loss since the stomata are open only during the cooler night-time. Succulent plants (cactus is an example) exhibit this arid-region characteristic.

We consider only C_3 plants in this work.

In Fig. 8.1 we sketch *photosynthetic capacity* curves (i.e., net CO_2 uptake, P, as a function of radiant flux, I(PAR)) for the leaves of typical C_3 and C_4 plants under conditions of unlimited water, natural CO_2 supply, and at the "optimal" temperature, T_m, for which photosynthesis is maximum. The proportionality between CO_2 assimilation and photosynthesis makes curves such as these the basis for analytical expressions of the relationship between plant productivity and available light.† Three features of the C_3 curve have particular significance for our work:

(1) The value, I_k, of insolation at which the net uptake of CO_2 is zero. Once again, at this value all the photoreduced CO_2 is needed to keep the leaf alive and is just equal to the respired CO_2. I_k is called the *compensation insolation* and represents the minimum light intensity at which leaves will grow. It varies from species to species and even within a given species according to the light and heat regimes to which the leaves have adapted. We will return to this important point below. Observations of the compensation insolation, I_k, are given in Table 8.1 for broad classes of plants (from Larcher, 1983, Table 3.6, p. 104).

† Note the units used in Fig. 8.1: on the ordinate, dm^2 = basal square decimeter of single-sided leaf area; and on the abscissa, klx = kilolux.

Table 8.1. *Observations of the photosynthetic reaction (natural CO_2 availability and optimal temperature)*[a]

Plant group	I_k [b] (total radiation) (klx)	I_{SL} [b] (total radiation) (klx)	P_{sm} [c] (mg CO_2 dm^{-2} h^{-1})
Herbaceous plants			
C_4 plants	1–3	over 80	30–80
C_3 plants			
Crops	1–2	30–80	20–45
Heliophytes (sun)	1–2	50–80	20–40
Sciophytes (shade)	0.2–0.5	5–10	4–20
Xerophytes (dry)			20–45
Grains and fodder grasses			15–35
Woody plants			
Tropical and subtropical trees			
Fruit trees			18–22
Forest canopy trees			12–24
Understory trees			5–10
Broad-leaved evergreens of warm tropics and sub-temperate zones			
Sun leaves			10–18
Shade leaves			3–6
Seasonally deciduous trees			
Sun leaves	1–1.5	25–50	15–25
Shade leaves	0.3–0.6	10–15	5–10
Conifers			
Winter deciduous			
Evergreen			5–18
Sun leaves	0.5–1.5	20–50	
Shade leaves	0.1–0.3	5–10	
Sclerophylls			5–15
Bamboos			5–10
Palms			6–10
Desert shrubs			6–20
Understory ferns	0.1–0.5	2–10	3–5
Mosses and lichens	0.4–2	10–20	0.5–3

[a] This table was prepared before the 1995 publication of Larcher's latest edition of *Physiological Plant Ecology*. The updated values presented in Larcher (1995, Table 2.4, p. 85, and Table 2.8, p. 98) are only quantitatively different and do not affect those conclusions of the present work which are based upon the data presented in this table.

[b] From Larcher, W., *Physiological Plant Ecology*, Table 3.6, p. 104, for single leaves, Copyright © 1983 Springer-Verlag GmbH & Co.; with kind permission from Springer-Verlag GmbH & Co., and from W. Larcher.

[c] From Larcher, W., *Physiological Plant Ecology*, Table 3.4, p. 94, Copyright © 1983 Springer-Verlag GmbH & Co.; with kind permission from Springer-Verlag GmbH & Co., and from W. Larcher.

Values for various tree species are given as a percentage of "full sunlight" by Baker (1950, Table 12, p. 143) while leaving the term "full sunlight" undefined.† I_k has an important variation with temperature because respiration and photosynthesis have different temperature dependencies. No *fixed* light value marks the threshold of existence of any species (Baker, 1950, p. 139). Leaves adapted to shade respire at a lower rate than do those adapted to strong light, hence the former reach the compensation value at a lower value of the light intensity (e.g., Larcher, 1975, Fig. 26, p. 41). However, the *relative* abilities of different species to survive under shade remains constant from one light environment to another (Baker, 1950, p. 140). This suggests that the photosynthetic behavior of a given species is relative to the incident light rather than being absolute and accounts for the reporting of most compensation insolations, I_k, as a percentage of the incident insolation, I_0, as we have seen in Table 3.9. Using results of this and later chapters, we prove in Appendix H that I_k/I_0 is a species constant.

(2) The value, I_{SL}, of insolation at which the net photosynthesis, P, can be considered to have reached its saturation limit, P_{sm}. I_{SL} is called the (CO_2) *saturation insolation*. It too varies from species to species, but because of the saturating behavior of P, estimation of I_{SL} from observations of P is problematic. Observations of the I_{SL} are given in Table 8.1. In Table 8.2 we give the average (from Table 8.1) observed I_{SL} for C_3 crops and for the sun leaves of deciduous trees and evergreen coniferous trees. We will assume that I_{SL} is the largest light intensity to which the plant has adapted, and therefore in regions with larger incident radiation we expect other species with higher I_{SL} to be more productive and thus to predominate.

(3) The light-saturated rate of net photosynthesis, P_s. Figure 8.2 (from Kramer and Kozlowski, 1960, Fig. 3.11, p. 83, after Müller, 1928, Fig. 2, p. 29) shows photosynthetic capacity curves for Arctic willow (*Salix glauca*) for unlimited water at three values of the ambient temperature, T_0. There we also see that photosynthesis saturates in weak light at low temperature and in much stronger light at high temperature. However, for a given light condition the maximum photosynthetic rate, P_s, does not increase monotonically with ambient temperature. Rather, it has a maximum, P_{sm}, at an intermediate, "optimum", value of ambient temperature, $T_0 = T_m$, as shown (for unlimited water) for *Pinus cembra* in Fig. 8.3 (from Kramer and Kozlowski, 1960, Fig. 3.10, p. 82, after Tranquillini, 1955, Fig. 6, p.169). Species-dependence of this optimal temperature, T_m, is illustrated by the relative curves of Fig. 8.4a (from Larcher, 1983, Fig. 3.35, p. 114), and some reported values of T_m are collected in Table 8.3. Adaptation of T_m by several degrees Celsius can occur over a few days in response to changing environmental conditions (Larcher, 1983, pp. 118–119).

† Using the definition of Gates (1980, p. 94), full sunlight = 107.6 klux, making the values of I_k in Table 3.9 much higher than those for trees in Table 8.1.

We have been assuming in this work that the leaf temperature, T_l, is that, T_m, for which the photosynthetic capacity is the species maximum as is shown in Fig. 8.4a, and furthermore, that the canopy sensible heat transfer is small enough that $T_l \approx T_0$.[†] Under these assumptions,

$$P_{sm} \equiv P_s(T_l = T_m) = \text{maximum light-saturated photosynthetic rate for an isolated leaf of a given species.}$$

Table 8.2. *Averages of observed photosynthetic reaction for isolated leaves (natural CO_2 availability and optimal temperature)*

Plant group	P_{sm} (g_s m^{-2} h^{-1})[a]	I_{SL}[b] (MJ(total) m^{-2} h^{-1})	$\varepsilon = P_{sm}/I_{SL}$ (g_s MJ(total)$^{-1}$)	Reference
C_3 crops[c]	1.63	2.07	0.79	Larcher (1983, Table 3.6, p. 104, Table 3.4, p. 94)
	(2.37)	(2.36)	(1.00)	Larcher (1995, Table 2.4, p. 85, Table 2.8, p. 98)
Deciduous trees[c]	0.69	0.94	0.73	Larcher (1983, Table 3.6, p. 104, Table 3.4, p. 94)
	(0.99)	(1.51)	(0.66)	Larcher (1995, Table 2.4, p. 85, Table 2.8, p. 98)
Coniferous trees[c]	0.58	0.79	0.73	Larcher (1983, Table 3.6, p. 104, Table 3.4, p. 94)
	(0.36)	(1.79)	(0.20)	Larcher (1995, Table 2.4, p. 85, Table 2.8, p. 98)
Sitka spruce[d]	0.55	0.52	1.06	Jarvis *et al.* (1976, Fig. 21, p. 229)
European beech[d]	0.33	0.32	1.03	Kramer and Kozlowski (1960, Fig. 3.9, p. 78)
Red oak[d]	0.28	0.46	0.61	Kramer and Decker (1944, Fig. 1, p. 352)
White oak[d]	0.26	0.46	0.57	Kramer and Decker (1944, Fig. 1, p.352)
Loblolly pine[d]	0.40[e]	0.60[f]	0.67	Kramer and Decker (1944, Fig. 1, p. 352), and Kramer and Clark (1947, Fig. 3, p. 55)
Creosote bush[d]	2.08	2.34	0.89	Ehleringer (1985, Fig. 7.8, p. 172)
Goethalsia[d]	0.38	0.44	0.86	Allen and Lemon (1976, Fig. 6, p. 290)
Arctic willow[d]	0.30	0.30	1.00	Kramer and Kozlowski (1960, Fig. 3.11, p. 83)
		Average	0.81 (0.78)	

[a] $g_s/g = 0.5$.
[b] $I(\text{total}) = 2I(\text{PAR})$.
[c] From Table 8.1. Updated, parenthetical values from Larcher (1995) make only a 4% difference in ε and are not used.
[d] Exponential fit to data from source.
[e] For seedlings.
[f] For individual needles.

[†] According to Larcher (1983, p. 119) natural selection adjusts T_m to the average climatic temperature, T_0, of the habitat.

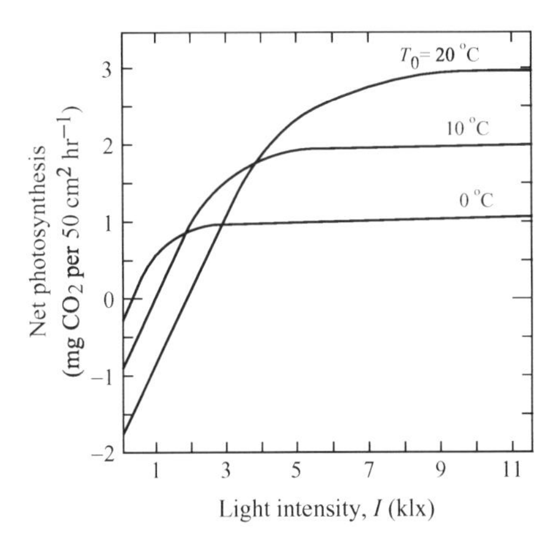

Fig. 8.2. Photosynthetic capacity of Arctic willow (water non-limiting, and ambient CO_2). (From Müller, D., Die Kohlensäureassimilation bei arktischen Pflanzen und die Abhängigkeit der Assimilation von der Temperatur, Fig. 2, p. 29, *Planta*, 6, Copyright © 1928 Springer-Verlag GmbH & Co. KG; as given by Kramer, P.J., and T.T. Kozlowski, *Physiology of Trees*, Fig. 3.11, p. 83, Copyright © 1960 McGraw-Hill Book Co., Inc.; with kind permission from Springer-Verlag GmbH & Co. KG and from T.T. Kozlowski.)

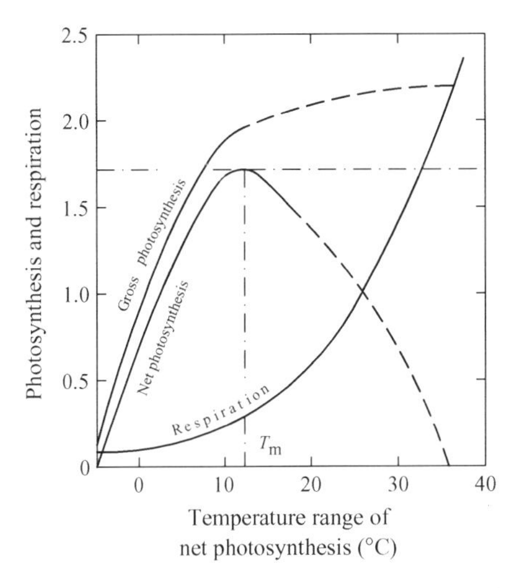

Fig. 8.3. Effects of ambient temperature on photosynthesis (*Pinus cembra*; water non-limiting). (From Tranquillini, W., Die Bedeutung des Lichtes und der Temperatur für die Kohlensäureassimilation von *Pinus cembre* Jungwachs an einem hochalpinen Standort, Fig. 6, p. 169, *Planta*, 46, Copyright © 1955 Springer-Verlag GmbH & Co. KG; as given by Kramer, P.J., and T.T. Kozlowski, *Physiology of Trees*, Fig. 3.10, p. 82, Copyright © 1960 McGraw-Hill Book Co., Inc.; with kind permission from Springer-Verlag GmbH & Co. KG, and from T.T. Kozlowski.)

The strength of the temperature dependence of plant productivity illustrated by the curves of Fig. 8.4a makes this behavior an important determinant of competitive advantage in establishing natural habitat. In such later discussions we will find it useful to have an approximate expression for the generalized form of these curves. To account

Table 8.3. *Optimum photosynthetic temperature for isolated leaves*

Plant type or species	Optimum photosynthetic temperature, T_m (°C)	Reference
Winter-deciduous trees of the temperate zone	15–25	Larcher (1983, Table 3.7, p. 115)
Pinus cembra	10–15	Tranquillini (1955)
Loblolly pine	30	Decker (1944)
Evergreen trees of tropics and subtropics	25–30	Larcher (1983, Table 3.7, p. 115)

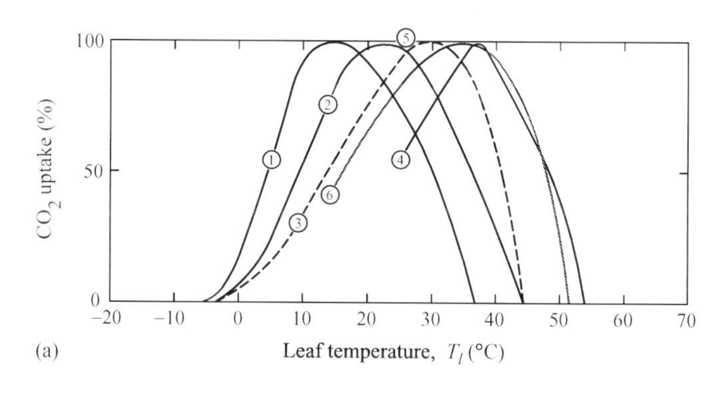

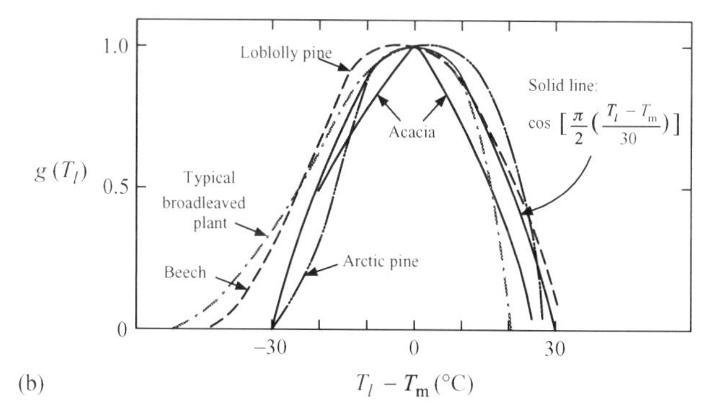

Fig. 8.4. Typical temperature dependence of net photosynthesis in woody plants (water non-limiting). (a) Comparative behavior of individual plants: 1, Arctic pine (Central Alps, 1850 m elevation; Larcher, 1983, Fig. 3.35, p. 114); 2, European beech (Central Europe; Larcher, 1983, Fig. 3.35, p. 114); 3, typical broadleaved plant (Gates, 1980, Fig. 3.13, p. 51); 4, acacia (Western Australia; Larcher, 1983, Fig. 3.35, p. 114); 5, loblolly pine (Kramer and Kozlowski, 1960), rising curve no. 2, falling curve no. 3, 100% for T_l = 20–30 °C; 6, creosote bush (Larcher, 1983, Fig. 3.40, p. 119), for day/night T_l = 35–25 °C. (After Larcher, W., *Physiological Plant Ecology*, corrected printing of 2nd Edition, Fig. 3.35, p. 114, Copyright © 1983 Springer-Verlag GmbH & Co. KG; with kind permission from Springer-Verlag GmbH & Co. KG, and from W. Larcher.) (b) Normalized behavior of various plants.

for suboptimal habitats in which $T_l \neq T_m$, we define

$P_s \equiv P_s(T_l \neq T_m)$ = suboptimal light-saturated photosynthetic rate for an isolated leaf of a given species.

We then generalize the light-saturated rate using the approximation

$$\frac{P_s}{P_{sm}} \equiv g(T_l) \approx \cos\left[\frac{\pi}{2}\left(\frac{T_0 - T_m}{30}\right)\right], \tag{8.1}$$

which is compared with the curves of Fig. 8.4a in Fig. 8.4b.

Values of P_{sm} are given in Table 8.1 for various land plants as reported by Larcher (1983, Table 3.4, p. 94).

B Photosynthetic capacity of an isolated leaf

Analytical expression

Assuming *biochemical* control, the rate of photosynthesis for an *isolated leaf* is usually expressed mathematically in terms of the rate of CO_2 assimilation through an equation in the form of the classical hyperbolic Michaelis–Menten equation (White *et al.*, 1968) applicable for enzymatic reactions. This dependence of photosynthetic rate on light intensity (the so-called *photosynthetic capacity* curve) has the form (Monteith, 1963, p. 106; Horn, 1971, p. 68; Gates, 1980, p.50)

$$P_t = P + P_r = \frac{P_s I}{I + k}, \tag{8.2}$$

in which for an isolated leaf

P_t = total rate of photosynthesis,
P = net rate of photosynthesis,
P_s = light-saturated rate of photosynthesis = photosynthetic capacity,
P_r = rate of respiration,
I = insolation, and
k = empirical "binding constant that measures the effectiveness of an isolated leaf in getting and processing photons" (Horn, 1971, p. ix).

Photosynthetic capacity curves are determined experimentally for a given species by isolating one or more leaves in a closed chamber called an assimilation chamber or "phytotron" with the leaf temperature, T_l, maintained at the species' optimum photosynthetic temperature, T_m, the CO_2 concentration maintained at its natural ambient value, and with non-limiting water supply; the intensity of PAR is controlled using artificial light while the mass flux of CO_2 is monitored continuously in the inflowing and outflowing air. Such observations yield the *net* rate of CO_2 assimilation, but since

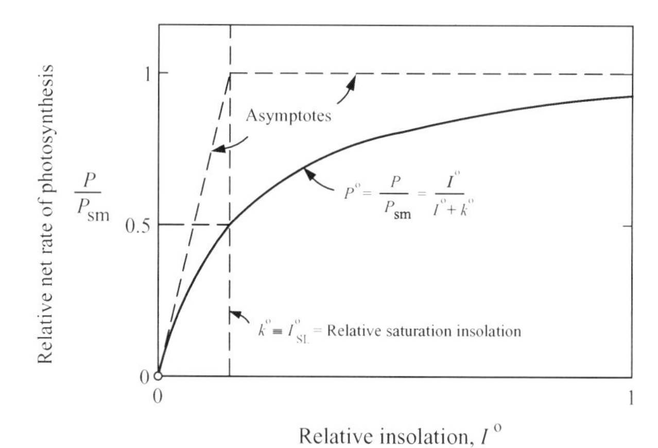

Fig. 8.5. Michaelian net rate of photosynthetic reaction (water non-limiting, optimal temperature, and ambient CO_2).

the ratio P_r/P_s is only approximately 10% of P_t/P_s for trees (Larcher, 1983, Tables 3.4, p. 95, Table 3.5, p. 97), we neglect it and rewrite Eq. 8.2 to estimate the *net* rate of photosynthesis, P/P_s. This gives

$$\frac{P}{P_{sm}} \approx \frac{P_t}{P_{sm}} = \frac{I}{I+k}, \tag{8.3}$$

where, because $T_l = T_m$ in these experiments, we replace P_s by its maximum value, P_{sm}. Normalizing all quantities on the right-hand side of Eq. 8.3 by the ambient "full sunlight" insolation, I_0, we define $I^o \equiv I/I_0$, and $k^o \equiv k/I_0$. Using this notation the behavior of Eq. 8.3 is sketched in Fig. 8.5.

With this normalization Horn (1971, p. 69) fits Eq. 8.3 to observations of photosynthesis in oak, beech, and pine using the maximum observed photosynthetic rate as P_{sm}. However, selection of the proper "full sunlight" radiance is problematic given observations of a process that saturates asymptotically with applied radiation. In response to this difficulty, we select as an objective definition of saturation insolation, I_{SL}, the intersection of the asymptotes of Eq. 8.3 as shown by the dashed lines in Fig. 8.5. We locate this point by evaluating the slope of Eq. 8.3 at $I = 0$ as

$$\left|\frac{dP}{dI}\right|_{I=0} = \left|\frac{k}{(I+k)^2}\right|_{I=0} = \frac{1}{k}, \tag{8.4}$$

from which we see that the saturation insolation $I_{SL} \equiv k$. Note that with this definition, we have effectively replaced the photosynthetic characteristic curve of Eq. 8.3 by its asymptotes, and therefore the saturation radiances derived in this work will be smaller than those reported in the literature and listed here in Tables 8.1 and 8.2.

Alternatively, since photosynthesis is *biophysically* controlled by the uptake of CO_2 through diffusive processes in the external leaf boundary layer and in the internal leaf protoplasm, it seems equally reasonable to fit observations with an *exponential* function of the form

$$\frac{P}{P_{sm}} = 1 - \exp(-I/k). \tag{8.5}$$

As before

$$\left|\frac{dP}{dI}\right|_{I=0} = \left|\frac{1}{k}\exp(-I/k)\right|_{I=0} = \frac{1}{k},$$

and intersection of the asymptotes again gives $I_{SL} \equiv k$ as is sketched in Fig. 8.6a. This reduces Eq. 8.5 to

$$\frac{P}{P_{sm}} = 1 - \exp(-I/I_{SL}). \tag{8.6}$$

Equation 8.6 is compared in Fig. 8.6b with the same observations used by Horn (1971, p. 69) where in the present case, the oak and pine photosynthetic rates have been normalized using the observed asymptotic values. Here we see that Eq. 8.6 fits all three species reasonably well using $k^o = I^o_{SL} = 0.11$, a value which corresponds

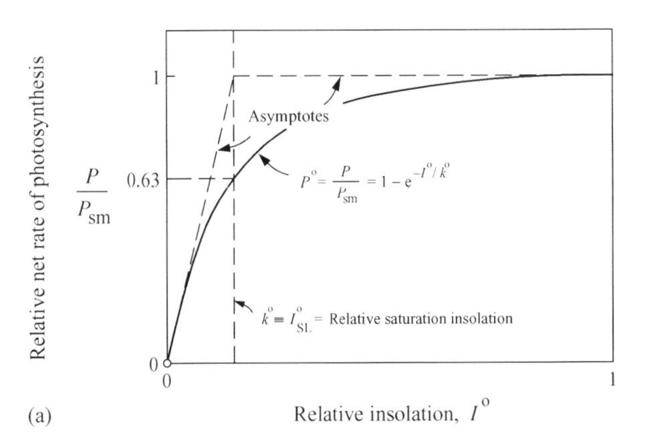

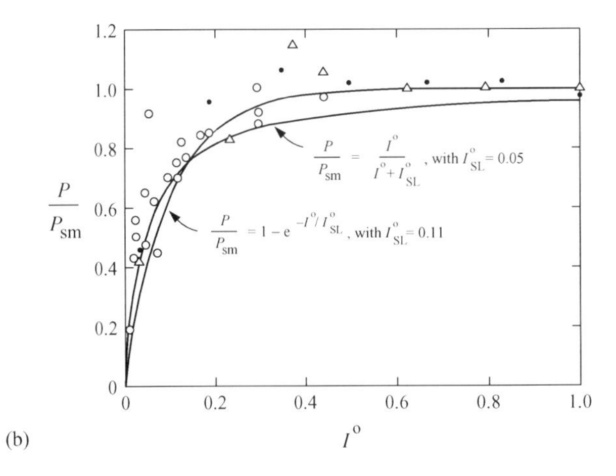

Fig. 8.6. Exponential net rate of photosynthetic reaction for leaf (water non-limiting, optimal temperature, and ambient CO_2). (a) Analytical form; (b) comparison with observations: ●, Eastern red oak (Kramer and Decker, 1944); ○, European beech (Boysen-Jensen, 1932, after Kramer and Kozlowski, 1960; △, Loblolly pine (Kramer and Clark, 1947).

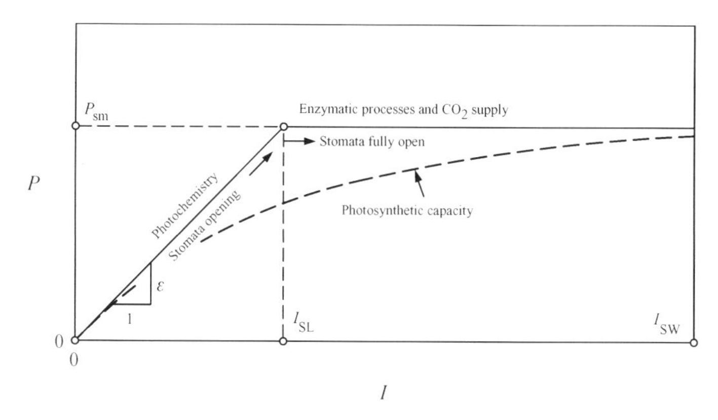

Fig. 8.7. Definition of the leaf light characteristic (unit basal C_3 leaf area, water non-limiting, optimal temperature, ambient CO_2).

to $I_{SL} \approx 11.9$ kilolux (i.e., 43 MJ (total) m^{-2} h^{-1}). However, according to the data of Table 8.1, this value is appropriate for the sun leaves of both deciduous and coniferous species only when multiplied by a factor of about 3, but we caution against direct comparisons of reported values because of the different definition used herein.

Leaf light characteristic

We believe the more appropriate physical basis of Eq. 8.6, plus the physical significance and objectivity of its asymptote intersection (see Appendix D), justifies its adoption here to represent the photosynthetic characteristic. Hereafter, we will discuss the photosynthetic capacity curve as though it were composed of its exponential asymptotes as sketched again in Fig. 8.7 under non-limiting conditions of water supply, and with $T_l = T_m$. We call these asymptotes the *leaf light characteristic*. Because CO_2 enters the leaf through the stomates, we must look at the principal mechanisms of stomatal control in order to understand further the physical significance of these asymptotes and their intersection.

C Light control of productivity of an isolated leaf

Carbon dioxide control circuit

We first look back briefly at the structure of the typical stomatal cavity as idealized in Fig. 4.27a for the C_3 plants and trees to which we will limit our consideration. The stomatal opening responds to (among other things) the difference of CO_2 concentration between the boundary layer on the leaf surface and the stomatal cavity. All else being constant, when this difference is large, the leaf is assimilating CO_2 faster than it can enter the leaf. Consequently, the stomatal opening increases, lowering the stomatal resistance and admitting more CO_2. This occurs under conditions of increasing

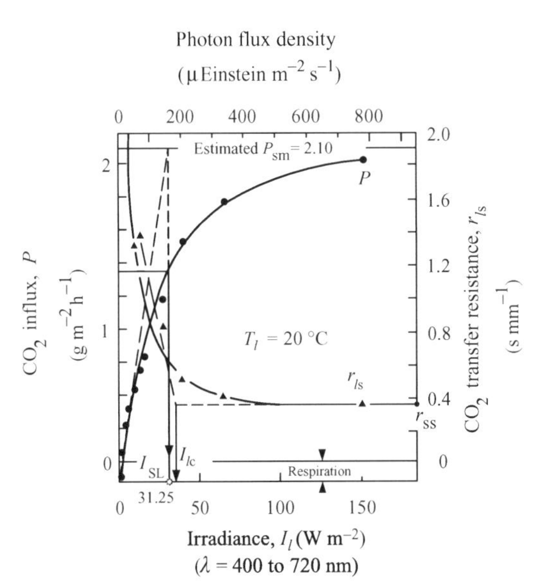

Fig. 8.8. Demonstration that stomates are effectively open at intersection of asymptotes of the photosynthetic capacity curve (*Picea sitchensis*; water non-limiting, and ambient CO_2). (From Ludlow, M.M., and P.G. Jarvis, Photosynthesis in Sitka spruce, I. General characteristics, Fig. 3, p. 934, *Journal of Applied Ecology*, 8, Copyright © 1971 Blackwell Science Ltd; with kind permission from Blackwell Science Ltd, Oxford, UK.)

light intensity (all else again being constant), and the reverse occurs as light intensity falls. With this understanding, and with water availability non-limiting, we see that the stomatal opening increases with increasing light intensity along the rising, photochemical asymptote of Fig. 8.7. At photosynthetic saturation (the intersection of the asymptotes in this idealization), the stomates are fully open, the stomatal resistance has its minimum value, and the assimilation rate becomes constant at its maximum value (consistent with the ambient concentration of CO_2) along the asymptote controlled by enzymatic processes (and hence temperature).

A piece of evidence supporting the use of the intersection of the asymptotes of the photosynthetic capacity curve to define the critical point of fully open stomates is given here in Fig. 8.8 (adapted from Ludlow and Jarvis, 1971, Fig. 3, p. 934). Observations of net CO_2 influx, P, and of stomatal resistance, r_{ls}, are presented there as a function of irradiance, I_l, for an isolated shoot of Sitka spruce (*Picea sitchensis*) undergoing the usual phytotron test. The saturation value, $P = P_{sm}$, is estimated at $P_{sm} = 2.10$ g m^{-2} h^{-1} as shown on the figure. Evaluating Eq. 8.4 at $I_l = I_{SL}$ gives

$$\frac{P}{P_{sm}}(I_l = I_{SL}) = 1 - 1/e = 0.632,$$

yielding $P(I_l = I_{SL}) = 1.33$ g m^{-2} h^{-1}. Entering Fig. 8.8 with this value locates I_{SL} as is shown by the arrows entering from the P scale.

The r_{ls} data points of Fig. 8.8 were fitted by Ludlow and Jarvis (1971) with the solid line shown, but over the range of observations could as well have been fitted with the dashed straight line segments shown. We use the intersection of these dashed segments to approximate the insolation, I_{lc}, at which the stomates first become fully open.

The proximity of I_{lc} to I_{SL} supports our use of the asymptote intersection of the exponential photosynthetic capacity curve as the point of effectively full stomatal opening.† In so doing it provides the major physical justification for our use of this intersection to define the normalizing "critical" insolation when characterizing the dimensionless photosynthetic capacity curve. We recognize that values of I_{SL} determined in this manner will be at variance with most of the existing observations, so we have developed our own set of I_{SL} values using observations, taken from the literature, of the photosynthetic characteristic of single leaves of eight separate species, and fitting each of these with an exponential function as described above. P_{sm} is normally measured in milligrams of CO_2 assimilated per unit of *stomated* leaf surface per unit of time at the optimum leaf temperature, $T_l = T_m$, and I_{SL} is measured in a variety of units of either total or shortwave incident radiant energy per unit of *illuminated* leaf area per unit of time. We have converted all energy fluxes to PAR (assumed to be one half the total radiation) in units of megajoules (i.e., MJ).

The species studied are the bottom eight entries in Table 8.2 (i.e., Sitka spruce through Arctic willow), and their experimentally determined photosynthetic capacity curves are from the sources in the last column of Table 8.2. The value of P_{sm} was determined in each case by estimating the saturation asymptote of the source curve and is listed in the first column of Table 8.2. For each species, several points were read from the source curve over the full range of P/P_{sm} and were used with Eq. 8.6 to obtain an average I_{SL} for that species. These values are listed in the second column of Table 8.2, and the resulting values of $\varepsilon = P_{sm}/I_{SL}$ are given in the third column.

The individual points transferred from the source curves and put in common units are displayed in Fig. 8.9. We note there that of those species studied, all but the creosote bush (*Larrea divaricata*) appear to cluster about a common exponential (i.e., the upper curve of Fig. 8.9) while creosote bush, the lower curve, reaches maximum productivity at a much higher insolation.

Also shown in Table 8.2 are average values of these parameters, taken from Table 8.1, for C_3 crops, and for the sun leaves of deciduous trees and evergreen coniferous trees, the three main vegetation types under consideration here.

Dry matter : radiation quotient

For purposes of interspecies comparison, the rate of photochemical production of plant biomass is commonly characterized by the proportionality between photosynthetic yield and available radiation in the region of opening stomates (see Fig. 8.7). Larcher (1975, p. 41) calls this the *dry matter : radiation quotient*, ε, commonly defined as:

$$\varepsilon \equiv \frac{\text{grams of dry biomass produced}}{\text{megajoules of intercepted radiant energy}}, \quad (g_s\ MJ^{-1}),$$

† This is another of our expedient assumptions. Larcher (1983, p. 93) points out that stomatal opening responds to the interplay of water potential, humidity, and temperature as well as to light, and that the stomates are completely open only rarely and briefly.

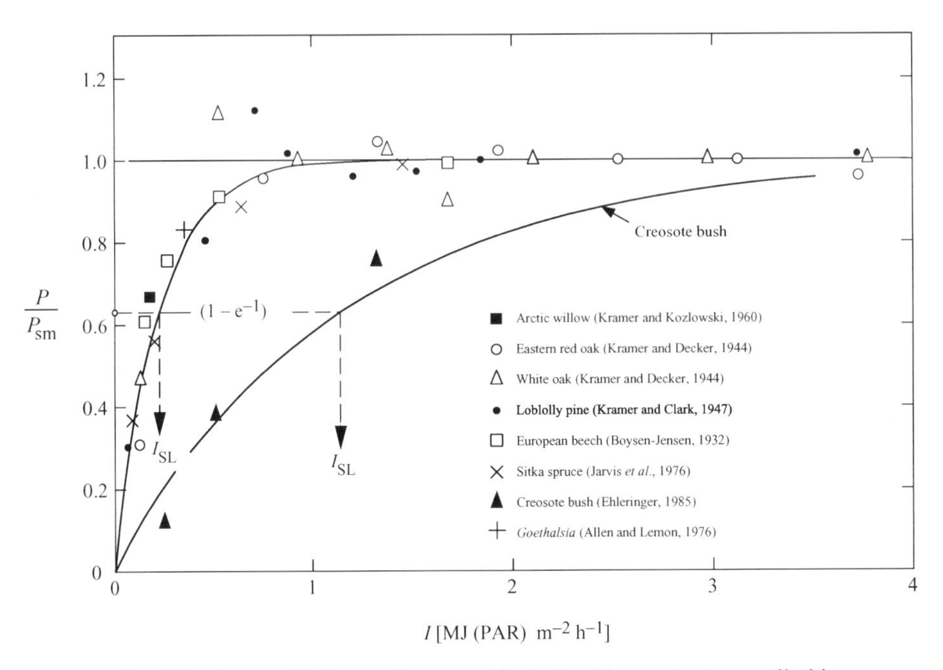

Fig. 8.9. Photosynthetic capacity curves for isolated leaves (water non-limiting, optimal temperature, and ambient CO_2).

in which the intercepted radiation used is sometimes short wave and sometimes total, while the mass of roots is usually excluded when ε is reported for whole trees. We recognize ε as the slope, P_{sm}/I_{SL}, of the photochemical asymptote of the leaf light characteristic presented in Fig. 8.7. However, in evaluating ε and in using values of ε from the literature, we must be careful of the units.

The estimates of P_{sm} given in the first column of Table 8.2 have been converted to the resulting mass of solid matter using the accepted average conversion factor

$$\vartheta = \frac{\text{g(solid matter)}}{\text{g(CO}_2\text{ assimilated)}} \equiv \frac{\text{g}_\text{s}}{\text{g}} \approx 0.5, \tag{8.7}$$

as given by Penning de Vries *et al.* (1974), and Ledig *et al.* (1976, *fide* Russell *et al.*, 1989b, p. 27).

The needles of some conifers have stomates on more than one side (see Chapter 2) in which case the the stomated and illuminated areas of the leaf are unequal. To accommodate such cases we introduce the additional modifier,

$$\eta_o = \frac{\text{stomated (i.e., assimilating) leaf area}}{\text{illuminated (i.e., projected) leaf area}}. \tag{8.8}$$

For example, loblolly pine (*Pinus taeda*) has stomates on all three needle surfaces (Kramer and Decker, 1944, p. 352), thus in direct light this species has $\eta_o \approx \frac{\text{perimeter}}{\text{arc-chord}} = 2.36$ (Raison *et al.*, 1992). The observed value of P_{sm} for this species was reported per unit of stomated area and thus was multiplied by $\eta_o = 2.36$ for listing in Table 8.2. For broad leaves $\eta_o = 1$.

With these modifiers we write

$$\varepsilon = \vartheta \frac{P_{sm}}{I_{SL}} \eta_o, \quad (g_s \ MJ^{-1}), \tag{8.9}$$

Unfortunately there are not many observations of ε for trees or even for woody perennials (Jarvis and Leverenz, 1983). Those which do exist have been assembled by Russell et al. (1989b, pp. 28–30) and are reproduced here in Table 8.4. Using the above dimensional modifiers as appropriate, we can estimate additional values of ε for isolated leaves from the observations of P_{sm} and I_{SL} tabulated in Table 8.2. The latter are plotted in Fig. 8.10 along with a solid line representing the average value of the single leaf *dry matter : radiation quotient*, ε. This average curve has the value $\langle \varepsilon \rangle = 0.81$ g_s MJ(total)$^{-1}$ which is consistent with the value $\varepsilon = 0.80$ g_s MJ(tot)$^{-1}$ found for the leaves of trees by Linder (1985). Other reported values are much higher, such as the "representative" value $\varepsilon = 2.2$ g_s MJ(tot)$^{-1}$ (Szeicz, 1974) and that for the leaves of C_3 plants of $\varepsilon = 1.4$ g_s MJ(tot)$^{-1}$ (Monteith, 1977).

It is important to understand that the species-independence of ε defines a biochemical limit to the assimilation rate of C_3 leaves of a given species. We call this function, $P_{sm} = \varepsilon I_{SL}$, the *biochemical assimilation capacity* for leaves of that species. It follows, as we will now see, that the function, $P_{sm} = \varepsilon I_0$, limits the C_3 leaf assimilation rate in a given radiational climate, I_0, thereby providing perhaps the primary pressure for selection of I_{SL} and hence species at a given I_0; we call this latter function the *climatic assimilation potential*.

Bioclimatic optimality with non-limiting water ("species control")

What is the relation of I_{SL} to the environmental insolation I_0 (i.e., insolation at the top of the crown) assuming that the availability of water is not limiting? We reason as follows:

(1) In Chapter 5 we saw that the temperature difference, $|T_l - T_0|$, driving the sensible heat transfer between the leaf and the ambient atmosphere is very small, so that for the purposes of this chapter we may assume $T_l \approx T_0$.

(2) *Because of the extreme sensitivity of biological productivity (i.e., photosynthetic rate) to leaf temperature (Fig. 8.4), we assume that the environmental temperature,* T_0, *will select a species such that* $T_m \approx T_0$.

(3) Consider the idealized photosynthetic capacity curve for species no. 1 shown as the solid line in Fig. 8.11a and having the saturating insolation I_{SL1}. If $T_0 \approx T_m$ but the environmental insolation, I_0, is such that $I_0 > I_{SL1}$, the system will operate at point no. 1, and there will be unused radiance which is not biologically productive with this species. While species no. 1 is maximally productive in this environment which is thus a *natural habitat* of the species, it is suboptimal in the evolutionary sense of the given environment's potential for

Table 8.4. *Estimates of the dry matter : radiation quotient, ε, for whole plants (water or nutrients not limiting; main period of growth; radiation is total intercepted)*

Plant name	Species	ε (g_s MJ^{-1})[a]	Reference
Agricultural C_3			
Rape	*Brassica napus*	1.1	Mendham *et al.*, 1981
Pigeon pea	*Cajanus cajan*	1.2	Hughes *et al.*, 1981
Pea	*Pisum sativum*	1.5	Heath and Hebblethwaite, 1985
Potato	*Solanum tuberosum*	1.4	Scott and Allen, 1978
Typical		1.4	Monteith, 1977
Woody C_3			
Monterey pine	*Pinus radiata*	0.9[b]	Linder, 1985
		1.45	Linder, 1985, plus Forrest and Ovington (1970), and Forrest (1973), *fide* Cannell (1982, p. 15), give: (root dry weight)/(total dry weight) = 0.38.
Scots pine	*Pinus sylvestris*	0.85[b]	Linder, 1985
		1.03	Linder, 1985, plus Mälkönen (1974); Albrektson (1980a, 1980b); Ovington (1957, 1959, 1961); *fide* Cannell (1982, pp. 63; 225; and 244), give: (root dry weight)/(total dry weight) = 0.24 (3 sites); 0.11 (15 sites); 0.31 (3 sites).
Black pine	*Pinus nigra*	0.85[b]	Linder, 1985
		1.06	Linder, 1985, plus Miller *et al.* (1980), *fide* Cannell (1982, p. 247), give: (root dry weight)/(total dry weight) = 0.20 (7 sites).
Sitka spruce	*Picea sitchensis*	0.85[b]	Linder, 1985
		1.05	Linder, 1985, plus Deans (1979, 1981); *fide* Cannell (1982, p. 242), give: (root dry weight)/(total dry weight) = 0.19 (1 site).
Eucalyptus	*Eucalyptus globulus*	0.45[b]	Linder, 1985
		0.56	Linder, 1985, plus Feller (1980); Rogers and Westman (1977, 1981), Westman and Rogers (1977a, 1977b); *fide* Cannell (1982, pp.15; 18), give: (root dry weight)/(total dry weight) = 0.10 (1 site *E. regnans*, and 1 site *E. obliqua*); 0.38 (1 site, *E. signata* and *E. umbra*).
Willow	*Salix viminalis*	1.4[b]	Cannell *et al.*, 1987

[a] Radiation is total intercepted.
[b] Above-ground biomass only (i.e., roots excluded).
Source: Russell, G., B. Marshall, and P.G. Jarvis, editors, *Plant Canopies: Their Growth, Form and Function*, Cambridge University Press, 1989, Table 2.1, p. 28 and following text; reprinted with permission of both Cambridge University Press and G. Russell.

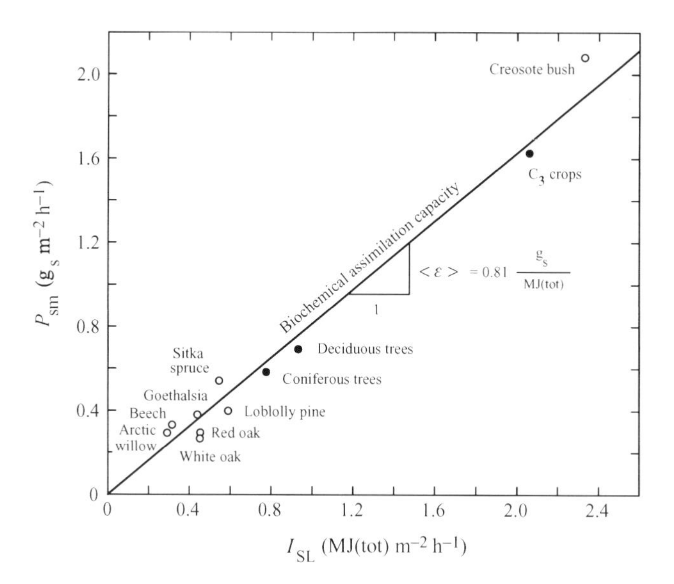

Fig. 8.10. Biochemical assimilation capacity of C_3 leaves. Data from Table 8.2. Note that these data have been updated by Larcher (1995, Table 2.4, p. 85 and Table 2.8, p. 98) since the original preparation of this figure. Use of the updated data here would result in greater scatter leading to $<\varepsilon> = 0.78$.

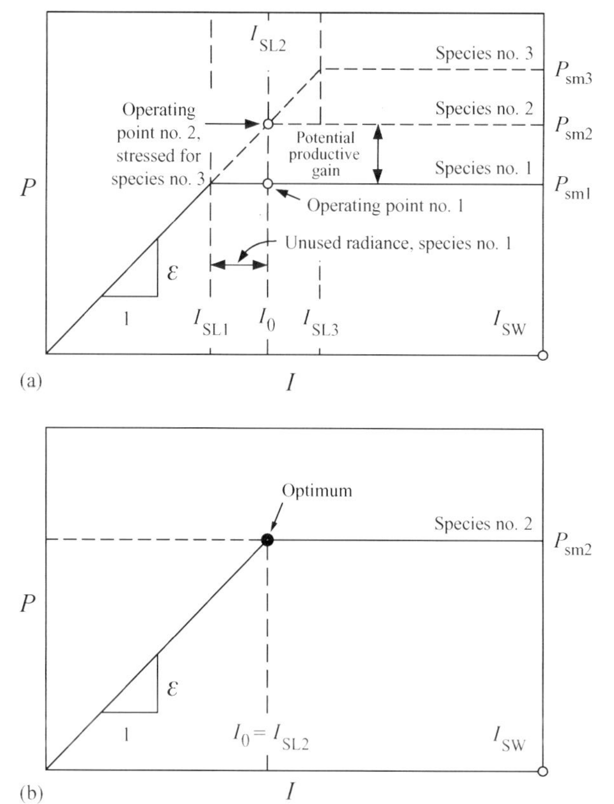

Fig. 8.11. Bioclimatic operating states under light control (unit basal leaf area, water non-limiting, optimal temperature, ambient CO_2). (a) Suboptimal; (b) optimal.

biomass production. Referring back to Fig. 8.10 now we note the direct proportionality between P_{sm} and I_{SL} (i.e., constant ε). Therefore (return to Fig. 8.11a), species having larger saturation insolation at this T_0 will be more productive (i.e., larger P_{sm}). We hypothesize that evolutionary pressure will move the biological system in this direction. In a single-species world we might expect this species to evolve or adapt biochemically to a new form having $I_{SL2} = I_0$ as is indicated by operating point no. 2 in Fig. 8.11a. However, in our multispecies world, we expect that a species no. 2, having the same T_m but larger P_{sm}, will take over (as indicated by the upper dashed curve) such that $I_{SL2} = I_0$, and it will operate at point no. 2. Species no. 2 is thus "optimal" for this climate; after Lindeman (1942) we will refer to it as the *climax* species, that is the species that is maximally productive there, and hence stable.[†‡]

(4) Continuing our consideration of Fig. 8.11a, if $T_0 = T_m$ for species no. 3 but $I_0 < I_{SL3}$, the operating point is again no. 2. Here the photosynthetic rate, P_{sm2}, is less than P_{sm3}, and at operating point no. 2 species no. 3 is not producing at capacity. Under such suboptimal operation, species no. 3 lives in an average state of stress due to insufficient light with associated increased risk of trauma during transient periods of lower insolation. Furthermore, plant support of such unproductive capacity is suboptimal because the constant and non-productive light requirements of respiration consume a greater proportion of I_0 than they do when I_0 has the saturation value, I_{SL}. Therefore, natural selection would seem strongly to favor a new species (species no. 2), having smaller I_{SL} at this I_0, such that $I_{SL2} = I_0$ as is shown by the lower dashed line in Fig. 8.11a. At operating point no. 2, this species will be just as productive as species no. 3 would be at this I_0, but optimally so. Species no. 2 would again qualify as a climax species for this environment.

(5) *We conclude that with ample water and a continuous range of species-specific* T_m, *the optimum relations between climate and species are* $T_0 \approx T_l = T_m$, *and* $I_0 = I_{SL}$, *for here* $P = P_{sm}$ *which is the maximum possible photosynthetic rate (as is shown in Fig. 8.11b). We refer to the suboptimal state* $I_{SL} > I_0$ *as "light-limited" and the suboptimal state* $I_{SL} < I_0$ *as "species-limited".*

[†] Note that we make this climax classification solely on the basis of full productive use of the environmental light resource provided, of course, that water is not limiting. To the first approximation (i.e., neglecting the stimulation of added assimilation capacity by the water-borne flux of added nutrients) productivity is independent of excess water supply beyond that needed to keep the stomates fully open.

[‡] These discussions of pressures for species substitution are simplistic when viewed from the viewpoint of species succession in natural ecosystems (cf. Colinvaux, 1973, pp. 550–572) where various other strategies are believed to be operative at different successional stages.

Plate 8.1. Norway spruce. (Photographed in Arnold Arboretum, Boston, by William D. Rich; Copyright © 2001 William D. Rich.)

D Water control of productivity of an isolated leaf

Transient case

Imagine a stand of trees growing under constant weather conditions which include adequate light (i.e., $I_0 \geq I_{SL}$), an *initially saturated* root-zone soil, and no further precipitation. Under these conditions the trees are initially unstressed by shortage of either light or water, and therefore the stomates are fully open. Accordingly, we expect leaf transpiration to occur initially at the potential rate $E_v = E_{ps}$. The net rate of biological productivity is shown in Appendix E to be directly proportional to the rate of transpiration. We thus expect the photosynthetic rate of at least the uppermost leaf initially to have the saturation value, $P = P_{sm}$. For subsequent times, it is helpful to understand the energetics of water movement in the plant from its reservoir in the root-zone soil up to the surface of the spongy mesophyll cells lining a stomatal cavity in this uppermost leaf where it undergoes an evaporative state change from liquid to vapor.

Water moves through the plant in the xylem under the influence of a potential gradient between xylem terminals in the root and the leaf. However, many crucial details of this extremely complex process remain controversial. For purposes of qualitative reasoning only, we will follow here the simplified engineering explanation offered by Hendricks and Hansen (1962).

Figure 8.12a gives a schematic representation of the energy grade line, and Fig. 8.12b shows typical values of the associated energy potentials. The pressure intensities, p_c, of fluid held in the soil pores by capillary forces are negative (i.e., the fluid is in tension) as are the related soil moisture ("capillary") potentials, $\psi_s = p_c/(\rho_f g) \equiv p_c/\gamma_f$, where ρ_f and γ_f are the mass density and the specific weight of the fluid respectively, and g is the gravitational constant. The root-zone capillary potential is related to the root-zone soil moisture concentration by the well-known relation from soil physics (cf. Eq. 6.30)

$$\psi_s = \psi_s(1)\, s^{-1/m}, \tag{8.10}$$

in which $\psi_s(1)$, the potential at saturation (i.e., $s = 1$), and m are hydraulic properties of the soil.

As time progresses in our imaginary transient experiment, transpiration and downward percolation in the soil reduce the average root-zone soil moisture concentration, s, and the associated root-zone soil moisture potential, ψ_s, becomes more negative. The withdrawal of moisture from the soil adjacent to the roots lowers the capillary potential there by the additional amount, h_s, needed to support flow to the root surfaces against the resistance of the soil. A positive osmotic potential, H_o, exists across the root cell membranes due to concentration differences between the soil solution and the sap solution (Kramer, 1969, pp. 150–173). This potential acts as a pump which may (seasonally at least) develop a rise in pressure head on the order of 1 to 2 atmospheres (Kramer, 1969, p. 158). On the plant side of the root walls there is a drop in potential due to friction losses in the roots (h_r), xylem (h_x), and leaves (h_l), while the height of the tree gives a positive gravitational potential, H_G. Across the leaf's spongy mesophyll cells, from the veins to the evaporation surfaces in the stomatal cavities, there is a positive capillary potential, $H_L \equiv \psi_l$, which can be sizeable depending upon the effective capillary diameter of the spaces between the mesophyll cells. In Fig. 8.12a these potentials are added to their respective elevations, z, at every point along the flow path to obtain the conventional "hydraulic grade line" (i.e., piezometric head line). When the hydraulic grade line lies below the elevation of the point in question, the moisture potential at that point is negative. As is shown in Fig. 8.12a, the fluid is under tension in much of the tree. Conventional wisdom has rejected tension-induced cavitation by

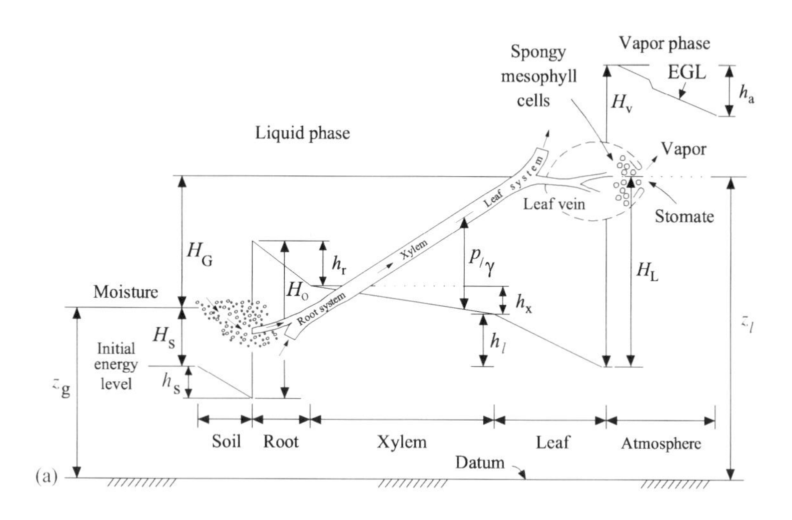

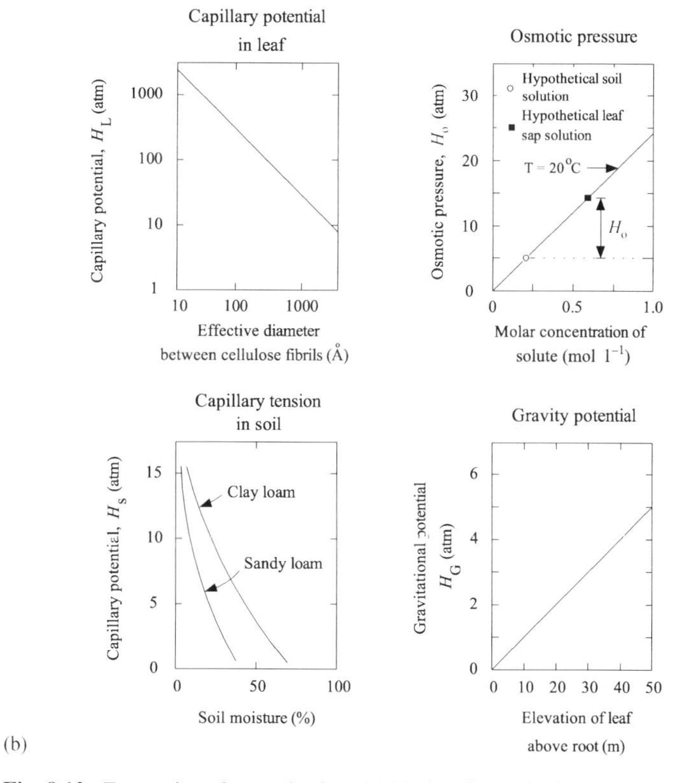

Fig. 8.12. Energetics of transpiration. (a) Hydraulic grade line; (b) energy potentials. (After Hendricks, D.W., and V.E. Hansen, Mechanics of evapo-transpiration, Fig. 3, p. 73 and Fig. 4, p. 74, *Proc. A.S.C.E., J. Irrigation and Drainage Div.*, 88, IR2, Copyright © 1962 American Society of Civil Engineers; with kind permission from the American Society of Civil Engineers.)

virtue of "cohesion" between adjacent fluid molecules (Briggs, 1950) and by "adhesion" between the fluid molecules and the elastic vessel walls (Slatyer, 1967, p. 212). However, more recent studies (cf. Milburn, 1993) demonstrate that xylem hydraulic conductivity may be substantially reduced by cavitation and the resulting formation of gas-filled passages. Holbrook and Zwieniecki (1999) hypothesize that these gas emboli are continuously repaired, perhaps diurnally (Zwieniecki and Holbrook, 1998), and that measured hydraulic conductivity is a "dynamic balance between the processes of damage and repair".[†]

Because the flow velocities are so small, the kinetic energy of the plant fluid is negligible, and the potentials of Fig. 8.12a can be summed to give H_f, the potential available to support friction loss, as

$$H_f = h_s + h_r + h_x + h_l = (H_o - H_G) + (\psi_l - |\psi_s(s)|), \tag{8.11}$$

in which $H_G \equiv z_l - z_g$.

As time progresses (without additional precipitation remember!), the magnitude of ψ_s increases sharply (Eq. 8.10), and the plant must work ever harder to pump moisture from the soil to the stomates at the atmospheric demand rate, E_{ps}. The values of H_o and H_G in Eq. 8.11 are fixed by the given plant, so to maintain constant H_f (and hence $E_v = E_{ps}$ and $P = P_s$), the leaf potential, ψ_l, must increase by the same amount as $|\psi_s|$.

At some ("critical") value of the falling soil moisture, $s = s_c$, the plant's available pumping capacity, $H_o + \psi_l$, can no longer overcome the opposing potentials, $|\psi_s| + H_G + H_f$, while keeping the flow at its maximum rate, E_{ps}. At subsequent times, without some change in the opposing potentials, the atmospheric demand would begin to desiccate the plant structure in order to make up the deficit in the plant's transpiration capacity. To prevent this, the plant has evolved a mechanism to keep evaporative demand more-or-less in balance with the falling transpiration capacity.

A biological sensor in the guard cells surrounding the stomatal opening senses the moisture potential, ψ_l, in the leaf. When ψ_l reaches the critical value, ψ_{lc} (corresponding to the critical soil moisture, s_c), the guard cells respond by reducing the stomatal opening and thereby putting the plant into a *stressed* (as opposed to relaxed) *physiological state*. This restriction to the vapor flow raises the vapor density within the stomatal cavity and thereby decreases the transpiration rate. The friction loss, H_f, then falls requiring (see Eq. 8.11) that $\Delta\psi_l < \Delta\psi_s$ for a given negative Δs. This decline in the transpiration rate for $\psi_l > \psi_{lc}$ implies a proportional decline in the rate of water-borne nutrient flux from the soil as well as in the rate of CO_2 assimilation. Thus there is a decrease in net biological productivity as the soil dries below the critical value s_c.

Figure 8.13 shows the typical variation of the rates of transpiration and of net photosynthesis for several crop and tree species as a function of ψ_l under conditions

[†] Zwieniecki *et al.* (2001) find that salt-sensitive gels in certain xylem membranes constantly shrink and swell as needed to regulate the water flow.

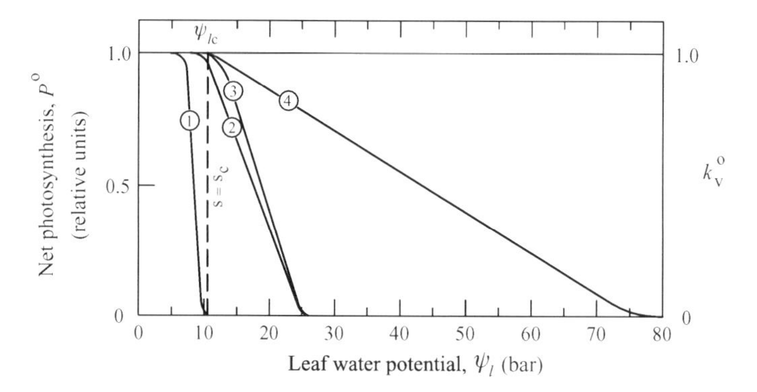

Fig. 8.13. Restriction of photosynthesis by leaf water stress (light non-limiting, optimal temperature, and ambient CO_2). 1, Annual vine; 2, English walnut; 3, Sitka spruce; 4, creosote bush. (Adapted from Larcher, W., *Physiological Plant Ecology*, corrected printing of 2nd Edition, Fig. 3.46, p. 123, Copyright © 1983 Springer-Verlag GmbH & Co. KG; with kind permission from Springer-Verlag GmbH & Co. KG, and from W. Larcher.)

of unlimited light as adapted from Larcher (1983, Fig. 3.46, p. 123). These ordinates are presented as normalized by their maximum rates such that

$$k_v^o \equiv \frac{k_v}{k_v^*}, \quad \text{and} \quad P^o \equiv \frac{P}{P_{sm}}. \tag{8.12}$$

Both normalized ordinates have the maximum value of unity while the stomates are fully open (see Chapters 6 and 8). We see in Fig. 8.13 that ψ_{lc} has virtually a common value of about 10 bars for the species considered.

Note in Fig. 8.13 that the stomatal control of transpiration is singularly weak in the creosote bush. This is probably due to low soil moistures characteristic of that plant's habitat, for at low soil moisture, $\Delta\psi_s/\Delta s$ is very large requiring a comparably large $\Delta\psi_l$ to accomplish a given reduction in H_f (i.e., in E_v).

Thus, in the manner just outlined for this case of transient drying, gradual stomatal closure permits the plant to continue production, albeit at an ever-decreasing rate, without desiccation but in a state of physiological stress until the drying soil forces complete shut-down of the stomates. Equation 8.11 can be used to transform the independent variable of Fig. 8.13 from ψ_l to s. With laminar flow in the soil pores and plant veins, friction losses are proportional to the velocity in these passages and we can write

$$H_f = k_L E_v, \tag{8.13}$$

in which k_L (sec) is the sum of the geometrically scaled potential-loss coefficients for the soil, root, xylem, and leaf passages. Equations 8.11 and 8.13 then give

$$E_v = \frac{\psi_l - |\psi_s| + H_o - H_G}{k_L}. \tag{8.14}$$

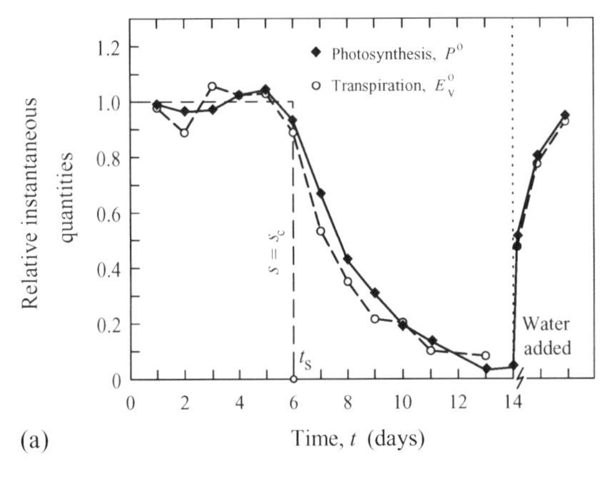

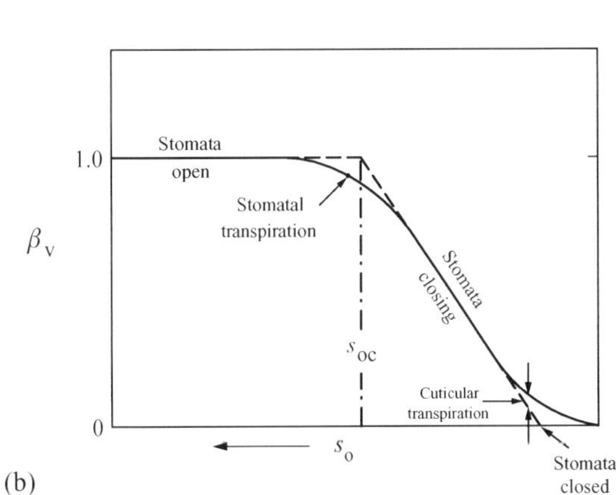

Fig. 8.14. Biological response to falling soil moisture (loblolly pine; light non-limiting, and ambient CO_2). (a) Observed transient transpiration and photosynthesis of loblolly pine (after Brix, H., The effect of water stress on the rates of photosynthesis and respiration in tomato plants and loblolly pine seedlings, Fig. 4, p. 15, *Physiologia Plantarum*, 15, Copyright © 1962 Munksgaard International Publishers Ltd., Copenhagen, Denmark; with kind permission from Munksgaard International Publishers Ltd); (b) generalized time-average leaf transpiration efficiency.

Using Eqs. 8.10 and E.12, Eq. 8.14 provides the desired coordinate transformation to

$$P^o = \frac{C_1}{k_L}\left[\psi_l - |\psi_s(1)|s^{-1/m} + H_o - H_G\right]. \tag{8.15}$$

With Eq. 8.15 each pair of values (P^o, ψ_l) for a particular species in Fig. 8.13 maps to a pair (P^o, s) having the transient form illustrated for a loblolly pine canopy in Fig. 8.14a. Note here that with falling root zone soil moisture, the photosynthetic rate remains constant at its maximum value until the critical transient soil moisture content, $s = s_c$, is reached and the stomates begin to close. This maximum value of photosynthesis corresponds to the maximum value of the canopy conductance, $k_v = k_v^*$ or $k_v^o = 1$. This transient case is of course the natural one where occasional additional precipitation produces soil moisture recharge which intermittently resets the decaying transpiration rate at a higher value $0 < E_v \le E_{pv}$. Note particularly in Fig. 8.14a the proportionality of P and E_v.[†]

[†] Here the proportionality results from control of both P and E_v by stomatal closure induced by the drying soil. In Appendix E we demonstrate approximate proportionality for the condition of open stomates.

Steady-state case

Time-averaging the succession of soil moisture transients characterizing a given climate, we define an imaginary steady-state which we assume to be representative of climatic behavior to a degree that becomes exact only at the limit in which variance of the weather vanishes. This steady-state approximation has yielded useful insights into the behavior of the climatic water balance (Eagleson, 1978f), and was introduced in Chapter 6 as the basis for analysis in this book. The time-averaged (i.e., *climatic*) root-zone soil moisture concentration is denoted by s_0. If we change the independent variable of Fig. 8.14a from the transient s to the climatic s_0, as in Fig. 8.14b, then the sketched function represents the relative productivity of that species as a function of soil moisture climate *provided that there is always enough light in this climate for photosynthesis at the rate* P *equivalent to* E_v. The insolation, I, at which s_0 attains its critical value, s_{0c}, and plant desiccation begins is an important bioclimatic parameter which, in comparison with I_{SL}, and the actual climatic insolation, I_0, determines plant habitat. In making this comparison through the radiation-driven transpiration rate we will be careful to deal with daylight-hour time periods and rates. For example, we use Eq. 6.46 to write, for a unit vegetated area

$$E_v^{\otimes} = \beta_v \, k_v^* \, E_{ps}^{\otimes}, \tag{8.16}$$

in which $E_{ps}^{\otimes} \equiv 2E_{ps}$ is the average daylight hour potential rate of evaporation[†] from a wet surface, and β_v is approximated from Eqs. 6.43 and 6.48 for the period following evaporative exhaustion of surface retention by retaining only the second term of Eq. C.6.[‡] That is

$$\beta_v \approx 1 - \exp\left[-\alpha\left(\frac{4\, n_e\, z_r}{\mathrm{M}\, k_v^*\, E_{ps}^{\otimes}} s_0\right)\right]. \tag{8.17}$$

Equation 8.17 is sketched as the solid line labeled "stomatal transpiration" in Fig. 8.14b.

Leaf water characteristic

Earlier in this chapter we examined the photosynthetic rate as a function of light when the water supply was not limiting. We called it the leaf light characteristic and sketched it in Fig. 8.7. We now need a means of considering the joint effects of light *and* water in order to specify the most productive unstressed state. To this end it will aid our physical reasoning to define a *leaf water characteristic* under conditions in which light is not limiting (i.e., $I \geq I_{SL}$).

We begin by demonstrating that the climatic root-zone soil moisture concentration, s_0, is inversely related to the ambient insolation, I, all else being constant. This

[†] That is, the time-averaged depth evaporated in one-half the time, making $m_{t_b'} E_{ps} \approx m_{t_b^{\otimes}} E_{ps}^{\otimes}$.
[‡] Note that in discussing the behavior of a single leaf, k_v^* and β_v can be less than unity only by virtue of stomatal closure. This contrasts with our use of k_v to describe the conductance of a multilayer canopy in which all stomates are assumed fully open.

Plate 8.2. Golden larch. (Photographed in Arnold Arboretum, Boston, by William D. Rich; Copyright © 2001 William D. Rich.)

variable transformation is important in that it allows us to present the falling leaf productivity of Fig. 8.14b as a result of increasing environmental insolation rather than of decreasing soil moisture, and thus facilitates comparison of the leaf water characteristic with the leaf light characteristic of Fig. 8.7. Using first Eq. C.17 and then the first-order approximation to Eq. E.1 we get, for a given moisture supply, the functional relationships[†]

$$s_o = f_1\left(E_{ps}^{\otimes -1}\right) = f_2(I^{-1}). \tag{8.18}$$

[†] Note that the function $P(I)$ given in Appendix E is different when carbon is supply-limited.

Thus, as I increases in a given precipitation climate, s_o will decrease due to increasing $E_{ps}^{\otimes}$. When the falling soil moisture causes $s_o \le s_{oc}$, progressive stomatal closure occurs, and β_v declines from unity. Provided I_{SL} is small enough to insure photosynthetic saturation, we can now define the *leaf water characteristic* for a unit leaf area of a C_3 species for a given soil, seasonal temperature and precipitation.

Referring to Fig. 8.15, as we increase the insolation, I, above I_{SL}, the stomates remain fully open under species control as discussed earlier (cf. Fig. 8.7), and the photosynthetic rate remains constant at P_{sm}.† Increase in the evaporation rate, E_v, is damped somewhat by virtue of increasing back radiation as shown in Eqs. E.1 and E.4. Nevertheless, with increasing I, the potential rate of evaporation, $E_{ps}^{\otimes}$, increases and s_o falls (cf. Eq. 8.18) non-linearly. With still larger I, the soil moisture falls to the critical value, $s_o = s_{oc}$, at which $I = I_{SW}$ and both the photosynthesis and evaporation rates come under water control with the stomates beginning to close in response to water stress. We define the *desiccation* or *water-critical insolation*, I_{SW}, at which stomatal closure begins by the intersection of the asymptotes of the $P(I)$ curve as is shown in Fig. 8.15.

It is interesting and important to note through Eq. 8.15 that with ψ_{lc} and H_o constant, decreasing P can be delayed to higher I (and hence to lower s_o) through a decrease in tree height as expressed through z_l. Here is a direct mechanism for control of plant height through water supply.

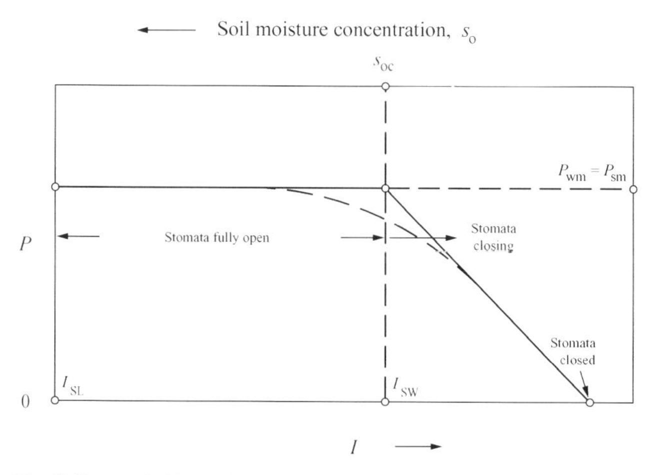

Fig. 8.15. Definition of the leaf water characteristic (unit basal leaf area of given species, unlimited saturation insolation, optimal temperature, ambient CO_2, given soil and precipitation).

† Assuming that increased transpiration maintains optimum leaf temperature.

E Joint control of leaf productivity by light and water

Bioclimatic optimality for an individual leaf

With the light and water characteristics expressed in the same coordinates we can combine them in order to consider the limits of their interaction for a given species and climate. We have three important values of the ambient insolation, I:

(1) I_0 = actual environmental insolation,

(2) $I_{SL} \equiv P_{sm}/\varepsilon$ = saturation or carbon-critical insolation for a leaf of given species, and

(3) I_{SW} = desiccation or water-critical insolation at which stomatal closure and the accompanying water stress begin.

Their *relative* values determine the feasibility, stability, and optimality of a given climate–vegetation system, as well as the direction of any evolutionary pressure. Although there are six permutations of the relative values of these three insolations, the feasible set is smaller when we consider a given species and observe a series of constraints:

Constraint no. 1: Productivity is stressless and is maximized in the given climate

Stresslessness fixes $P_{sm} \leq P_{wm}$, and productivity maximization fixes $I_{SL} \leq I_0$, placing the maximum stressless productivity of this species at

$$I_{SL} \leq I_0 \leq I_{SW}, \tag{8.19}$$

as is demonstrated in the sketch of Fig. 8.16. This range of I_0 defines the *feasible radiational habitat* of the given species for the given soil and water supply. A few comments about this figure are in order:

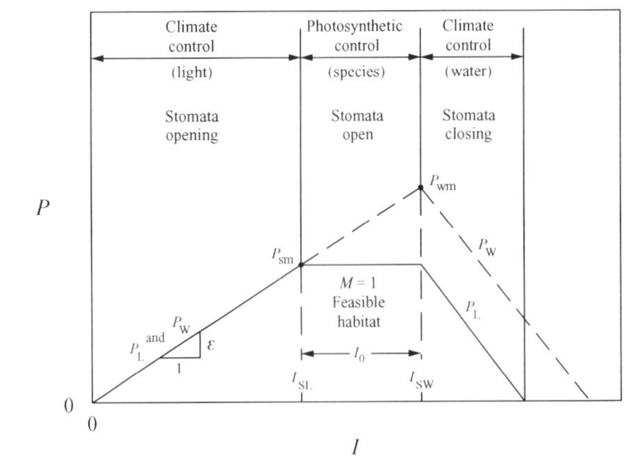

Fig. 8.16. Jointly controlled leaf photosynthesis (fixed species, soil, and precipitation). Solid line, behavior of given species as controlled by light saturation; dashed line, climatic potential behavior realizable only by a different species having $P_{sm} \geq P_{wm}$.

The solid line represents the *actual* photosynthesis, P_L, of the leaf given by its actual light characteristic as influenced by the moisture supply of the given climate, while the dashed line represents the *potential* water-limited photosynthesis of the leaf, P_W, as influenced only by the climatic moisture supply and the stomatal opening as though light saturation did not occur. The jointly controlled leaf photosynthetic characteristic follows the form of the light characteristic in Fig. 8.7, until photosynthetic saturation at $I = I_{SL}$. The amount of light controls P_L for $I \leq I_{SL}$, and the species, through its photosynthetic saturation, P_{sm}, controls P_L for $I_{SL} < I \leq I_{SW}$. In this latter range, the stomates are fully open, there is excess water since $P_W > P_L$, and thus the canopy cover is complete at $M = 1$. At $I = I_{SW}$, the stomates begin to close under contol of the available water and both the actual and the water-limited potential photosynthesis of the individual leaf fall accordingly. At $I_{SL} = I = I_{SW}$ decline of the climatic water supply will be felt through decrease of the canopy cover, M, leaving $P_{sm} = P_{wm}$. Additional constraints follow from optimality considerations as we now see.

To the extent that $I_0 > I_{SL}$, the given species is climatically suboptimal and hence is "selectively unstable" in the sense that it does not make maximum productive use of the available radiation (cf. Fig. 8.11a) and there will be selective pressure to do so. To the extent that $I_0 = I_{SL}$ but $I_0 < I_{SW}$, the species is photosynthetically optimal and stable, even though it does not make maximum productive use of the available water, since there is no resulting selective pressure. Such a system is *light-limited* because with larger I_0 a species with correspondingly larger I_{SL} and P_{sm} may be substituted. Accordingly, for optimal conditions we add additional constraints.

Constraint no. 2: Efficiency of insolation utilization is maximum

Maximum energy efficiency requires $I_{SL} = I_0$ (as discussed above) and with Eq. 8.19 limits the stable space to the climax criterion,

$$I_{SL} = I_0 \leq I_{SW}. \tag{8.20}$$

Constraint no. 3: Productive potential of given climate–soil combination is fully utilized

Production to the full extent allowed by the available soil moisture requires $I_0 = I_{SW}$ and with Eq. 8.20 further reduces the stable space to the climatically optimum subset,

$$I_{SL} = I_0 = I_{SW}. \tag{8.21}$$

Bioclimatic optimality for a canopy of individual leaves

In order to understand the role of canopy cover in leaf bioclimate, we now expand our scale from the individual leaf to an imaginary canopy composed of a *single leaf layer* covering fraction, M, of the surface. For the purpose of these discussions we use the

Plate 8.3. Nikko fir. (Photographed in Arnold Arboretum, Boston, by William D. Rich; Copyright © 2001 William D. Rich.)

energy balance as written in the simple form of Eq. 5.5.[†] Incorporating the Bowen ratio estimator of Eq. 5.38 we have, for the daylight hours

$$R_n = (1 \pm \boldsymbol{R}_b)\,\lambda E_T, \tag{8.22}$$

in which $\boldsymbol{R}_b$ (through its dependence on k_v^*) is a function of plant species. During the daylight hours, the net radiation is related to the ambient insolation, I, by the definition

$$R_n \equiv (1 - \rho_T)\,I - q_b, \tag{8.23}$$

where ρ_T = spectral reflectance (i.e., albedo) of canopy, and q_b is the net longwave back radiation as is given empirically by Eq. B.12.

Together Eqs. 8.22 and 8.23 approximate the daylight hour balance of energy fluxes *for a unit vegetated area* in which we replace the nominal (i.e., 24 hour) evaporation rate, E_T, by the actual daylight-hour rate, $E_v^{\otimes}$, giving

$$I = \frac{1}{1 - \rho_T}\left[(1 \pm \boldsymbol{R}_b)\,\lambda E_{ps}^{\otimes} + q_b\right], \tag{8.24}$$

where $E_v^{\otimes}$ is linearly related to the photosynthetic rate by a climate- and species-dependent coefficient derived in Appendix E, and where the $\pm$ refers to sensible heat rejection (+) or acceptance (−) by the canopy.

Substituting Eqs. 6.46, 6.66, and 8.16 into Eq. 8.24 gives

$$I = \frac{1}{1 - \rho_T}\left[(1 \pm \boldsymbol{R}_b)\frac{\lambda V_e \beta_v}{m_{t_b^{\otimes}} M} + q_b\right]. \tag{8.25}$$

At the critical condition $I = I_{SW}$, $s_o = s_{oc}$, and $\beta_v = 1$. We then rewrite Eq. 8.25 for this single leaf layer,

$$I_{SW} = \frac{1}{1 - \rho_T}\left[(1 \pm \boldsymbol{R}_b)\frac{\lambda V_e}{m_{t_b^{\otimes}} M} + q_b\right], \tag{8.26}$$

in which, to the first approximation Eqs. 6.67 and 6.68 become independent of the soil properties (cf. Eq. 6.65 and Table 6.4) and define, for deep water tables

$$V_e \approx m_h\left[1 - \frac{\overline{h_o}}{m_h} + \frac{\Delta S}{m_\nu m_h}\right]. \tag{8.27}$$

We see from Eq. 8.26 that the larger the moisture supply rate, $V_e/m_{t_b^{\otimes}}$, the larger will be the critical radiance, I_{SW}.

It is helpful to remember the relationship between M and I_{SW} as the average available soil moisture changes.

(1) Start with the situation $I_0 < I_{SW}$ signifying that there is excess available moisture which insures that $M = 1$, and that the stomates are fully open.

[†] For the purpose of calculation in Chapter 9, we will use the Penman–Monteith formulation, Eq. 5.28.

(2) Now decrease the rate, $V_e/m_{t_b^{\otimes}}$, at which soil moisture is available (all other environmental variables remaining constant), until the system arrives at the critical average state, $I_{SW} = I_0$, with the stomates still open, and $M = 1$.

(3) Continue to decrease $V_e/m_{t_b^{\otimes}}$ and, since $I_{SW} < I_0$, there are only two alternatives: (a) the stomates close to reduce the rate of moisture use while maintaining $M = 1$ (but placing the plant in a state of stress), or (b) the stomates remain fully open (and the plant remains unstressed) while the canopy cover declines just enough to keep $V_e/(m_{t_b^{\otimes}} M)$ constant, $I_{SW} = I_0$, and thereby maintain the critical water supply for each leaf. Adhering to our basic premise of a stressless average state, the latter alternative is the only feasible operating state. However, to get there it is necessary for individual plants to die off through transient pursuit of alternative (a). As the most sensitive plant succumbs to stress, the additional soil moisture thereby made available reduces the stress in those plants still living.[†]

The insignificance of percolation to groundwater under these conditions of critical soil moisture[‡] means that whenever the average operating state is $I_0 = I_{SW}$, groundwater can be generated during the growing season only due to finite variance of the climatic properties, or during the dormant season due to $E_v^{\otimes} \equiv 0$. This demonstrated insensitivity of I_{SW} to s_{oc} may be the physical basis for the previous observation by Denmead (1973) that "transpiration is almost independent of soil conditions".[§]

In summary, our hypothesis of an *optimal bioclimatic state* or *climax state* for the individual leaf is founded on the concept that any average operating state which requires stomates to be less than fully open (i.e., $I_{SW} < I_{SL}$) is stressful to the plant and will be selected out naturally. It proposes that in a given climate–soil, *natural selection and/or adaptation leads to* $T_l = T_m = T_0$ *at* $I_0 = I_{SL} \leq I_{SW}$, *at which point* $P = P_{sm}$. We will refine these conditions after acounting for the principal interactions among the leaves in a real canopy.

F Bioclimatic optimality for the canopy-average leaf

Assumptions

Scaling up from photosynthetic capacity tests on an individual leaf to the lumped behavior of a full canopy demands that we make assumptions about: (1) the representativeness of the tested leaf's light and water characteristics; and (2) the manner of canopy response to the conditions of bioclimatic optimality for an individual leaf which were explicated in the previous sections:

[†] Such benefit of the group through seemingly altruistic self-sacrifice by the individual plant is not a negation of the "selfish organism" assumption of natural selection (cf. Dawkins, 1989) but is instead a reflection of variance in the transient responses of the individuals in the group.

[‡] See Table 6.4 for estimates of the terms in the water balance leading to reducing Eq. 6.67 to the form of Eq. 8.24.

[§] Once again we caution that this excludes significant soil moisture supply from groundwater which is considered as a special case in Eq. 6.72.

Plate 8.4. Scrub pine. (Photographed in Arnold Arboretum, Boston, by William D. Rich; Copyright © 2001 William D. Rich.)

(1) Since all leaves tested from various tree and C_3 crop species appear to have a common value of ε (cf. Fig. 8.10), we are led to the enabling, but gross, approximation that *all productive leaves of a given species have identical light capacity curves at the same temperature*. Under such circumstances, P_{sm} is the only variable light characteristic, changing with both species and operating temperature. Of course, Table 8.1 seems to belie this assumption with its demonstration of different photosynthetic characteristics for "sun" leaves and "shade" leaves of the same species at optimal temperature. We acknowledge this by further assuming that only the sun leaves are significant contributors to the *net* productivity which yields seeds, and hence only these leaves are the determinants of plant survival strategy.

(2) The canopy is comprised primarily of sun leaves which we assume to be distributed uniformly throughout the crown. Clearly, all leaves cannot have bioclimatically optimal incident radiance at all daylight moments during the growing season. The incident insolation, I, changes with time, and the leaf insolation, I_l, depends upon both I and position within the crown (cf. Eq. 3.53). Therefore, we assume that *bioclimatic optimality of the canopy is achieved only in the space–time average* .

Canopy-averaged insolation†

Using the "hat", $\widehat{\cdot}$, to distinguish *crown-averaged* values from those of a single leaf, the above assumptions lead to the *effective canopy saturation insolation*, $\widehat{I_{SL}}$, for a crown of a given species remaining unchanged from the individual leaf at

$$\widehat{I_{SL}} \equiv I_{SL} = \frac{P_{sm}}{\varepsilon}, \tag{8.28}$$

While the comparable *effective canopy environmental insolation*, $\widehat{I_l}$, is obtained by averaging the actual insolation, I_l, over the crown volume.

For *cylindrical multilayers*, the insolation, I_l, incident upon a leaf decays with depth into the canopy according to Eq. 3.53 which we average over the canopy depth to get

$$\frac{\widehat{I_l}}{I_0} \equiv f_I(\kappa L_t) = \int_0^1 \exp(-\kappa L_t \xi) \, d\xi = \frac{1 - e^{-\kappa L_t}}{\kappa L_t}. \tag{8.29}$$

For *tapered multilayers* (e.g., cones or hemispherical segments), the compensation light intensity is assumed to exist for $z = h_s$ at all radii, leading again to Eq. 8.29. Invoking our important earlier finding that $\kappa = \beta$ (cf. Eq. 3.52), we then write *for all crown shapes*

$$\frac{\widehat{I_l}}{I_0} \equiv f_I(\beta L_t) = \frac{1 - e^{-\beta L}}{\beta L_t}. \tag{8.30}$$

Equation 8.30 is plotted over the important range of βL_t in Fig. 8.17. Over this range Eq. 8.30 yields the approximate mean, $\langle f_I(\beta L_t) \rangle = 0.36$.

For *monolayers* of course, the incident insolation acts undiminished on all leaves giving

$$\frac{\widehat{I_l}}{I_0} \equiv f_I(\beta L_t) = 1. \tag{8.31}$$

Similarly, with I_{SW} determined by the uppermost leaf in the canopy using Eq. 8.26, we define the effective *leaf desiccation insolation*, $\widehat{I_{SW}}$, for the entire canopy to be

$$\widehat{I_{SW}} \equiv f_I(\beta L_t) I_{SW}. \tag{8.32}$$

† Although generally overlooked, these canopy-average reductions of the insolation incident at canopy-top are the effective insolations for use with "big leaf" canopy models.

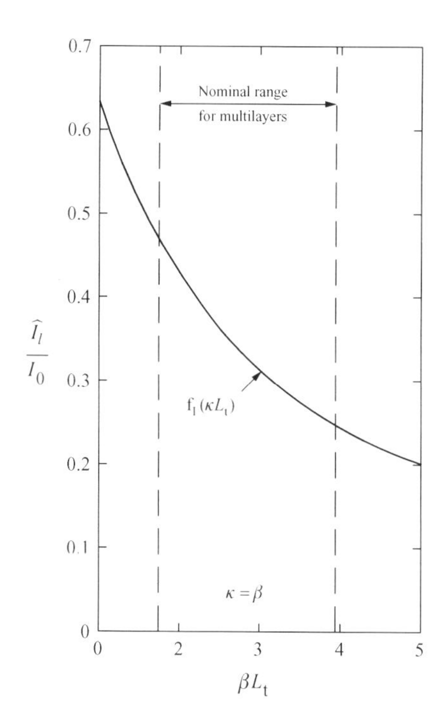

Fig. 8.17. Canopy-average insolation for multilayer crowns. "Exact" for cylinders, assumes homogeneity at constant radius for cones.

With these averagings, we have created a fictitious canopy in which *every leaf* satisfies Eq. 8.28 and is exposed to the canopy environmental insolation, $\widehat{I_l} \equiv f_I(\beta L_t) I_0$, in which I_0 is the daylight-hour average environmental insolation during the growing season.

Bioclimatic optimality for a canopy of crowns

The environmental conditions yielding maximum stressless productivity of a *given species* define its *feasible habitat*. For the canopy of crowns these conditions follow our prior reasoning for the canopy of individual leaves and consist of the three hypotheses:

Heat

$$T_0 \approx T_l = T_m; \tag{8.33}$$

Light

$$\widehat{I_{SL}} \le \widehat{I_l}, \tag{8.34}$$

in which the equality, together with Eqs. 8.28 and 8.30, defines the canopy *climatic assimilation potential*, $P_{sm} = \varepsilon f_I(\beta L_t) I_0$, while the degree of inequality signifies species instability and measures the productivity pressure for species substitution; and finally, to ensure no water stress,

Water

$$\widehat{I_l} \le \widehat{I_{SW}}, \tag{8.35}$$

Plate 8.5. Cedar among oaks. Left to right: scarlet oak, white oak, and Eastern red cedar. (Photographed in Holden Arboretum, Cleveland, by William D. Rich; Copyright © 2001 William D. Rich.)

or, using Eqs. 8.32, 8.34, and 8.35,

$$I_{SL} \leq f_I(\beta L_t) I_{SW}, \tag{8.36}$$

in which I_{SW} is given by Eq. 8.26 with V_e given by Eq. 8.27.

The constraints on canopy bioclimatic space are as given for the individual leaf in Eqs. 8.19 through 8.21 by letting $I_0 \rightarrow \widehat{I_l}$, $I_{SW} \rightarrow \widehat{I_{SW}}$, and $\widehat{I_{SL}} \rightarrow \widehat{I_{SL}}$:

Species feasibility and productive optimality

$$\widehat{I_{SL}} \leq \widehat{I_l} \leq \widehat{I_{SW}}, \tag{8.37}$$

Species stablility

$$\widehat{I_{SL}} = \widehat{I_l} \leq \widehat{I_{SW}}, \tag{8.38}$$

Climatic productive optimality

$$\widehat{I_{SL}} = \widehat{I_l} = \widehat{I_{SW}}. \tag{8.39}$$

The feasible and productively optimal bioclimatic space for single-species canopies is illustrated by the prism sketched in Fig. 8.18a using the boundaries provided by Eq. 8.37. The stable subset of this region is shown by the cross-hatched sloping surface according to Eq. 8.38, and the productively optimal climate for a given species is defined from Eq. 8.39 by the heavy edge of this latter surface. Note in this figure that the vertical coordinate is a species parameter since $I_{SL} \equiv P_{sm}/\varepsilon$, while the two horizontal coordinates define the environmental controls of light and water (including the soil). We see clearly from this that the range of feasible species increases with increasing

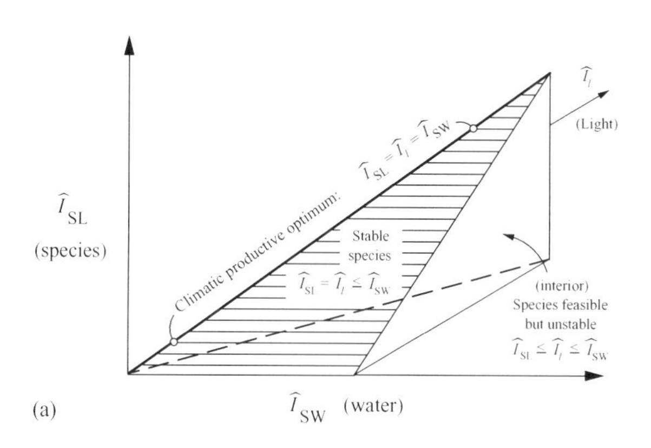

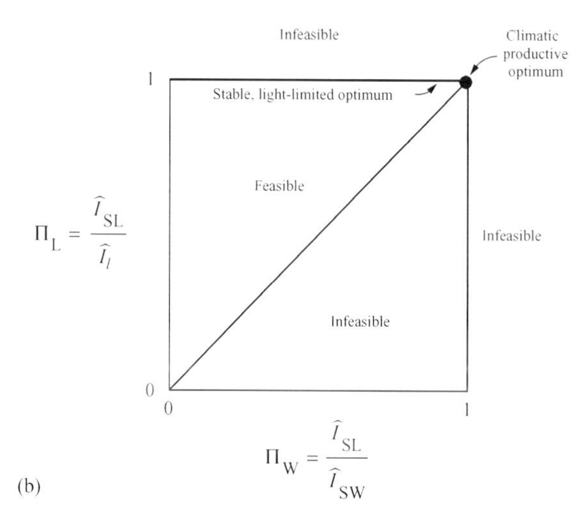

Fig. 8.18. Feasible canopy bioclimatic space. (a) In one species and two climate dimensions; (b) in two dimensionless species/climate coordinates.

availability of light and water while the climate range becomes more restricted. The inverse is true in resource-poor climates.

It is perhaps easier to examine this space by reducing its dimensionality through the use of dimensionless variables. Dividing Eq. 8.37 by $\widehat{I_{SL}}$ and inverting each term we have

$$1 \geq \Pi_L \geq \Pi_W, \tag{8.40}$$

in which the dimensionless habitat variables are each a ratio of species-to-climate parameters according to their prior definitions for light:

$$\Pi_L \equiv \frac{\widehat{I_{SL}}}{\widehat{I_l}} = \frac{P_{sm}/\varepsilon}{f_I(\beta L_t) I_0}, \tag{8.41}$$

and for water:

$$\Pi_W \equiv \frac{\widehat{I_{SL}}}{\widehat{I_{SW}}} = \frac{I_{SL}}{f_I(\beta L_t) I_{SW}} = \frac{P_{sm}/\varepsilon}{\dfrac{f_I(\beta L_t)}{1-\rho_T}\left[(1 \pm \boldsymbol{R}_b)\dfrac{\lambda V_e}{m_{t_b^{\otimes}} M} + q_b\right]}. \tag{8.42}$$

The feasible canopy bioclimatic space of Fig. 8.18a is shown in these dimensionless coordinates in Fig. 8.18b where the feasible prism transforms into a planar triangle, the stable surface called the climatic climax becomes the uppermost boundary, and the climatic optimum reduces to a point.

The hydrology of the stable bioclimatic state

We gain hydrologic insight into the subset of stable bioclimatic states through a plot of Π_W vs. the vegetation state variable, M, for $\Pi_L = 1$ as is given in Fig. 8.19. A few important observations are in order:

(1) Wherever $M < 1$, the soil moisture is at its critical value and therefore $\widehat{I_l} = \widehat{I_{SW}}$.

(2) The condition, $\Pi_W = 1$, $M \leq 1$, is called the *climatic climax condition*. There maximum productive use is made of the environmental resources, light and water.

(3) For $\Pi_W = 1$, $M < 1$, the climax condition is productively *water-limited* in that canopy productivity will rise with water availability due to increasing M.

(4) For $\Pi_W < 1$, $M = 1$, there is excess water and the climax condition is productively *light-limited*. However, there is a magnitude of this water excess beyond which productivity is nitrogen-limited. This occurs due to soil flooding which restricts oxygen concentration and thereby causes anerobic activity producing toxic compounds that disrupt nitrogen turnover (Larcher, 1983, p. 86). We do not consider the limits imposed by soil flooding in this work.

(5) The condition $\Pi_W < 1$, $M < 1$,[†] is a feasible natural habitat but is unstable in the long term due to $\widehat{I_l} > \widehat{I_{SL}}$. We call it the *sub-climax* condition.

[†] Note that in this representation of bioclimatic space, the branch for $\Pi_W \leq 1$ is shared by both the stable light-limited climax and the limiting unstable sub-climax.

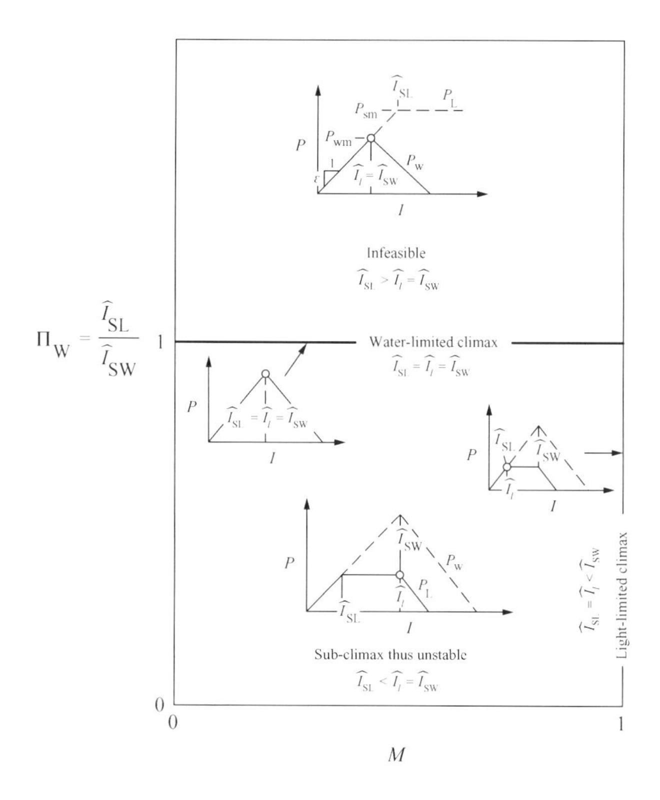

Fig. 8.19. Canopy bioclimatic space (M is a surrogate for rate of moisture availability, $V_e/m_{t_b \otimes}$).

(6) The situation $\Pi_W > 1$ is presumed to be biologically infeasible because the stomates are partially closed at any $\widehat{I_l}$ and thereby put the plant under stress. Any such overdeveloped capacity (such as might be present following rapid climate change) would rapidly die back and be replaced by species with productive capacity matching the environmental potential so that $\widehat{I_l} = \widehat{I_{SL}} = \widehat{I_{SW}}$.[†‡]

(7) As the above conditions are formulated, we note that the soil hydraulic properties appear solely in the surface runoff fraction, $e^{G(0)\ \ 2\sigma(0)^{3/2}}$, of V_e as given by Eqs. 6.67 and 6.68, and even disappear entirely with our approximation of V_e in Eq. 8.27. Soil hydraulic properties are indeed important in controlling stomatal closure. However, at the critical soil moisture state, s_{oc}, which determines I_{SW}, the stomates remain fully open (but at incipient closure). The soil is therefore only weakly influential (cf. Table 6.4) *as long as the water table is deep*. Nevertheless, the literature emphasizes the importance of plant–soil water relations (e.g., Slatyer, 1957a, 1957b; Gardner, 1960; Eyre,

† Note the role that advected heat (i.e., negative $\boldsymbol{R}_b$) from the bare soil fraction can have in increased transpiration and thus in reducing $\widehat{I_{SW}}$ (cf. Eq. 8.26), and the opportunity for regulation of I_{SW} by β through its effect upon ρ_T. Note also that such increased water use will bring increased nitrogen flux with resulting increase in $\widehat{I_{SL}}$ and in βL_t, the latter of which produces decreased $\widehat{I_l}$.

‡ Restriction of the availability of carbon and of soluble nutrients in the soil will be considered briefly in Chapter 10.

Plate 8.6. European black pine. (Photographed in Arnold Arboretum, Boston, by William D. Rich; Copyright © 2001 William D. Rich.)

1968; Zahner, 1968), perhaps an indication of the pervasiveness of water table influence.[†]

(8) There may be multiple species, which we will call *competing species*, that meet these climax conditions in the same environment. In such cases, we hypothesize that the stable or climax species will be determined by which of the competing species has the greatest annual production of above-ground, seed-producing, biomass in the given environment. We deal with this issue in Chapter 10.

In Appendix H we use these results to show that βL_t and hence I_k/I_0 (cf. Eqs. 7.3 and 7.4), are *species constants* under climax conditions. Observations for a wide range of communities, from desert scrub to rainforest, are compared with these conditions in Chapter 9.

[†] Refer to Salvucci and Entekhabi (1995, corrected 1997) in order to find V_e under water table influence.

9

Natural habitats and climax communities

The separate heat, light, and water propositions of the bioclimatic optimality hypothesis are tested against the characteristics of observed canopies and their environments. On the basis of the limited evidence presented all three propositions are strongly supported. Seven of the nine vegetation communities tested appear to be in their hypothesized natural habitat, and only two are more than 6 percent removed from the hypothesized climax state.

A Review of habitat constraints

In Chapter 8 we propose a set of conditions relative to environmental heat, light, and water that must be satisfied for a given environment to support a given species in an unstressed (and thus feasible) state. These conditions, $\widehat{I_{SL}} \leq \widehat{I_l} \leq \widehat{I_{SW}}$, and $T_0 \approx T_m$, define the set of *feasible natural habitats* for the particular species without regard to issues of stability or competition.

We further hypothesize in the previous chapter that in a feasible habitat, natural selection and/or adaptation will lead to a stable (i.e., *climatic climax*) vegetative canopy that maximizes productivity, thereby restricting the habitat space according to $\widehat{I_{SL}} = \widehat{I_l} = \widehat{I_{SW}}$ when water is limiting and to $\widehat{I_{SL}} = \widehat{I_l} < \widehat{I_{SW}}$ when light is limiting.

Our hypothesis contains separate propositions for heat, light, and water which we will now test against the characteristics of observed (assumed stable) canopies and their environments. This implements the instruction of Billings (1985, p. 12) that to explain species distribution we should seek tolerance ranges which make prediction of geographical ranges possible. Note however, the data have been taken from the literature, they are sparse, and they are used in combinations from heterogeneous sources; thus any conclusions drawn therefrom can only be tentative.

Plate 9.1. Creosote bush desert community. (*Larrea tridentata* photographed in the Mohave Desert, Nevada, by Robert H. Webb; courtesy of Robert H. Webb and the U.S. Department of the Interior.)

B Proposition no. 1: Heat

Statement

The average leaf temperature, $\widehat{T}_l$, has the value, T_{m}, at which the rate of photosynthesis of the given species is maximum. Under the assumption (cf. Chapter 5) that *on average* the transpiration process is closely isothermal, the average leaf temperature approximates the growing season average atmospheric temperature at screen height, T_0. This has been noted by Larcher (1983, pp. 119–120) and leads to our heat proposition

$$T_0 \approx \widehat{T}_l = T_{\mathrm{m}}. \tag{9.1}$$

Test

In testing Eq. 9.1 we are limited to the species for which experimental determinations of T_{m} are available *and* for which the geographic location (and hence T_0) of their stable habitat is also known. Values of T_{m} are given for several species in Fig. 8.4a taken from Larcher (1983, Fig. 3.35, p. 114), and the natural North American habitats of many species are given by Burns and Honkala (1990, vols. 1 and 2). With these data, the test is possible for only three species.

Loblolly pine

The only exact match of species between these two data sources is for loblolly pine. Figure 8.4a gives for loblolly pine, 23 °C $\leq T_m \leq$ 30 °C, while Baker and Langdon (1990, Fig. 1, p. 497) present the natural habitat for loblolly pine in North America indicated by the shaded area of Fig. 9.1. Average July atmospheric temperatures, T_0, for the period 1941–70 are given in the table accompanying Fig. 9.1 as taken from U.S.

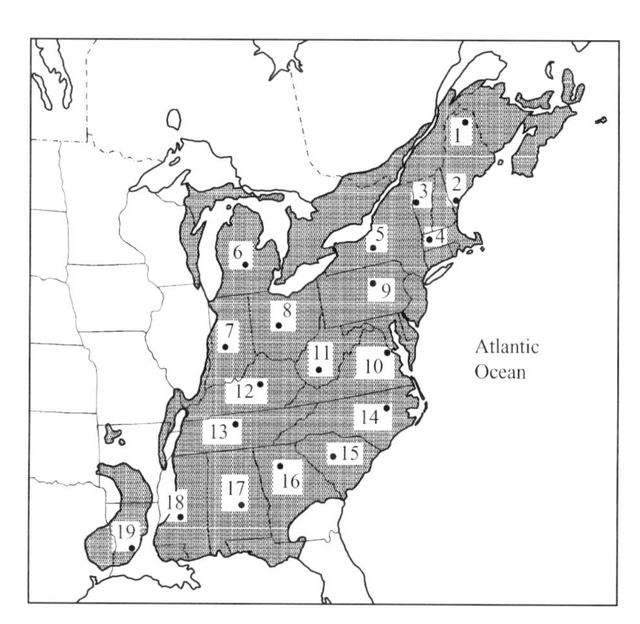

Fig. 9.1. Natural North American habitat of loblolly pine. (From Baker and Langdon, 1990, Fig. 1, p. 497; courtesy of the U.S. Department of Agriculture.)

Station	Location	$\bar{T}_0{}^a$ (°C)	Station	Location	$\bar{T}_0{}^a$ (°C)
1	Caribou, Maine	18.3	10	Richmond, Virginia	25.5
2	Portland, Maine	20.0	11	Beckley, West Virginia	21.1
3	Burlington, Vermont	21.0	12	Lexington, Kentucky	24.5
4	Pittsfield, Massachusetts	19.9	13	Nashville, Tennessee	26.5
			14	Raleigh, North Carolina	25.3
5	Syracuse, New York	22.0	15	Columbia, South Carolina	27.3
6	Lansing, Michigan	21.6			
7	Indianapolis, Indiana	23.9	16	Atlanta, Georgia	25.6
8	Columbus, Ohio	23.1	17	Montgomery, Alabama	27.2
9	Williamsport, Pennsylvania	22.7	18	Jackson, Mississippi	27.6
			19	Alexandria, Louisiana	27.2
				Average	23.7

[a] T_0 = (Normal July Daily Maximum + Normal July Daily Minimum) / 2; 1941–70.

Weather Service climatological records at nine stations within this habitat. The stations are located in the cities shown on Fig. 9.1 and are chosen to cover the geographical range of the habitat.

American beech

Here we must compare two different varieties of beech. Tubbs and Houston (1990, Fig. 1, p. 326) give the natural habitat for American beech as is shown in Fig. 9.2, and U.S. Weather Service average July atmospheric temperatures, T_0, throughout the habitat are again listed in the caption. Figure 8.4a records $T_m = 23\,^\circ$C for European beech.

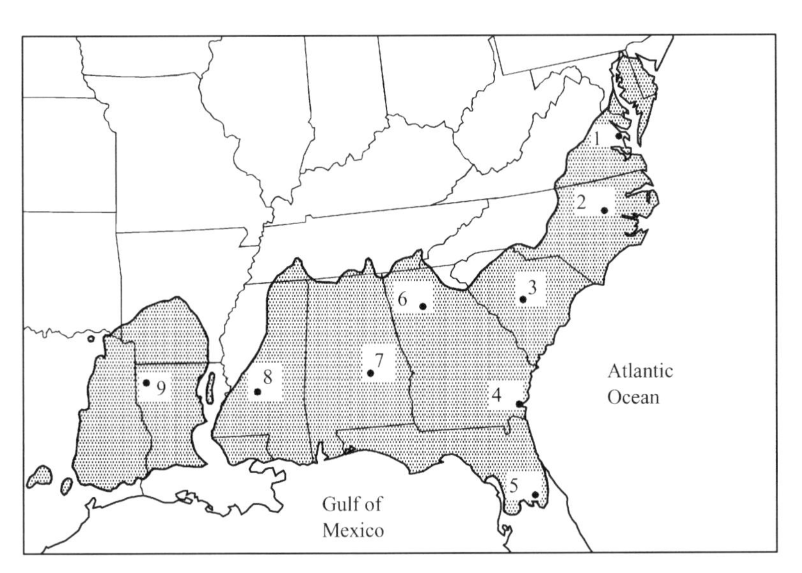

Fig. 9.2. Natural North American habitat of American beech. (From Tubbs and Houston, 1990, Fig. 1, p. 326; courtesy of the U.S. Department of Agriculture.)

Climatological station	Location	$\bar{T}_0{}^a$ (°C)
1	Richmond, Virginia	25.5
2	Raleigh, North Carolina	25.3
3	Columbia, South Carolina	27.3
4	Savannah, Georgia	26.9
5	Lakeland, Florida	27.5
6	Atlanta, Georgia	25.6
7	Montgomery, Alabama	27.2
8	Jackson, Mississippi	27.6
9	Shreveport, Louisiana	28.5
	Average	26.8

[a] T_0 = (Normal July Daily Maximum + Normal July Daily Minimum) / 2; 1941–70.

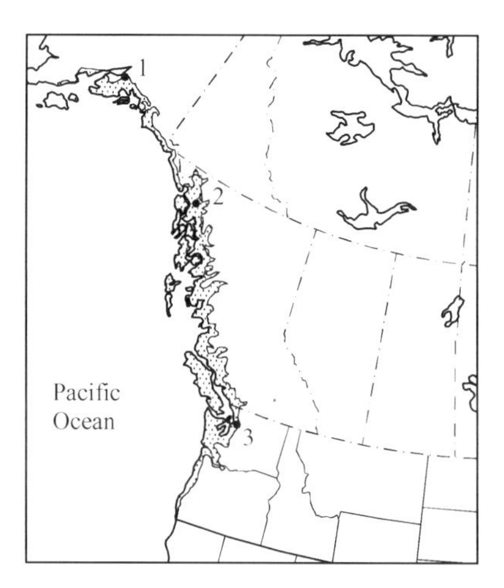

Fig. 9.3. Natural North American habitat of Sitka spruce. (From Harris, 1990, Fig. 1, p. 260; courtesy of the U.S. Department of Agriculture.)

Station	Location	$\bar{T}_0{}^a$ (°C)
1	Anchorage, Alaska	14.4
2	Juneau, Alaska	13.2
3	Seattle–Tacoma, Washington	18.0
	Average	15.2

$^a T_0$ = (Normal July Daily Maximum + Normal July Daily Minimum) / 2; 1941–70.

Sitka spruce

In this last case, we are forced to make this comparison between two species of cold-weather conifer, Arctic pine for T_m (cf. Fig. 8.4a where $T_m = 15\,^\circ$C) and Sitka spruce for T_0 (cf. Harris, 1990, Fig. 1, p. 260). The North American habitat and average July temperatures within the Sitka spruce habitat are given in Fig. 9.3.

The result of this limited testing of our heat proposition is summarized in Fig. 9.4 where the lines indicate the range of observation and the plotted circles the averages. The data strongly support our proposition that $T_0 \approx \widehat{T_l} = T_m$. We cannot say whether this agreement is the result of natural selection of the species for the particular environment or of adaptation of the species to the environmental conditions. In either case, the result is the same.

C Proposition no. 2: Light

Statement

The minimum average leaf insolation, I_{SL}, for photosynthetic saturation (i.e., fully open stomates) is given by the photosynthetic characteristic of the species as P_{sm}/ε which, under climax (i.e., stable) conditions, must be identical to the incident environmental

Table 9.1. *Test of solar radiation proposition*

Number	Species	P_{sm}[a] (g_s m^{-2} h^{-1})	I_{SI}[b] (MJ m^{-2} h^{-1}) (PAR)	Climate[c]	Φ[d] (degrees)	I_0[e] (MJ m^{-2} h^{-1}) (PAR)	β	L_t	βL_t	$f_I(\beta L_t)$	$\frac{f_I(\beta L_t) I_0}{I_{SL}}$	References
Cold												
1	Arctic willow	0.30	0.15	Finland, 1/2	60	0.56	0.45[f]	6.75[g]	3.04	0.31	1.15	Kramer and Kozlowski (1960, Fig. 3.11, p. 83).
2	Sitka spruce	0.55	0.26	Scotland, 1/2	55	0.60	0.51[f]	5.14[h]	2.62[f,i]	0.35	0.81	Jarvis *et al.* (1976, Fig. 21b, p. 229).
3	European beech	0.33	0.16	Norway, 1/2	60	0.56			3.19[j]	0.30	1.05	Kramer and Kozlowski (1960, Fig. 3.9, p. 78).
Temperate												
4	Eastern red oak	0.28	0.23	1/2	43–32	0.73			2.60[j]	0.35	1.11	Kramer and Decker (1944, Fig. 1, p. 352).
5	White oak	0.26	0.23	1/2	45–30	0.72			2.60[j]	0.35	1.10	Kramer and Decker (1944, Fig. 1, p. 352).
6	Loblolly pine	0.40[k]	0.30	1/2	35–27	0.75			2.58[j]	0.35	0.88	Kramer and Clark (1947, Fig. 3, p. 55).
Hot												
7	Creosote bush	2.08	1.17	Mojave desert, clear	35–20	1.24	0.37[l]	1.00[m]	0.37	0.84	0.89	Ehleringer (1985, Fig. 7.8, p. 172).
8	Goethalsia	0.38	0.22	Costa Rica	10	0.62[n]	0.79[o]	3.30[o]	2.61	0.36	1.02	Allen and Lemon (1976, Fig. 6, p. 290).
										Average	1.00	

[a] From data, using conversion factor, g_s g^{-1} = 0.5 (see Eq. 8.7). [b] From Fig. 7.15 at $P/P_s = 1 - 1/e$, shortwave only.
[c] "1/2" signifies the average of "very clear sky" and "overcast" sky radiances.
[d] Single entry from map, using site of phytotron test; Temperate climate double entry from North American natural range (Burns and Honkala, 1990), Hot climate range from map in MacMahon (1988, Fig. 8.18, p. 252).
[e] Daylight Hour Average from Fig. 3.16. [f] From Table 2.3 assuming $\kappa = \beta$. [g] From Table 2.2. [h] $L_t = \beta L_t / \beta$. [i] $a_w = \beta L_t$; average for all spruce, Table 2.2.
[j] From Baker (1950, Table 12, p. 143); see Table 3.8. [k] 0.17 g_s m^{-2} (stomated area) h^{-1} × 2.36 m^2 (stomated area) / m^2 (illuminated area) (Raison *et al.*, 1992).
[l] For Eucalyptus assuming $\kappa = \beta$ (see Table 2.2). [m] Whittaker and Likens (1975) as given by Larcher (1983, Table 3.13, p. 151).
[n] Allen and Lemon (1976, p. 298). [o] Allen and Lemon (1976, p. 288).

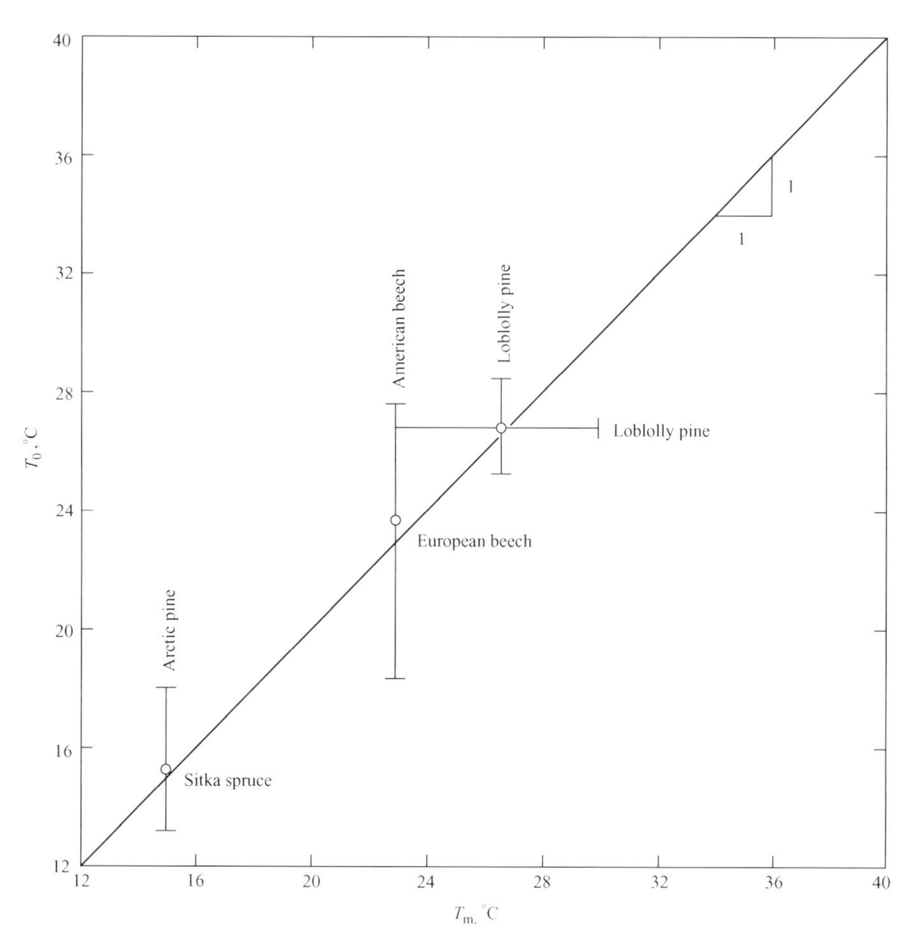

Fig. 9.4. Limited test of heat proposition: $T_0 = T_m$.

insolation as averaged both in time over the growing season and in space over the depth of the canopy.[†] That is

$$f_I(\beta L_t) I_0 = \frac{P_{sm}}{\varepsilon} \equiv I_{SL}, \tag{9.2}$$

or, in dimensionless form,

$$\Pi_L \equiv \frac{\widehat{I_{SL}}}{\widehat{I_l}} = \frac{I_{SL}}{f_I(\beta L_t) I_0} = 1. \tag{9.3}$$

Test

Data for the test of this proposition are collected from the literature for eight species native to a wide range of climates and are summarized here in Table 9.1.

[†] In Fig. 8.10 we see that for C_3 plants, an average value of ε is $\langle \varepsilon \rangle = 0.81\ g_s\ MJ(tot)^{-1}$.

(1) The values of P_{sm} are obtained by scaling from the source photosynthetic capacity curves for individual leaves as determined from phytotron experiments. The values of I_{SL} are then obtained by fitting Eq. 8.6 to pairs of values scaled from these curves.

(2) Estimation of the effective growing season average environmental insolation presents a more difficult problem. We use the "total light" curves of Fig. 3.10 to estimate the daylight-hour average PAR. For all climates save the Mojave desert and the rainforest, we assume the annual average to be the mean of the values for "very clear sky" and "overcast sky" and signify this with the notation "1/2". For the desert we use the "very clear sky" value unmodified by cloudiness, while for the rainforest we use the reported site observation. The latitude at which Fig. 3.10 is entered to obtain I_0 is an estimate of that at which the phytotron test was performed in cases where the natural range of the species was unknown (i.e., the "cold" climates). In the other cases the ranges are known from the referenced maps, and their limiting latitudes are used. The corresponding irradiances are then averaged to get I_0.

(3) Conversion of irradiance from leaf to canopy (cf. Eqs. 8.29–8.31) requires an estimate of the light decay coefficient, κL_t.[†] For species nos. 3 through 6 this is found using Baker's (1950, Table 12, p. 143) estimates of light intensity at the compensation point (cf. Table 3.8) along with Eq. 3.48.[‡] In the other cases, we make use of our finding that $\kappa = \beta$ (see Chapter 3) and obtain separate estimates of β and of L_t from the literature as is described in the notes beneath Table 9.1.

The two outside members of Eq. 9.2 are compared in Fig. 9.5 to test the solar radiation proposition. The ratio, $\Pi_L = \frac{I_{SL}}{f_I(\beta L_t)I_0}$, has a mean value of 1.00 as is shown in the last column of Table 9.1, and a standard deviation of $\pm 12\%$. Considering all the approximation involved in arriving at the individual values, this remarkable agreement of the average with that hypothesized in Eq. 9.3 must be regarded as somewhat fortuitous.

This very significant result is highlighted in Fig. 9.6 by comparison of the (small sample) distribution of $f_I(\beta L_t)I_0/I_{SL}$ with the functional depiction of this ratio at leaf scale for the case of light control. In this representation of the distribution, the circle is the mean, the line shows the range, and the box includes all values within $\pm$ one standard deviation of the mean. The scatter of this small sample about its mean, $\Pi_L = 1$, appears biased very slightly toward $\Pi_L > 1$ which is opposite to the direction we would expect if the communities were photosynthetically underdeveloped (i.e., $\widehat{I_l} > \widehat{I_{SL}}$). Discarding this tendency and disallowing the stressed circumstance, $\widehat{I_l} < \widehat{I_{SL}}$, we accept the hypothesis $\Pi_L = 1$.

These results strongly support this proposition.

[†] Note that all but one of the species tested here (i.e., Sitka spruce) appear to have homogeneous cylindrical crowns as far as the distribution of leaf area is concerned.

[‡] For loblolly pine we use Baker's lodgepole pine and for pinyon–juniper we use either Baker's Northern white cedar or Baker's lodgepole pine as detailed in a footnote in Chapter 7, Section E.

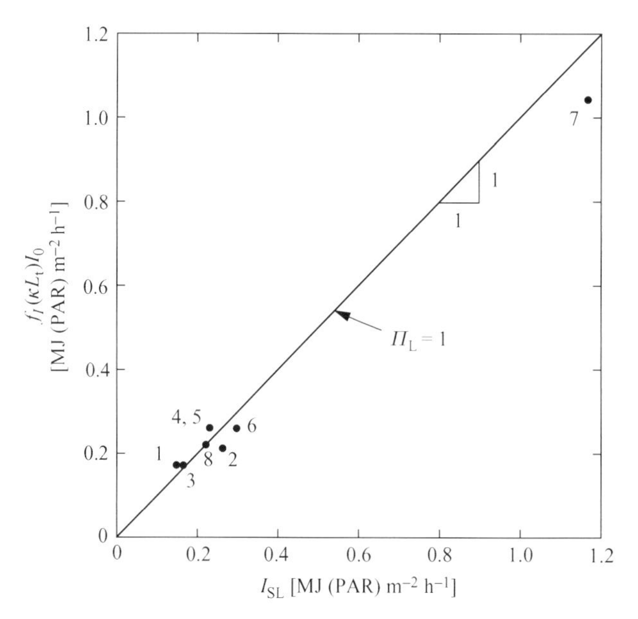

Fig. 9.5. Test of light proposition: $f_I(\beta L_t)I_0 = I_{SL}$. 1, Arctic willow; 2, Sitka spruce; 3, European beech; 4, Eastern red oak; 5, white oak; 6, loblolly pine; 7, creosote bush; 8, *Goethalsia*. (Data from Table 9.1.)

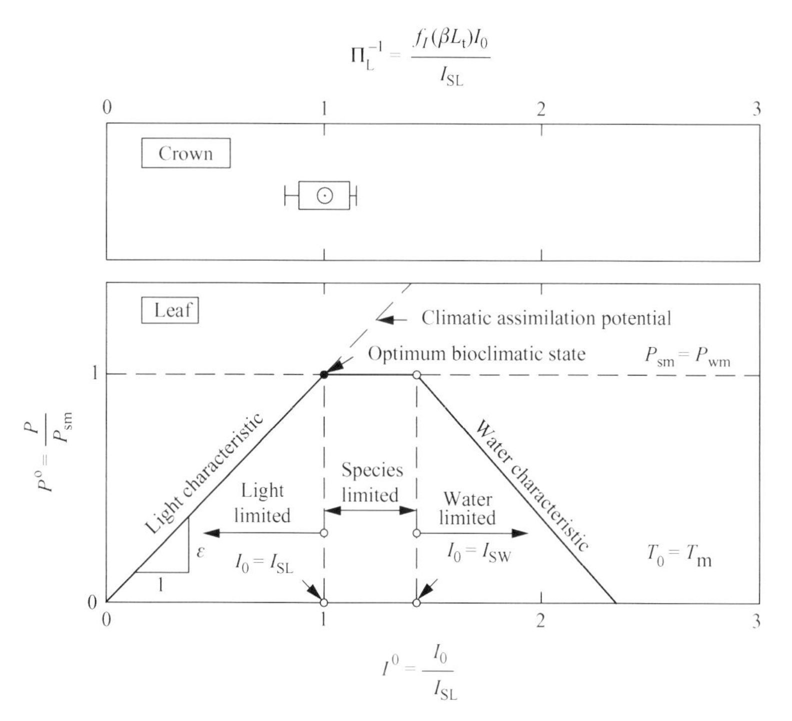

Fig. 9.6. Functional view of light proposition and the supporting evidence (fixed species, soil, and precipitation). (Data from Table 9.1.)

D Proposition no. 3: Water

Statement

In order for the leaf to assimilate at its maximum rate, P_{sm}, the water supply must be such as to insure that $I_{SL} \leq I_{SW}$. At canopy scale this becomes (cf. Chapter 8)

$$\frac{P_{sm}}{\varepsilon} \equiv I_{SL} \leq f_I(\beta L_t) I_{SW}. \tag{9.4}$$

To express I_{SW} in terms of the canopy water supply we use the energy equation in which the latent heat term, λE_T, is derived from the Penman–Monteith equation (Eq. B.1). We work with a unit of *vegetated* surface so that $E_T \equiv E_v$, and then use Eq. 6.46 to write, for the crown, at $\beta_v = 1$

$$E_T = E_v = k_v E_{ps} = k_v^* E_{ps}. \tag{9.5}$$

Replacing k_v and E_{ps} through Eqs. 5.29 and B.1 respectively, Eq. 9.5 becomes

$$\lambda E_v = \frac{R_n + D_p}{1 + \frac{\gamma_o}{\Delta}\left(1 + \frac{\widehat{r_c}}{r_a}\right)}, \tag{9.6}$$

in which the D_p term incorporates the Bowen ratio. We expand R_n according to Eq. B.3, evaluate D_p empirically from Eq. B.18, and note from Eqs. 6.66 and 9.5 that the "true" (i.e., daylight-hour) transpiration is controlled by the available soil moisture to be

$$E_v^{\otimes} = \frac{V_e}{m_{t_b^{\otimes}} M}, \tag{9.7}$$

all leading to the desiccation insolation

$$I_{SW} \equiv \frac{1}{1 - \rho_T}\left\{\frac{\lambda V_e}{m_{t_b^{\otimes}} M}\left[1 + \frac{\gamma_o}{\Delta}\left(1 + \frac{\widehat{r_c}}{r_a}\right)\right] + q_b\left[1 - \frac{(1 - S_r)}{0.85}\right]\right\}. \tag{9.8}$$

In Eq. 9.8:

(1) V_e, the moisture supply to the root zone, is approximated using Eq. 6.65 with $\beta_v = 1$ (for deep water tables and with term estimates from Table 6.4) as†

$$V_e \approx m_h\left[1 - \frac{\overline{h_0}}{m_h} + \frac{\Delta S}{m_\nu m_h}\right]; \tag{9.9}$$

(2) $m_{t_b^{\otimes}}$, the daylight-hour time between storms, is (Eq. 6.63)

$$m_{t_b^{\otimes}} \approx \frac{1}{2} m_{t_b} - \frac{\eta_o L_t h_0 / 2}{E_{ps}^{\otimes}} = \frac{1}{2} m_{t_b'}; \tag{9.10}$$

† There are obviously many common and much more complex cases in which the root-zone moisture is supplied totally or in part from below and a very different approximation of Eq. 6.65 is called for. Accurate estimation of V_e is at once the most difficult and the most important task in formulating the water limit to vegetation growth, and the devil is in the details.

Plate 9.2. Subalpine fir and Engelmann spruce forest. (Photographed at Bow Lake in Banff National Park, Alberta, by William D. Rich; Copyright © 2001 William D. Rich. Tentative species identification by Arthur L. Fredeen from the photograph.)

(3) $\frac{\widehat{r_c}}{r_a}$, the crown average resistance ratio, is (Eq. 7.25)

$$\frac{\widehat{r_c}}{r_a} = (1 - M)\left(\frac{r_c}{r_a}\right)_{M \to 0} + M\left(\frac{r_c}{r_a}\right)_{M = 1}; \tag{9.11}$$

(4) q_b, the net longwave back radiation, is (Eq. B.12)

$$q_b = (1 - 0.80N)\left(0.245 - 0.145 \times 10^{-10} T_{0K}^4\right); \tag{9.12}$$

and

(5) S_r, the atmospheric saturation ratio (i.e., fractional relative humidity) at screen height is obtained from station meteorological records.

The water proposition has both a water-limited branch,

$$\Pi_W \equiv \frac{I_{SL}}{f_I(\beta L_t) I_{SW}} = 1, \quad M \leq 1, \tag{9.13}$$

and a light-limited branch,

$$\Pi_W \equiv \frac{I_{SL}}{f_I(\beta L_t) I_{SW}} < 1, \quad M = 1. \tag{9.14}$$

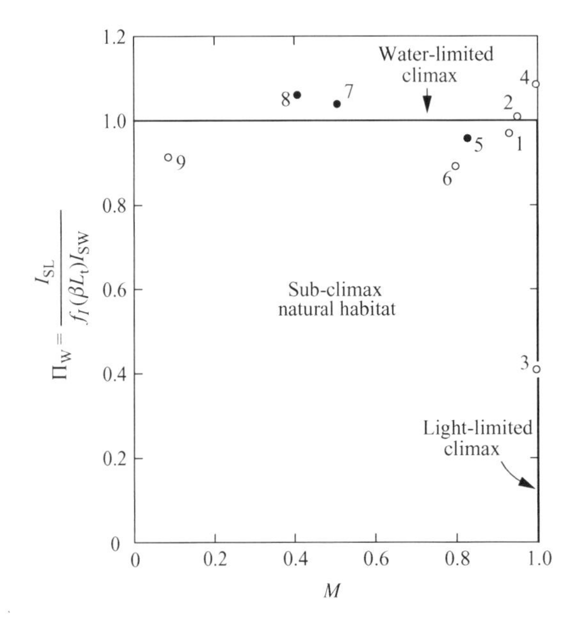

Fig. 9.7. Test of water proposition: $I_{SL} = f_I(\beta L_t)I_{SW}$ when $M \leq 1$ and $I_{SL} < f_I(\beta L_t)I_{SW}$ when $M = 1$. 1, Beech; 2, oak; 3, *Goethalsia* (rainforest); 4, Sitka spruce; 5, ponderosa pine; 6, loblolly pine; 7, pinyon–alligator juniper; 8, pinyon–Utah juniper; 9, creosote bush (desert). (Data from Table 9.2; solid points are Beaver Creek watersheds.)

Test

We have already supported this proposition by the test in Fig. 6.15 using the extraordinarily complete data set for the three pinyon–juniper and ponderosa pine communities at Beaver Creek, Arizona. Here we expand the test to a wider range of habitats using the less direct formulation in terms of desiccation insolation as is called for in Eq. 9.13 and as prescribed in Eq. 9.8.

We test the proposition further in Fig. 9.7 (as was suggested in Fig. 8.19) by examining the observed variation of the ratio, $\Pi_W \equiv \frac{I_{SL}}{f_I(\beta L_t)I_{SW}}$, with the canopy density, M, for stands of nine different species covering the range from the desert shrub (*Larrea divaricata*) to the rainforest (*Goethalsia meiantha*). The supporting data are presented and their sources are identified in Table 9.2. In interpreting this figure remember that I_{SW} is the value of the growing season average insolation, I_0, at which the available soil moisture has that critical value which is just sufficient to keep the stomates at incipient closure at the given temperature. Therefore, given verification of the heat and light propositions as just presented, and given the strong verification of the critical water balance (cf. Eq. 6.15), the degree to which $I_{SL} = f_I(\beta L_t)I_{SW}$ for $M < 1$ must be viewed as a test of the energy balance.

As noted in Chapter 6, evaluation of V_e in I_{SW} is problematic. This is particularly true for the interception loss and the carryover storage for precipitation-fed soil moisture, and is even more uncertain where soil moisture is recharged from the water table. Among the data presented, the carryover storage is best known for the ponderosa pine and pinyon–juniper watersheds of Beaver Creek (Flagstaff, Arizona) which were studied in Chapter 6 and reported in Tables 6.3 and 6.4. However, the value of I_{SL} was not available for ponderosa pine or for either of the pinyon–junipers. On the other hand, I_{SL} is known for the other six species (see Table 8.2), but *all* components of V_e are unknown at the sites where I_{SL} was measured. Thus in all nine cases, completion of the data set required either outright estimation of one or more parameters or the use of observations of the given species at other sites. This must be kept in mind when evaluating Fig. 9.7.

The parenthetical superscripts used in Table 9.2 and defined in the listing accompanying that table, identify the source and estimators of the parameter values used in this test. Most should be self-explanatory; however a few deserve expanded commentary due to the uncertainty of their estimates:

(1) Canopy cover (M). This variable is seldom reported by the foresters and plant physiologists responsible for the bulk of useful data. The studies by Whittaker (1966) of natural beech and oak stands in the Great Smoky Mountains of Tennessee which we utilize heavily here, report a small percentage, less than 10%, of light transmitted to the forest floor. We assume this to reflect the structure of the *canopy* rather than that of the crown, and take M for these canopies to be 1 minus the percentage of light transmitted.

(2) Interception (h_0). An interception depth, $h_0 = 0.1$ cm, was assumed retained on the horizontal projection (i.e., βL_t) of all flat leaves and on all sides (i.e., $\eta_0 L_t$) of needle leaves. Note from Table 9.2 and Eq. 6.52 that with this assumption the Sitka spruce canopy gets soil moisture recharge from only 2% of the rainfall. This small V_e leads to $\frac{I_{SL}}{f_I(\beta L_t) I_{SW}} = 5$ for the spruce which is off the scale of Fig. 9.7. There must be another moisture source for this canopy.

(3) Carryover moisture storage (ΔS). The presence of soil moisture in storage at either the beginning ($+\Delta S$) or end ($-\Delta S$) of the growing season is known for only the Beaver Creek watersheds (Salvucci, 1992) and is reported here in Table 6.3. However, the seasonal asynchronism of heat (i.e., growing season) and water (i.e., rainy season) present at Beaver Creek is similar to that found for the Sitka spruce at Spokane, Washington. We see this in Fig. 9.8 where the monthly precipitation is much higher just before than during the growing season at Flagstaff and at Spokane. In contrast we note in Fig. 9.8 that for loblolly pine at Savannah, Georgia the monthly rainfall builds to a peak at the conclusion of the growing season raising the likelihood of a negative carryover moisture.

In addition to carryover soil moisture storage, Douglas fir (Waring and Running, 1978), and Scots pine (Waring *et al.*, 1979) are known to store up to 1.5 mm of water per day in the sapwood of the tree. This mechanism may be an evolutionary response to the asynchronous seasonality just discussed and also enables available soil moisture to be carried over either to or from the growing season.

In moist habitats such as those of the Sitka spruce, drip from fog condensation and direct absorption of condensation by the foliage (cf. Stone, 1957; Goodell, 1963) may provide the missing moisture supplement.

Table 9.2. *Test of water proposition*[a]

Number	Species	Site	Latitude (°N)	T_0 (°C)	m_τ (d y^{-1} : h d^{-1})	E_{ps}[24] (cm d^{-1})	m_h (cm)	m_ν	m_{t_b} (d)	M	ρ_T	S_r[20]	ΔS (cm)
1	Beech	Knoxville, TN	36	31.0[20]	202[28] : 12[4]	0.36[34]	0.98[7]	63.0[7]	3.21[8]	0.93[9]	0.25[3]	0.62[36]	unknown
2	Oak	Allentown, PA	41	23.4[20]	183[28] : 12[4]	0.37[34]	0.97[7]	61.0[7]	3.00[2]	0.95[9]	0.25[3]	0.52[36]	unknown
3	*Goethalsia*	Turrialba, Costa Rica	10	23.0[14]	365[4] : 12[4]	0.34[27]	0.68[23]	365.0[4]	0.90[4]	1.00[4]	0.25[3]	0.71[37]	unknown
4	Sitka spruce	Scotland[b]	47	27.0[20]	106[28] : 12[4]	0.41[34]	0.44[7]	22.0[7]	4.82[8]	1.00[9]	0.25[3]	0.80[38]	unknown[c]
5	Ponderosa pine	Flagstaff, AZ	34	14.2[2]	169[2] : 12[4]	0.31[2]	0.70[33]	31.8[8]	5.32[33]	0.83[2]	0.25[3]	0.34[36]	+21.0[2]
6	Loblolly pine	Savannah, GA	32	27.0[20]	78[28] : 12[4]	0.48[34]	1.46[7]	25.0[7]	3.12[8]	0.80[25]	0.25[3]	0.60[36]	unknown[d]
7	Pinyon–alligator juniper	Flagstaff, AZ	34	15.9[2]	182[2] : 12[4]	0.32[2]	0.75[33]	31.7[2]	5.74[33]	0.51[2]	0.25[3]	0.34[36]	+4.70[2]
8	Pinyon–Utah juniper	Flagstaff, AZ	34	18.5[2]	215[2] : 12[4]	0.35[2]	0.63[33]	36.8[8]	5.86[33]	0.41[2]	0.25[3]	0.34[36]	+6.13[2]
9	Creosote bush	Las Vegas, NV[e]	37	11.0[20]	60[15] : 12[4]	0.38[35]	0.45[7]	8.0[8]	7.50[8]	0.09[16]	0.35[26]	0.39[38]	unknown

No.	I_{SL} (MJ(tot) m^{-2} h^{-1})	N[41]	η_o	$\beta = \kappa$	L_t	βL_t	$f_I(\kappa L_t)$	$\frac{\hat{r}_c}{r_a}$	$\overline{h}_o$[13] cm	$\lambda E_v^{\otimes}$[40] (MJ m^{-2} h^{-1})	q_b^*[39] (MJ m^{-2} h^{-1})	$\frac{q_b^*}{\lambda E_v^{\otimes}}$	I_{SW}[46] (MJ m^{-2} h^{-1})	R_b[19]	$\frac{I_{SL}}{f_I(\beta L_t) I_{SW}}$
1	0.32[1]	0.41	1.00	0.73[42]	4.4[29]	3.19[5]	0.30[11]	1.15[44]	0.42	0.70	0.13	0.19	1.10	−0.19	0.97
2	0.46[1]	0.37	1.00	0.53[42]	4.9[29]	2.60[5]	0.36[11]	0.89[44]	0.36	0.85	0.10	0.12	1.26	−0.15	1.01
3	0.44[1]	0.50[4]	1.00	0.79[12]	3.3[12]	2.61[10]	0.35[11]	1.95[44]	0.36	2.73	0.20	0.07	3.07	−0.73	0.41
4	0.52[1]	0.73	2.00[9]	0.51[30]	5.1[30]	2.62[30]	0.35[11]	0.76[45]	0.43	0.88	0.14	0.16	1.36	−0.30	1.09
5	0.60[6]	0.24	2.50[21]	0.28[42]	6.3[29]	1.77[5]	0.47[11]	0.51[43]	0.47	0.93	0.07	0.07	1.33	−0.17	0.96
6	0.60[1]	0.36	2.36[22]	0.52[42]	5.0[29]	2.58[5]	0.36[11]	0.81[31]	0.59	1.28	0.12	0.09	1.87	−0.12	0.89
7	0.60[6]	0.24	2.50[21]	0.52[42]	5.0[29]	2.58[5]	0.36[11]	1.31[43]	0.43	1.14	0.07	0.06	1.60	−0.07	1.04
8	0.60[6]	0.24	2.50[21]	0.42[42]	5.4[29]	2.27[5]	0.40[11]	1.32[43]	0.33	1.00	0.07	0.07	1.42	+0.13	1.06
9	2.34[1]	0.22	1.00	0.37[17]	1.0[18]	0.37[10]	0.84[11]	0.75[43]	0.10	1.88	0.09	0.05	3.03	−0.22	0.92

[a] Parenthetical superscript numbers refer to the source listing given in the notes below.
[b] Aberdeen, Scotland for I_{SL}; Spokane, WA for climate.
[c] We assume $\Delta S = +0.5 P_d$ as was observed for ponderosa pine at Flagstaff, AZ.
[d] We suspect a positive ΔS due to the seasonal rainfall (cf. Fig. 9.8).
[e] Las Vegas for $T_0^{(20)}$, Ely for precipitation (Appendix F).

Sources:

1. Table 8.2. 2. Williams and Anderson (1967).

3. Ross (1981, Fig. 112, p. 332). Equation 9.8 is formulated for unit vegetated area, thus we use $\rho_T = 0.25$ for "dense plant stands" rather than $\rho_T = 0.16$ at "tree top" (Iqbal, 1983, Table 9.4.1, p. 289).

4. Estimated.

5. βL_t from Table 3.8: using lodgepole pine for loblolly, Northern white cedar for pinyon–Utah juniper, and lodgepole pine for pinyon–alligator juniper as suggested by sketches of needle structure (cf. Zim and Martin, 1987, pp. 25, 36, 37, and 39).

6. Assumed identical to loblolly pine. 7. Appendix F. 8. $m_{t_b} = m_\tau / m_v$.

9. Whittaker (1966, Table II, p. 107); average of [1 – % light penetration].

10. Product of separate β and L_t estimations. 11. Eq. 8.27 or Fig. 8.20. 12. Allen and Lemon (1976, p. 288).

13. Eq. 6.52 with $h_0 = 0.1$ cm in all cases. 14. Allen and Lemon (1976, p. 289).

15. MacMahon (1988, Fig. 8.5, p. 236), shows most favorable season to be Jan–Feb.

16. MacMahon (1988, Table 8.2, p. 240). 17. For Eucalyptus with $\beta = \kappa$, (Whitford *et al.*, 1995).

18. Desert shrub (Larcher, 1983, Table 3.13, p. 151). 19. This "apparent" Bowen ratio is calculated from Eq. 5.5. See text for discussion.

20. U.S. Weather Service (1974). 21. Waring (1983). 22. Raison *et al.* (1992a). 23. Whitford *et al.* (1995).

24. Gross seasonal rate (i.e., total for 24-h day during growing season).

25. Rogerson (1967), average of ten stands in Oxford, MS.

26. Ehleringer (1985, pp. 168–169). 27. Budyko (1977, Map 34). 28. Fig. 3.11. 29. Fig. 7.4.

30. Table 2.2 for $\kappa = \beta$ and $\beta L_t = a_w$, average for all spruce; $L_t = \beta L_t / \beta$.

31. As in Note 43 but divided by 2.36 to account for parallel flux from other stomated needle surfaces (cf. McNaughton and Jarvis, 1983; Jarvis and McNaughton, 1986).

32. Average for maple; Table 2.4. 33. Salvucci (1992); Table 6.4. 34. Kohler *et al.* (1959, Plate 2).

35. Farnsworth *et al.* (1982, Map 2) gives May–Oct. $E_{ps} = 1.14$ m; Kohler *et al.* (1959, Plate 2) gives full-year $E_{ps} = 1.83$ m; thus, winter $E_{ps} = 0.69$ m/180 days.

36. July @ 1 pm. 37. Allen and Lemon (1976, Table III, p. 287). 38. January @ 1 pm.

39. $q_b^* \equiv q_b[1 - (\frac{1-S_r}{0.85})]$ with q_b from Eq. B.12 using daily minimum (i.e., night-time) radiation temperature.

40. $E_v^\otimes \equiv \frac{2V_e}{M m_{t_b'}}$ with V_e and $m_{t_b'}$ from Eqs. 6.67 and 6.63 respectively. The factor "2" reflects evaporation during daylight hours only.

41. U.S. Weather Service (1974) using $N = 1 - \%$ possible sunshine. 42. $\beta L_t / L_t$. 43. Eq. 7.25 and the data in the caption of Fig. 7.9.

44. Eq. 7.13. 45. Fig. 7.6 for $\bar{d}/t = 0.006$. 46. Eq. 9.8

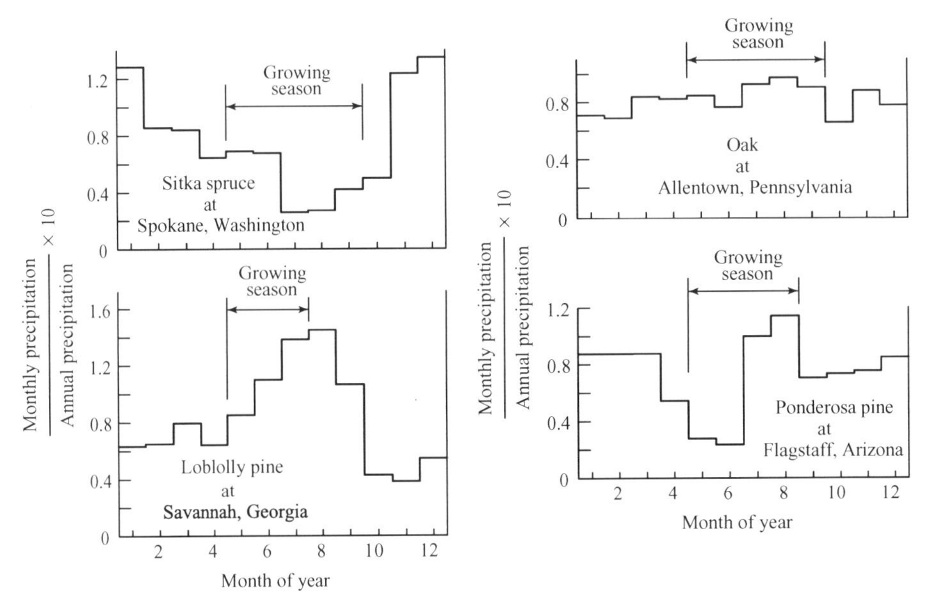

Fig. 9.8. Typical relationships of growing season to rainy season.

We resolve this uncertainty for the Sitka spruce rather arbitrarily by relying on the similarity (cf. Fig. 9.8) of seasonal phases at the beginning of the growing season and making the *assumption* that for the spruce ΔS is the same +50% of the dormant season precipitation as is observed for the ponderosa pine (cf. Table 6.3). This assumption gives $\Delta S = +17$ cm for the spruce, making $\frac{I_{SL}}{f_I(\beta L_t) I_{SW}} = 1.09$ as is plotted in Fig. 9.7.

(4) Light decay coefficient (κL_t). Although admittedly inaccurate, Baker's (1950) values of κL_t were used for matching species. Otherwise, separate estimates of $\beta = \kappa$ and of L_t were made using Tables 2.2 and 2.3. The Beaver Creek communities have a continuous variation with topographic elevation (Baker, 1982) from ponderosa at the top, through pinyon–alligator juniper down to pinyon–Utah juniper and finally to desert shrub at the bottom. Not knowing κL_t for either pinyon or juniper, we use the physical resemblance of the needle structure to represent the pinyon–alligator juniper as pinyon using Baker's (1950) κL_t for lodgepole pine, and to represent the pinyon–Utah juniper as juniper using Baker's (1950) κL_t for Northern white cedar (see Zim and Martin, 1987, pp. 36 and 39 for drawings of the respective needle structures).

(5) Wet surface potential evaporation (E_{ps}). The gross annual water surface evaporation was estimated for the given location from climatological maps and then reduced to the appropriate season using the ogive distribution of Salvucci (1992) presented here as Eq. B.13.

(6) Sensible heat flux (H). Because of the large uncertainty in V_e and hence in λE_T, we are uncertain of the actual values of $\boldsymbol{R}_b$, and the values of $\boldsymbol{R}_b$ estimated in Table 9.2 using Eq. 5.5 are termed "apparent" Bowen ratios. The water balance components are best known for the pinyon–juniper canopies where we have no evidence of sapwood storage capability and ΔS was observed. Indeed, the "apparent" Bowen ratios for these canopies *are* closely zero and are assumed to be "actual". All other apparent $\boldsymbol{R}_b$ are negative indicating an apparent transfer of sensible heat from the atmosphere to the vegetation in augmentation of the net radiation. This is likely for the moist canopies such as oak, beech, *Goethalsia*, and Sitka spruce, and even for the loblolly and ponderosa although the actual magnitudes may be different due to the water balance uncertainties discussed above. The driest, creosote bush canopy, has a winter growing season and may well be accepting heat as the negative $\boldsymbol{R}_b$ indicates. In any event, with the exception of the rainforest (i.e., *Goethalsia*), and beech, the apparent $\boldsymbol{R}_b$ satisfy the criterion $\boldsymbol{R}_b < O(10^{-1})$ which is consistent with our finding in Fig. 9.4 that $T_0 \approx \widehat{T_l} = T_m$.

The results are presented in Fig. 9.7 where we see that according to this proposition and the assumptions discussed above, seven of the nine communities are in their natural habitat. Note in particular that only the loblolly pine fails to satisfy closely the conditions for climax communities specified by Eqs. 9.13 and 9.14. It seems possible that *buffered stress tolerance* (see Baker, 1950, pp.140–143; Whittaker, 1975, p. 179) may be responsible for this deviation. Our reasoning follows.

For simplicity and convenience, we have developed the habitat constraints using the approximation that random variables are replaced by their mean values. However, this formulation does not account for the fact that while *in the average* a species may satisfy the climax conditions, the distributed nature of the climatic variables guarantees that *in some years* the species will be under stress in this habitat. The ability of certain species to survive the runs of stressful seasons characteristic of climatic variability is often called the *tolerance* of that species (cf. Baker, 1950, pp. 140–143) and is a function of both species and habitat. To survive, the species must be in a *mean state* that falls short of optimum productivity by an amount providing a margin of safety to accommodate the resource shortages of extreme seasons. Such a stable but sub-climax state appears to be one of so-called *buffered tolerance* (Whittaker, 1975, p. 45) in which the departure (i.e., the buffering) from climax conditions depends upon the tolerance that the particular species has for stress. It is interesting to note that Baker (1950, Fig. 12, p. 143) lists lodgepole (similar to loblolly) pine among the "intolerant" species. This classification is consistent with the behavior demonstrated by this species in the habitat plot of Fig. 9.7. The degree of buffering provided is probably controlled by the population density, M. That is, for a given value of the ordinate, $\frac{I_{SL}}{f_I(\beta L_t) I_{SW}}$, the amount of water available, on the average, to each leaf can be increased by decreasing the canopy cover, M.

Plate 9.3. Nurselog. Fallen fir tree nourishes new growth in the Olympic National Park, Washington. (Photograph by William D. Rich; Copyright © 2001 William D. Rich.)

Further approximation of the water-limited climax condition

For the climatic climax, $\Pi_W = 1$, Eqs. 9.4 and 9.8 give

$$\frac{(1-\rho_T)P_{sm}/\varepsilon}{f_I(\beta L_t)} = \lambda E_v^{\otimes}\left\{\left[1+\frac{\gamma_o}{\Delta}\left(1+\frac{\widehat{r_c}}{r_a}\right)\right]+\frac{q_b}{\lambda E_v^{\otimes}}\left[1-\frac{(1-S_r)}{0.85}\right]\right\}. \tag{9.15}$$

Table 9.2 shows that the second square-bracketed term on the right-hand side of Eq. 9.15 may be neglected with respect to the first, leaving

$$\frac{(1-\rho_T)P_{sm}/\varepsilon}{f_I(\beta L_t)} \approx \lambda E_v^{\otimes}\left\{\left[1+\frac{\gamma_o}{\Delta}\left(1+\frac{\widehat{r_c}}{r_a}\right)\right]\right\}, \tag{9.16}$$

which is helpful to our reasoning. Note that, except for minor ρ_T variation with β, the left-hand side of Eq. 9.16 is a species constant as is the minimum resistance ratio on the right-hand side. Climate is contained within $E_v^{\otimes}$ and $\frac{\gamma_o}{\Delta}$. Two significant observations flow from this:

(1) Figure 7.3 shows that the resistance ratio for a given species varies strongly with β without significant sacrifice of optimality. Thus, a given species can maintain maximum productivity (i.e., maintain $\Pi_W = 1$) over a range of climates by varying β while simultaneously varying L_t so as to keep βL_t at the species-constant value. This ability may account for some of the scatter of β around the theoretical value, $\beta = \kappa$, observed in Fig. 3.18.

Plate 9.4. Baldcypress swamp community. (Photograph of the Okefenokee Swamp in Georgia by F. Eugene Hester; Copyright © 1996 F. Eugene Hester.)

(2) Nutritional deficiency decreases the leaf area, L_t. If β increases to maintain the species-constant βL_t, the resistance ratio increases (cf. Fig. 7.3) making $\Pi_W < 1$. With desert and rainforest soils notoriously short of nitrogen, can this be the reason for the sub-climax position of the creosote bush and the *Goethalsia* in Fig. 9.7?

E Conclusion

On the basis of the evidence presented, all three propositions are strongly supported, and their combination, as presented in Fig. 9.7, defines both the feasible natural habitat of species with given I_{SL} and βL_t, and the climax limits thereto.

10

Net primary productivity and ecotones

The photosynthetic capacity of an isolated leaf is expanded to canopy scale in order to obtain the annual above-ground *carbon demand*. This demand is then normalized by that of a groundcover monolayer of the same species to obtain the *productive gain* of the canopy structure subject to the availability of adequate CO_2 and nutrients. The annual canopy gain in above-ground *carbon supply* is estimated using the flux-gradient relationship.

The independent variable of normalized productivity is the *canopy absorption index*, βL_t, which from optical optimality is a surrogate for tree species. For cylindrical multilayer crowns of given leaf angle and without local carbon recycling, there is a critical $\beta L_t = \widehat{\beta L_t}$ separating species productively limited by carbon demand from those limited by atmospheric carbon supply, and at which the productive gain has its global maximum. For tapered multilayer crowns the atmospheric carbon supply exceeds demand at all values of βL_t.

Collected productivity observations from the literature support the theory over a wide range of plant communities from desert shrub to rainforest with the wet communities clustering around $\widehat{\beta L_t}$.

Normalizing the productivity by the local *climatic productive potential* places all species dependence in a new function of βL_t that displays a global maximum over the range of βL_t corresponding to that observed in nature and thereby confirming our fundamental hypothesis that selection favors increasing productivity.

Equating the productivity of unit plant areas for three competing species is shown to predict well the latitude of the mixed forest ecotone of eastern North America.

A Canopy biomass production

Introduction

Green plants use solar energy to synthesize organic compounds from CO_2 and water and must take up inorganic nutrients, primarily nitrogen, from the soil to manufacture these organic components. The available solar energy creates a *demand* for CO_2, water,

Plate 10.1. Collection of pines. Left to right: Himalayan pine, Cevennes pine, and Japanese red pine. (Photographed in Arnold Arboretum, Boston, by William D. Rich; Copyright © 2001 William D. Rich.)

and nitrogen in order for the plant to fulfill its productive potential. Here we assume a nutrient supply that is non-limiting and climatic climax conditions (cf. Chapter 8) that guarantee light saturation and sufficient water supply to keep the stomates fully open. We now formulate the production of canopy biomass in terms of carbon supply and demand.

Carbon balance

The accumulated green plant biomass (plus any net transfer of organic carbon from the green plant compartment to other compartments within the ecosystem) is termed the *net primary productivity* or *NPP* of the community (Bormann and Likens, 1979, p. 16), and is normally quoted on an annual basis. Larcher (1983, p. 146) expresses this carbon budget as

$$\mathrm{NPP} = B + C + D, \qquad (10.1)$$

where B = additions to plant biomass, C = plant matter consumed by grazers, and D = plant matter shed as detritus.

The remainder of the annually assimilated carbon is lost back to the atmosphere through the open stomates in a process, R, called *respiration*. The *gross primary productivity*, or *GPP*, is then

$$\text{GPP} = \text{NPP} + R. \tag{10.2}$$

With NPP measured in grams of dry solid matter (i.e., g_s) and GPP measured in grams of CO_2 assimilated, we write the ratio

$$\frac{\text{NPP}}{\text{GPP}} = \frac{1}{1 + \dfrac{R}{\text{NPP}}} \equiv \vartheta, \quad (g_s\, g(CO_2)^{-1}, \quad \text{or just } g_s\, g^{-1}), \tag{10.3}$$

which was introduced here in Chapter 8. It is a necessary empirical quantity for our work and is variable among plant types. Larcher (1983, p. 149) gives $\vartheta = 0.40\text{–}0.67\ g_s\, g^{-1}$ for temperate forests, and $\vartheta \approx 0.25\ g_s\, g^{-1}$ for moist–warm tropical forests. Furthermore, data presented by Larcher (1983, Table 3.12, p. 149) show that for these two types of deciduous trees at least, the value of ϑ is less than 10% different if the contributions of the roots are omitted from both R and NPP in Eq. 10.3.† Here we will not attempt to include the variability of ϑ with plant type and will instead use the single value

$$\vartheta = 0.50\ g_s\, g^{-1}, \tag{10.4}$$

reported by Penning de Vries *et al.* (1974) and by Ledig *et al.* (1976, *fide* Russell *et al.*, 1989b, p. 27). Consistent with this approximation then, we may also take the chosen value of ϑ to represent g_s(above-ground) g^{-1}. We will find this alternate interpretation convenient because the usual field observations of forest productivity omit the root biomass.

To formulate canopy NPP we begin at the leaf level following the developments of Chapter 8. Maximization of *plant* NPP is taken to be the driving force behind the natural selection processes establishing the monocultures with which we deal on the basis of the assumption that NPP is proportional to seed productivity and hence is a surrogate for survival probability, all else being constant. This proportionality may vary with species.

Crown-average leaf photosynthetic capacity

In Chapter 8 we considered the process of photosynthesis for an isolated leaf and found the photosynthetic capacity per unit of one-sided leaf area to be adequately represented by Eq. 8.6. When modified for non-optimal ambient temperature using Eq. 8.1 (remember, we assume $T_l \approx T_0$), and for leaf angle, Eq. 8.6 is written

$$\frac{P_D}{g(T_0)\, \vartheta\, P_{sm}\, \beta} = 1 - \exp\left[-\frac{I(\xi)}{I_{SL}}\right], \tag{10.5}$$

† The data compiled by Cannell (1982) indicate that between 65% and 87% of total net primary production is above-ground production.

Plate 10.2. Tupelo. The tupelo or sourgum is the large rounded crown in this image. (Photographed in Holden Arboretum, Cleveland, by William D. Rich; Copyright © 2001 William D. Rich.)

in which (using our alternate interpretation of ϑ) P_D is the rate of above-ground carbon demand in g_s per unit of one-sided leaf area, P_{sm} is the maximum light-saturated photosynthetic rate for an isolated leaf in $g(CO_2)$ per unit of basal leaf area,[†] β is the ratio of basal to one-sided leaf areas, and $I(\xi)$ is the insolation on a leaf at depth, ξ, in a crown of similar leaves for a situation unlimited by CO_2 availability.[‡]

In Chapter 9 we showed (cf. Eq. 9.3 and Fig. 9.5) that for maximum productivity

$$I_{SL} = f_I(\kappa L_t) I_0, \tag{10.6}$$

whereupon, Eq. 10.5 becomes

$$\frac{P_D}{g(T_0)\,\vartheta\, P_{sm}\,\beta} = 1 - \exp\left[-\frac{1}{f_I}\frac{I(\xi)}{I_0}\right], \tag{10.7}$$

[†] P_{sm} will change with changing ambient CO_2 concentration.

[‡] Here, "unlimited" implies that the natural, ambient CO_2 concentration is available at every leaf surface.

where f_I is a shorthand representation of (cf. Eq. 8.27)

$$f_I(\kappa L_t) \equiv \frac{1 - e^{-\kappa L_t}}{\kappa L_t}. \tag{10.8}$$

We have seen earlier that $I(\xi)/I_0$ captures the decay of insolation with crown depth through (cf. Eq. 7.1)

$$\frac{I(\xi)}{I_0} = \exp(-\kappa L_t \xi). \tag{10.9}$$

Substituting Eq. 10.9 in Eq. 10.7 and averaging over the depth of the crown, we get the dimensionless above-ground carbon demand, $\widehat{P_D}$, of the average leaf in the crown

$$\frac{\widehat{P_D}}{g(T_0)\,\vartheta\, P_{sm}\,\beta} = 1 - \int_0^1 \exp\left(-\frac{1}{f_I}\, e^{-\kappa L_t \xi}\right) d\xi. \tag{10.10}$$

To perform this integration, we make the substitution†

$$e^{-\kappa L_t \xi} = u, \tag{10.11}$$

whereupon Eq. 10.10 becomes

$$\frac{\widehat{P_D}}{g(T_0)\,\vartheta\, P_{sm}\,\beta} = 1 + \frac{1}{\kappa L_t} \int_{1/f_I}^{(e^{-\kappa L_t})/f_I} \frac{e^{-t}}{t} dt. \tag{10.12}$$

From Abramowitz and Stegun (1964, p. 228) we recognize the exponential integral

$$E_1(z) = \int_z^\infty \frac{e^{-t}}{t} dt \qquad (|\arg z| < \pi), \tag{10.13}$$

so that Eq. 10.12 becomes

$$\frac{\widehat{P_D}}{g(T_0)\,\vartheta\, P_{sm}\,\beta} = 1 - \frac{1}{\kappa L_t}\left[E_1\left(\frac{1}{f_I} e^{-\kappa L_t}\right) - E_1\left(\frac{1}{f_I}\right)\right], \tag{10.14}$$

and the arguments, z, of $E_1(z)$ are both functions of the light extinction coefficient, κL_t. In Appendix H we show that κL_t is a species constant.

For $0 \le z \le 1$, Abramowitz and Stegun (1964, p. 229, no. 5.1.11) give

$$E_1(z) = -\gamma - \ln(z) - \sum_{n=1}^{\infty} \frac{(-1)^n z^n}{n \cdot n!}, \tag{10.15}$$

in which

$$\gamma \equiv \text{Euler's constant} \approx 0.577, \tag{10.16}$$

and for $1 \le z < \infty$, Abramowitz and Stegun (1964, p. 231, no. 5.1.54) give the approximation

$$E_1(z) = \frac{1}{z\, e^z}\left[\frac{z^2 + a_1 z + a_2}{z^2 + b_1 z + b_2} + \varepsilon(z)\right], \tag{10.17}$$

† Professor David Benney, personal communication.

Plate 10.3. Cedar of Lebanon. (Photographed in Arnold Arboretum, Boston, by William D. Rich; Copyright © 2001 William D. Rich.)

where

$$a_1 = 2.334733, \qquad b_1 = 3.330657,$$
$$a_2 = 0.250621, \qquad b_2 = 1.681534,$$
$$|\varepsilon(z)| < 5 \times 10^{-5}.$$

Canopy annual carbon demand

To expand the rate of carbon demand from an average unit of one-sided leaf area to the full canopy we convert $\widehat{P_{\mathrm{D}}}$ to a unit of horizontal land area by multiplying first by L_{t}, the total one-sided leaf area per unit of crown basal area, and then by the canopy cover, M. To expand from an instantaneous demand, $\widehat{P_{\mathrm{D}}}$, to an annual total demand, $\widehat{P_{\mathrm{T}}}$, we must further multiply $\widehat{P_{\mathrm{D}}}$ by the length of the growing season, m_τ. Eq. 10.14 is then

$$\widehat{P_{\mathrm{T}}} = \mathrm{g}(T_0)\,\vartheta\, P_{\mathrm{sm}}\, M m_\tau \frac{\beta}{\kappa}\left\{\kappa L_{\mathrm{t}} - \left[E_1\left(\frac{1}{f_I}\,\mathrm{e}^{-\kappa L_{\mathrm{t}}}\right) - E_1\left(\frac{1}{f_I}\right)\right]\right\}. \qquad (10.18)$$

We now compare this above-ground annual canopy carbon demand with that of a groundcover monolayer of this species having $\beta = 1, \ L_t = 1$, which has the above-ground annual carbon demand

$$p_D = g(T_0)\,\vartheta\,P_{sm}\,M m_\tau. \tag{10.19}$$

We use Eq. 10.19 to rewrite Eq. 10.18

$$\frac{\widehat{P_T}}{p_D} = \frac{\widehat{P_T}}{g(T_0)\,\vartheta\,P_{sm}\,M m_\tau} = f_D\left(\kappa L_t, \frac{\beta}{\kappa}\right), \tag{10.20}$$

for the light environment determining κL_t, and in which $f_D(\kappa L_t, \frac{\beta}{\kappa})$ is the potential *productive gain* of the canopy structure subject to the availability of adequate CO_2, water, and nutrients.[†] We note that this gain is linearly proportional to β/κ which suggests a selective pressure to maximize this ratio. In Chapter 3 we argued that *for optical optimality* this maximum should be $\beta/\kappa = 1$, and we presented supporting observational evidence (cf. Fig. 3.19). On these bases we continue to let $\beta = \kappa$ whereupon

$$f_D(\beta L_t) = \left\{\beta L_t - \left[E_1\left(\frac{1}{f_I}\,e^{-\beta L_t}\right) - E_1\left(\frac{1}{f_I}\right)\right]\right\}, \tag{10.21}$$

which vanishes for $\beta L_t = 0$ and theoretically is unbounded[‡] as $\beta L_t \to \infty$.

To the first approximation, $\widehat{P_T}$ equals the above-ground *potential net primary productivity*, NPP, (subject to available carbon, water, and nutrients), and we write finally[§]

$$\frac{\text{NPP}}{p_D} = \frac{\text{NPP}}{g(T_0)\,\vartheta\,P_{sm}\,M m_\tau} = f_D(\beta L_t). \tag{10.22}$$

Equation 10.22 is plotted as curve (a) in Fig. 10.1 where $f_D(\beta L_t)$ may be viewed alternatively as the above-ground dimensionless *carbon demand function* for the given light environment. Note that $f_D(\beta L_t)$ satisfies all the conditions for optical optimality for a given species as given in Chapter 3[††] and is independent of all other aspects of climate as well as of crown shape. Note also that for the "diffuse" monolayer, $\beta L_t = 1$, $f_D(\beta L_t) < 1$, as compared with the groundcover monolayer (i.e., $\beta = 1, L_t = 1$) where NPP $= p_D$ as is shown by curve (f) in Fig. 10.1.

[†] Note that the ability of the plant to use light depends upon the supply of nitrogen (Sharkey, 1985), while its ability to use CO_2 depends upon the adequacy of the water supply to maintain fully open stomates as specified by $I_{SL} \le f_I(\kappa L_t) I_{SW}$. This latter condition is expressed in Eq. 10.20 through the separable variable, M.

[‡] Actually of course, there is an upper limit to βL_t in the neighborhood of $\beta L_t = 4$ to 5 imposed by compensation light requirements (cf. Table 3.9).

[§] Remember that here P_{sm} has the units of $g(CO_2)\ m^{-2}\ h^{-1}$.

[††] With the exception of ρ_{PAR} which is insensitive to β for these $h_\otimes$, and is thus fixed by the leaf texture of the species.

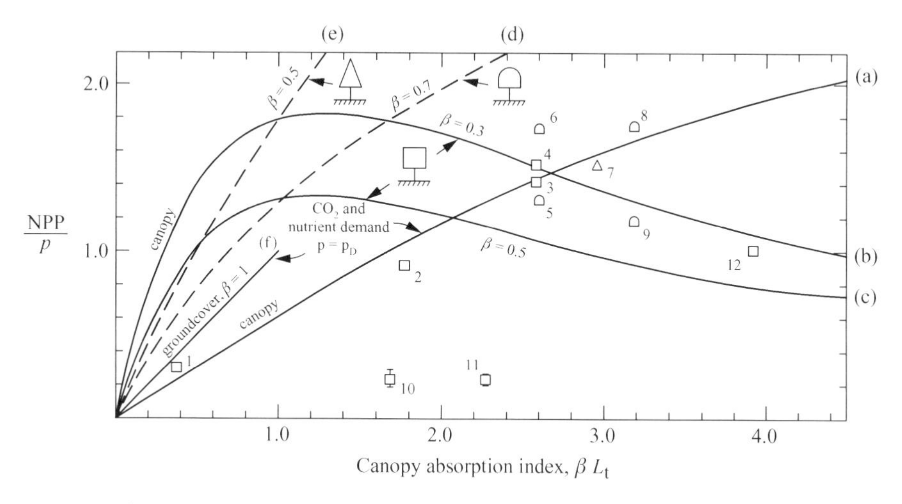

Fig. 10.1. Productive gain of C_3 multilayer canopies. (a) CO_2 demand (Eqs. 10.21 and 10.22); (b) atmospheric CO_2 supply, cylindrical crown with $\beta = 0.3$ (Eqs. 10.30 and 10.31); (c) atmospheric CO_2 supply, cylindrical crown with $\beta = 0.5$ (Eqs. 10.30 and 10.31); (d) atmospheric CO_2 supply, hemispherical crown with $\beta = 0.7$ (Eq. 10.33); (e) atmospheric CO_2 supply, conical crown with $\beta = 0.5$ (Eq. 10.34); (f) CO_2 demand, groundcover with $\beta = 1.0$, $\mathrm{NPP}/p_{\mathrm{d}} = L_{\mathrm{t}}$. Data points from Table 10.1: 1, Creosote bush, Nye Co., Nevada; 2, ponderosa pine, Tucson, Arizona; 3, loblolly pine, Tucson, Arizona; 4, oak, Great Smoky Mountains, Tennessee; 5, rainforest, Kade, Ghana; 6, rainforest, Yangambi, Congo; 7, red spruce, Great Smoky Mountains, Tennessee; 8, beech, Eastern Europe; 9, beech, Great Smoky Mountains, Tennessee; 10, pinyon–alligator juniper, Flagstaff, Arizona; 11, pinyon–Utah juniper, Flagstaff, Arizona; 12, sugar maple, Great Smoky Mountains, Tennessee.

Canopy annual atmospheric carbon supply for cylindrical multilayers in a closed canopy

Let the downward rate of atmospheric CO_2 supply to a unit basal area of the canopy's ensemble of stomata be Q. It is regulated by the average eddy diffusivity of mass, K_{c}, and the vertical gradient of the CO_2 concentration, $\mathrm{d}c/\mathrm{d}z$, within the crown as given in Chapter 4 (cf. Eq. 4.116) by the flux–gradient relationship[†]

$$Q = \rho\, K_{\mathrm{c}} \frac{\mathrm{d}c}{\mathrm{d}z}. \tag{10.23}$$

Invoking the Reynolds analogy, we replace K_{c} by its equivalent, K_{m}, for momentum, as derived in Chapter 4. For $m = 1/2$ and $n = 2$, Eq. A.24 gives for cylindrical multilayer crowns the spatial average diffusivity over the crown depth

$$\widehat{K_{\mathrm{m}}} = K_{\mathrm{m}}(0)\, \frac{1 - \mathrm{e}^{-\beta L_{\mathrm{t}}}}{\beta L_{\mathrm{t}}}, \tag{10.24}$$

[†] We are assuming neutral atmospheric stability in the long-term average because it is unwise to calculate mass fluxes using eddy momentum diffusivities under conditions of atmospheric instability (Jarvis *et al.*, 1976, pp. 224–225).

in which $K_m(0)$ is the diffusivity at the top of the canopy, $\xi = 0$. Precise evaluation of the CO_2 concentration gradient is problematic. We will approximate it by assuming that its average over the crown depth increases with increasing foliage density in the manner

$$\frac{\widehat{dc}}{dz} = \frac{c_0 - c_s}{h - h_s}(1 - e^{-\beta L_t}), \tag{10.25}$$

in which c_0 and c_s represent the concentrations at the uppermost and lowermost leaves respectively. For homogeneous canopies it follows that

$$h - h_s = k_c L_t, \tag{10.26}$$

in which k_c is a proportionality coefficient with the units of meters.

Multiplying the instantaneous, unit leaf area rate of Eq. 10.23 by L_t, M and m_τ, and using Eqs. 10.24, 10.25 and 10.26, the annual total above-ground supply of carbon to the canopy per unit of horizontal land area becomes, in terms of potential above-ground solid matter†

$$\widehat{Q_S} = \vartheta \rho K_m(h) \frac{c_0 - c_s}{k_c} M m_\tau \frac{(1 - e^{-\beta L_t})^2}{\beta L_t}. \tag{10.27}$$

As we did earlier for carbon demand, we now compare this above-ground canopy carbon supply with the above-ground supply to the monolayer, $L_t = 1$. For this diffuse monolayer we have the annual above-ground carbon supply

$$p_S = \vartheta \rho K_m(0) \frac{c_0 - c_s}{k_c} M m_\tau \frac{(1 - e^{-\beta})^2}{\beta}. \tag{10.28}$$

Assuming $c_0 - c_s$ is the same for the monolayer and the full crown, we may factor Eq. 10.28 from Eq. 10.27 to obtain the annual canopy gain in above-ground carbon supply†

$$\frac{\widehat{Q_S}}{p_S} = f_S(\beta, L_t), \tag{10.29}$$

in which

$$f_S(\beta, L_t) \equiv \frac{\beta}{\beta L_t}\left[\frac{1 - e^{-\beta L_t}}{1 - e^{-\beta}}\right]^2. \tag{10.30}$$

Without carbon recycling, $\widehat{Q_S}$ equals the potential above ground NPP (subject to available light, water, and nutrients of course) and we rewrite Eq. 10.29

$$\frac{\text{NPP}}{p_S} = f_S(\beta, L_t), \tag{10.31}$$

which is plotted in comparison with Eq. 10.22 in Fig. 10.1 as solid curves (b) for $\beta = 0.30$, and (c) for $\beta = 0.50$, representative values for deciduous and evergreen

† It is the elimination of unknown proportionality constants such as $(c_0 - c_s)/k_c$ which motivates presenting these supply–demand comparisons in terms of dimensionless ratios.

trees respectively (cf. Fig. 3.18). Alternatively, we may call $f_S(\beta, L_t)$ the above-ground dimensionless *carbon supply function*. Note that the maximizing behavior of Eq. 10.31 is characteristic of turbulent vertical mass flux in the closed canopy of homogeneous cylindrical multilayer crowns.

With vanishing canopy, the concentration gradient at the monolayer surface is maximized and it seems reasonable to assume that the monolayer carbon supply, p_s, adjusts to meet the monolayer carbon demand, p_D. With $p_s = p_D$, we may view Fig. 10.1 as a comparison of carbon supply and demand under the same conditions.

We see from Eq. 10.24 that $\widehat{K_m} \rightarrow 0$ as βL_t gets large due to internal resistance of the foliage elements, while from Eq. 10.25, $\widehat{\frac{dc}{dz}} \rightarrow 0$ as $\beta L_t \rightarrow 0$ due to the decline of sink strength. Thus the concentration gradient controls the turbulent flux at low βL_t while the diffusivity controls that flux at large βL_t with a maximum in between. In Chapter 7 we found a minimum in crown resistance which, together with Eqs. 5.29 and 5.30, dictated a maximum in *plant* transpiration rate[†] at intermediate values of βL_t.

Note that for $\beta = 0.3$, Fig. 10.1 shows the dimensionless CO_2 supply rate to maximize at $\beta L_t \approx 1.2$, while we found in Fig. 7.3 that at this same β, the dimensionless transpiration rate, k_v^*, maximizes (i.e., $\widehat{r_c/r_a}$ minimizes) at $\beta L_t \approx 1.95$. Equations 10.23 and 10.24 apply equally to water vapor and to CO_2, so the difference in the peak of the two flux curves must lie with differences in the vertically averaged concentration gradients of the two quantities. Indeed, the source concentration of water vapor is assumed at saturation in *all* stomates throughout the crown, while the sink concentration of CO_2 in the air is presumed to decline with depth into the crown.

The critical canopy absorption index for cylindrical multilayers without carbon recycling

We see in Fig. 10.1 that for cylindrical multilayers there is an intermediate, *critical canopy absorption index*, $\widehat{\beta L_t}$, separating regions of productivity control by carbon demand ($\beta L_t < \widehat{\beta L_t}$) as limited by leaf photochemistry, from regions of productivity control by carbon supply ($\beta L_t > \widehat{\beta L_t}$) as limited by the biophysics of turbulent diffusion within the canopy. We estimate $\widehat{\beta L_t}$, which varies inversely with β, by equating the normalized productivities of Eqs. 10.22 and 10.31 to obtain

$$\frac{\beta}{(1 - e^{-\beta})^2} = \frac{\widehat{\beta L_t}}{\left(1 - e^{-\widehat{\beta L_t}}\right)^2} f_D(\widehat{\beta L_t}). \tag{10.32}$$

Equation 10.32 is plotted as the solid line in Fig. 10.2 in comparison with the optically optimal foliage state for several species (dashed hyperbolas) presented earlier

[†] Note that maximization of the *plant* moisture flux rate, $E_v \equiv k_v^* E_{ps}$, through maximization of k_v^*, does not affect the *canopy* moisture flux rate, $M E_v = M k_v^* E_{ps}$, in a water-limiting climate, because there the product, $M k_v^*$, is a climatic constant (cf. Eq. 6.66). However, because this product is invariant in a given water-limiting situation, maximizing k_v^* means minimizing M, and the canopy biomass will decline (cf. Eq. 10.22). Thus evidence such as Fig. 7.11 in support of maximum k_v^* supports the concept of the "selfish gene" (Dawkins, 1976).

in Fig. 7.1. Species lying to the left of the $\widehat{\beta L_t}$ curve will be carbon demand-controlled while those to the right will be carbon supply-controlled. With all the caveats put upon both the theory and the observations displayed in Fig. 10.2, we should be cautious about drawing species-specific conclusions therefrom. Nevertheless, we see in this figure that the averages of Tables 2.2–2.4 for each of the three vegetation classes, when assessed collectively, seem to support an operating state close to $\widehat{\beta L_t}$. Furthermore, it is apparent from the dashed lines of Fig. 10.2 that some species, such as ponderosa pine, may be limited by the demand for carbon. For example, in Fig. 10.2 we see that ponderosa pine is carbon demand-controlled at all β, while Norway spruce and sugar maple are carbon supply-controlled at all β.

Atmospheric carbon supply for tapered multilayers

For tapered multilayer crowns (e.g., cones and hemispheres) the potential atmospheric carbon supply increases monotonically with βL_t at least until solid-body resistance comes into play. Retaining the above assumptions concerning concentration gradient, and locating the reference monolayer at the crown surface, we replace only Eq. 10.24 in Eq. 10.23 to obtain for:

Hemispherical crowns (using Eq. 7.17)

$$\frac{\text{NPP}}{p_s} = \left[\frac{0.55 + 0.052\,\beta L_t + 0.008\,(\beta L_t)^2}{0.55 + 0.052\,\beta + 0.008\,\beta^2}\right]\frac{1 - e^{-\beta L_t}}{1 - e^{-\beta}}, \tag{10.33}$$

and

Conical crowns (using Eq. 7.19)

$$\frac{\text{NPP}}{p_s} = \left[\frac{0.605 + 0.108\,\beta L_t + 0.042\,(\beta L_t)^2}{0.605 + 0.108\,\beta + 0.042\,\beta^2}\right]\frac{1 - e^{-\beta L_t}}{1 - e^{-\beta}}. \tag{10.34}$$

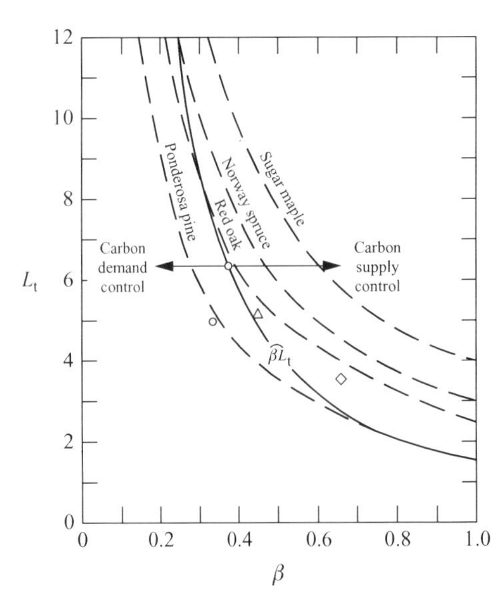

Fig. 10.2. Critical canopy absorption index without carbon recycling. Dashed lines from Baker's I_k/I_0 (1950, Table 12, p. 143) using $\kappa = \beta$; Table 3.9. Plotted points are averages of observations of L_t, and average of β and κ from Tables 2.2–2.4. ○, Deciduous; △, evergreen; ◇, crops and grasses.

Plate 10.4. Collection of spruces. (Photographed in Arnold Arboretum, Boston, by Beverly G. Eagleson.)

Equations 10.33 and 10.34 are plotted in Fig. 10.1 as dotted curves (d) and (e) respectively using values of β representative of the hemispherical crowns common to rainforests (i.e., $\beta = 0.70$), and conical evergreens (i.e., $\beta = 0.50$).

Atmospheric carbon supply for tapered monolayers

In Eq. 7.19 and Fig. 4.25a, we saw the homogeneous, conical, multilayer crown of low foliage density to have an eddy viscosity that increases monotonically with βL_t. Incorporating the vertical concentration gradient through Eq. 10.26, the dimensionless CO_2 supply curve for such crowns is shown in Fig. 10.1 as curve (e). As this crown of given β increases in foliage density (say under the need to retain heat), the leaves (probably needles in the assumed cold climate) concentrate in a monolayer.

In Eq. 7.21 and Fig. 4.25b we described the eddy viscosity of dense conical monolayers as increasing monotonically but weakly with βL_t. It seems reasonable to assume that the CO_2 concentration gradient under this increasing density also behaves as in Eq. 10.26, in which case the dimensionless carbon supply for the dense conical monolayer will be proportional to curve (e) of Fig. 10.1 as are the eddy viscosities of the high density monolayer and the homogeneous multilayer in Figs. 4.25b and 4.25a respectively. This leaves the estimated dimensionless CO_2 supply curve for the conical monolayer higher than the dimensionless CO_2 demand curve at all βL_t, and the same situation obtains for the hemispherical monolayer as well. These qualitative estimates are not shown on Fig. 10.1.

Before discussing the very revealing Fig. 10.1, a few comments concerning the other major ingredients of production are in order.

Canopy water supply and demand

For photosynthesis, the plant requires water in order to: (1) maintain turgor and hence to keep the stomates open for CO_2 assimilation; (2) maintain the moisture potential in the protoplasm at the level necessary for cell metabolism; and (3) supply hydrogen and oxygen for use in photosynthetic chemical reactions (Larcher, 1983, p. 121). In addition, production of plant tissue requires a supply of nitrogen and other water-soluble nutrients, resident in the soil but entrained and transported to the plant by the through-flow of transpired water. In this work we focus on the plant operating state in which the stomates are fully open (cf. Chapter 8), and concern ourselves with the water supply required to maintain this state where we assume that the moisture potential meets the requirements of cell metabolism, and that the photochemical water use is a negligible fraction of the soil water through-flow induced by the climatic transpiration potential.

In Eq. 5.30 we defined the open-stomate conductance of a unit of vegetated surface as

$$k_v^* = \frac{E_v}{E_{ps}}, \tag{10.35}$$

so that the water use rate of the entire canopy becomes

$$ME_v = Mk_v^* E_{ps}. \tag{10.36}$$

In Chapter 6 we found that when k_v^*, a function of species and temperature, satisfies

$$k_v^* > \frac{V_e}{m_{t_b'} E_{ps}}, \tag{10.37}$$

in which V_e is the average available interstorm soil moisture, and $m_{t_b'}$ is the net interstorm duration, the canopy adjusts its water demand to equal the climatic supply by setting $M < 1$.[†] Otherwise, $M = 1$, and there is an excess of water supply which runs off through the natural surface water and/or groundwater drainage systems. In this way, through self-regulation, the canopy water supply always equals or exceeds its water demand.

Plant nitrogen (and other nutrient) demand

Nitrogen is an important plant nutrient whose demand often exceeds supply thereby limiting plant productivity. Nitrogen demand is proportional to carbon demand

[†] Actually of course, the ecological pressure for this adjustment arises from the *individual* plants through the demise of the weakest ones due to stress induced by water shortage.

Plate 10.5. Riverside community. This community along the Ayeyarwady (Irrawaddy) River in Myanmar (Burma) contains a giant hemispherical tree of unknown species within a grove of Royal palms. (Photograph by Peter S. Eagleson.)

stoichiometrically, and therefore the dimensionless nitrogen demand, N_D/n_D, is identical to the dimensionless carbon demand, $f_D(\beta L_t)$, as given by Eq. 10.21.

Plant nitrogen (and other nutrient) supply

Trees take up inorganic nitrogen from the soil, incorporate it into organic compounds from which proteins and other nitrogen compounds are synthesized, and return it to the soil in organic form primarily through the falling of leaves and fruits (Larcher, 1983, pp. 175–180). For plants having a relatively high root density, most of the mineral nitrogen in the root zone is available for uptake quickly, perhaps within one or two days, provided the nitrogen is in the form of nitrate (van Keulen *et al.*, 1989, p. 91). Thus, if the nitrogen reservoir is non-limiting, one of the expedient basic assumptions of this work, the nitrogen supply to the plant is proportional to the transpiration (van Keulen *et al.*, 1989, p. 91). Nitrogen carried into the plant by the soil water accumulates as structural protein in the plant dry matter. As a component of chlorophyll, nitrogen acts to aid in the assimilation of CO_2. Many studies have demonstrated, both in annual C_3 crops (cf. van Keulen *et al.*, 1989, p. 85) and in trees (Mooney *et al.*, 1978, for eucalyptus; and Larcher, 1983, Fig. 3.5.1, p. 127, for spruce), that the maximum rate of CO_2 assimilation displays a strong linear relationship to leaf nitrogen concentration over a wide range of the latter. Here then is a selection pressure for maximizing the flux rate of the solvent water through the plant.

We begin our analysis of nitrogen supply by assuming the reservoir of soluble nitrates in the soil to be large enough to impose no limit on their solution in the soil water. The concentration of nitrate in the soil water will then be fixed by the kinetics of the water uptake which is under the control of the climate, soil, and vegetation.

The plant water supply equals the plant water demand which is given by Eq. 10.35 as

$$E_v = k_v^* E_{ps}. \tag{10.38}$$

Under the above assumption, the plant nitrogen supply, N_s, is then

$$N_s = C_N \, k_v^* \, E_{ps}, \tag{10.39}$$

in which C_N is a constant for given climate, soil, and vegetation. A diffuse monolayer under these same conditions will have the the nitrogen supply

$$n_s = C_N \, k_{vm}^* \, E_{ps}, \tag{10.40}$$

in which k_{vm}^* is the monolayer conductance. With Eq. 5.29, the dimensionless supply of nitrogen *and all other minerals* is then

$$\frac{N_S}{n_S} = \frac{k_v^*}{k_{vm}^*} = \frac{1 + \dfrac{\Delta}{\gamma_0} + \left(\widehat{\dfrac{r_c}{r_a}}\right)_m}{1 + \dfrac{\Delta}{\gamma_0} + \widehat{\dfrac{r_c}{r_a}}}. \tag{10.41}$$

Unfortunately, our big-leaf model for the resistance ratio, r_c/r_a, does not hold at the monolayer limit, $L_t = 1$. However:

(1) *For cylindrical crowns* we infer from Fig. 7.3 that irrespective of β, $(\widehat{r_c/r_a})_m$ must become large as $L_t \to 1$. As noted above, there is a selection pressure to maximize N_S/n_S and hence to minimize $\widehat{r_c/r_a}$. Thus, for the common observed range of β (i.e., $0.3 \le \beta \le 0.5$), $(\widehat{r_c/r_a})_{min} = 0.34$ at $\beta L_t = 1.8$, and $(\widehat{r_c/r_a})_{min} = 0.81$ at $\beta L_t = 2.5$ (cf. Fig. 7.3), and for normal temperatures, $1 + \Delta/\gamma_0 \approx 3$ (cf. Table 5.1). The denominator of Eq. 10.41 therefore increases modestly with βL_t while the numerator stays constant. It follows then that $N_S/n_S > 1$ for all βL_t.

(2) *For tapered crowns* we infer from Fig. 7.5 that the denominator of Eq. 10.41 decreases with βL_t, and that therefore $N_S/n_S > 1$ for all βL_t and increases with βL_t.

Unfortunately, quantitative estimation of N_S/n_S is problematic, although nitrogen sufficiency seems probable at all βL_t at least for the infinite reservoir assumed here.

Plate 10.6. Northern red oak and white oak. Stand of Northern red oak on the left and white oak on the right. (Photographed by William D. Rich in Arnold Arboretum, Boston; Copyright © 2001 William D. Rich.)

Conclusions for atmospheric carbon supply

Figure 10.1 yields many conclusions for this case of minimum carbon supply:

(1) Potential carbon supply equals or exceeds carbon demand for full local carbon recycling and atmospheric carbon replaces that exported.

(2) The CO_2 (and nutrient) demand is independent of crown shape.

(3) The CO_2 supply is dependent upon crown shape.

(4) For crown shapes producing a rough canopy top (e.g., tapered shapes such as cones and hemispherical segments) CO_2 supply exceeds CO_2 demand for all values of βL_t for both multilayer and monolayer foliage. Therefore, the demand function governs for these crowns at all βL_t.

(5) For cylindrical crowns only, the CO_2 supply rate maximizes at low βL_t making supply equal demand at a critical intermediate value of βL_t denoted $\widehat{\beta L_t}$,

which varies inversely with β. At low βL_t for such crowns the relative canopy resistance, $\widehat{r_c/r_a}$, gets very large (cf. Fig. 7.3) making the canopy conductance and hence the transpiration rate correspondingly small. These are conditions for low water use and regions of low water supply should favor such plants. We have seen (cf. Chapter 8) that in water-limited situations, water demand (i.e., "use") is maintained equal to water supply (at given βL_t) through adjustment of the canopy density, M.

(6) For $\beta L_t < \widehat{\beta L_t}$, CO_2 supply exceeds demand for all crown shapes and the demand function governs the productivity which increases with increasing βL_t. All else being constant, all crown shapes should coexist in this range of βL_t.

(7) For $\beta L_t > \widehat{\beta L_t}$, CO_2 demand exceeds supply for cylindrical crowns, and the supply function governs the productivity[†] which decreases with increasing βL_t. For tapered crowns the demand function continues to govern. All else being constant, tapered crowns will bring maximum dimensionless productivity in this range of βL_t. Without supplemental CO_2 supply such as by regular fog or rain (cf. Wilson, 1948), or by the augmented mixing provided by wind-induced leaf motion, *only* tapered crowns will follow the CO_2 demand curve in this region.

(8) For $\beta L_t = \widehat{\beta L_t}$, the dimensionless productivity of cylindrical crowns has its global maximum.

(9) For $\beta L_t \leq 1$, a monolayer groundcover, curve (f), has greater productivity than a monolayer canopy.

(10) For $1 < \beta L_t < 1.7$, a canopy has no productive advantage over the monolayer groundcover, $\beta L_t = 1$. Can it be that this window of βL_t is vacant of natural canopies?

(11) For all βL_t, $\widehat{r_c/r_a}$ will be minimized in order to maximize the supply of nutrients to each plant.

(12) For $\beta L_t < \widehat{\beta L_t}$, where CO_2 supply exceeds demand, selection pressure is on the photochemistry to develop more efficient assimilation mechanisms.

(13) For $\beta L_t \geq \widehat{\beta L_t}$, where CO_2 demand exceeds supply, selection pressure is on crown shape to develop the tapered form that facilitates vertical fluxes.

(14) Water-limited communities have a more efficient CO_2 flux path and thus a higher CO_2 supply curve than those for the closed canopies plotted on Fig. 10.1. Therefore, we expect water-limited communities to be CO_2 demand-limited. Under such an excess of supply over demand, natural selection has produced more efficient photochemical processes: the C_4 process

[†] NPP limitation through foliage-induced CO_2 flux constraints seems to be a new finding. Larcher (1983, p. 85) states that CO_2 supply is often a yield-limiting factor for both land and water plants but he does not identify the constraint. Later in the same work (Larcher, 1983, p. 154) he writes, "On the *continents* it is almost always the water supply that limits yield, though other factors are nutrient deficiency and . . . shortening of the production period due to cold."

(found in sugar cane, maize, millet, and many savanna grasses; cf. Larcher, 1983, p. 79), that operates in a biophysically similar manner to the C_3 plants but generates much higher productivities, and the CAM process (found in succulents such as cactus) that is fundamentally different by virtue of its nocturnal assimilation of CO_2. Note that the demand-NPP as given by Eq. 10.18 contains the product, $P_{sm} M$. For C_3 and C_4 plants whose stomates are open during the day, Eqs. E.9 and 6.46 give $P_{sm} \propto E_v \propto E_{ps}$, while for *water-limited* systems, Eq. 6.70 gives $M \propto E_{ps}^{-1}$. Thus under water limitation $P_{sm} M = \text{constant}$. This removes the selection pressure for increasing P_{sm} in either C_3 or C_4 plants in the presence of excess carbon supply when water is limiting. We can then expect $M < 1$ climax communities to have species in the range $\beta L_t < \widehat{\beta L_t}$. Stated in another way, $\beta L_t < \widehat{\beta L_t}$ is a *necessary* condition for water limitation. It is not a sufficient condition since it specifies no climate.

(15) Following the above, species for which $\beta L_t \geq \widehat{\beta L_t}$ will be light limited. For light-limited communities, $M \equiv 1$, and the NPP as given by Eq. 10.18 remains sensitive to changes in P_{sm}. This preserves the selection pressure to close the potential supply–demand productivity gap. We expect that climax communities will have closed this gap resulting in $M = 1$ climax communities having species in the range $\beta L_t \geq \widehat{\beta L_t}$.

(16) Items 14 and 15 above lead heuristically to $\widehat{\beta L_t}$ as a necessary condition separating the species of water-limited ($M < 1$) and light-limited ($M = 1$) communities.

Productivity observations

In Table 10.1 we have collected productivity observations for a small set of vegetation communities from a variety of sources as described in the table's footnotes. As was the case in Chapter 9, our emphasis in the data collection has been to test the theory over the range of communities from desert to rainforest rather than to assemble masses of observations for the commonly studied systems. Table 10.1 with its accompanying notes is self-explanatory except that in plotting the points on Fig. 10.1 we have used symbols evoking the crown shape according to our best estimate. That is, squares are used for assumed cylindrical crowns, triangles for assumed conical crowns, and semicircles for assumed hemispherical crowns. Tubbs and Houston (1990, p. 329) describe American beech as having a "round-topped head"; photographs of red spruce (Blum, 1990, Fig. 3, p. 254) show a conical form; Fechner (1990, p. 243) describes the blue spruce as "pyramidal" in form; Odum (1975, Fig. 7-10, p. 197) pictures the rainforest canopy-top as a set of spherical segments; and Godman *et al.* (1990, Fig. 2, p. 80) show the sugar maple to have a cylindrical crown. Other than the beech, spruce, and rainforest, we assume all crowns to be cylindrical. In making the comparison

Table 10.1. *Observations of canopy productivity*[a]

Number	Species	Site	NPP[b] ($g_s\ m^{-2}\ y^{-1}$)	P_{sm} ($g_s\ m^{-2}\ h^{-1}$)	M	m_τ ($d\ y^{-1}$: $h\ d^{-1}$)	β	L_t	βL_t	$\frac{NPP}{P_{sm}\, m_\tau\, M}$
1	Creosote bush	Nye Co., NV	39.8[1]	2.08[2]	0.09[3]	60:12[4]	0.37[5]	1.0[6]	0.37	0.30
2	Ponderosa pine	Tucson, AZ	61,200[7]	0.40[8]	0.83[9]	169:12[10]			1.77[11]	0.91
3	Loblolly pine	Tucson, AZ	435[12]	0.40[8]	0.80[4]	80:12[13]	0.52[39]	5.0[38]	2.58[14]	1.41
4	Oak	Gt. Smoky Mts., TN	963[15]	0.28[2]	0.95[16]	202:12[13]	0.53[39]	4.9[38]	2.60[11]	1.49
5	Rainforest	Kade, Ghana	2,188[17]	0.38[18]	1.00[19]	365:12[20]	0.79[21]	3.3[21]	2.60	1.31
6	Rainforest	Yangamb., Congo	2,884[22]	0.38[18]	1.00[19]	365:12[20]	0.79[21]	3.3[21]	2.60	1.73
7	Red spruce	Gt. Smoky Mts., TN	950[23]	0.55[2]	0.92[24]	85:14.5[25]	0.51[26]	5.10[26]	2.62[26]	1.52
8	Beech	Eastern Europe	1,300[27]	0.33[2]	0.93[28]	202:12[13]			3.19[11]	1.75
9	Beech	Gt. Smoky Mts., TN	883[29]	0.33[2]	0.93[28]	202:12[13]			3.19[11]	1.19
10	Pinyon–alligator juniper	Flagstaff, AZ	0.19 to 0.25[30]	0.40[8]	0.52[31]	182:12[31]			1.68[32]	0.25 to 0.19
11	Pinyon–Utah juniper	Flagstaff, AZ	0.20 to 0.26[30]	0.40[8]	0.41[31]	215:12[31]			2.27[32]	0.26 to 0.20
12	Sugar maple	Gt. Smoky Mts., TN	1,340[33]	0.69[34]	0.96[35]	202:10[36]	0.35[37]		3.91[11]	1.00

[a] Parenthetical superscript numbers refer to the source listing given in the notes below.

[b] These values are *above-ground* and do not include roots. From Table 8.4 we see that if roots were included the total NPP would be 10% to 40% higher.

Sources:

1. Rosenzweig (1968, Table 1, p. 69, Code A) 2. Table 8.2. 3. Average of five sites in southern California (McMahon, 1988, Table 8.2, p. 240). 4. Table 9.2. 5. Eucalyptus (Table 2.3). 6. Desert shrub (Larcher, 1983, Table 3.13, p. 151). 7. Cannell (1982, p. 311). 8. Loblolly pine (Table 8.2). 9. Williams and Anderson (1967). 10. Salvucci (1992, Table 4.5). 11. Table 3.9. 12. Whittaker and Marks (1975, Table 4–6, pp. 80–81). 13. Fig. 3.11. 14. Lodgepole pine (Table 3.9). 15. Whittaker (1966, Table VI, p. 111, average of Sample Numbers 20, 21, 27, and 28). 16. Whittaker (1966, Table II, p. 107, average of Sample Numbers 20, 21, 27, and 28). 17. Nye (1961). 18. *Goethalsia* (Table 8.2). 19. Assumed. 20. Costa Rica; Allen and Lemon (1978, pp. 298, 300). 21. Allen and Lemon (1978, p. 288). 22. Bartholomew *et al.* (1953). 23. Whittaker (1966, Table II, p. 107, average of Sample Numbers 29 and 30). 24. Whittaker (1966, Table I, p. 106), from average light penetration. 25. Fig. 3.11 for cool multilayers; white pine and red spruce cohabit (Moore, 1917); Table 3.5 with full daylight length due to high elevation. 26. Table 2.3, $\kappa = \beta$, and $\beta L_t = a_w$, average for all spruce; $L_t = \beta L_t/\beta$. 27. Drozdov (1971, Table 1, p. 55). 28. Whittaker (1966, Table II, p. 107, average of Sample Numbers 25 and 26). 29. Average of Bray (1964) and Whittaker (1966, Table II, p. 107, sites 15, 25, 26). 30. Clary *et al.* (1974). 31. Table 6.3. 32. From physical resemblance of needle structure (cf. Zim and Martin, 1987, pp. 36 and 39) we represent the pinyon–alligator juniper as pinyon using Baker's (1950) κL_t for lodgepole pine, and we represent the pinyon–Utah juniper as juniper using Baker's (1950) κL_t for Northern white cedar. 33. Cannell (1982, p. 248). 34. Table 8.2: Average for deciduous trees. 35. Whittaker (1966, Table II, p. 107, Sample Number 24). 36. Fig. 3.11 with estimated shortening of day length due to low elevation and northern exposure. 37. Table 3.7 for maple. 38. Fig. 7.4. 39. $\beta L_t/L_t$.

between theory and observation, we keep in mind that the former assumes atmospheric carbon supply and contains other major approximations:

(1) The form of the βL_t-dependence of the CO_2 concentration gradient has been reasoned rather than derived (and only for $M = 1$), and this form is crucial to the shape of the supply curve. While analogy with the derived canopy conductance, k_v^*, demonstrates that the supply curve clearly must peak, our location, $\widehat{\beta L_t}$, of the important supply–demand equality must be approximate.

(2) We have in effect assumed that the annual net increase of above-ground biomass is a constant percentage of the annual assimilated CO_2 for all forest types. This expedient permits us to use observed annual above-ground biomass production to evaluate the theory but it limits us to qualitative conclusions.

Note in Fig. 10.1 that except for the two pinyon–juniper canopies, all observations are in reasonable agreement with the theory. In particular, we see a cluster of data points in the vicinity of the global optimum where for cylindrical crowns with $\beta = 0.3$, supply equals demand. Furthermore, beyond this point the only canopies following the demand curve are two apparently tapered crowns as predicted. Productivity observations in the range $\beta L_t < \widehat{\beta L_t}$ are few but as is shown in Fig. 10.1 and Table 10.1, they support $\beta L_t < \widehat{\beta L_t}$ being water-limited with $M < 1$ and $\beta L_t \geq \widehat{\beta L_t}$ being light-limited with $M = 1$.

It is particularly interesting to note that the literature finds C_3 trees to be carbon supply-limited at the current atmospheric CO_2 concentration (Larcher, 1983, p. 85; Sharkey, 1985; Gunderson and Wullschleger, 1994; Drake *et al.*, 1997) which would imply a direct response to anthropogenic CO_2 increases. Here we find, by both theory and limited observation, that carbon is limiting only beyond a critical basal leaf area, $\widehat{\beta L_t}$, and then only for non-tapered crowns.

The general agreement of observation with theory supports our earlier tentative conclusion that with an adequate nutrient reservoir, the supply of nutrients will not limit productivity. However, the soil at the Beaver Creek site of the pinyon–juniper canopies is described (Williams and Anderson, 1967) as a "very stony clay" derived from basalt and volcanic cinders, rich in exchangeable calcium and magnesium, but low in potassium. The nitrogen content ranges from 0.17 to 0.02 percent. It seems possible that the productivity of these soils may be limited by the available reservoir of either potassium or nitrogen.

In evaluating the comparisons of observations with theory we must remember that we have not developed a theoretical predictor for m_τ, the length of the growing season, and have estimated this important factor using a somewhat uncertain empirical correlation (cf. Fig. 3.11). Season length is limited by complicated biochemical processes beginning with bud-break in the spring which is commonly forecast through some measure of accumulated heat (Valentine, 1983; Lechowitz, 1984). The season ends in the autumn with leaf senescence. Transpiration (Gee and Federer, 1972) and photosynthesis (Mitchell, 1936) cease as chlorophyll and other nutrients are removed from the leaves before they fall. Our estimates of m_τ may introduce considerable error into the observational ordinates of Fig. 10.1.

Plate 10.7. Colorado spruce. (Photograph by William D. Rich in Arnold Arboretum, Boston; Copyright © 2001 William D. Rich.)

Verification of species productivity maximization

Underlying the whole of this work is the assumption that the natural selective pressure on the individual plant is to maximize its reproductive potential and hence its annual biomass and proportional seed production. Equation 10.22 gives us a means of testing this assumption if we move all species dependence to the right-hand side.

The focus of attention when deriving Eq. 10.22 was carbon demand by a given species. Therefore, we integrated the falling light intensity over the saturating photosynthetic capacity curve with depth into the canopy in order to obtain the total assimilation as a function of the species-constant maximum assimilation rate, P_{sm}. To introduce the climate supportive of this carbon demand we use Eq. 8.31 (confirmed in Fig. 8.11), and the bioclimatic light optimum of Eq. 9.2 (confirmed in Fig. 9.5), along with the optical optimum Eq. 3.52 (confirmed in Fig. 3.19) to replace P_{sm} by the *climatic assimilation potential*

$$P_{sm} = \varepsilon f_I(\beta L_t) I_0, \tag{10.42}$$

in which $\varepsilon = 0.81\ g_s MJ(tot)^{-1}$ as found in Fig. 8.10 and P_{sm} is in $g_s\,m^{-2}\,h^{-1}$. Equation 10.42 allows us to factor additional species dependence out of the denominator on the left-hand side of Eq. 10.22. In addition, we use the thermal optimality of Eq. 8.30 (confirmed in Fig. 9.4) to write

$$g(T_0) = g(T_m) \equiv 1. \tag{10.43}$$

The canopy cover, M, is a complex function of βL_t through the effect of canopy resistance on transpiration (cf. Eqs. 5.29, 5.30, and 6.66). We avoid this complication here by restricting Eq. 10.22 to conditions of non-limiting water supply, $M = 1$, thereby making the resulting NPP a *potential* maximum in that no limitations to the supply of either water or carbon dioxide are considered.

With these substitutions Eq. 10.22 is rewritten to put all species dependence on the right-hand side yielding the maximum potential (i.e. light-limited) productivity in a given climate

$$\frac{\text{NPP}}{\varepsilon I_0\, M\, m_\tau} = f_D(\beta L_t) f_I(\beta L_t). \tag{10.44}$$

Note that the quantity by which NPP is normalized in Eq. 10.44 now represents the productivity of a groundcover monolayer having $\beta = 1,\ L_t = 1$.

Equation 10.44 is plotted in Fig. 10.3 where we see a rising limb for small βL_t due to dominance of $f_D(\beta L_t)$. In this range the low foliage density allows relatively undiminished light penetration so that each additional leaf contributes heavily to increased

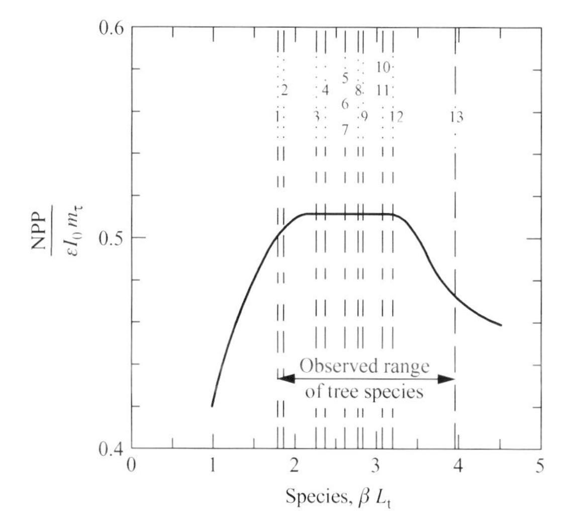

Fig. 10.3. Evidence for maximization of the light-limited productivity of trees. Observed range from Baker's I_k/I_0 (1950, Table 12, p. 143) using $\kappa = \beta$; Table 3.9.
1, Ponderosa pine; 2, Scots pine; 3, Northern white cedar; 4, tamarack; 5, lodgepole pine; 6, Douglas fir; 7, red oak; 8, hackberry; 9, Engelmann spruce; 10, Norway spruce; 11, Eastern hemlock; 12, beech; 13, sugar maple.

canopy productivity. For large βL_t, the high foliage density causes heavy light decay and the falling average light intensity, $f_I(\beta L_t)$, dominates the increasing leaf surface so that canopy productivity falls. Importantly, there is an intermediate range of βL_t in which the dimensionless productivity has the almost constant *global maximum* with the approximate value

$$\left(\frac{\text{NPP}}{\varepsilon I_0 m_\tau}\right)_{\max} \approx 0.51. \tag{10.45}$$

Note that at least for $\beta = 0.3$ as plotted on Fig. 10.1, $\widehat{\beta L_t} = 2.65$ lies in the center of the productivity peak of Fig. 10.3.

In Appendix H we show that $\beta L_t = \kappa L_t$ is a species constant, and in Table 3.9 we relate κL_t to species through Eq. 3.53 and the compensation point observations of Baker (1950). This allows us to associate βL_t with specific species and to draw the vertical dashed lines on Fig. 10.3 representing the full range of species observed by Baker (1950). Note that 10 of the 13 species lie in the range of the global maximum productivity. We take this as observational confirmation of natural selection pressure to maximize survival probability as represented by its surrogate, NPP.

Extreme climates may dictate extreme βL_t which are locally but not globally optimum productively. For example, hot, dry growing seasons demand low βL_t to reduce water use (e.g. $\beta L_t = 0.37$ for creosote bush; not plotted in Fig. 10.3), and cold but wet growing seasons demand high βL_t to conserve heat (e.g. $\beta L_t = 3.9$ for sugar maple).

B Ecotones

Definition

An *ecotone* is a transition zone between different plant communities (Colinvaux, 1973). Because they represent marginal conditions, we expect ecotones to be sensitive to changing climate (Arris and Eagleson, 1994). Therefore, a necessary condition for the utility of the present work in anticipating the effects of future climate change is the ability of our analyses to locate principal existing ecotones.

The proposition

Let us assume that in the battle for survival, the *plant* with the highest productivity wins due to its superior seed production. Let us also assume, for purposes of this illustration, that the gradient of climatic factors controlling productivity of a given species is expressed primarily through latitude. Extremes of latitude should depress productivity due to the temperature control of photosynthesis leaving a preferred intermediate latitude at which the productivity is maximum. In the competition between two monocultures, the geographical productivity variation is as sketched in Fig. 10.4 (Arris and Eagleson, 1994, Fig. 3, p. 4). Under these assumptions the local productivity maximum shifts species where the two curves of Fig. 10.4 cross thereby marking the ecotone location.

A test

As is shown in Fig. 10.5, the forest formations of eastern North America display ecotones which are approximately latitudinal. The shaded region embracing the Great Lakes is the broad, mixed forest transition (i.e., ecotone) between the evergreen boreal (primarily spruce) forest and the deciduous (primarily oak) forest. The thinner shaded band represents a second mixed forest ecotone between the deciduous and southern pine (primarily loblolly).

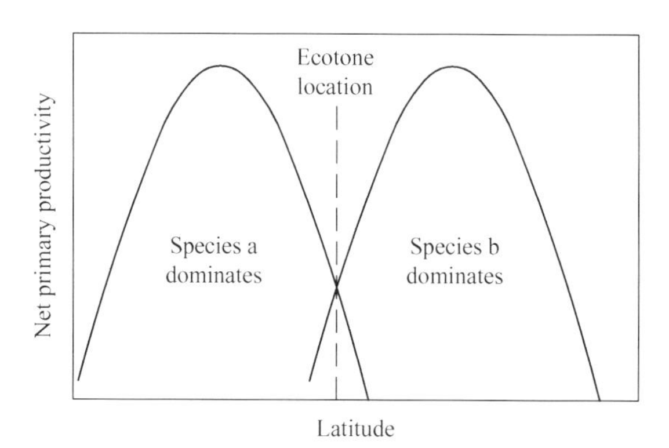

Fig. 10.4. Hypothetical relationship of productivity and latitude for two competing vegetation types. (From Arris and Eagleson, 1994, Fig. 3, p. 4; with permission of the American Geophysical Union.)

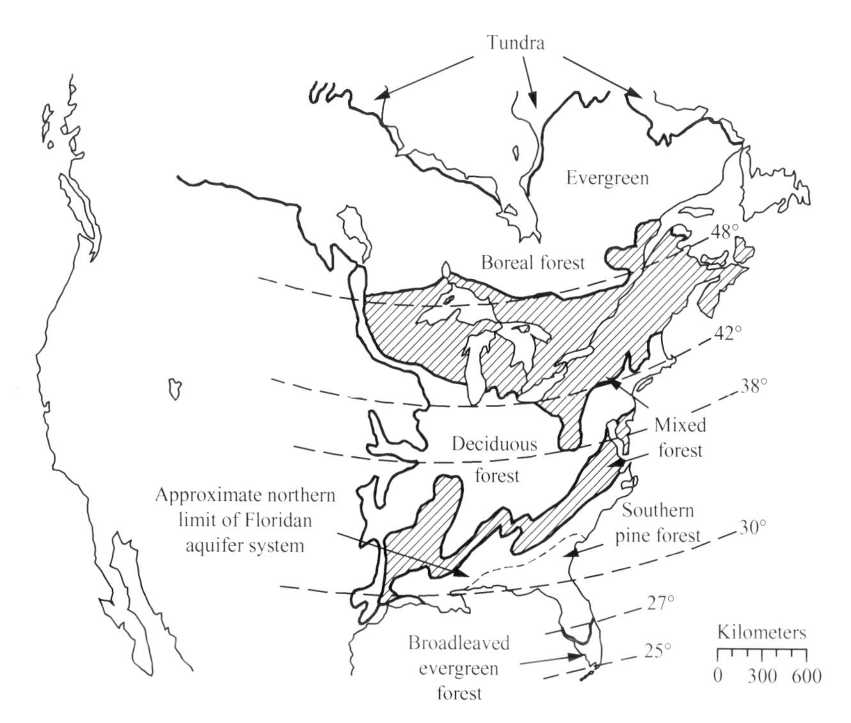

Fig. 10.5. Forest formations of eastern North America. Map from Little (1971); vegetation boundaries from Eyre (1968); aquifer boundary from Sun and Weeks (1991).

To compare our ecotone postulate with the observations of Fig. 10.5, we return to Eq. 10.20 rewritten to represent an *individual plant* as

$$\frac{\widehat{P_T}}{M} = f_D(\beta L_t)\, g(T_0)\, \vartheta\, P_{sm}\, m_\tau, \quad (g_s\, m^{-2}\, y^{-1}). \tag{10.46}$$

The terms of Eq. 10.46 are evaluated as a function of latitude in eastern North America for spruce, oak and loblolly pine as explained and listed in Table 10.2. The reference growing season atmospheric temperature at each latitude is taken as the average daily maximum July temperature which is presented in Fig. 10.6 for locations in eastern North America from climatic records of the U.S. Weather Service (1974). A common value of β is assumed from Table 10.2, and Eq. 10.46 is plotted vs. latitude in Fig. 10.7.

Examining Fig. 10.7 we find Eq. 10.46 to predict spruce as being the most productive of the three species in the observed boreal forest zone and oak as the most productive in the deciduous forest zone. The interchange of productive order among the three species takes place clearly within the observed broad mixed forest ecotone. At variance with the theory is the predicted strong superior productivity of oak in the observed southern pine forest zone. Clearly some influence is present here in nature that the theory omits. Arris and Eagleson (1989a) found this same behavior and have suggested nutrient supply as is indicated in the following discussion paraphrased from their work.

Table 10.2. *Productivity vs. latitude in eastern North America*

Φ (° N)	$T_l = T_0$ (°C) (Fig. 10.5)	$g(T_0)$ (Fig. 8.4b)	P_{sm} $g_s\, m^{-2}\, h^{-1}$ (Table 8.2)	m_τ $d\, y^{-1} : h\, d^{-1}$ (Fig. 3.11) (Table 3.5)	$f_D(\beta L_t)$ (Table 3.9) (Eq. 10.21)	NPP/M $10^2\, g_s\, m^{-2}\, y^{-1}$ (Eq. 10.22)
Spruce ($\beta L_t = 2.91$) ($\beta = 0.51$) (*Table 10.1*)						
50	17	0.98(1)	0.55	100 : 16	1.37	11.81
45	25	0.88(1)	0.55	85 : 15.5	1.37	8.74
40	29	0.77(1)	0.55	15 : 15	1.37	1.30
Oak ($\beta L_t = 2.60$) ($\beta = 0.53$) (*Table 10.1*)						
50	17	1.00(2)	0.28	156 : 16	1.40	9.78
45	25	1.00(2)	0.28	173 : 15.5	1.40	10.51
40	29	1.00(2)	0.28	189 : 15	1.40	11.11
35	31.5	0.98(2)	0.28	202 : 14.5	1.40	11.25
30	32.5	0.97(2)	0.28	212 : 14	1.40	11.28
Loblolly pine ($\beta L_t = 2.58$) ($\beta = 0.52$) (*Table 10.1*)						
50	17	0.97(3)	0.40	122 : 16	1.40	10.60
45	25	1.00(3)	0.40	110 : 15.5	1.40	9.60
40	29	0.98(3)	0.40	98 : 15	1.40	8.07
35	31.5	0.96(3)	0.40	85 : 14.5	1.40	6.63
30	32.5	0.95(3)	0.40	74 : 14	1.40	5.51

1. Artic pine: $T_m = 15$ °C. 2. Beech: $T_m = 23$ °C. 3. Loblolly pine: $T_m = 26$ °C.

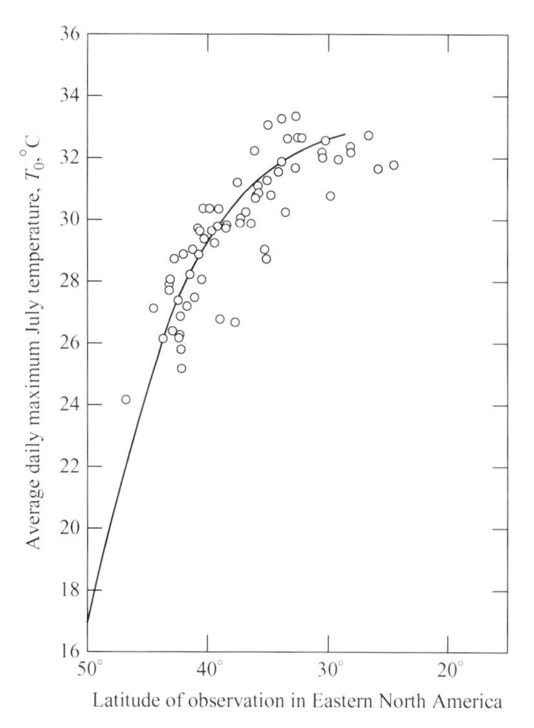

Fig. 10.6. Average maximum July temperatures in eastern North America. (From records of U.S. National Weather Service.)

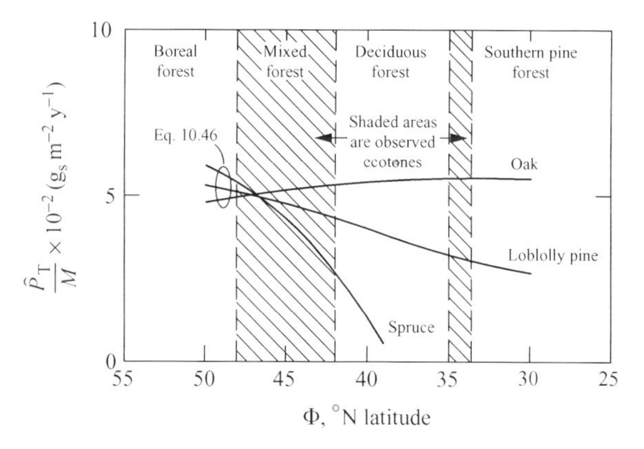

Fig. 10.7. Ecotone location in eastern North America. (Calculations summarized in Table 10.2.)

Note in Fig. 10.5 how this narrow ecotone separating the deciduous and the southern pine forests takes a NE–SW direction along the approximate northern boundary of the Floridan aquifer system. This marks the beginning of the sandy alluvial formation terminating in the Florida peninsula. The coarse texture of sandy soils leads to rapid percolation of soil moisture and to better soil aeration which causes rapid decomposition of organic matter. The small fraction of clay-sized soil particles coupled with the low organic content dramatically reduces the surface area available to hold nutrients in the soil against the leaching effects of percolation (Brady, 1974; Spurr and Barnes, 1980). This is particularly true for deciduous trees which, in contrast to evergreens, contribute a concentrated pulse of nutrients to the soil in the spring when the autumn leaf fall decomposes. Evergreen leaves are shed more-or-less continuously and have both chemical and structural differences which ensure a more gradual return of nutrients to the soil with less likelihood of loss through leaching from the root zone (Monk, 1966; Spurr and Barnes, 1980; Chabot and Hicks, 1982).

Monk (1966) and Chabot and Hicks (1982) suggest that evergreen species may dominate in areas of low nutrient availability for biological reasons also. Evergreen leaves tend to contain smaller quantities of key nutrients which reduces the amount of replacement nutrient required in the soil. Evergreens also have more efficient internal nutrient cycling which results in a higher carbon gain per unit of nutrient turned over (Chabot and Hicks, 1982).

As a result of the above, it seems likely that differential nutrient requirements may make the loblolly pine dominate the oak in the sandy soils characterizing the southern forest zone.

11

Summary, speculations, and opportunities

Through theoretical study of the fluxes of solar radiation, momentum, heat, water, carbon dioxide, and nutrients, we have distilled and identified (in many cases for the first time) some of the opportunities for natural selection to maximize plant productivity and hence the probability of its reproductive success by shaping the form and function of natural vegetation canopies. Advancement toward this goal has required many assumptions and simplifications, the most serious of which are perhaps: (1) our severing of the feedback path from canopy to atmosphere in order to consider the vegetation as being passive, (2) our formulating the biophysical dynamics in terms of long-term and large-scale space–time averages over the growing season rather than by averaging the solution to space- and time-dependent dynamic equations, and (3) our assumption of non-limiting nutrient supply. Where possible, we have tested our resulting hypotheses against observations from the literature. Although these observations are limited in number and are often the combined results of different investigators, they cover a wide range of community types from desert shrub to tropical rainforest.

A Summary

Problem simplification

Interaction between the atmosphere and vegetated land surface involves the exchange of mass, momentum, and energy between the two climate components in processes which have characteristic time-scales varying from seconds (turbulent wind gusts) to decades or centuries (life span of individual trees). In seeking first-order understanding of the natural selection of vegetation form and function we have chosen a priori to focus our attention at the time-scale of the growing season and to treat the canopy as a steady-state passive responder to climatic forcing without feedback from canopy to atmosphere. Additional major assumptions are:

(1) Coefficients of variation of climatic variables in both time and horizontal space are small enough to allow writing dynamic equations in terms of the climatic

averages over the growing season for the local vegetation community (Chapter 4).

(2) Atmosphere is neutrally stable in this climatic average (Chapter 4).

(3) Flux divergences of heat and water vapor above the canopy are negligible in this climatic average (Chapter 5).

(4) The vegetal state in this climatic average is stressless in the interests of plant health and productivity, and thus the stomates are in an effectively open condition (Chapter 8).

Bioclimatic optimality

Our principal thesis here has been the existence of a relentless evolutionary pressure toward a bioclimatically optimal vegetal state assumed to be that of maximum probability of reproductive success which we equate to maximum seed productivity and hence, by proportion, to maximum biomass productivity (Chapter 10). From the biophysics we have identified the following major optimization opportunities for exercise of this selective pressure.

Optical optimality

We propose that the optimum foliage arrangement is such as to extract the maximum photosynthetically useful energy from the radiation beam.[†]

This requires that the reflection of PAR is minimized, the upper leaf surfaces are fully illuminated, the lowest leaf in the canopy has the minimum radiation intensity needed for metabolic sufficiency (i.e., the *compensation* intensity), and that the basal area of the average leaf be in full shadow. We show (cf. Figs. 3.19–3.20) that these conditions result in equality of the canopy extinction coefficients for momentum and insolation (i.e., $\kappa = \beta$) and the selection of optimum leaf angle by the solar altitude (cf. Eq. 3.51 and Fig. 3.15, curve (b)).

We further propose that the climatic insolation, $\widehat{I}_l$, *on the average leaf in the crown is just equal to the minimum insolation,* I_{SL}, *for fully open stomates to insure stable systems in which crown carbon assimilation is maximized (cf. Fig. 8.11).*[†]

This requires $I_{SL} \equiv \frac{P_{sm}}{\varepsilon} = \widehat{I}_l \equiv f_I(\beta L_t) I_0$, where $f_I(\beta L_t)$ averages I_0 over the canopy (cf. Eq. 8.29). It is supported over a wide range of canopies in Figs. 8.10 and 9.5,[†] and defines the crown *climatic assimilation potential*, $P_{sm} = \varepsilon f_I(\beta L_t) I_0$, determining the maximum crown CO_2 assimilation rate for a given C_3 species and climate.

Mechanical optimality

We propose that with optical optimality in homogeneous cylindrical multilayer crowns, there is a particular leaf angle for each species at which the canopy conductance of water vapor and CO_2 *is maximum thereby defining an optimum foliage state.*

[†] We believe this to be a new finding.

With stomates fully open, the resistive network for turbulent transfer in the canopy from stomatal cavities to the free atmosphere is shown adequately to approximate canopy total resistance, r_c, by a function solely of the interleaf layer atmospheric resistance, r_i,† which is a function of βL_t (shown in Appendix H to be a species constant†). The relative canopy resistance, $\widehat{r_c}/r_a$, is thus derived as a function of both β and L_t. For homogeneous cylindrical crowns $\widehat{r_c}/r_a$ has a minimum,† and thus the vertical turbulent fluxes are maximum at a particular leaf angle for each species. For tapered crowns $\widehat{r_c}/r_a$ declines monotonically with L_t for a given β.† The dependence of canopy conductance, k_v^*, on $\widehat{r_c}/r_a$ and T_0 provides selective pressure toward tapered crowns in climates where the fluxes of carbon, nitrogen, or heat may limit productivity.†

Thermal optimality

We propose that during the growing season, the average leaf temperature, T_l, *equals the photosynthetically optimum temperature,* T_m, *and approximates the average screen-height atmospheric temperature,* T_0, *thereby making the canopy–atmosphere system isothermal on the average.*

We confirm the finding of earlier investigators by our limited demonstration in Fig. 9.4 that $T_0 = T_m$. Minor environmental temperature regulation is provided in the short term by temporary modulations of leaf angle, while major genetic regulation occurs through both foliage arrangement and crown shape.‡ We are concerned here only with the genetic mechanisms. Some speculation is in order:

(1) Very limited evidence is presented (cf. Fig. 3.15) that in quite hot climates the average leaf angle tends to produce upward specular reflection probably as a means of most efficiently reflecting heat-producing longwave radiation back to space. Hot climates appear to favor tapered multilayer crowns perhaps because for these the canopy conductance increases monotonically with increasing leaf area (up until the onset of significant solid body resistance) and thus maximizes the flux of sensible heat from plant to atmosphere.

(2) Similarly limited evidence is presented (cf. Fig. 3.15) that in cold climates the average leaf angle tends to produce horizontal specular reflection which may be most efficient in retaining heat-producing longwave radiation within the canopy. Cold climates appear to favor needle leaves in monolayer arrangement probably because these best inhibit heat transfer to the atmosphere from within the crown. Crown shape in these climates is typically conical with the monolayer surface nearly normal to the radiation beam in order to minimize reflection of NIR.

† We believe this to be a new finding.

‡ Among the genetic mechanisms is a tapered multilayer crown which, as we have seen in Chapter 7, insures a maximum canopy conductance for given β and L_t, and hence maximum transfer of latent and sensible heat, CO_2, and dissolved nutrients.

Hydrological optimality

We propose that at the end of the average growing season, water-limited communities (i.e., $M < 1$*) achieve a critical moisture state at which the individual plant is at incipient stress and the crown-average insolation is* I_{SW}*, the maximum insolation for open stomates as given by the leaf water characteristic (cf. Fig. 8.15, Eqs. 6.67 and 8.26).*†

This insures that the stomates remain open for carbon assimilation the maximum possible time and is borne out by the water balance data presented in Fig. 6.15.† Joint achievement of the optimum foliage state and the critical moisture state defines the optimum canopy state which compares well with limited observations (cf. Fig. 7.11).†

Nutritional optimality

We propose that the optimum foliage state will be that giving maximum canopy conductance, because that will provide maximum water-borne nutrient flux from soil to plant.

This requires that $\widehat{r_c}/r_a = (\widehat{r_c}/r_a)_{min}$ for cylindrical multilayers as is illustrated in Fig. 7.3 which, along with the conditions of optical optimality, selects the species and fixes the leaf area index in this crown shape for given leaf angle (cf. Fig. 7.4). For given species having the tapered crowns which are favored at high leaf area, the solar-determined leaf angle along with optical optimality fixes the leaf area index and thus the resistance ratio as given in Fig. 7.5. Strong support of this proposition is obtained from the detailed study of three semi-arid canopies summarized in Fig. 7.11.†

We further propose that tapered multilayer crowns are indicated at high basal leaf area in order to insure that through high canopy conductance, carbon supply satisfies the light-driven carbon demand.

This is borne out by the productivity data presented in Fig. 10.1.†

Climax conditions

We propose that the maximum habitat insolation at which stomates can remain open without water stress, I_{SW}*, is always equal to or greater than the minimum insolation for open stomates,* I_{SL}*, as given by the leaf light characteristic.*

Along with optical optimality this proposition insures that the stomates are always fully open and that for $M < 1$, all available moisture is used. As given in Chapter 8, this defines the water-limited climax condition, $\frac{P_{sm}}{\varepsilon} = f_I(\beta L_t) I_{SW}$ for $M < 1$, and the light-limited climax condition $\frac{P_{sm}}{\varepsilon} \leq f_I(\beta L_t) I_{SW}$ for $M = 1$, which receive strong support for a variety of canopies and habitats in Fig. 9.7.†

Productivity

The net primary productivity of a multilayer canopy is derived in Chapter 10 and normalized by that of a groundcover monolayer of the same species. Derived using

† We believe this to be a new finding.

the optimality conditions cited above, this ratio is called the productivity *gain* of the canopy. It is shown to agree quite well with observations from the literature over a wide range of the basal leaf area, βL_t, which is shown in Appendix H to be a species constant.† Limited by carbon *demand*, productivity gain rises with increasing βL_t until, for non-tapered crowns, a critical value, $\widehat{\beta L_t}$, is reached beyond which the foliage density limits atmospheric carbon *supply* to the leaves. The zone of productive limitation by carbon demand appears to contradict implications in the literature that C_3 plants are always carbon supply-limited. It is reasoned that $\beta L_t < \widehat{\beta L_t}$, and $\beta L_t \geq \widehat{\beta L_t}$ are necessary conditions for water-limited and light-limited communities respectively. For tapered crowns we show that carbon supply is non-limiting for all species.†

Confirmation of productivity maximization

The (normalized) net primary productivity is rederived for light limitation, also in Chapter 10, in terms of a new function of βL_t that incorporates all the species-dependence and is shown there to have a maximum over a broad, intermediate range of the species-specific parameter βL_t. Values of βL_t calculated from observations of the relative compensation light intensity, I_k/I_0, for a wide range of tree species are concentrated in this range of global-maximum productivity thereby confirming our underlying assumption of productivity maximization as a surrogate goal of the natural selection process.† Climate extremes beget βL_t extremes that differently constrain productivity to a reduced maximum.

B Speculations

Synthesized selection

With the above mechanisms in mind, how might we synthesize the selection of the most productive species in a given habitat?

We begin with temperate climates and the cylindrical multilayer crowns that seem to predominate there:

(1) At the given latitude, Φ, the mean solar altitude, $h_{\otimes}$, determines the mean leaf angle, $\cos^{-1}\beta$, through optical optimality by Eq. 3.51. This equates the canopy decay coefficients for light and momentum giving $\kappa = \beta$.

(2) Mechanical optimality gives L_t using Eq. 7.14 and bioclimatic optimality fixes P_{sm} through Eq. 8.41.

(3) The optimum canopy state (k_v^*, M) for this β, L_t, and habitat is determined from the critical moisture state through Eqs. 6.66 and 7.31 as is illustrated in Figs. 7.10 and 7.11.

(4) The season length, m_τ, is estimated empirically from Fig. 3.11 and Table 3.6.

† We believe this to be a new finding.

(5) Departure, $g(T_l \approx T_0)$, from optimum canopy temperature, T_m, is estimated empirically from Fig. 8.4b.
(6) The maximum productivity is now given by Eq. 10.18 with proportional reduction, as may be called for by carbon supply limitation, obtained from Fig. 10.1.

There is implicit evidence, presented here in Fig. 7.11, that the canopy conductance is maximized at least for the three water-limited communities tested. However, mechanical optimality leads theoretically to deciduous β being higher than the β for pine as is shown in Fig. 7.4, while observationally the reverse is found as is shown in Fig. 3.18.

Evidence is clear (cf. Fig. 3.15) that various climate extremes may dictate sacrifice of the optically optimal β apparently for the control of canopy temperature and/or water use achieved through energy retention or reflection. In other cases, excess sensible heat or nutrient-poor soil may call for the increased canopy fluxes obtained only with tapered crowns thereby sacrificing mechanical optimality. In all such cases β is again estimated from $h_{\otimes}$ but this time using the empirical curve (a) of Fig. 3.15. In the absence of a minimum canopy resistance ratio, we must now resort to trial calculation of the productivities of assumed species:

(1) Assume a species thus fixing the compensation light intensity, I_k, from Table 3.9 and yielding L_t.
(2) Proceed as for cylindrical multilayers beginning with no. 3 and repeat for alternative species in the same habitat until the productivity calculated in no. 6 is maximized.

A response to global warming

To the extent that global warming increases the intensity and frequency of "severe" storms while the seasonal precipitation remains unchanged, the runoff component of the water budget will increase, the available soil moisture, V_e, will fall as will the desiccation insolation, I_{SW} (cf. Eq. 9.8), and existing climax communities will go into stress. These stressed species will die off and be replaced by others with smaller I_{SL} (and hence smaller P_{sm}) in order to recover the optimum state, $\Pi_W = 1$, all at the cost of reduced productivity. Of course accompanying increase in the total seasonal precipitation will counter this tendency.

Acclimation to changing ambient CO_2 concentration

Studies of the response of isolated leaves or seedlings to increases in the concentration of CO_2 in the ambient atmosphere (cf. Gunderson and Wullschleger, 1994) reveal a pattern of initial increase in CO_2 assimilation rate for a year or two followed by a return to approximately pre-change rates. This is refered to as *acclimation* of the plant to permanently changed conditions without permanent change in its response. The mechanisms for this behavior are not well understood.

The developments of Chapter 10 offer the following possible scenario for acclimation by trees to increased concentration of atmospheric CO_2 *provided the trees are under CO_2 supply limitation, and provided the nitrogen supply is adequate to support additional foliage*:[†]

(1) Increase in ambient CO_2 concentration initially brings added productivity since it removes the CO_2 supply constraint.

(2) Over a growing season (or longer), the increased productivity causes an increase in the crown absorption index, βL_t, which restores the CO_2 supply limitation (cf. Fig. 10.1) and productivity falls.

C Opportunities

The first-order relationships presented here provide a simple, albeit approximate, analytical basis for sensitivity analyses and probabilistic forecasting in the manner exploited by the author in studies of the water balance (cf. Eagleson, 1978g). Accordingly, the following extensions seem particularly promising:

(1) Both the productivity formulation and its comparison with field observations of NPP suffer from the expedient assumption that nutrient availability is non-limiting (cf. Chapter 10). It would be helpful to make the NPP forecasts probabilistic by incorporating into Eq. 10.18 the natural variability in nutrient supply as well as that of the important independent climate variables such as season length, temperature, and available soil moisture. It would then be possible to evaluate the accuracy of Eq. 10.18 better when compared with reported observations, to gain a theoretical understanding of observed interanual variability of primary production (cf. Knapp and Smith, 2001), and to make sensitivity studies of climate change.

(2) The photosynthetic capacity curve used in generating the productivity function (Eq. 10.18) as well as the habitat conditions (Eqs. 9.13–9.14) should be generalized to incorporate variable ambient CO_2 concentration in order to investigate the effects of this climate change (cf. LaDeau and Clark, 2001).

Certain of the fundamental components of productivity remain empirically defined and are treated here in an intellectually unsatisfactory manner which might benefit from a fresh look. Particularly troublesome are the length of the growing season with its multitude of constraints and influences, and the leaf angle for tapered crowns which we have concluded to respond to the thermal need to retain or reject longwave radiation.

Finally, far more work is needed in careful testing of our results against observation before we can claim understanding of the true basis for natural selection in trees.

[†] Drake *et al.* (1997, Table 1, p. 612) collected 12 long-term studies of 8 field-grown species (crops such as wheat and cotton, and non-forest ecosystems such as prarie, grassland and wetland) and found only a 3% increase in leaf area following a doubling of the CO_2.

Appendices

A

Effect of crown shape on flow in canopy

1 EDDY VISCOSITY FOR MULTILAYER FOLIAGE ($M = 1$)

Introduction

In Chapter 4 we derived the vertical distribution of atmospheric eddy viscosity within the crown of leafy plants for closed canopies (i.e., $M = 1$) of homogeneous circular cylinders (Eq. 4.40, $m = 1/2$), and for a nearly conical spruce (Eq. 4.102, $m = 1/2$, $\Psi = 1$). These crowns were assumed to have homogeneous foliage density which is high areally and low volumetrically. Using the terminology of Horn (1971), such crowns are often referred to as being *multilayer*. To compare the effect of multilayer crown geometry on the turbulent mixing we first derive the equations for a hemispherical crown and for a pure conical crown, both again at $M = 1$.

Keeping in mind the requirement for compensation light intensity at the lowest leaf level at all radii, any crown which does not fill its circumscribing cylinder must be non-homogeneous in leaf area density, that is $a_t = a_t(\xi, r)$. We will deal with this variation in an approximate fashion by first averaging radially. In other words, we assume $a_t(\xi)$ to be homogeneous and isotropic in the radial direction and deal with slices, $\Delta\xi$, of the circumscribing cylinder which are each homogeneous over the full cylindrical crossection (cf. Fig. A.1b).

Hemispherical multilayer crowns

In the definition sketch, Fig. A.1a, the shaded area defines a hemispherical segment with base diameter $2r$. A larger segment of basal diameter, b, is shown as a representation of the full crown. Such a segment is probably a better representation of an actual tree crown than is the full hemisphere. Eshbach (1936, p. 2–43) gives the volume, $\forall_s$, of a

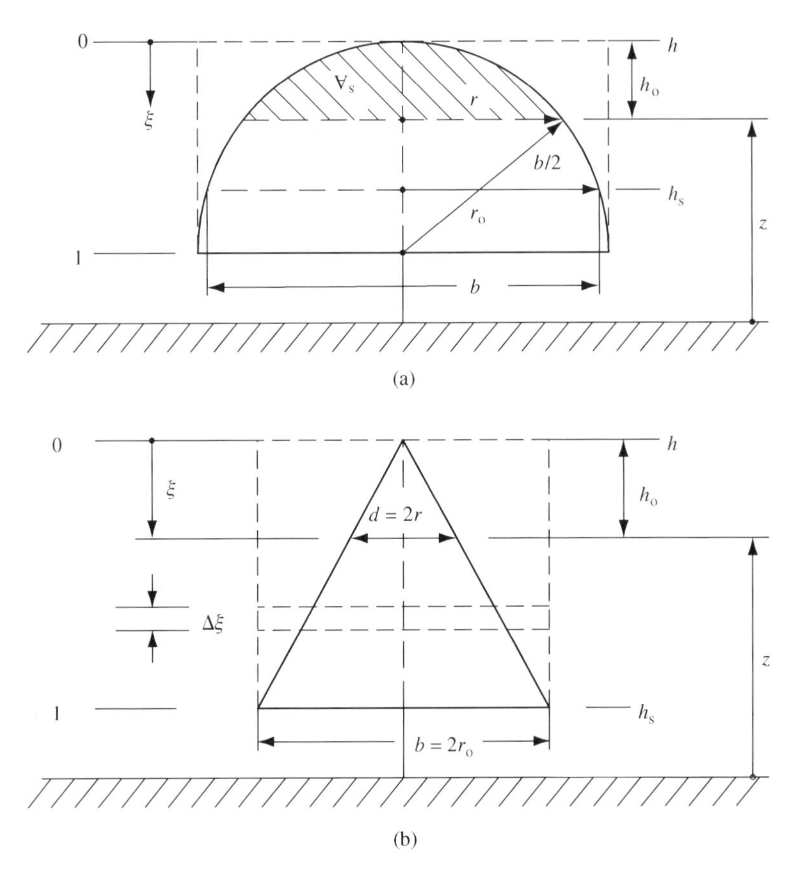

Fig. A.1. Definition sketches for multilayers. (a) Hemispherical; (b) conical.

spherical segment as

$$\forall_s = \frac{\pi}{6} h_o \left(3r^2 + h_o^2\right), \tag{A.1}$$

in which geometry allows the radius, r, to be written

$$r^2 = r_o^2 - (r_o - h_o)^2 = 2r_o h_o - h_o^2. \tag{A.2}$$

Examining a horizontal slice of unit thickness located at depth $\xi \equiv \frac{h-z}{h-h_s} = \frac{h_o}{h-h_s}$, and using Eq. A.2, the foliage area density, $a_t(\xi)$, is

$$a_t(\xi) = \frac{\pi r^2 a_{cr}}{\pi \left(\frac{b}{2}\right)^2} = 4\left(\frac{r}{b}\right)^2 a_{cr} = 4\left(\frac{h - h_s}{b}\right)^2 \left[2\left(\frac{r_o}{h - h_s}\right)\xi - \xi^2\right] a_{cr}, \tag{A.3}$$

where

$$a_{cr} = \text{crown one-sided } \textit{leaf area density} = \frac{\text{one-sided leaf area}}{\text{crown volume}}.$$

For the special case of $h_o = h - h_s$, $r = b/2$, and Eq. A.3 becomes

$$a_t(\xi) = \frac{2\left(\dfrac{r_o}{h - h_s}\right)\xi - \xi^2}{2\left(\dfrac{r_o}{h - h_s}\right) - 1} a_{cr}. \tag{A.4}$$

Applying Eq. 4.15 to Eq. A.4

$$\int_0^1 a_t(\xi)\,d\xi \equiv L_t = \frac{\left[\dfrac{r_o}{h - h_s} - \dfrac{1}{3}\right]}{2\left(\dfrac{r_o}{h - h_s}\right) - 1} a_{cr}, \tag{A.5}$$

which, together with Eq. A.4, gives

$$a_t(\xi) = L_t \left[\frac{2\left(\dfrac{r_o}{h - h_s}\right)\xi - \xi^2}{\left(\dfrac{r_o}{h - h_s}\right) - \dfrac{1}{3}}\right], \tag{A.6}$$

in which the homogeneity function appears as the factor

$$a(\xi) = \left[\frac{2\left(\dfrac{r_o}{h - h_s}\right)\xi - \xi^2}{\left(\dfrac{r_o}{h - h_s}\right) - \dfrac{1}{3}}\right]. \tag{A.7}$$

We approximate the shear distribution within the three-dimensional shape by assuming that Eq. 4.17 yields the areal average shear stress at any elevation. Then using Eq. A.7 in Eq. 4.17 under the restriction that $\varepsilon \equiv \frac{r_o}{h - h_s} \geq 1$, gives the momentum flux variation

$$\frac{\tau_f(\xi)}{\tau_f(0)} = \exp\left[-n\beta L_t \left(\frac{\varepsilon\xi^2 - \dfrac{\xi^3}{3}}{\varepsilon - \dfrac{1}{3}}\right)\right]. \tag{A.8}$$

To match velocity gradients at $\xi = 0$, the interference function for the hemisphere is again

$$s(\xi) = B\xi. \tag{A.9}$$

As we did for the cone in Eqs. 4.87–4.90, we can now find the velocity distribution

$$\frac{u(\xi)}{u_0} = \exp\left[-\gamma L_t \left(\frac{\varepsilon\xi^2 - \dfrac{\xi^3}{3}}{\varepsilon - \dfrac{1}{3}} - B_0\xi\right)\right], \tag{A.10}$$

where again $B_0 \equiv \frac{B}{\gamma L_t}$. Equating velocity gradients at $\xi = 0$ gives $B_0 = -1$. Then, with $m = 1/2$ and $\Psi = 1$ (see Eq. 4.111 and Fig. 4.26b), Eq. 4.37 gives the eddy viscosity variation for the arbitrary hemispherical segment having a single base

$$K_{\mathrm{m}}^{\mathrm{o}} \equiv \frac{K_{\mathrm{m}}(\xi)}{K_{\mathrm{m}}(0)} = \frac{\exp\left[-\gamma L_{\mathrm{t}}\left(\dfrac{3\varepsilon\xi^2 - \xi^3}{3\varepsilon - 1} - \xi\right)\right]}{1 + \dfrac{6\varepsilon\xi - 3\xi^2}{3\varepsilon - 1}}. \tag{A.11}$$

The limiting cases of Eq. A.11 are of particular interest. For $\varepsilon = 1$ we have the hemisphere and Eq. A.11 becomes

$$K_{\mathrm{m}}^{\mathrm{o}} = \frac{\exp\left[-\gamma L_{\mathrm{t}}\left(\dfrac{3\xi^2}{2} - \dfrac{\xi^3}{2} - \xi\right)\right]}{1 + 3\xi - \dfrac{3\xi^2}{2}}, \tag{A.12}$$

and for $\varepsilon \gg 1$ we have a "palm tree" or an umbrella-like savanna tree and Eq. A.11 gives

$$K_{\mathrm{m}}^{\mathrm{o}} = \frac{\exp[-\gamma L_{\mathrm{t}}(\xi^2 - \xi)]}{1 + 2\xi}, \tag{A.13}$$

both of which are plotted in Fig. 4.24a.

Conical multilayer crowns

Figure A.1b gives a definition sketch for the conical crown which has the volume

$$\forall_{\mathrm{co}} = \frac{\pi r^2}{3} h_{\mathrm{o}}, \tag{A.14}$$

where

$$\frac{h_{\mathrm{o}}}{h - h_{\mathrm{s}}} \equiv \xi = \frac{2r}{b}. \tag{A.15}$$

Again examining a horizontal slice of unit thickness located at $\xi = \frac{h_{\mathrm{o}}}{h - h_{\mathrm{s}}}$, the canopy foliage area density, $a_{\mathrm{t}}(\xi)$, is

$$a_{\mathrm{t}} = \frac{\pi r^2(1)\, a_{\mathrm{cr}}}{\pi\left(\dfrac{b}{2}\right)^2(1)} = 4\left(\frac{r}{b}\right)^2 a_{\mathrm{cr}} = \xi^2 a_{\mathrm{cr}}. \tag{A.16}$$

Then

$$\int_0^1 a_{\mathrm{t}}(\xi)\, \mathrm{d}\xi \equiv L_{\mathrm{t}} = \frac{a_{\mathrm{cr}}}{3}, \tag{A.17}$$

and therefore

$$a_{\mathrm{t}}(\xi) = L_{\mathrm{t}}(3\xi^2) = L_{\mathrm{t}}\, a(\xi), \tag{A.18}$$

giving the homogeneity function

$$a(\xi) = 3\xi^2. \tag{A.19}$$

Using Eq. A.19 in Eq. 4.17 gives the areal average momentum flux variation

$$\frac{\tau_f(\xi)}{\tau_f(0)} = \exp\left(-n\beta L_t \xi^3\right). \tag{A.20}$$

Choosing the interference function again as $s(\xi) = B\xi$, the velocity variation is

$$\frac{u(\xi)}{u_0} = \exp\left[-\gamma L_t\left(\xi^3 - B_0\xi\right)\right], \tag{A.21}$$

in which $B_0 \equiv \frac{B}{\gamma L_t}$. Equating the velocity gradients at $z = 0$ again gives $B_0 = -1$. From Eq. 4.111 and Fig. 4.24b, $\Psi = 1$, and with $m = 1/2$, Eq. 4.37 gives for the cone

$$K_m^o = \frac{\exp\left[\gamma L_t(\xi - \xi^3)\right]}{1 + 3\xi^2}, \tag{A.22}$$

which is also plotted in Fig. 4.24a.

Canopy-average eddy viscosity for multilayers

The vertical spatial average atmospheric eddy viscosity, $\widehat{K_m^o}$, is defined by

$$\widehat{K_m^o} \equiv \widehat{\frac{K_m(\xi)}{K_m(0)}} \equiv \int_0^1 \frac{K_m(\xi)}{K_m(0)} d\xi. \tag{A.23}$$

(1) Cylindrical crown (i.e., homogeneous canopy; $M = 1$)

Equations 4.40 and A.23 give the lumped eddy viscosity within the homogeneous (i.e., cylindrical) crown as

$$\widehat{K_m^o} \equiv \frac{1 - \exp\left[-\left(\frac{1-m}{m}\right)\gamma L_t\right]}{\left(\frac{1-m}{m}\right)\gamma L_t}. \tag{A.24}$$

(2) Hemispherical crown ($M = 1$)

Equations A.12 and A.13 do not seem to be integrable analytically, but we wish to compare values of $\widehat{K_m^o}$ for different crown shapes. Accordingly, we integrate Eqs. A.12 and A.13 numerically for selected values of γL_t and fit the result empirically with a polynomial over the practical range, $0 \le \gamma L_t \le 4$. With $mn = 1$, and $\gamma \equiv \beta$, the fitted functions are:

full hemisphere ($\varepsilon = 1$):

$$\widehat{K_m^o} = 0.55 + 0.052(\beta L_t) + 0.008(\beta L_t)^2, \tag{A.25}$$

and

limiting hemispherical segment ($\varepsilon \gg 1$):

$$\widehat{K_m^o} = 0.54 + 0.086(\beta L_t) + 0.012(\beta L_t)^2, \tag{A.26}$$

which are displayed in Fig. 4.25a where the plotted points represent the numerically integrated values.

(3) Conical crown ($M = 1$)

Equation A.22 does not seem analytically integrable either. Following the same numerical procedure

$$\widehat{K_{\mathrm{m}}^{\mathrm{o}}} = 0.605 + 0.108(\beta L_{\mathrm{t}}) + 0.042(\beta L_{\mathrm{t}})^2, \tag{A.27}$$

which is also shown in Fig. 4.25a. The values of $\widehat{K_{\mathrm{m}}^{\mathrm{o}}}$ at the common value $\beta L_{\mathrm{t}} = 1.5$ are added to Fig. 4.24a as vertical dashed lines. Once again we call attention to the quite profound qualitative difference in these functions for differing crown shape.

2 EDDY VISCOSITY FOR MONOLAYERS OF LOW FOLIAGE DENSITY ($M = 1$)

We first consider the case of monolayers having foliage volume densities low enough that flow passes readily through the crown. In this case the primary momentum absorption is internal to the crown and results from the drag of the individual foliage elements as we have seen for leafy multilayers:

Cylindrical monolayer at low foliage density

All relationships for the homogeneous cylindrical monolayer are identical to those already derived for the case of the homogeneous cylindrical multilayer. However, for the monolayer, $h - h_{\mathrm{o}} = t$.

Hemispherical monolayer at low foliage density

The hemispherical monolayer is illustrated in the definition sketch Fig. A.2a and its notation is given in Fig. A.1a. Examining a horizontal slice at depth $\xi = \frac{h_{\mathrm{o}}}{h - h_{\mathrm{s}}}$, and

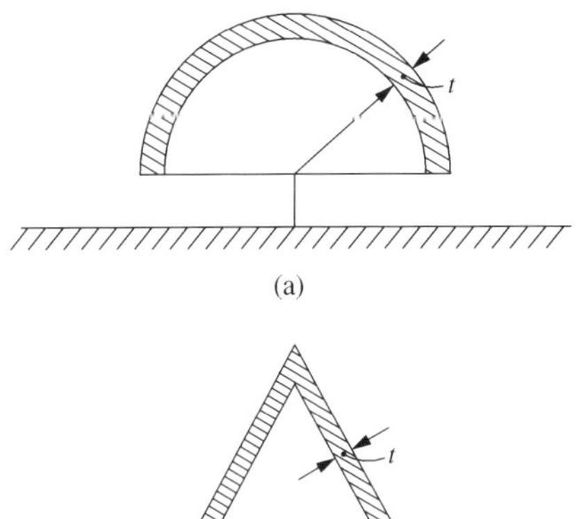

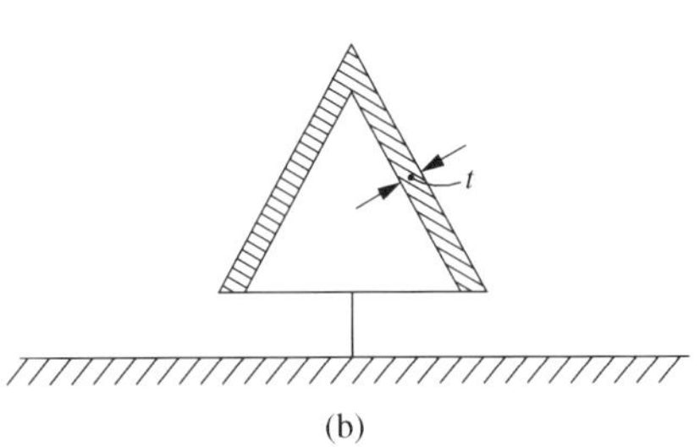

Fig. A.2. The monolayer crown. (a) Hemispherical; (b) conical.

using Eq. A.2, the foliage area density, $a_t(\xi)$, is approximately

$$a_t(\xi) \approx \frac{2\pi r t(1) a_{cr}}{\pi r_o^2(1)} = \frac{t}{r_o}(2\xi - \xi^2)^{1/2} a_{cr}. \tag{A.28}$$

From the definition of foliage area density, and using Eq. A.28

$$\int_0^1 a_t(\xi)\,d\xi \equiv L_t = \frac{t}{r_o} \cdot \frac{\pi}{4} a_{cr}, \tag{A.29}$$

which together with Eq. A.28 gives

$$a_t = L_t \left[\frac{4}{\pi}(2\xi - \xi^2)^{1/2} \right], \tag{A.30}$$

in which the homogeneity function appears as the factor

$$a(\xi) = \left[\frac{4}{\pi}(2\xi - \xi^2)^{1/2} \right]. \tag{A.31}$$

Using Eq. A.31 in Eq. 4.17 gives the momentum flux variation

$$\frac{\tau_f(\xi)}{\tau_f(0)} = \exp\left[-n\beta L_t \frac{4}{\pi} \int_0^{\xi} (2y - y^2)^{1/2}\,dy \right], \tag{A.32}$$

which can be integrated (Gradshteyn and Ryzhik, 1980, p. 82, no. 2.262-1 and p. 81, no. 2.261) to obtain

$$\frac{\tau_f(\xi)}{\tau_f(0)} = \exp\left\{ -n\beta L_t \frac{2}{\pi} \left[\frac{\pi}{2} - \arcsin(1 - \xi) - (1 - \xi)(2\xi - \xi^2)^{1/2} \right] \right\}. \tag{A.33}$$

The interference function will again be taken as in Eq. A.9, which gives the velocity distribution

$$\frac{u(\xi)}{u_0} = \exp\left\{ -\gamma L_t \frac{2}{\pi} \left[\frac{\pi}{2} - \arcsin(1 - \xi) - (1 - \xi)(2\xi - \xi^2)^{1/2} - B_0 \xi \right] \right\}, \tag{A.34}$$

in which $B_0 = \frac{mB}{\frac{2}{\pi}\gamma L_t}$. Equating velocity gradients at the canopy top and setting $\Psi = 1$ (Eq. 4.111 and Fig. 4.26b) establishes that $B_0 = -\frac{\pi}{2}$. Taking $m = 1/2$ then allows us to write

$$K_m^o = \frac{\exp\left\{ -\gamma L_t \frac{2}{\pi} \left[\frac{\pi}{2}(1 - \xi) - \arcsin(1 - \xi) - (1 - \xi)(2\xi - \xi^2)^{1/2} \right] \right\}}{1 + \frac{4}{\pi}(2\xi - \xi^2)^{1/2}}. \tag{A.35}$$

Conical monolayer at low foliage density

The definition sketch for the conical monolayer is given in Fig. A.2b with notation in Fig. A.1b, and defines $\xi \equiv \frac{h-z}{h-h_s} = \frac{d}{b}$. The foliage area density is written

$$a_t(\xi) = \frac{\pi dt\left[1+\frac{1}{4}\left(\frac{b}{h-h_s}\right)^2\right]^{1/2} a_{cr}}{\pi\left(\frac{b}{2}\right)^2} = 4\frac{t}{b}\left[1+\frac{1}{4}\left(\frac{b}{h-h_s}\right)^2\right]^{1/2} a_{cr}\xi. \tag{A.36}$$

Applying Eq. 4.15 to Eq. A.36

$$\int a_t(\xi)\,d\xi \equiv L_t = 2\frac{t}{b}\left[1+\frac{1}{4}\left(\frac{b}{h-h_s}\right)^2\right]^{1/2} a_{cr}, \tag{A.37}$$

which together with Eq. A.36 gives

$$a_t(\xi) = 2\xi L_t, \tag{A.38}$$

in which the homogeneity function appears as the factor

$$a(\xi) = 2\xi. \tag{A.39}$$

Using Eq. A.39 in Eq. 4.17 gives the momentum flux variation

$$\frac{\tau_f(\xi)}{\tau_f(0)} = \exp\left(-n\beta L_t\xi^2\right). \tag{A.40}$$

Once again the interference function is taken as $s(\xi) = B\xi$, giving the velocity distribution

$$\frac{u(\xi)}{u_0} = \exp\left[-\gamma L_t\left(\xi^2 - B_0\xi\right)\right], \tag{A.41}$$

where $B_0 = \frac{mB}{\gamma L_t}$. Equating velocity gradients at $\xi = 0$, and setting $\Psi = 1$ gives $B_0 = -1$.

With $m = 1/2$ the eddy viscosity variation is

$$K_m^o = \frac{\exp\left[-\gamma L_t(\xi^2 - \xi)\right]}{1+2\xi}, \tag{A.42}$$

which is identical to that found for the multilayer hemispherical segment with $\varepsilon \equiv \frac{r_o}{h-h_s} = \frac{r_o}{h_o} \gg 1$. Once again these monolayer eddy viscosities are compared for the typical value $\gamma L_t = 1.5$ in Fig. 4.24b.

3 EDDY VISCOSITY FOR MONOLAYERS OF HIGH FOLIAGE DENSITY ($M = 1$)

We next consider the case of monolayers of high foliage area density in non-cylindrical crowns. In this case, the high solid density of the crown surface forces the bulk of the flow to go around the crown rather than through it. The drag element then becomes the crown shape itself and our flow field becomes the canopy rather than the interior of the crown. In this case, the interference drag results from the separation between adjacent crowns at a given elevation rather than from varying separation of neighboring needles and/or shoots within the crown. This peculiar foliage area distribution is found primarily in the conifers of cold climates where the high density serves to maintain sufficient needle warmth for photosynthesis. We will confine our treatment of high densities to conical monolayers.

Conical monolayer at high foliage volume density

At the upper limit of foliage volume density, the crown becomes solid. However, since the foliage must remain photosynthetically active, its elements must lie in the surface of the cone where they have access to light. This coincidence of leaf surface and crown surface allows us to express the one-sided leaf area index in terms of the geometry of the crown (see Fig. A.1b) as

$$L = \frac{1}{2}\frac{\pi \frac{b}{2}\left[\left(\frac{b}{2}\right)^2 + (h - h_s)^2\right]^{1/2}}{\pi (b/2)^2} = \frac{1}{2}\left[1 + 4\left(\frac{h - h_s}{b}\right)^2\right]^{1/2}. \tag{A.43}$$

For $L > 1$, the error in letting $L \approx (h - h_s)/b$ is less than 10%. Continuing the approximation $L \approx L_t$, Eq. A.43 then gives

$$L_t \approx L \approx \left(\frac{h - h_s}{b}\right). \tag{A.44}$$

Making the crude assumption that the substrate shear is negligible, Eq. 4.9 is written as

$$f_D(z) = \frac{C_D \rho u^2 d \Delta z}{\pi (b/2)^2 \Delta z} = \tau_f a_D(z), \tag{A.45}$$

so that Eq. 4.8 becomes

$$\frac{d\tau_f}{\tau_f} = 4\frac{d}{b^2}\, dz. \tag{A.46}$$

Alternatively, since $\frac{d}{b} = \xi$, and $\xi \equiv \frac{h-z}{h-h}$, Eq. A.46 can be written

$$\frac{d\tau_f}{\tau_f} = -4\left(\frac{h - h_s}{b}\right)\xi\, d\xi, \tag{A.47}$$

or, with the approximation of Eq. A.44,

$$\frac{d\tau_f}{\tau_f} = -4L_t \xi\, d\xi. \tag{A.48}$$

Introducing the variable of integration, x, Eq. A.48 yields

$$\frac{\tau_f(\xi)}{\tau_f(0)} = \exp\left(-4L_t \int_0^{\xi} x\, dx\right) = \exp\left(-2L_t \xi^2\right). \tag{A.49}$$

Using the interference function, Eq. 4.87, the shear stress is again as in Eq. 4.88, and we can write the velocity variation

$$\frac{u_{\mathrm{f}}(\xi)}{u_0} = \exp\left(-2mL_{\mathrm{t}}\xi^2 + mB\xi\right). \tag{A.50}$$

Equating velocity gradients inside and outside the canopy at $\xi = 0$, and using Eq. 4.111, one obtains

$$mB = -\frac{u_*}{ku_0}\left(\frac{h-h_{\mathrm{s}}}{h-d_0}\right) \equiv -\gamma L_{\mathrm{t}}\psi. \tag{A.51}$$

Assuming Fig. 4.26b applies, $\Psi = 1$, and Eq. A.51 becomes

$$mB = -\gamma L_{\mathrm{t}} = -mn\beta L_{\mathrm{t}}. \tag{A.52}$$

With the single-sided foliage area equal to half the lateral crown surface area in this limiting case, $n = 2$, and $\beta \approx \frac{2}{\pi}$, reducing Eq. A.52 to

$$B = -\frac{2}{\pi}L_{\mathrm{t}}. \tag{A.53}$$

At the high crown shape Reynolds numbers associated with even moderate wind speeds, $m = 1/2$ (Table 7.1). Thus from Eqs. 4.37, A.49, and A.50

$$K_{\mathrm{m}}^{\mathrm{o}} = \frac{K_{\mathrm{m}}(\xi)}{K_{\mathrm{m}}(0)} = \frac{\exp\left[L_{\mathrm{t}}\left(-\xi^2 + \frac{2}{\pi}\xi\right)\right]}{1+\pi\xi}, \tag{A.54}$$

which is compared with the low density monolayers for the value $\gamma L_{\mathrm{t}} = 1.5$ in Fig. 4.24b. Note for the conical monolayer that the change in drag behavior induced by the high foliage volume density causes much less vigorous vertical mixing than was present in the low density case.

4 CANOPY AVERAGE EDDY VISCOSITY FOR MONOLAYERS

Equations A.35, A.42, and A.54 are integrated numerically to obtain the spatial average, $\widehat{K_{\mathrm{m}}^{\mathrm{o}}}$, of the eddy viscosities over the crown depth. Respectively, this yields:

Hemispherical monolayer at low foliage density

$$\widehat{K_{\mathrm{m}}^{\mathrm{o}}} = 0.51 + 0.04(\beta L_{\mathrm{t}}) + 0.0016(\beta L_{\mathrm{t}})^2, \tag{A.55}$$

Conical monolayer at low foliage density

$$\widehat{K_{\mathrm{m}}^{\mathrm{o}}} = 0.54 + 0.086(\beta L_{\mathrm{t}}) + 0.012(\beta L_{\mathrm{t}})^2, \tag{A.56}$$

which is identical to that found for the hemispherical multilayer with $\varepsilon \equiv \frac{r_{\mathrm{o}}}{h-h_{\mathrm{s}}} = \frac{r_{\mathrm{o}}}{h_{\mathrm{o}}} \gg 1$ in Eq. A.26, and

Conical monolayer at high foliage density

$$\widehat{K_m^o} = 0.454 + 0.009L_t + 0.002L_t^2. \quad (A.57)$$

Cylindrical monolayer

The monolayer cylindrical crown is just a shallow multilayer cylinder and its $\widehat{K_m^o}$ has been given as Eq. A.24.

All monolayer values of $\widehat{K_m^o}$ are presented in Fig. 4.25b as a function of βL_t where they may be compared with the multilayer values given in Fig. 4.25a.

5 RESISTANCE RATIO

Homogeneous hemispherical multilayer ($M = 1$)

Referring to Fig. A.1a, Eqs. A.1 and A.2, and with $\xi \equiv \frac{h_o}{r_o}$, the volume of the spherical segment of height h_o is

$$\forall_s = \frac{\pi}{6} h_o \left[6r_o h_o - 2h_o^2\right] = \pi r_o^3 \left(\xi^2 - \frac{\xi^3}{3}\right), \quad (A.58)$$

which can be closely approximated (see Fig. A.3) by

$$\frac{\forall_s}{\pi r_o^3} \approx 0.7\xi^{1.85}. \quad (A.59)$$

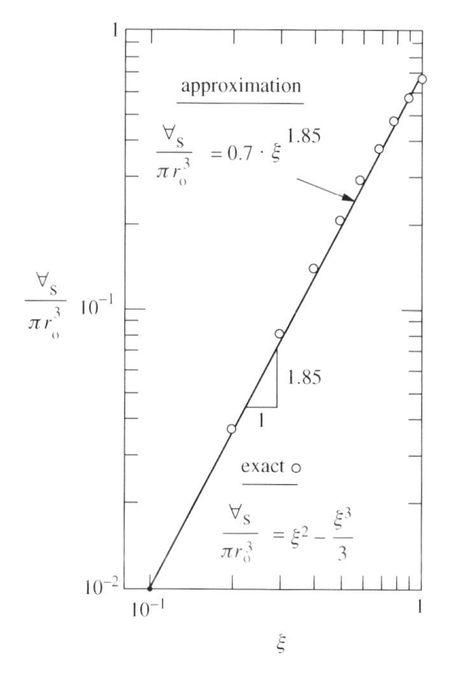

Fig. A.3. Volume of hemispherical segment.

With the hemisphere volume defined as

$$\forall_{so} = \frac{2}{3}\pi r_o^3, \quad (A.60)$$

Eq. A.59 is written

$$\frac{\forall_s}{\forall_{so}} \approx 1.05\xi^{1.85}. \quad (A.61)$$

With the foliage distributed homogeneously throughout the hemispherical crown volume, and assuming the topmost leaf layer to diffuse directly into the free atmosphere, we discretize the canopy into $L_t - 1$ atmospheric "layers" of volume

$$\Delta\forall_s = \frac{\forall_{so}}{L_t - 1}, \quad (A.62)$$

each containing an equal foliage area. The bottom of the topmost layer lies at $\xi = \xi_1$, so that Eqs. A.61 and A.62 give

$$\Delta\forall_s \equiv \forall_s(\xi_1) = 1.05\forall_{so}\xi_1^{1.85}, \quad (A.63)$$

and

$$\Delta\xi_1 \equiv \xi_1 = \left[\frac{1}{1.05}\right]^{1/1.85}\left[\frac{1}{L_t - 1}\right]^{1/1.85} = \left[\frac{0.95}{L_t - 1}\right]^{0.54} \quad (A.64)$$

is the thickness of the uppermost layer. If we now let $\forall_s(\xi_2) = 2\Delta\forall_s$, we get

$$\forall_s(\xi_2) = 1.05\forall_{so}\xi_2^{1.85} = \frac{2\forall_{so}}{L_t - 1}, \quad (A.65)$$

or

$$\xi_2 = \left[\frac{(2)(0.95)}{L_t - 1}\right]^{0.54}, \quad (A.66)$$

and the second layer down has the thickness

$$\Delta\xi_2 = \xi_2 - \xi_1 = \left[\frac{0.95}{L_t - 1}\right]^{0.54}[2^{0.54} - 1^{0.54}]. \quad (A.67)$$

Continuing in this fashion, the bottom layer has thickness

$$\Delta\xi_{L_t-1} = \xi_{L_t-1} - \xi_{L_t-2} = \left[\frac{0.95}{L_t - 1}\right]^{0.54}\left[(L_t - 1)^{0.54} - (L_t - 2)^{0.54}\right]. \quad (A.68)$$

Using Eqs. 4.38 and 4.40, Eq. 4.165 gives the interleaf layer resistance as

$$r_{i,n} = \frac{h - h_s}{ku_*(h - d_0)}\Delta\xi_n/\widehat{K_m^o}, \quad (A.69)$$

and using Eq. 4.111, Eq. A.69 is

$$r_{i,n} = \frac{\gamma L_t \psi u_0}{u_*^2}\Delta\xi_n/\widehat{K_m^o}. \quad (A.70)$$

As before for the cylindrical crown, the $r_{i.n}$ are arrayed in parallel series to give the reciprocal of the canopy resistance as

$$\frac{1}{r_c} = \frac{1}{r_{i.1}} + \frac{1}{r_{i.1} + r_{i.2}} + \cdots + \frac{1}{r_{i.1} + r_{i.2} + r_{i.3} + \cdots + r_{i.L_t-1}}, \tag{A.71}$$

but in this case the $r_{i.n}$ are unequal due to the varying thickness of the atmospheric layers containing equal foliage area. Using Eq. A.67, Eq. A.71 becomes

$$\frac{1}{r_c} = \frac{u_*^2 \widehat{K_m^o}}{\gamma L_t \psi u_0} \left[\frac{L_t - 1}{0.95}\right]^{0.54} \times \left\{\frac{1}{1^{0.54}} + \frac{1}{1^{0.54} + (2^{0.54} - 1^{0.54})} + \cdots + \frac{1}{(L_t - 1)^{0.54}}\right\}, \tag{A.72}$$

or, with Eq. 4.143

$$\frac{r_a}{r_c} = \left(\frac{u_*}{ku_0}\right)^2 \frac{\widehat{K_m^o}}{\gamma L_t \psi} \left[\frac{L_t - 1}{0.95}\right]^{0.54} \sum_{n=1}^{L_t-1} \frac{1}{n^{0.54}}. \tag{A.73}$$

From Fig. 4.26b we see that $\psi \approx 1$, and that $(\frac{u_*}{ku_0}) \approx 1$ for a homogeneous hemisphere. Also for these leafy trees, $mn = 1$, so Eq. A.73 reduces to the desired resistance ratio

$$\frac{r_c}{r_a} = 0.97 \frac{\beta L_t}{\widehat{K_m^o}(L_t - 1)^{0.54}} \left(\sum_{n=1}^{L_t-1} \frac{1}{n^{0.54}}\right)^{-1}, \tag{A.74}$$

in which $\widehat{K^o}$ is given by Eq. A.25. Equation A.74 is plotted in Fig. 7.5 for the common range of β, and we notice that there is an important difference between the resistance ratio for the hemispherical and cylindrical crowns.

Homogeneous conical multilayer

Referring to Fig. A.1b, the volume of the conical crown is

$$\forall_{co} = \frac{1}{3} \cdot \frac{\pi}{4} b^2 (h - h_s). \tag{A.75}$$

With the foliage distributed homogeneously throughout this volume, and continuing to assume the topmost leaf layer to diffuse directly into the free atmosphere, we discretize the canopy into $L_t - 1$ atmospheric "layers" each containing an equal foliage volume of magnitude

$$\Delta\forall_{co} = \frac{\forall_{co}}{L_t - 1} = \frac{\pi b^2}{12(L_t - 1)}(h - h_s). \tag{A.76}$$

The crown volume above level z is

$$\forall_{cr}(z) = \frac{\pi d^2}{12}(h - z). \tag{A.77}$$

Using the definition

$$\xi \equiv \frac{h - z}{h - h} = \frac{d}{b}, \tag{A.78}$$

Eq. A.77 becomes

$$\forall_{cr}(\xi) = \frac{\pi b^2}{12}(h - z)\xi^2 = \forall_{co}\xi^3. \tag{A.79}$$

As we did earlier for the hemisphere, the thickness, $\Delta\xi_1$, of the topmost atmospheric layer follows from Eqs. A.76 and A.79 as

$$\Delta\xi_1 = \xi_1 = \left[\frac{\forall_c(\xi_1)}{\forall_{co}}\right]^{1/3} = \left[\frac{1}{L_t - 1}\right]^{1/3}, \tag{A.80}$$

while that of the second atmospheric layer is

$$\Delta\xi_2 = \xi_2 - \xi_1 = \left[\frac{2\forall_c(\xi_1)}{\forall_{co}}\right]^{1/3} - \left[\frac{\forall_c(\xi_1)}{\forall_{co}}\right]^{1/3} = \left[\frac{1}{L_t - 1}\right]^{1/3}[2^{1/3} - 1^{1/3}], \tag{A.81}$$

and so on until the bottom atmospheric layer is

$$\Delta\xi_{L_t-1} = \xi_{L_t-1} - \xi_{L_t-2} = \left[\frac{(L_t - 1)\forall_c(\xi_1)}{\forall_{co}}\right]^{1/3} - \left[\frac{(L_t - 2)\forall_c(\xi_1)}{\forall_{co}}\right]^{1/3}, \tag{A.82}$$

or,

$$\Delta\xi_{L_t-1} = \left[\frac{1}{L_t - 1}\right]^{1/3}\left[(L_t - 1)^{1/3} - (L_t - 2)^{1/3}\right]. \tag{A.83}$$

Using Eqs. 4.38 and 4.40, Eq. 4.165 gives the interleaf atmospheric resistance, $r_{i,n}$, as shown for the hemisphere in Eq. A.70. The $r_{i,n}$ are again arrayed in parallel series to arrive at the reciprocal of the canopy resistance as given by Eq. A.71. The $r_{i,n}$ are unequal due to the varying thickness of the atmospheric layers containing equal foliage area. Using Eqs. A.80–A.83, Eq. A.71 becomes

$$\frac{1}{r_c} = \frac{u_*^2\widehat{K_m^o}(L_t - 1)^{1/3}}{\gamma L_t \psi u_0} \times \left\{\frac{1}{1^{1/3}} + \frac{1}{1^{1/3} + (2^{1/3} - 1^{1/3})} + \cdots + \frac{1}{(L_t - 1)^{1/3}}\right\}, \tag{A.84}$$

or, with Eq. 4.143

$$\frac{r_a}{r_c} = \left(\frac{u_*}{ku_0}\right)^2 \frac{\widehat{K_m^o}(L_t - 1)^{1/3}}{\gamma L_t \psi} \cdot \sum_{n=1}^{L_t-1}\frac{1}{n^{1/3}}. \tag{A.85}$$

From Fig. 4.30b we see that $\psi = 1$, and that $(\frac{u_*}{ku_0}) \approx 0.9$ for a conical spruce. Also, for these leafy trees, $mn = 1$, so Eq. A.85 reduces to the desired resistance ratio

$$\frac{r_c}{r_a} = 1.23\frac{\beta L_t}{\widehat{K_m^o}(L_t - 1)^{1/3}} \cdot \left(\sum_{n=1}^{L_t-1}\frac{1}{n^{1/3}}\right)^{-1}, \tag{A.86}$$

in which $\widehat{K_{\mathrm{m}}^{\mathrm{o}}}$ is given by Eq. A.27. Equation A.86 is plotted in Fig. 7.5 for the common range of β, and we notice there the same important difference between the resistance ratio for the conical and cylindrical crowns that we previously found between the hemispherical and cylindrical crowns.

Conical monolayer of low foliage density

Referring to the definition sketch for conical monolayers in Fig. 7.6a, we have

$$\xi \equiv \frac{h_o}{h - h_s} = \frac{d}{b}, \tag{A.87}$$

and the foliage is contained within a layer of thickness, t, parallel to the cone's lateral surface. The fraction of the cone within which the foliage is distributed (homogeneously) is thus

$$\forall_{\mathrm{cm}} = \frac{\pi b^2 t}{4}\left[1 + 4\left(\frac{h - h_s}{b}\right)^2\right]^{1/2} = A_{\mathrm{cm}} t, \tag{A.88}$$

in which the surface area, A_{cm}, of the monolayer is

$$A_{\mathrm{cm}} = \frac{\pi b^2}{4}\left[1 + 4\left(\frac{h - h_s}{b}\right)^2\right]^{1/2}. \tag{A.89}$$

Once again we assume that the topmost leaf layer diffuses directly into the free atmosphere, and we discretize the canopy into $L_t - 1$ atmospheric layers each containing an equal foliage volume of

$$\Delta\forall_{\mathrm{cm}} = \frac{\pi b^2 t}{4} \frac{\left[1 + 4\left(\frac{h - h_s}{b}\right)^2\right]^{1/2}}{L_t - 1}. \tag{A.90}$$

The crown volume above level z is

$$\forall_{\mathrm{cm}}(\xi) = \frac{\pi d b t}{4}\left[1 + 4\left(\frac{h - h_s}{b}\right)^2\right]^{1/2} \xi. \tag{A.91}$$

The bottom of the topmost layer lies at $\xi = \xi_1$, so that Eqs. A.90 and A.91 give

$$\Delta\forall_{\mathrm{cm}} = \forall_{\mathrm{cm}}(\xi_1) = \frac{\pi d_1 b t}{4}\left[1 + 4\left(\frac{h - h_s}{b}\right)^2\right]^{1/2} \xi_1, \tag{A.92}$$

and

$$\Delta\xi_1 \equiv \xi_1 = \left[\frac{1}{L_t - 1}\right]^{1/2} \tag{A.93}$$

is the thickness of the uppermost layer. The second layer down from the top has $\forall_{\mathrm{cm}}(\xi_2) = 2\Delta\forall_{\mathrm{cm}}$, so that Eqs. A.92 and A.93 give

$$\frac{\pi d_2 b t}{4}\left[1 + 4\left(\frac{h - h_s}{b}\right)^2\right]^{1/2} \xi_2 = 2\frac{\pi d_1 b t}{4}\left[1 + 4\left(\frac{h - h_s}{b}\right)^2\right]^{1/2} \xi_1, \tag{A.94}$$

or

$$\xi_2 = \left(\frac{2}{L_t - 1}\right)^{1/2}, \tag{A.95}$$

so that the second layer down has the thickness

$$\Delta(\xi_2) = \xi_2 - \xi_1 = \left(\frac{1}{L_t - 1}\right)^{1/2} (2^{1/2} - 1^{1/2}). \tag{A.96}$$

Continuing in this way, the bottom layer has thickness

$$\Delta\xi_{L_t-1} = \xi_{L_t-1} - \xi_{L_t-2} = \left[\frac{1}{L_t - 1}\right]^{1/2} \left[(L_t - 1)^{1/2} - (L_t - 2)^{1/2}\right]. \tag{A.97}$$

Using Eqs. 4.38 and 4.40, Eq. 4.165 gives the interleaf atmospheric resistance, $r_{i,n}$, as is shown for the hemispherical multilayer in Eq. A.70. The unequal $r_{i,n}$ are again arrayed in parallel series to arrive at the reciprocal of the canopy resistance as given by Eq. A.71. Using Eqs. A.93 to A.97, Eq. A.71 becomes

$$\frac{1}{r_c} = \frac{u_*^2 \widehat{K_m^o} (L_t - 1)^{1/2}}{\gamma L_t \psi u_0} \times \left\{ \frac{1}{1^{1/2}} + \frac{1}{1^{1/2} + (2^{1/2} - 1^{1/2})} + \cdots + \frac{1}{(L_t - 1)^{1/2}} \right\}, \tag{A.98}$$

or, with Eq. 4.143

$$\frac{r_a}{r_c} = \left(\frac{u_*}{ku_0}\right)^2 \frac{\widehat{K_m^o}(L_t - 1)^{1/2}}{\gamma L_t \psi} \cdot \sum_{n=1}^{L_t-1} \frac{1}{n^{1/2}}. \tag{A.99}$$

Again using Fig. 4.26b, we see that $\psi = 1$, and $(\frac{u_*}{ku_0}) \approx 0.9$ for a conical spruce. Also, for these leafy trees, $mn = 1$, so Eq. A.99 reduces to the desired resistance ratio

$$\left(\frac{r_c}{r_a}\right)_s = 1.23 \frac{\beta L_t}{\left(\widehat{K_m^o}\right)_s (L_t - 1)^{1/2} N(L_t)}, \tag{A.100}$$

in which

$$N(L_t) \equiv \left(\sum_{n=1}^{L_t-1} \frac{1}{n^{1/2}}\right). \tag{A.101}$$

In Eq. A.100, the subscript "s" on the resistance ratio and on the eddy viscosity signifies they are for the case of "sparse monolayer" in which the wind blows freely among the leaves or needles. Evaluating the diffusivity through Eq. A.56, Eq. A.100 is plotted as the lower dashed line in Fig. 7.7 for the common value, $\beta = 0.5$. We notice there that $\left(\frac{r_c}{r_a}\right)_s$ declines monotonically with L_t just as it does for the conical and hemispherical multilayers (see Fig. 7.5).

Conical monolayer of limiting foliage density

As the foliage volume density approaches unity, that is as the crown becomes solid, flow through the crown ceases and the drag becomes that of the solid crown shape; all the active foliage elements lie in the surface of the crown, and the (active) foliage area index is defined by the crown geometry. Spruce is a typical example. In this case turbulent diffusion to and from the foliage does not pass through other interleaf layers. Rather, it passes through the intercrown atmosphere over a distance which averages, for the distributed elements of the conical surface

$$\overline{\Delta z} = \frac{2}{3}(h - h_s). \tag{A.102}$$

From Eq. 4.165, neglecting r_{cs}

$$r_c = r_{ci} = r_i = \frac{\overline{\Delta z}}{\widehat{K_m(z)}}. \tag{A.103}$$

Using Eqs. 4.38 and A.102, Eq. A.103 becomes

$$r_c = \frac{\frac{2}{3}(h - h_s)}{ku_*(h - d_o)\widehat{K^o_m}}. \tag{A.104}$$

Equations 4.143 and A.104 give the desired resistance ratio as

$$\left(\frac{r_c}{r_a}\right)_d = \frac{2}{3}\left(\frac{ku_o}{u_*}\right)\left(\frac{h - h_s}{h - d_o}\right)\frac{1}{\left(\widehat{K^o_m}\right)_d}, \tag{A.105}$$

where the subscript "d" on the resistance ratio and on the eddy viscosity signifies they are for the limiting case of "dense monolayer" in which the wind blows around the solid crown. In the absence of observations of the bulk drag parameter for the case of solid cones, we assume $\frac{u_*}{ku_o} = 1$ (see Fig. 4.26b), and use Eq. 4.51 to write Eq. A.105

$$\left(\frac{r_c}{r_a}\right)_d = \frac{2}{3}\left[\frac{n\beta L_t}{1 - \frac{n\beta L_t \exp(-n\beta L_t)}{1 - \exp(-n\beta L_t)}}\right]\frac{1}{\left(\widehat{K^o_m}\right)_d}. \tag{A.106}$$

For the solid cone, $n = 2, \beta = \frac{2}{\pi}$, and the diffusivity is given by Eq. A.57. Equation A.106 is plotted as the upper dashed line in Fig. 7.7 for the common value, $\beta = 0.5$. We see there that the sparse and dense conical monolayers have opposing variations with L_t. All else being constant, Eq. A.100 must apply for very small L_t, while Eq. A.106 applies for very large L_t. A weighting function to define r_c/r_a for the practical, intermediate case is given in Chapter 7 as Eq. 7.24.

B

Estimation of potential evaporation from wet simple surfaces[†]

1 ANNUAL AVERAGE FOR WEAKLY SEASONAL CLIMATES

The Penman equation

In Chapter 5 we wrote the Penman equation for potential rate of evaporation from wet surfaces, E_{ps}, in the form (cf. Eq. 5.21)

$$\lambda E_{ps} = \frac{1}{1 + \gamma_o/\Delta}(R_n + D_p), \tag{B.1}$$

in which the drying power, D_p, is defined to be (cf. Eq. 5.22)

$$D_p \equiv \frac{\rho c_p}{r_a}\frac{e_s}{\Delta}(1 - S_r), \tag{B.2}$$

and S_r is the atmospheric saturation ratio (i.e., fractional relative humidity) at screen height.

Net radiation

The net radiation, R_n, is defined as the difference

$$R_n \equiv q_i - q_b, \tag{B.3}$$

in which the *net* incoming solar radiation, q_i, is given by

$$q_i \equiv (1 - \rho_T)I_0, \tag{B.4}$$

in which ρ_T is the surface albedo (all wavelengths).

[†] The material of this Appendix is adapted from the author's unpublished lecture notes for the Tenth Annual John R. Freeman Memorial Lecture entitled "Climate, Soil, and the Water Balance: A Framework for their Analytical Coupling", MIT, Cambridge, MA, 1977.

To estimate the *net* longwave back radiation, q_b, we begin[†] by considering the longwave incoming radiation from a clear sky,

$$I'_{lw} = \varepsilon_a \sigma T_{0K}^4, \tag{B.5}$$

in which ε_a is the atmospheric emissivity, and T_{0K} is the near-surface atmospheric temperature in degrees Kelvin. Many empirical expressions have been developed to reflect the dependence of ε_a on vapor pressure. We choose that of Swinbank (1964) which includes the dependence of vapor pressure on temperature to give

$$I'_{lw} = 0.971 \times 10^{-10} T_{0K}^4 - 0.245, \quad (\text{cal cm}^{-2}\ \text{min}^{-1}), \tag{B.6}$$

where T_{0K} is measured at screen height above the vegetal surface.

Water surfaces will reflect some longwave radiation which we accommodate through the longwave reflectivity, ρ_l. The *net* incoming longwave radiation is then

$$I^*_{lw} = (1 - \rho_l) I'_{lw}. \tag{B.7}$$

The outgoing or "back" longwave radiation is

$$I_{lw} = \varepsilon_0 \sigma T_{sK}^4, \tag{B.8}$$

in which E_s is the surface emissivity ($\varepsilon_0 = 0.97$ for water and $\varepsilon_0 = 1$ for other surfaces), and T_{sK} is the surface temperature in degrees Kelvin.

The *net* longwave back radiation *under clear skies*, R_c, is then

$$R_c = I_{lw} - I^*_{lw}. \tag{B.9}$$

Finally, the effect of cloud cover on the longwave radiation exchange between atmosphere and surface is approximated by writing the net longwave back radiation

$$q_b = (1 - k_l N) R_c, \tag{B.10}$$

which becomes, using Eqs. B.9, B.8, B.7, and B.6, in cal cm^{-2} min^{-1}

$$q_b = (1 - k_l N)\left[\varepsilon_0 \sigma T_{sK}^4 - (1 - \rho_l)\left(0.971 \times 10^{-10} T_{0K}^4 - 0.245\right)\right], \tag{B.11}$$

in which

N = fraction of the sky covered by clouds,
k_l = reduction factor < 1,
ρ_l = longwave albedo of the surface,
ε_0 = longwave emissivity of the surface, and
σ = Stefan–Boltzmann constant $= 0.826 \times 10^{-10}$ cal cm^{-2} min^{-1} K^{-4}.

Here, we apply Eq. B.1 using time-averaged variables to estimate the annual average evaporation rate. However, because of the non-linearity of Eq. B.1 with respect to

[†] This development of net longwave back radiation follows that of *Snow Hydrology* (U.S. Army Corps of Engineers, 1956, pp. 156–161).

temperature, the estimation degenerates with increasing climatic seasonality. Consequently, we restrict this approach to climates having weak seasonality and make the following additional approximations:

(1) $k_l = 0.80$ (U.S. Army Corps of Engineers, 1956),
(2) $\rho_l = 0$,
(3) $T_0 = T_s$ (assumed isothermal evaporation),
(4) $\overline{T^4} \approx \overline{T}^4$ (overbars indicate long-term time average), and
(5) $\varepsilon_0 \approx 1.0$.

Introducing the above and remembering that all quantities are now *time-averaged*, Eq. B.11 reduces to

$$q_b = (1 - 0.80\,N)\left(0.245 - 0.145 \times 10^{-10} T_{0K}^4\right), \tag{B.12}$$

in cal cm^{-2} min^{-1}, and is in terms of readily available meteorological data.

The drying power

Estimation of D_p is problematic due to the need to evaluate the atmospheric resistance, r_a, which is a function of both wind speed and surface roughness (cf. Eq. 4.141). It is useful to divide Eqs. B.12 and B.2 to obtain the dimensionless ratio

$$\frac{q_b}{D_p} = \frac{(1 - 0.80\,N)\left(0.245 - 0.145 \times 10^{-10} T_{0K}^4\right)}{\frac{c_p}{r_a}\frac{e_s}{\Delta}(1 - S_r)}, \tag{B.13}$$

which can be written

$$\frac{q_b}{D_p} = \left(\frac{(1 - 0.80\,N)}{c_p/r_a}\right)\left[\frac{\left(0.245 - 0.145 \times 10^{-10} T_{0K}^4\right)}{e_s/\Delta}\right]\left(\frac{1}{1 - S_r}\right). \tag{B.14}$$

As is shown in Fig. B.1, the saturation vapor pressure, e_s, can be approximated by

$$\frac{e_s}{T_s} = 6989 - 51.87\,T_{sK} + 0.0966\,T_{sK}^2, \quad (\text{dyne cm}^{-2}\,\text{K}^{-1}) \tag{B.15}$$

over the range $0\,^\circ\text{C} \leq T_s \leq 34\,^\circ\text{C}$, from which

$$\frac{e_s}{\Delta} \equiv \frac{e_s}{de_s/dT_{sK}} = \frac{T_{sK}\left(6989 - 51.87\,T_{sK} + 0.0966\,T_{sK}^2\right)}{6989 - 103.74\,T_{sK} + 0.29\,T_{sK}^2}, \quad (\text{K}). \tag{B.16}$$

Assuming $T_s = T_0$, the second factor, $F_2 = \frac{0.245 - 0.145 \times 0^{-10}\,T_{0K}^4}{e_s/\Delta}$, of Eq. B.14 will vary by less than 25% over the nominal range of T_0. However, the dimensionless third factor,

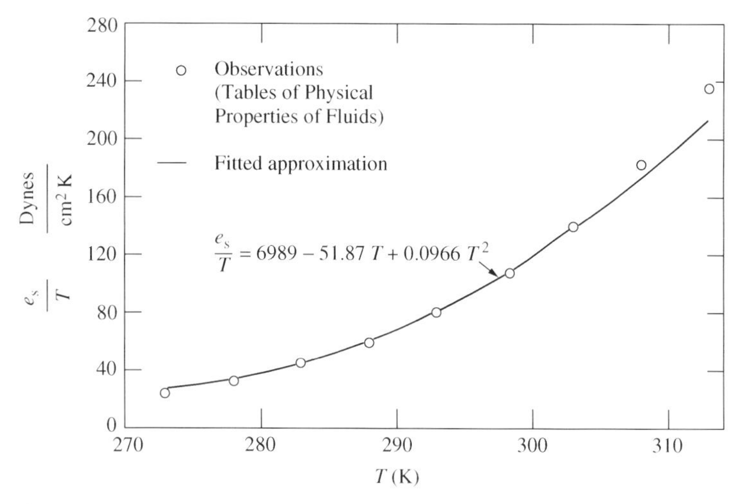

Fig. B.1. Saturated vapor pressure vs. temperature.

$F_3 \equiv \frac{1}{1-S_r}$, will vary by 150% over the accompanying range of relative humidity. The first factor of Eq. B.14 will be randomly variable according to cloud cover, wind speed, and surface roughness. We thus expect, for given r_a and N, that

$$\frac{q_b}{D_p} \approx f(S_r). \tag{B.17}$$

We now explore Eq. B.17 empirically.

Van Bavel (1966, p. 456) has pointed out that

> Potential evaporation can be defined for any situation in terms of the appropriate meteorological variables and the radiative and aerodynamic properties of the surface. When the surface is wet and imposes no restriction upon the flow of water vapor, the potential value is reached.

Accordingly, we use the readily available observations of "free water surface" evaporation[†] (cf. Farnsworth *et al.*, 1982) in developing our estimation procedure for simple surfaces such as bare soil and leaves.

Jensen and Haise (1963, Table 3, pp. 30–31) tabulated the observed mean monthly values of insolation, I_0, in many different climates as taken from weekly U.S. Weather Bureau records. The resulting annual averages are plotted in Fig B.2 where the least-squares linear fit to the observations is given by the dashed line. For comparison, the annual average clear sky insolation is also shown as calculated by Budyko (1958). From Jensen and Haise (1963) we select those sites for which the presence of ice for a significant fraction of the year may be neglected, and we estimate annual water

[†] "Free water surface" evaporation is that which occurs from a thin film of water having insignificant heat storage and thus it closely represents the potential evaporation from adequately watered "simple" natural surfaces such as individual leaves and bare soil (Farnsworth *et al.*, 1982, p. 4).

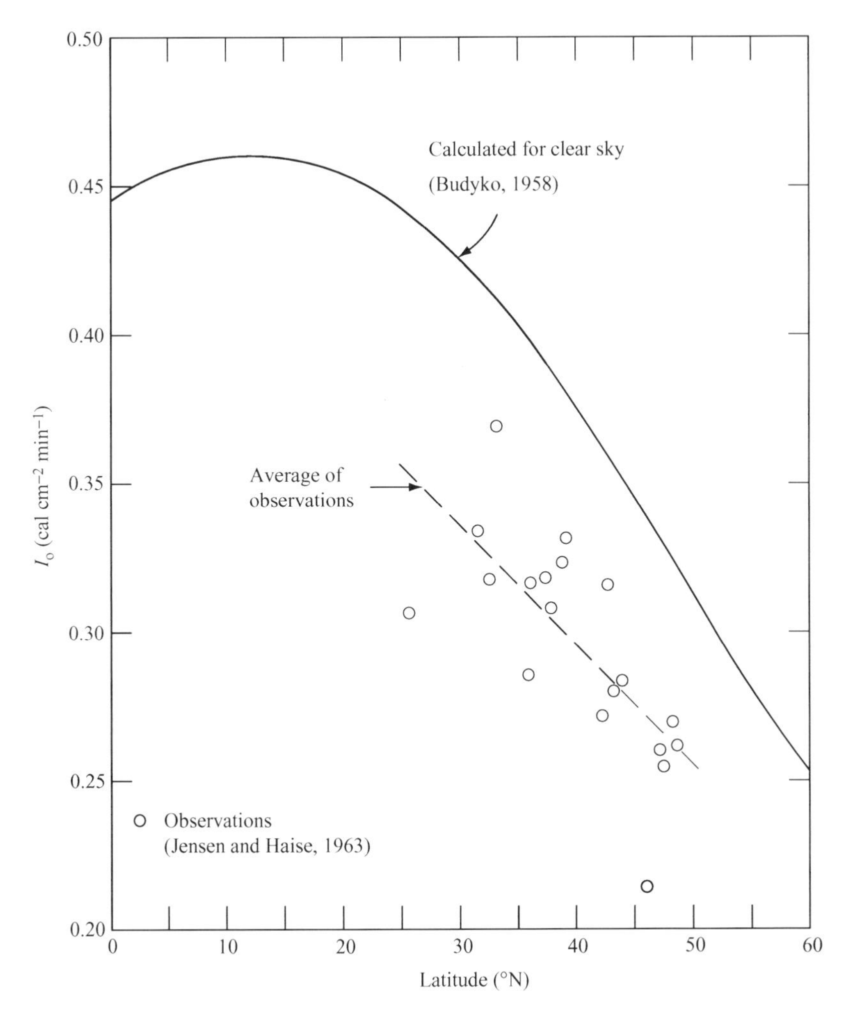

Fig. B.2. Annual average surface insolation.

surface evaporation at these sites from the contour map of annual "lake" evaporation† (presented by Kohler *et al.*, 1959, Plate 2). For these water surfaces we assume an annual average reflectance, $\rho_T = 0.05$ (U.S. Army Corps of Engineers, 1956). The associated meteorological variables, N, S_r, and T_o at the chosen sites are averaged for the year from the values given by the U.S. National Weather Service in their Annual Climatological Summary using the station closest to the site. Finally, the parameter γ_0/Δ is given as a function of T_0 in Table 5.1.

With these assembled data (listed in Table B.1), Eq. B.1 is solved for D_p using the observed annual evaporation at each of the nine sites, and the relationship of q_b/D_p

† The term "lake" evaporation has been used by Kohler *et al.* (1959) with the same meaning as "free water surface" evaporation (cf. Farnsworth *et al.*, 1982, p. 4).

Table B.1. *Estimation of potential evaporation for water surfaces (negligible ice)*

Site location	Φ (°N)	I_0[a] (ly min^{-1})	ρ_r[b]	T_0[c] (°C)	N[c]	u_2[c] (m s^{-1})	S_r[c]	E_{ps}[d] (observed) (m y^{-1})	q_i[e] (ly min^{-1})	q_b[f] (ly min^{-1})	R_n[g] (ly min^{-1})	$1+\frac{\gamma_0}{\Delta}$[h]	D_p[i] (ly min^{-1})	E_{ps}[k] (calculated) (m y^{-1})	Reference
1 Phoenix, AZ	33.4	0.371	0.05	21.3	0.20	2.7	0.43	1.82	0.352	0.114	0.238	1.44	0.061	1.92	Jensen and Haise (1963)
2 Davis, CA	38.4	0.310	0.05	15.7	0.30	3.9	0.52	1.27	0.295	0.109	0.186	1.59	0.045	1.36	Jensen and Haise (1963)
3 Fresno, CA	36.8	0.317	0.05	16.8	0.25	2.8	0.43	1.52	0.301	0.114	0.187	1.56	0.084	1.48	Jensen and Haise (1963)
4 Stillwater, OK	36.1	0.286	0.05	15.6	0.40	5.5	0.50	1.45	0.272	0.104	0.168	1.59	0.095	1.26	Jensen and Haise (1963)
5 Brownsville, TX	25.9	0.307	0.05	23.2	0.40	5.2	0.77	1.47	0.292	0.090	0.202	1.40	0.033	1.42	Jensen and Haise (1963)
6 Fort Worth, TX	32.8	0.320	0.05	18.6	0.40	4.8	0.63	1.45	0.304	0.095	0.209	1.50	0.039	1.46	Jensen and Haise (1963)
7 Midland, TX	31.9	0.334	0.05	17.7	0.25	4.5	0.50	1.86	0.317	0.113	0.204	1.53	0.121	1.55	Jensen and Haise (1963)
8 Spokane, WA	47.6	0.257	0.05	8.5	0.40	3.9	0.42	0.98	0.244	0.105	0.139	1.89	0.072	0.98	Jensen and Haise (1963)
9 Lake Cochituate, MA	42.5	0.286	0.05	10.7	0.33	4.9	0.70	0.59	0.200	0.111	0.089	1.78	0.031	0.63	Hamon (1961)

[a] Jensen and Haise (1963).
[b] Eagleson (1970, Fig. 3-10, p. 37).
[c] U.S. Weather Service Annual Climatological Summary, 1941–70 (at nearest station).
[d] Kohler *et al.* (1959, Plate 2).
[e] $q_i = (1 - \rho_r)I_0$. [f] Eq. B.6. [g] $R_n = q_i - q_b$. [h] Table 5.1. [i] Eq. B.1. [k] Eqs. B.1 and B.12.

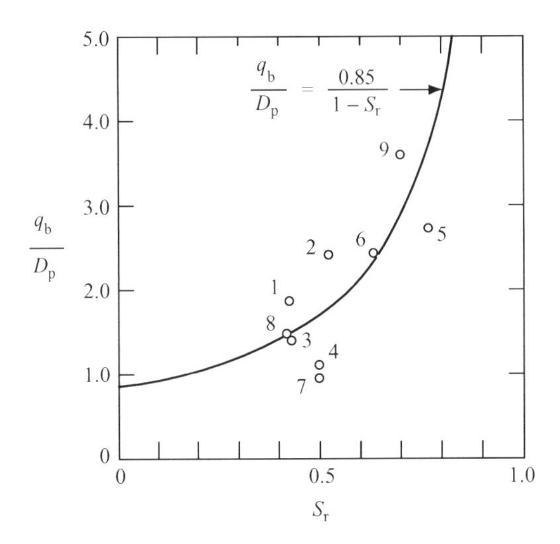

Fig. B.3. Empirical estimation of "drying power". Site numbers refer to Table B.1.

to S_r is plotted in Fig. B.3. In agreement with our prior reasoning (refer to Eq. B.17), we see that over the included range of variables their functional relationship may be approximated empirically by

$$\frac{q_b}{D_p} = \frac{0.85}{1 - S_r}. \tag{B.18}$$

Using Eqs. B.3, B.4, B.12, and B.18 in Eq. B.1 gives finally

$$\lambda E_{ps} = \frac{1}{1 + \gamma_o/\Delta}\left\{(1 - \rho_T)I_0 - q_b\left[1 - \frac{(1 - S_r)}{0.85}\right]\right\}. \tag{B.19}$$

Goodness of fit

Because of the secondary importance of D_p in Eq. B.1, the errors inherent in Eq. B.18 (due to our omission of dependencies on wind speed and surface roughness) are reduced when it is used in Eq. B.19 to estimate E_{ps}. This is shown in Fig. B.4 where E_{ps} as given by Eqs. B.1 and B.18 is compared with the observed value for each of the nine water surfaces of Table B.1. Equations B.19 and B.12 provide the estimates of E_{ps} in this work.

2 GROWING SEASON AVERAGE FOR STRONGLY SEASONAL CLIMATES[†]

In strongly seasonal climates we can use the approach of the previous section by dividing the year into two or more discrete seasons in each of which the climatic variables

[†] This material is for use in Chapters 6, 7, and 9.

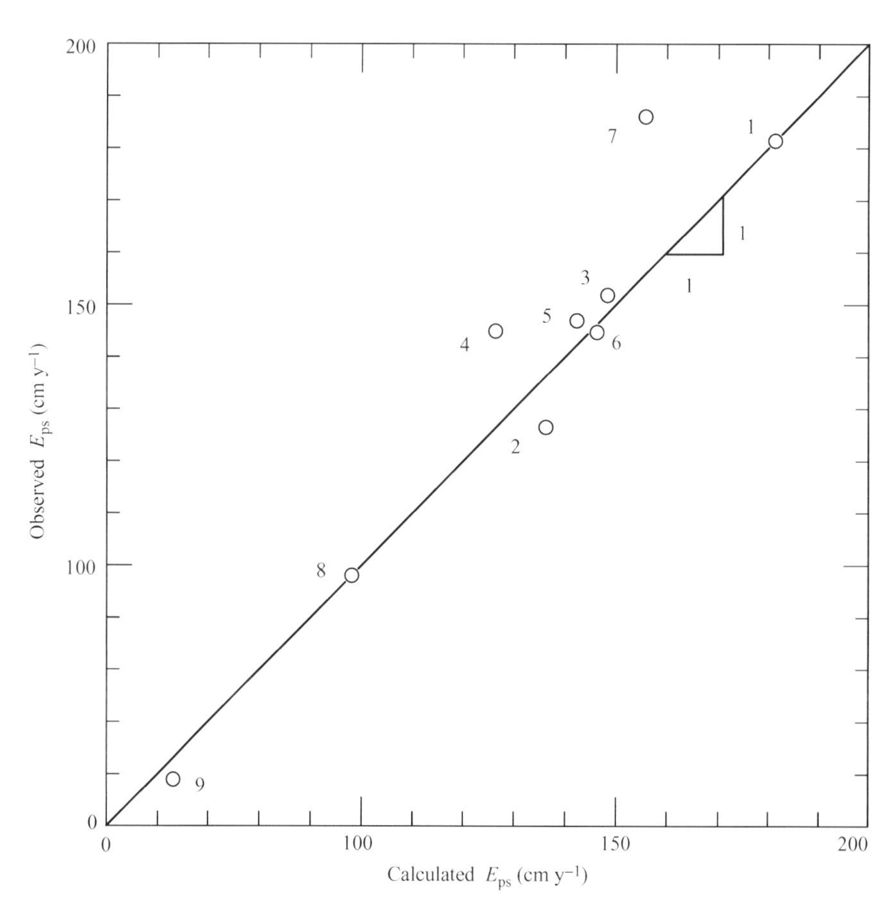

Fig. B.4. Free water surface evaporation: goodness of fit. Site numbers refer to Table B.1.

are assumed to be only weakly variable. In this work we use potential evaporation as a base for estimating annual transpiration, so we are concerned with climatic variables which are time averaged over only the growing season.†

In Table B.2 we present the data for 18 forested watersheds within the Beaver Creek basin of Arizona studied by Baker (1982, 1986), Salvucci (1992), and others. The atmospheric temperatures listed are the average for the growing season, and along with the other average variables are used with Eq. B.19 to calculate the potential free water surface evaporation, $E_{ps\tau}$, during the growing season, τ.

Cooley (1970) presents the cumulative percentage of annual water surface evaporation as a function of Julian day as observed in Arizona, and Salvucci (1992) fitted this with an ogive curve which may be approximated well (except for its cold-weather tails)

† Estimation of the growing season is considered briefly in Chapter 3.

Table B.2. *Estimation of free water surface potential evaporation for Beaver Creek watersheds, Arizona*

	Growing season													Dormant season
Watershed number	S_r[a]	N[b]	T_0[a,b,c] (°C)	Start[d] (Julian day)	Stop[d] (Julian day)	I_0[e,f] (ly min^{-1})	ρ_T[g]	$1+\frac{\gamma_0}{\Delta}$[h]	q_b[i] (ly min^{-1})	R_n[k] (ly min^{-1})	D_p[l] (ly min^{-1})	$E_{ps\tau}$[m] (cm d^{-1})	% Total[h]	E_{psd}[o] (cm d^{-1})
1	0.51	0.43	18.5	130	284	0.315	0.18	1.51	0.093	0.165	0.054	0.35	65	0.14
2	0.50	0.43	19.3	126	287	0.315	0.18	1.48	0.092	0.166	0.054	0.35	68	0.13
3	0.50	0.43	17.7	125	287	0.315	0.18	1.53	0.093	0.165	0.055	0.34	68	0.13
4	0.55	0.43	16.4	141	278	0.315	0.18	1.57	0.095	0.163	0.050	0.33	59	0.14
5	0.55	0.43	15.8	144	276	0.315	0.18	1.58	0.095	0.163	0.050	0.32	56	0.14
6	0.56	0.43	15.5	144	276	0.315	0.18	1.60	0.095	0.163	0.049	0.32	56	0.14
7–18	0.58	0.43	14.2	151	271	0.315	0.18	1.64	0.097	0.161	0.048	0.31	65	0.08

[a] Baker (1982).
[b] NOAA data for Flagstaff, AZ, as reported by Salvucci (1992, Table B.2).
[c] Baker (1986).
[d] Interpolated from Williams and Anderson (1967).
[e] Φ estimated from topographic maps.
[f] Fig. B.2, average of observations at $\Phi = 35°$ N.
[g] Kondratyev (1969, Table 7.5, p. 425): assumed 50% grass @ $\rho_T = 0.22$, and 50% forest @ $\rho_T = 0.14$.
[h] Table 5.1. [i] Eq. B.6. [k] Eq. B.3. [l] Eq. B.18. [m] $(\tau/365)$ Eq. B.19. [n] Eq. B.20.
[o] $E_{psd} = \frac{\tau E_{ps\tau}}{365-\tau}\left(\frac{1}{\%\ \mathrm{tot}} - 1\right)$.

by the linear relation

$$\sum \%E_{ps} = \begin{cases} 0, & 0 \leq t_J \leq 65, \\ 100\left(\frac{t_J - 65}{300 - 65}\right), & 65 \leq t_J \leq 300, \\ 100, & 300 \leq t_J \leq 365, \end{cases} \tag{B.20}$$

in which t_J = Julian day. Using Eq. B.20 and the tabulated season length, we find $E_{ps\tau}$ as a percentage of the total annual E_{ps}, and from this we calculate the free water surface potential evaporation, E_{psd}, during the dormant season. Listed in Table B.2, these values are useful in the water balance calculations of Chapters 6, 7, and 9.

C

Water balance equations

1 AVERAGE INTERSTORM BARE SOIL EXFILTRATION

We take the expected value of the integral, over the full range of the randomly variable time between storms, of the bare soil evaporation rate, e_s, the smaller of E_{ps} and f_e^* in Fig. 6.6. We simplify the notation somewhat by using the shorthand, E_p, to represent the potential transpiration rate, E_{ps}, for bare soil. The result is

$$
\begin{aligned}
E[E_{sj}] \approx \frac{E_p}{\alpha}\Bigg\{ & \frac{\gamma[\kappa_o, \lambda_o h_o]}{\Gamma(\kappa_o)} - \left[1 + \frac{\alpha h_o/E_p}{\lambda_o h_o}\right]^{-\kappa_o} \frac{\gamma[\kappa_o, \lambda_o h_o + \alpha h_o/E_p]}{\Gamma(\kappa_o)} \mathrm{e}^{-BE} \\
& + \left\{1 - \frac{\gamma[\kappa_o, \lambda h_o]}{\Gamma(\kappa_o)}\right\} \cdot \Bigg\{1 - \mathrm{e}^{-BE-\alpha h_o/E_p}\left[1 + Mk_v^* + (2B)^{1/2}E - w/E_p\right] \\
& + \mathrm{e}^{-CE-\alpha h_o/E_p}\left[Mk_v^* + (2C)^{1/2}E - \frac{w}{E_p}\right] + (2E)^{1/2}\mathrm{e}^{-\alpha h_o}\Bigg[\gamma\left(\frac{3}{2}, CE\right) \\
& - \gamma\left(\frac{3}{2}, BE\right)\Bigg]\Bigg\} + \left[1 + \frac{\alpha h_o/E_p}{\lambda_o h_o}\right]^{-\kappa_o} \frac{\gamma[\kappa_o, \lambda_o h_o + \alpha h_o/E_p]}{\Gamma(\kappa_o)} \Bigg\{(2E)^{1/2} \\
& \times \left[\gamma\left(\frac{3}{2}, CE\right) - \gamma\left(\frac{3}{2}, BE\right)\right] + \mathrm{e}^{-CE}\left[Mk_v^* + (2C)^{1/2}E - w/E_p\right] \\
& - \mathrm{e}^{-BE}\left[Mk_v^* + (2B)^{1/2}E - w/E_p\right]\Bigg\}\Bigg\},
\end{aligned}
\tag{C.1}
$$

where the dimensionless coefficients B and C are defined by

$$
B \equiv \frac{1 - M}{1 + Mk_v^* - w/E_p} + \frac{M^2 k_v^* + (1 - M)w/E_p}{2(1 + Mk_v^* - w/E_p)^2}, \tag{C.2}
$$

and

$$C \equiv \frac{1}{2}(Mk_v^* - w/E_p)^{-2}. \tag{C.3}$$

In the above:

α = reciprocal of mean time, m_{t_b}, between storms (s^{-1}),
$\gamma[\,]$ = incomplete gamma function (dimensionless),
$\Gamma[\,]$ = gamma function (dimensionless),
κ_o = shape parameter of gamma distribution of storm depth (dimensionless),
λ_o = scale parameter of gamma distribution of storm depth (cm^{-1}),
$\psi[1]$ = saturated soil matrix potential (cm suction),
ϕ_e = dimensionless exfiltration diffusivity,
d = soil diffusivity index (dimensionless),
h_o = surface retention capacity (cm),
m = soil pore size distribution index (dimensionless),
n_e = soil effective porosity (dimensionless),
s_o = effective soil moisture concentration (dimensionless),
$E_p \equiv E_{ps}$ = bare soil potential evaporation rate ($cm\,s^{-1}$),
k_v^* = unstressed canopy conductance $\equiv E_{pv}/E_{ps}$ (dimensionless),
M = vegetated surface fraction (dimensionless),
w = rate of capillary rise ($cm\,s^{-1}$),
E = bare soil evaporation effectiveness (dimensionless), and
$K(1)$ = saturated hydraulic conductivity ($cm\,s^{-1}$).

Defining the *bare soil evaporation efficiency*, β_s, in the manner of Eagleson (1978d),

$$\beta_s = \frac{E[E_{sj}]}{\alpha^{-1}E_{ps}}. \tag{C.4}$$

Again letting $E_p \equiv E_{ps}$ and using Eq. C.1, Eq. C.4 becomes

$$\begin{aligned}
\beta_s \approx\ & \frac{\gamma(\kappa_o, \lambda_o h_o)}{\Gamma(\kappa_o)} - \left[1 + \frac{\alpha h_o/E_p}{\lambda_o h_o}\right]^{-\kappa_o} \frac{\gamma[\kappa_o, (\lambda_o h_o + \alpha h_o/E_p)]}{\Gamma(\kappa_o)}\, e^{-BE} \\
& + \left[1 - \frac{\gamma(\kappa_o, \lambda_o h_o)}{\Gamma(\kappa_o)}\right]\left\{1 - e^{-BE-(\alpha h_o/E_p)}\left[1 + Mk_v^* + E(2B)^{1/2} - \frac{w}{E_p}\right]\right. \\
& + e^{-CE-(\alpha h_o/E_p)} \cdot \left[Mk_v^* + E(2C)^{1/2} - \frac{w}{E_p}\right] + (2E)^{1/2} e^{-\alpha h_o/E_p} \\
& \left.\times \left[\gamma\left(\frac{3}{2}, CE\right) - \gamma\left(\frac{3}{2}, BE\right)\right]\right\} + \left[1 + \frac{\alpha h_o/E_p}{\lambda_o h_o}\right]^{-\kappa_o} \frac{\gamma[\kappa_o, (\lambda_o h_o + \alpha h_o/E_p)]}{\Gamma(\kappa_o)} \\
& \times \left\{(2E)^{1/2}\left[\gamma\left(\frac{3}{2}, CE\right) - \gamma\left(\frac{3}{2}, BE\right)\right] + e^{-CE}\left[Mk_v^* + E(2C)^{1/2} - \frac{w}{E_p}\right]\right. \\
& \left. - e^{-BE}\left[Mk_v^* + E(2B)^{1/2} - \frac{w}{E_p}\right]\right\}.
\end{aligned} \tag{C.5}$$

2 MOISTURE-CONSTRAINED TRANSPIRATION

The time to stress is

$$t_s/2 = t^* + \frac{1}{\alpha}\left\{\frac{n_e s_o z_r \alpha}{M k_v^* E_{ps}} + \frac{wE}{k_v^* E_{ps}}\left(\frac{1}{M} - 1\right)(B - C) + CE(1 - M) + \frac{E}{k_v^*}\left(\frac{1}{M} - 1\right)\left[(2B)^{1/2} - (2C)^{1/2}\right] + BE\left(M - 1 - \frac{1}{Mk_v^*} + \frac{1}{k_v^*}\right)\right\}, \tag{C.6}$$

in which B and C are defined above and E is given in Eq. 6.38.

3 AVERAGE INTERSTORM STRESSED TRANSPIRATION

Maintaining the shorthand, $E_p \equiv E_{ps}$, the expected value of the integral of the rate of stressed interstorm transpiration is

$$E[E_{vj}] = \int_0^{h_o} f(h)\,dh\left\{k_v^* E_p \int_0^{t_s'+h/E_p} t_b f(t_b)\,dt_b + \int_{t_s'+h/E_p}^{\infty}\left[t_s' + \frac{h}{E_p}\right] f(t_b)\,dt_b\right\} + k_v^* E_p \int_{h_o}^{\infty} f(h)\,dh\left\{\int_0^{t_s'=h_o/E_p} t_b f(t_b)\,dt_b + \int_{t_s'+h_o/E_p}^{\infty}\left[t_s' + \frac{h_o}{E_p}\right] f(t_b)\,dt_b\right\}, \tag{C.7}$$

which integrates to

$$E[E_{vj}] = \frac{k_v^* E_p}{\alpha}\left\{\frac{\gamma[\kappa_o, \lambda h_o]}{\Gamma(\kappa_o)} - e^{-\alpha t_s'}\left(1 + \frac{\alpha}{E_p}\right)^{-\kappa_o} \cdot \frac{\gamma[\kappa_o, (\lambda + \alpha/E_p)h_o]}{\Gamma(\kappa_o)} + \left[1 - e^{-\alpha(t_s'+h_o/E_p)}\right]\left[1 - \frac{\gamma(\kappa_o, \lambda h_o)}{\Gamma(\kappa_o)}\right]\right\}. \tag{C.8}$$

4 APPROXIMATE SOLUTIONS OF THE WATER BALANCE EQUATION

To begin we divide the behavioral range of s_o into two distinct regions separated by the intersection of the two limiting asymptotes:

Arid climates (soil control, $s_o \to 0$)

When the climate–soil parameter, E (cf. Eq. 6.38), satisfies $E \le 2/\pi$, the bare soil evaporation is under control of the soil properties. We then let that evaporation be given by the arid climate asymptote of Eq. 6.40 which has the form (cf. Fig. 6.8)

$$\beta_s = \left[\frac{\pi E}{2}\right]^{1/2} = E_o s_o^{(d+2)/2}, \tag{C.9}$$

in which

$$E_o \equiv \left[\frac{\alpha n_e K(1)\psi(1)\phi_e}{m E_{ps}^2}\right]^{1/2}. \tag{C.10}$$

Because this asymptote of β_s has the value $\beta_s = 1$ at $E = \frac{2}{\pi}$, we can use Eq. C.9 to define the upper limit of soil-controlled equilibrium soil moisture as

$$s_o = E_o^{-2/(d+2)}. \tag{C.11}$$

We now use Eqs. 6.52 and C.9 in Eq. 6.54 to get an approximation to the water balance under soil-controlled conditions:

$$\begin{aligned} 1 - e^{-G-2\sigma^{3/2}} - \frac{\overline{h_o}}{m_h} + \frac{\Delta S}{m_\nu m_h} + \frac{m_\tau K(1)}{m_{P_\tau}}\left[1 + \frac{3/2}{mc-1}\right]\left[\frac{\psi(1)}{z_w}\right]^{mc} \\ = \frac{m_{t_b} E_{ps}}{m_h}\left[(1-M)E_o s_o^{1+d/2} + Mk_v^* \beta_v\right] + \frac{m_\tau K(1)}{m_{P_\tau}} s_o^c \end{aligned} \tag{C.12}$$

in which the left-hand side represents moisture supply, S_W, to the root zone, and the right-hand side represents the exhaustion of that moisture by bare soil evaporation, transpiration, and percolation.

As $s_o \to 0$, $G \to G(0)$, and $\sigma \to \sigma(0)$. The data of Table 6.3 show that for s_o less than about $s_o = 0.25$, the last (percolation) term on the right-hand side of Eq. C.12 vanishes. Under these dry condition and with a deep, homogeneous soil, we expect that climatic depression of the water table, z_W, will make the capillary rise term vanish as well. Finally, since $m_{t_b} \gg m_{t_r}$ in dry climates we have

$$m_{P_\tau}/m_h \equiv m_\tau/\left(m_{t_b} + m_{t_r}\right) \approx m_\tau/m_{t_b}, \tag{C.13}$$

and Eq. C.12 can be reduced in dry climates to

$$1 - e^{-G(0)-2\sigma(0)^{3/2}} - \frac{\overline{h_o}}{m_h} + \frac{\Delta S}{m_\nu m_h} = \frac{m_{t_b} E_{ps}}{m_h}\left[(1-M)E_o s_o^{1+d/2} + Mk_v^* \beta_v\right], \tag{C.14}$$

in which

$$G(0) \equiv \omega K(1)/2, \tag{C.15}$$

and

$$\sigma(0) \equiv \left[\frac{5 n_e \lambda^2 K(1)\psi(1)\phi_i(d,0)}{6\pi \delta m \kappa_o^2}\right]^{1/3}. \tag{C.16}$$

Equation C.14 gives a *dry climate approximation* for the soil moisture state variable:

$$s_o^{1+d/2} = \frac{1}{(1-M)E_o}\left\{\frac{m_h}{m_{t_b} E_{ps}}\left[1 - e^{-G(0)-2\sigma(0)^{3/2}} - \frac{\overline{h_o}}{m_h} + \frac{\Delta S}{m_\nu m_h}\right] - Mk_v^* \beta_v\right\}. \tag{C.17}$$

Wet climates (climate control, $s_o \to 1$)
With plentiful soil moisture, β_v and β_s both approach unity, and Eq. 6.54 becomes

$$1 - e^{-G-2\sigma^{3/2}} - \frac{\overline{h_o}}{m_h} + \frac{\Delta S}{m_\nu m_h} + \frac{m_\tau K(1)}{m_{P_\tau}}\left[1 + \frac{3/2}{mc-1}\right]\left[\frac{\psi(1)}{z_w}\right]^{mc}$$
$$= \frac{m_{t_b} E_{ps}}{m_h}\left[(1-M) + Mk_v^* + \frac{K(1)}{E_{ps}} s_o^c\right]. \quad (C.18)$$

Equations. C.14 and C.18 are identical when the bare soil evaporation asymptotes meet as specified by Eq. C.11. For the catchments of Table 6.3 at least, this will occur at $s_o \approx 0.75$, at and above which the final bracketed term on the right-hand side of Eq. C.18 (i.e., the percolation) dominates that side of the water balance. Further assuming negligible capillary rise and $s_o \to 1$, we have $G \to G(1)$, and $M \to 1$. For $m_{t_r}/m_{t_b} \ll 1$, and since $G(1) \equiv 2G(0)$, Eq. C.18 yields a *wet climate approximation* for the soil moisture state variable:

$$s_o^c \approx \frac{m_h}{m_{t_b} K(1)}\left[1 - e^{-2G(0)} - \frac{\overline{h_o}}{m_h} + \frac{\Delta S}{m_\nu m_h}\right] + \left[1 + \frac{3/2}{mc-1}\right]\left[\frac{\psi(1)}{z_w}\right]^{mc} - k_v^*. \quad (C.19)$$

Intermediate moisture states
For intermediate moisture states meeting the condition s_o less than about $s_o = 0.75$, the water balance is approximately quadratic in $s_* \equiv s_o^{1+d/2}$. We see this by first noting that to be quadratic, the relationship between the exponents of s_o in Eq. C.12 must be

$$c = 2\left[\frac{d+2}{2}\right], \quad (C.20)$$

while through soil physics, Brooks and Corey (1966) relate these parameters according to $c = 2d - 1$ (cf. Eq. 6.34).

Equations 6.34 and C.20 are equal at $c = 5$, and are plotted in Fig. C.1 where dashed horizontal lines represent the nominal values of c for the classic soil types as given in Table 6.2. We note from this plot that the water balance equation does indeed approximate a quadratic in $s_* \equiv s_o^{1+d/2}$ for silty loam soils provided the above constraint on s_o is met and provided the surface runoff parameters have the fixed values, $G(s_o) = G$, and $\sigma(s_o) = \sigma$. With those approximations, the only positive solution of Eq. C.12 is

$$s_* \equiv s_o^{1+d/2} = \frac{-B_* + [B_*^2 - 4A_*C_*]^{1/2}}{2A_*}, \quad (C.21)$$

where

$$A_* \equiv \frac{K(1)}{E_{ps}}, \quad (C.22)$$

$$B_* \equiv (1-M)E_o, \quad (C.23)$$

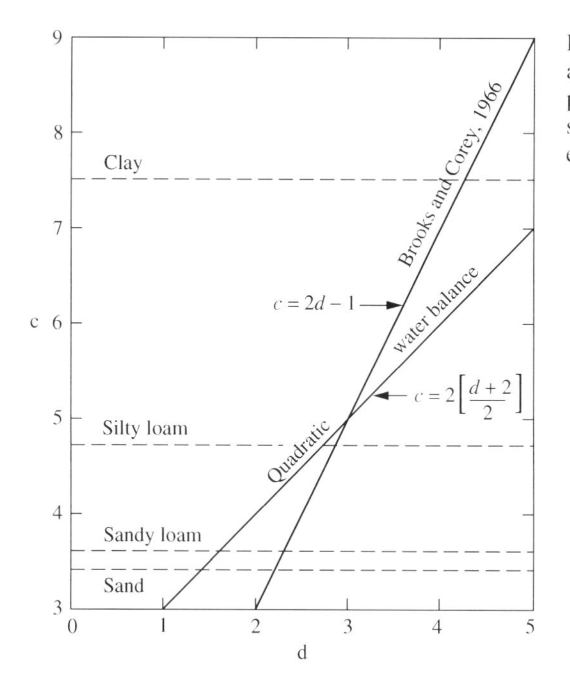

Fig. C.1. Quadratic approximation of soil properties for analytical solution of water balance equations.

in which E_o is given by Eq. C.10, and

$$C_* \equiv M k_v^* \beta_v - \frac{m_h}{m_{t_b} E_{ps}}\left[1 - e^{-G-2\sigma^{3/2}} - \frac{\overline{h_o}}{m_h} + \frac{\Delta S}{m_v m_h}\right] + \frac{K(1)}{E_{ps}}\left[1 + \frac{3/2}{mc-1}\right]\left[\frac{\psi(1)}{z_w}\right]^{mc}. \tag{C.24}$$

What constant values of G and σ should we use in this analytical determination of the equilibrium soil moisture? Using their dry soil approximation when $M = 0$, and their wet soil approximation when $M = 1$, we suggest for intermediate cases a linear interpolation for the surface runoff component of Eq. C.24 in the form

$$e^{-G-2\sigma^{3/2}} \approx (1 - M)\, e^{-G(0)-2\sigma(0)^{3/2}} + M\, e^{-2G(0)}. \tag{C.25}$$

5 ONE-DIMENSIONAL DIFFUSION OF SOIL MOISTURE

The diffusion equation

Assuming constant diffusivity, D, the one-dimensional diffusion equation is written (Crank, 1956, p. 26)

$$\frac{\partial s}{\partial t} = D\frac{\partial^2 s}{\partial z^2}, \tag{C.26}$$

in which t is time and z represents distance from the surface into the diffusive medium. Note that z is not associated with the vertical as there is no gravitational effect

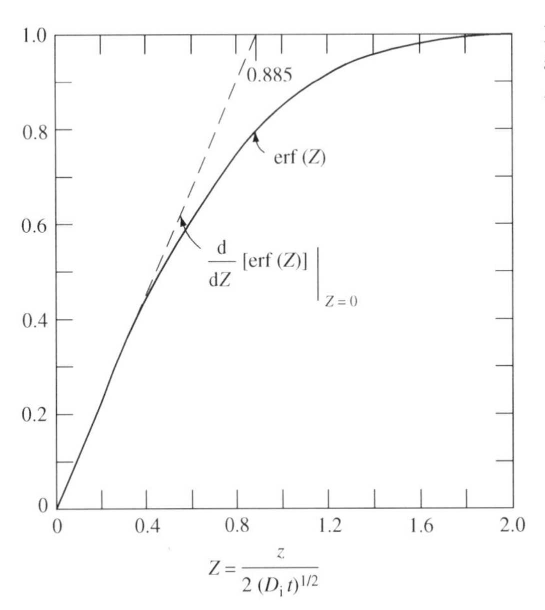

Fig. C.2. The error function and its slope. (From Abramowitz and Stegun, 1964, Table 7.1, p. 310.)

considered here. We will solve Eq. C.26 under the initial and boundary conditions

$$\begin{Bmatrix} s = 1;\ z = 0, t > 0 \\ s = s_5;\ z > 0, t = 0 \end{Bmatrix}. \tag{C.27}$$

The solution is (Crank, 1956, p. 30, Eq. 3.13)

$$\frac{s-1}{s_5-1} = \text{erf}\left[\frac{z}{2(Dt)^{1/2}}\right] \equiv \text{erf}[Z], \tag{C.28}$$

which is plotted in Fig. C.2 using tabulated values from Abramowitz and Stegun (1964, Table 7.1, p. 310). The asymptote of this function evaluated at $Z = 0$ is also given in the above-referenced table and is plotted as the dashed line on Fig. C.2. The intersection of the asymptotes of erf $[Z]$ locates the (dimensionless) *characteristic penetration depth* of this diffusion as $Z = 0.885$.

D

Characterization of exponential decay

Consider the exponential decay function

$$A = A_0\,\mathrm{e}^{-kt/\tau}, \tag{D.1}$$

in which k is a dimensionless decay parameter, and τ is a normalizing parameter carrying the same units as t. Since $A/A_0 \to 0$ as $t \to \infty$, the choice of τ, which should be physically significant, is somewhat arbitrary. The convention in physics utilizes the average, $\frac{\bar{t}}{\tau}$, where

$$\frac{\bar{t}}{\tau} = \int_0^\infty \mathrm{e}^{-kt/\tau}\,\mathrm{d}\left(\frac{t}{\tau}\right) = \frac{1}{k}, \tag{D.2}$$

at which point Eq. D.1 gives

$$\frac{A}{A_0} = \mathrm{e}^{-1}, \tag{D.3}$$

commonly called the "e-folding" value of the dimensionless dependent variable, A/A_0.

Differentiating Eq. D.1 with respect to t/τ gives

$$\frac{\mathrm{d}(A/A_0)}{\mathrm{d}(t/\tau)} = -k\,\mathrm{e}^{-kt/\tau}, \tag{D.4}$$

which gives the slope of the function at $\frac{t}{\tau} = 0$ as

$$\frac{\Delta(t/\tau)}{\Delta(A/A_0)} = \frac{1}{k}. \tag{D.5}$$

The tangent to A/A_0 at the origin thus intersects the horizontal axis at the average of the independent variable as we see in the sketch of Fig. D.1. This is the intersection

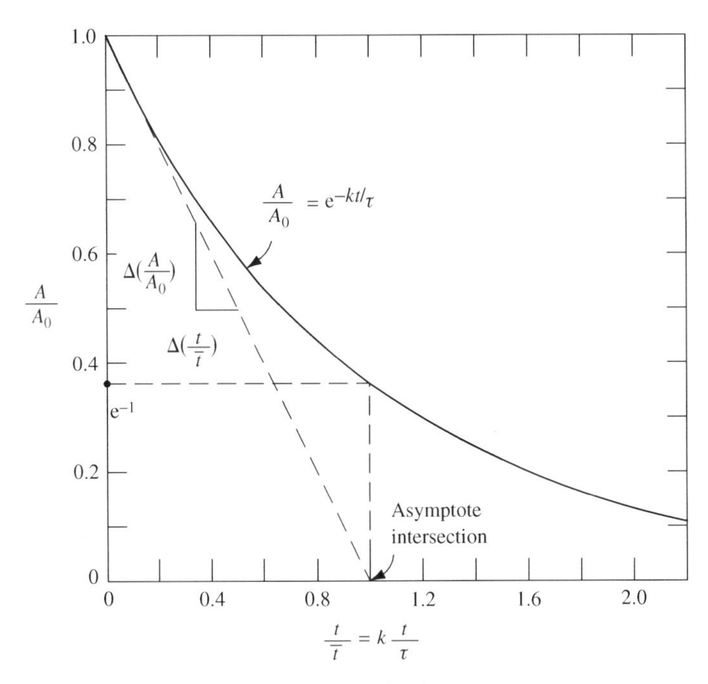

Fig. D.1. Characterization of exponential decay.

of the asymptotes of the exponential function and provides the physical justification for using this point to characterize such a function.

E

Transpiration as a productivity surrogate

The canopy

In Appendix B we used the energy equation to derive the potential evaporation from a wet simple surface in Eq. B.19. In terms of average rates over the growing season, this is

$$\lambda E_{ps} = \frac{1}{1+\gamma_o/\Delta}\left\{(1-\rho_T)I_0 - q_b\left[1-\frac{(1-S_r)}{0.85}\right]\right\}. \tag{E.1}$$

We must carefully bookkeep the dimensions: $\lambda \approx 585$ cal g_w^{-1}, the environmental insolation, I_0, is in cal cm^{-2} per unit time, and the potential evaporation, E_{ps}, is in cm per unit time.

For the forest communities of Table 9.2 we find that

$$\frac{q_b\left[1-\dfrac{(1-S_r)}{0.85}\right]}{\lambda E_{ps}[1+\gamma_o/\Delta]} \leq 0.13, \tag{E.2}$$

and therefore, to the first approximation, the back radiation from the wet surface may be neglected and Eq. E.1 reduces to

$$I_0 = \frac{\lambda(1+\gamma_o/\Delta)}{1-\rho_T}E_{ps}. \tag{E.3}$$

At the critical moisture condition, $\beta_v = 1$, and from Eq. 6.46

$$E_{ps} = \frac{E_v}{k_v^*}. \tag{E.4}$$

Equations E.3 and E.4 give, *for a horizontal unit of vegetated surface*,

$$I_0 = \left(\frac{1 + \gamma_0/\Delta}{1 - \rho_T}\right)\frac{\lambda E_v}{k_v^*}. \tag{E.5}$$

As long as carbon utilization by the forest is limited by the light-driven demand (cf. Chapter 10), we have from Eq. 10.20 at $\beta = \kappa$,

$$\text{NPP} = g(T_l)P_{sm}Mm_\tau f_D(\beta L_t), \tag{E.6}$$

in which

$g(T_l) = 1$ under optimum conditions (cf. Fig. 8.4),
$M =$ canopy cover (i.e., vegetated fraction),
$m_\tau =$ growing season length in h y^{-1},
$P_{sm} =$ maximum light-saturated photosynthetic rate for an isolated leaf in units of g_s m^{-2} (basal leaf area) h^{-1},
NPP $=$ net primary productivity in g_s m^{-2} (horizontal area) y^{-1}.

Written in terms of the canopy-average insolation, and with $g(T_l) = 1$, P_{sm} is given by Eqs. 8.9 and 9.3 as

$$P_{sm} = \varepsilon f_I(\beta L_t)I_0, \tag{E.7}$$

where

$\varepsilon =$ (g_s of dry biomass produced)/(MJ of intercepted radiant energy)
$= 0.81$ g_s MJ(tot)$^{-1}$, and
$I_0 =$ environmental insolation in MJ(tot) m^{-2} (basal canopy area) h^{-1}

The functions $f_I(\beta L_t)$ and $f_D(\beta L_t)$ are defined in Eqs. 8.30 and 10.21 respectively. Finally, using Eqs. E.5 and E.7, Eq. E.6 is written in dimensionless form:

$$\frac{\text{NPP}}{\varepsilon\lambda m_\tau M E_v} = F = \frac{1}{k_v^*} f_I(\beta L_t) f_D(\beta L_t)\left(\frac{1 + \gamma_0/\Delta}{1 - \rho_T}\right). \tag{E.8}$$

in which

$\lambda =$ latent heat of vaporization $= 597.3$ cal g^{-1} (at 0 °C).

In Eq. E.8 the product, $m_\tau MF$, is a constant for a given climate and species.

An isolated leaf

Again using Eqs. E.5 and E.7, the productivity, P_s, of an isolated horizontal leaf can be written

$$P_s = g(T_l)P_{sm} = g(T_l)\frac{(1+\gamma_o/\Delta)}{1-\rho_T}\varepsilon\lambda E_v, \tag{E.9}$$

where for the single leaf, $f_I(\beta L_t) \equiv 1$, $r_c/r_a \equiv 0$, and therefore $k_v^* \equiv 1$. This demonstrates the linear dependence of single-leaf productivity on the transpiration rate in which the proportionality constant is a function primarily of the leaf temperature.

Comparison with canopy observations

In order to compare Eq. E.8 with the behavior of natural canopies it is desirable to have observations of the two principal variables, NPP and E_v. Rosenzweig (1968) presents a collection of NPP observations from a variety of communities and while E_v, the transpiration rate for a unit vegetated area, is not measured at these sites (as it never is for lack of appropriate sensors) he estimates the values of actual evapotranspiration, E_T, using the methods of Thornthwaite and Mather (1957). We use pairs of Rosenzweig's "observations" covering the full moist–arid climatic range as the plotted points in Fig. E.1. The data for this comparison are given in Table E.1.

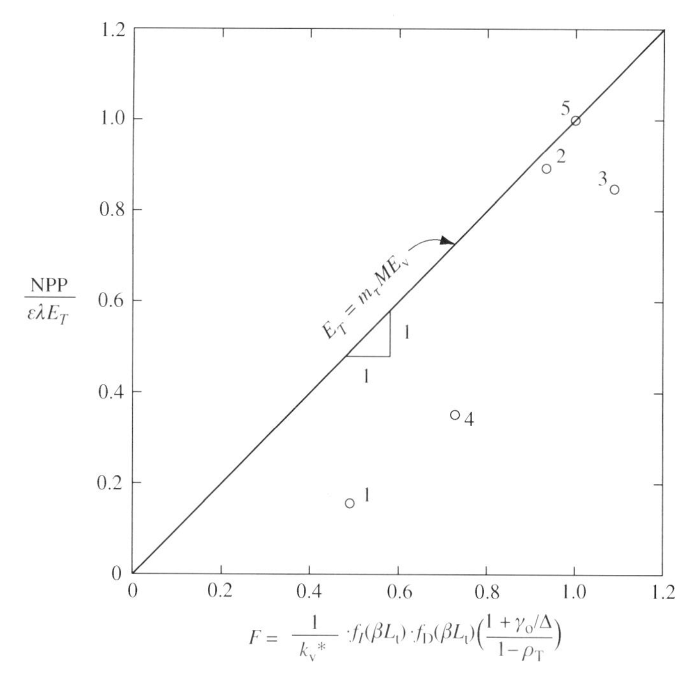

Fig. E.1. Test of NPP vs. E_v linearity. 1, Creosote bush; 2, tropical forest; 3, chestnut–oak; 4, loblolly pine; 5, beech.

Table E.1. *Annual NPP as a linear function of transpiration rate*

Number	Vegetation type	NPP (g_s m^{-2} y^{-1})	E_T (mm y^{-1})	$\frac{NPP}{\varepsilon \lambda E_T}$ dimensionless	$\beta L_t{}^a$	$f_I{}^a$	$f_D{}^b$	$T_0{}^a$	$\frac{\gamma_0}{\Delta}{}^c$	$\rho_T{}^a$	$\frac{1+\frac{\gamma_0}{\Delta}}{1-\rho_T}$	$k_v^*{}^d$	M^a	m_τ dimensionless	F^e	References for NPP and E_T
1	Creosote bush	40[7]	126[7]	0.16	0.37	0.84	0.22	21	0.44	0.35	2.22	0.84	0.09	0.08	0.49	Odum (1959)
2	Tropical forest	2188[8]	1230[8]	0.89	2.61	0.35	1.44	32	0.25	0.25	1.67	0.88	1.00	0.50	0.84	Nye (1961)
3	Chestnut–oak	1400[9]	832[9]	0.85	2.60	0.36	1.44	31.5	0.26	0.25	1.68	0.88	0.95	0.25	0.99	Whittaker (1966)
4	Loblolly pine	380[10]	525[10]	0.36	2.58	0.36	1.42	27	0.32	0.25	1.76	0.83	0.80	0.11	0.73	Whittaker (1963)
5	Beech	977[11]	5[11]	1.00	3.19	0.30	1.68	27	0.32	0.25	1.76	0.88	0.93	0.28	1.00	Whittaker (1966)

[a] Table 9.2.
[b] Eq. 10.21 with $\beta = \kappa$ (cf. Fig. 10.1).
[c] Table 5.1.
[d] Fig. 7.9.
[e] Eq. E.8.

Note that as the climate gets more moist, the annual transpiration from a unit of horizontal area, $m_\tau M E_v$, grows in magnitude approaching the annual evapotranspiration, E_T, in the limit. We plot $\frac{\text{NPP}}{\varepsilon \lambda E_T}$[†] versus the right-hand side of Eq. E.8 in Fig. E.1 where Eq. E.1 is represented by the 45° limit-line along which $E_T = m_\tau M E_v$. Since $E_T > m_\tau M E_v$ always, we expect the "observations" to lie below the line (which they do), and to be farther below the line the drier the climate (which they may be). Not much more can be said about this very limited test of the NPP vs. E_v linearity proposed by Eq. E.8 except to remind the reader of the many approximations involved.[‡]

It is interesting to speculate that the degree of agreement of these evapotranspiration data with the proposed linear transpiration relation may be due to compensating approximations. On the one hand, we neglected (as being small) the back radiation term in Eq. E.1 which makes E_{ps} smaller than it should be and results in the left-hand side of Eq. E.8 being too large. This error will be largest in the hot, dry climates. On the other hand, the left-hand side of Eq. E.8 has been evaluated from observations of E_T rather than E_v, and since $E_T > E_v$, makes the left-hand side of Eq. E.8 too small. This error will also be largest in the hot, dry climates.

[†] Note that the dimensions of the constituent variables of this parameter are adjusted from those given in Table E.1 such that the resulting plotted parameter is dimensionless. With NPP in $g_s\ m^{-2}\ y^{-1}$; E_T in mm y^{-1}; $\varepsilon = 0.81\ g_s\ \text{MJ(tot)}^{-1}$; $\lambda = 585$ cal g_w^{-1}; and there being 23.8×10^4 MJ cal^{-1} this parameter must be divided by 1.99 to be dimensionless.

[‡] Further evidence of the NPP vs. E_v proportionality is presented in Fig. 8.14a in the form of a comparison of measured photosynthesis and transpiration during a drying transient for loblolly pine.

F

Climatology of station storm rainfall in the U.S.: Parameters of the Poisson rectangular pulses model

Table F.1 is an index of 75 first-order stations of the U.S. Weather Service in the continental United States containing the station identification number, name, period of record, percentage of record useable, location, elevation, and mean annual precipitation (Hawk and Eagleson, 1992, Table 7-5, pp. 277–290).

In Table F.2 are tabulated eight parameters of the Poisson rectangular pulses model of station rainfall averaged over each month of the year from the historical records of hourly precipitation for each of the 75 stations identified in Table F.1. The column headings of the table translate into the notations used in this work as follows:

Column	*Definition*
1	Station identification number
2	Latitude of station (northern hemisphere)
3	Longitude of station
4	Month of year (January — 1)
5	Number of independent storms per month (m_ν)
6	Storm duration, hours (m_{t_r})
7	Storm depth, mm (m_h)
8	Storm intensity, mm h^{-1} (m_i)
9	Time between storms, hours (m_{t_b})
10	Intensity–duration correlation ($\rho[i, t_r]$)
11	Shape index of storm depth distribution (κ_o)
12	Scale parameter of storm depth distribution, mm^{-1} (λ_o)

Table F.1. *Index of stations*

State	Station ID	Station Name	Years	Percentage of records usable	Latitude (°N)	Longitude (°W)	Elevation (m)	P_A (m)
Alabama								
	33	Huntsville	1958–88	96	34:39	86:46	189	1.41
	34	Montgomery	1948–88	98	32:18	86:24	67	1.29
	67	Mobile	1958–88	94	30:41	88:15	64	1.65
Arizona								
	1	Flagstaff	1950–88	99	35:08	111:40	2137	0.56
	2	Tucson	1948–88	98	32:08	110:57	786	0.29
	55	Phoenix	1948–88	95	33:26	112:01	338	0.19
Arkansas								
	40	Fort Smith	1948–88	98	35:20	94:22	137	1.03
California								
	3	San Diego	1948–88	98	32:44	117:10	3	0.25
	4	San Francisco	1948–88	98	37:37	122:23	3	0.51
	75	Los Angeles	1948–88	98	33:56	118:24	34	0.31
Colorado								
	5	Denver	1948–88	98	39:46	104:52	1609	0.39
Connecticut								
	51	Bridgeport	1948–88	98	41:10	73:08	3	1.04
Florida								
	6	Fort Myers	1960–88	95	26:35	81:52	6	1.34
	7	Jacksonville	1948–88	98	30:30	81:42	9	1.32
	68	Key West	1957–88	96	24:33	81:45	12	1.02
	69	Daytona Beach	1942–88	98	29:44	81:02	6	1.39
Georgia								
	8	Athens	1957–88	95	33:57	83:19	244	1.26
	9	Atlanta	1948–88	98	33:39	84:26	308	1.23
	56	Savannah	1948–88	98	32:08	81:12	15	1.25
Idaho								
	35	Boise	1948–88	98	43:34	116:13	866	0.30
Illinois								
	10	Chicago	1948–88	96	41:44	87:46	189	0.91
	11	Peoria	1948–88	98	40:40	89:41	198	0.90
Indiana								
	12	Indianapolis	1948–88	98	39:44	86:16	241	1.01
Iowa								
	13	Des Moines	1948–88	98	41:32	93:39	293	0.80
Kansas								
	42	Dodge City	1948–88	98	37:46	99:58	786	0.53
Kentucky								
	52	Paducah	1948–88	94	37:06	88:36	101	1.13
Louisiana								
	45	Baton Rouge	1947–88	97	30:32	91:09	18	1.43
Maine								
	49	Portland	1948–88	98	43:39	70:19	3	1.11
Massachusetts								
	14	Boston	1948–88	98	42:22	71:02	6	1.09

(*continued*)

Table F.1. (*cont.*)

State	Station ID	Station Name	Years	Percentage of records usable	Latitude (°N)	Longitude (°W)	Elevation (m)	P_A (m)
Michigan								
	66	Sault Ste Marie	1948–88	98	46:28	84:22	219	0.86
Minnesota								
	15	Duluth	1948–88	98	46:50	92:11	436	0.77
	16	Internat. Falls	1948–88	98	48:34	93:23	360	0.62
Mississippi								
	48	Meridian	1948–88	98	32:20	88:45	88	1.37
Missouri								
	44	Springfield	1948–88	98	37:14	93:23	387	1.05
Montana								
	36	Billings	1948–88	98	45:48	108:32	1088	0.37
	64	Glasgow	1957–88	97	48:13	106:37	695	0.27
Nebraska								
	41	Norfolk	1948–88	98	41:59	97:26	472	0.63
	54	Valentine	1948–88	98	42:52	100:33	789	0.46
Nevada								
	17	Ely	1948–88	98	39:17	114:51	1908	0.24
New Hampshire								
	18	Mt. Washington	1948–88	98	43:12	71:30	107	0.94
New Mexico								
	19	Albuquerque	1947–88	97	35:03	106:37	1618	0.21
New York								
	53	Buffalo	1948–88	98	42:56	78:44	216	0.96
	70	Rochester	1948–88	98	43:07	77:40	168	0.80
North Carolina								
	20	Raleigh	1948–88	98	35:52	78:47	116	1.05
	71	Cape Hatteras	1957–88	95	35:16	75:33	3	1.39
North Dakota								
	37	Bismarck	1948–88	98	46:46	100:46	503	0.39
	65	Williston	1948–88	98	48:11	103:38	579	0.35
Ohio								
	31	Cincinnati	1948–83	92	39:09	84:31	232	1.03
	32	Cleveland	1948–88	98	41:25	81:52	235	0.92
Oklahoma								
	43	Oklahoma City	1947–88	97	35:24	97:36	390	0.82
Oregon								
	46	Salem	1948–88	98	44:55	123:01	61	1.03
	61	Eugene	1948–88	98	44:07	123:13	110	1.20
	62	Astoria	1953–88	99	46:09	123:53	3	1.73
Pennsylvania								
	21	Allentown	1948–88	98	40:39	75:26	119	1.23
	22	Philadelphia	1948–88	100	39:53	75:14	3	1.06
South Carolina								
	23	Columbia	1948–88	95	33:57	81:07	64	1.23
	73	Charleston	1948–88	98	32:47	79:56	3	1.17

Table F.1. *(cont.)*

State	Station		Years	Percentage of records usable	Latitude (°N)	Longitude (°W)	Elevation (m)	P_A (m)
	ID	Name						
South Dakota								
	38	Huron	1948–88	98	44:23	98:13	390	0.49
Tennessee								
	24	Knoxville	1948–88	98	35:48	84:00	290	1.17
Texas								
	25	Abilene	1940–88	95	32:26	99:41	544	0.59
	59	El Paso	1942–88	96	31:48	106:24	1195	0.21
	60	Corpus Christi	1947–88	97	27:46	97:30	12	0.75
	74	Brownsville	1942–88	96	25:54	97:26	6	0.67
Utah								
	57	Salt Lake City	1948–88	98	40:47	111:57	1286	0.40
	58	Milford	1948–88	97	38:26	113:01	1533	0.23
Vermont								
	50	Burlington	1948–88	98	44:28	73:09	101	0.86
Virginia								
	26	Richmond	1948–88	98	37:30	77:20	49	1.10
	27	Roanoke	1948–88	97	37:19	79:58	351	1.01
	72	Norfolk	1952–88	96	36:54	76:12	6	1.13
Washington								
	39	Spokane	1948–88	98	47:38	117:32	719	0.43
	63	Seattle	1964–88	96	47:27	122:18	137	0.94
West Virginia								
	28	Charleston	1948–88	98	38:22	81:36	311	1.07
Wisconsin								
	29	Greenbay	1948–88	98	44:29	88:08	207	0.72
	30	Lacrosse	1948–69	95	43:52	91:15	198	0.74
Wyoming								
	47	Sheridan	1948–88	98	44:46	106:58	1201	0.37

Table F.2. *Poisson rectangular pulse parameters*

Station ID	Latitude (°N)	Longitude (°W)	Month	m_ν	m_{t_r} (h)	m_h (mm)	m_i (mm h^{-1})	m_{t_b} (h)	$\rho[i, t_r]$	κ_0	λ_0 (mm^{-1})
1	35.133	−111.667	1	3	34	16.21	0.547	198	−0.19	0.7289	0.0450
1	35.133	−111.667	2	4	16	12.12	0.628	134	0.20	0.4764	0.0393
1	35.133	−111.667	3	4	25	12.19	0.590	141	−0.19	0.7108	0.0583
1	35.133	−111.667	4	3	18	9.98	0.584	182	−0.08	0.5148	0.0516
1	35.133	−111.667	5	3	9	5.11	0.673	207	−0.18	0.7717	0.1511
1	35.133	−111.667	6	2	9	6.78	1.170	323	−0.21	0.2876	0.0424
1	35.133	−111.667	7	3	66	19.29	0.779	135	−0.28	0.8170	0.0423
1	35.133	−111.667	8	9	5	7.09	1.848	68	−0.15	0.4081	0.0576
1	35.133	−111.667	9	4	11	9.85	1.124	141	−0.13	0.3145	0.0319
1	35.133	−111.667	10	3	17	13.73	0.931	215	−0.14	0.4829	0.0352
1	35.133	−111.667	11	3	18	14.06	0.751	206	0.02	0.4758	0.0338

(continued)

Table F.2. (*cont.*)

Station ID	Latitude (°N)	Longitude (°W)	Month	m_ν	m_{t_r} (h)	m_h (mm)	m_i (mm h^{-1})	m_{t_b} (h)	$\rho[i, t_r]$	κ_o	λ_o (mm^{-1})
1	35.133	−111.667	12	3	22	15.88	0.642	190	0.14	0.4954	0.0312
2	32.017	−110.950	1	2	19	8.41	0.594	257	−0.31	0.9850	0.1172
2	32.017	−110.950	2	2	13	6.92	0.858	274	−0.15	0.5744	0.0830
2	32.017	−110.950	3	3	8	5.36	0.578	213	0.05	0.6567	0.1225
2	32.017	−110.950	4	1	7	4.50	0.810	388	−0.20	0.7963	0.1769
2	32.017	−110.950	5	1	3	2.44	0.763	540	−0.03	0.6044	0.2474
2	32.017	−110.950	6	1	3	3.53	1.271	405	−0.21	0.4745	0.1344
2	32.017	−110.950	7	4	44	14.65	0.814	129	−0.33	0.6688	0.0456
2	32.017	−110.950	8	8	3	5.94	2.102	80	−0.09	0.4825	0.0812
2	32.017	−110.950	9	4	4	8.10	1.994	159	−0.11	0.4443	0.0549
2	32.017	−110.950	10	2	14	11.32	0.999	336	−0.16	0.3647	0.0322
2	32.017	−110.950	11	2	9	7.18	0.876	311	−0.18	0.7892	0.1098
2	32.017	−110.950	12	2	15	8.76	0.775	258	−0.22	0.6217	0.0709
3	32.733	−117.167	1	3	26	15.04	0.737	185	−0.19	0.6544	0.0435
3	32.733	−117.167	2	3	17	11.58	0.746	173	−0.11	0.6813	0.0588
3	32.733	−117.167	3	4	13	9.81	0.772	147	−0.09	0.4953	0.0505
3	32.733	−117.167	4	3	11	5.89	0.547	208	−0.06	0.4669	0.0793
3	32.733	−117.167	5	1	10	3.15	0.381	415	−0.17	0.2980	0.0948
3	32.733	−117.167	6	1	1	0.85	0.363	458	0.44	0.3505	0.4148
3	32.733	−117.167	7	0	15	1.02	0.400	999	−0.29	0.9880	0.9698
3	32.733	−117.167	8	0	27	6.91	0.746	999	−0.21	0.2345	0.0340
3	32.733	−117.167	9	1	6	3.62	0.710	570	−0.06	0.3151	0.0869
3	32.733	−117.167	10	2	4	3.46	0.762	319	0.03	0.4202	0.1216
3	32.733	−117.167	11	2	17	11.78	0.797	232	−0.13	0.6428	0.0545
3	32.733	−117.167	12	2	17	11.30	0.769	212	−0.13	0.5337	0.0472
4	37.617	−122.383	1	3	48	31.04	0.718	140	−0.10	0.6753	0.0218
4	37.617	−122.383	2	4	22	16.30	0.757	109	−0.04	0.6197	0.0380
4	37.617	−122.383	3	4	28	15.98	0.631	122	−0.10	0.4102	0.0257
4	37.617	−122.383	4	2	21	10.13	0.590	199	−0.22	0.6987	0.0690
4	37.617	−122.383	5	1	8	3.68	0.472	333	−0.11	0.3346	0.0909
4	37.617	−122.383	6	1	3	2.33	0.479	622	0.39	0.2797	0.1202
4	37.617	−122.383	7	0	7	2.08	0.477	999	−0.30	0.6131	0.2946
4	37.617	−122.383	8	0	2	1.24	0.508	603	0.32	0.4738	0.3817
4	37.617	−122.383	9	1	3	4.50	0.887	555	0.41	0.2422	0.0538
4	37.617	−122.383	10	2	15	11.60	0.705	290	0.06	0.1979	0.0171
4	37.617	−122.383	11	3	32	19.59	0.728	162	−0.15	0.6435	0.0328
4	37.617	−122.383	12	3	46	25.43	0.674	162	−0.18	0.5554	0.0218
5	39.767	−104.867	1	4	9	3.03	0.331	164	−0.10	0.5199	0.1717
5	39.767	−104.867	2	5	5	2.88	0.427	112	0.16	0.4830	0.1679
5	39.767	−104.867	3	6	9	4.79	0.495	98	0.00	0.4082	0.0851
5	39.767	−104.867	4	6	11	7.46	0.711	102	−0.06	0.4295	0.0576
5	39.767	−104.867	5	7	10	8.55	0.865	85	−0.04	0.2998	0.0351
5	39.767	−104.867	6	7	6	5.90	1.126	88	−0.11	0.2658	0.0451
5	39.767	−104.867	7	7	5	6.05	1.551	88	−0.17	0.4132	0.0683
5	39.767	−104.867	8	9	2	3.97	1.466	74	0.05	0.2707	0.0682
5	39.767	−104.867	9	5	6	5.35	0.995	122	−0.08	0.4275	0.0799
5	39.767	−104.867	10	4	8	6.27	0.704	166	0.08	0.4752	0.0757
5	39.767	−104.867	11	4	7	4.49	0.535	145	0.09	0.5632	0.1253

Table F.2. (*cont.*)

Station ID	Latitude (°N)	Longitude (°W)	Month	m_ν	m_{t_r} (h)	m_h (mm)	m_i (mm h^{-1})	m_{t_b} (h)	$\rho[i, t_r]$	κ_0	λ_0 (mm^{-1})
5	39.767	−104.867	12	5	6	3.08	0.406	136	0.16	0.3102	0.1008
6	26.583	−81.867	1	4	4	9.49	2.558	156	−0.06	0.5166	0.0545
6	26.583	−81.867	2	4	9	11.12	2.177	149	−0.16	0.6078	0.0546
6	26.583	−81.867	3	4	3	16.10	4.276	154	0.12	0.3924	0.0244
6	26.583	−81.867	4	3	3	9.40	3.171	235	−0.14	0.7349	0.0782
6	26.583	−81.867	5	3	22	25.06	3.204	179	−0.26	0.6648	0.0265
6	26.583	−81.867	6	6	31	37.74	2.553	86	−0.31	0.6346	0.0168
6	26.583	−81.867	7	11	9	17.87	4.675	56	−0.26	0.6526	0.0365
6	26.583	−81.867	8	25	1	6.37	6.371	27	0.00	0.4422	0.0694
6	26.583	−81.867	9	7	18	26.36	4.020	75	−0.12	0.4134	0.0157
6	26.583	−81.867	10	5	9	14.88	2.608	137	−0.22	0.4836	0.0325
6	26.583	−81.867	11	3	4	11.78	2.741	220	−0.03	0.4428	0.0376
6	26.583	−81.867	12	3	4	10.31	2.778	221	−0.17	0.5393	0.0523
7	30.500	−81.700	1	8	6	9.70	1.491	87	0.05	0.4942	0.0509
7	30.500	−81.700	2	8	5	11.31	2.161	72	−0.06	0.4087	0.0361
7	30.500	−81.700	3	6	7	13.94	1.832	100	0.01	0.4730	0.0339
7	30.500	−81.700	4	6	4	11.87	2.953	104	−0.06	0.3205	0.0270
7	30.500	−81.700	5	6	7	14.11	2.677	100	−0.18	0.4392	0.0311
7	30.500	−81.700	6	10	7	14.41	2.832	63	−0.16	0.3533	0.0245
7	30.500	−81.700	7	15	3	10.19	3.269	45	−0.09	0.5521	0.0542
7	30.500	−81.700	8	14	4	13.27	3.451	45	−0.07	0.2751	0.0207
7	30.500	−81.700	9	11	8	17.39	2.995	55	−0.12	0.2904	0.0167
7	30.500	−81.700	10	6	10	14.47	1.602	98	−0.06	0.2674	0.0185
7	30.500	−81.700	11	6	5	7.82	1.618	108	−0.04	0.4577	0.0586
7	30.500	−81.700	12	7	6	9.36	1.339	99	0.08	0.4600	0.0491
8	33.950	−83.317	1	10	8	12.39	1.205	65	0.19	0.5345	0.0400
8	33.950	−83.317	2	9	7	11.68	1.405	61	0.14	0.6079	0.0500
8	33.950	−83.317	3	10	6	13.57	1.783	64	0.12	0.5750	0.0400
8	33.950	−83.317	4	7	7	13.74	1.803	87	−0.03	0.4787	0.0300
8	33.950	−83.317	5	9	4	12.36	2.510	71	0.04	0.4082	0.0300
8	33.950	−83.317	6	6	10	15.06	2.184	100	−0.22	0.3328	0.0200
8	33.950	−83.317	7	9	6	13.81	3.072	72	−0.18	0.5065	0.0300
8	33.950	−83.317	8	8	7	11.89	2.984	87	−0.17	0.5861	0.0400
8	33.950	−83.317	9	6	8	13.38	1.715	102	−0.05	0.3850	0.0200
8	33.950	−83.317	10	5	9	14.55	1.856	122	−0.12	0.5041	0.0300
8	33.950	−83.317	11	6	11	15.07	1.478	108	−0.10	0.7950	0.0500
8	33.950	−83.317	12	9	7	10.77	1.190	71	0.28	0.5057	0.0400
9	33.650	−84.433	1	9	8	11.96	1.214	68	0.18	0.5354	0.0400
9	33.650	−84.433	2	10	6	10.85	1.472	57	0.13	0.4587	0.0400
9	33.650	−84.433	3	6	20	22.34	1.516	97	−0.22	0.8372	0.0300
9	33.650	−84.433	4	7	8	15.31	1.907	90	−0.08	0.5643	0.0300
9	33.650	−84.433	5	8	6	11.35	2.148	78	−0.09	0.5297	0.0400
9	33.650	−84.433	6	7	8	10.98	2.210	84	−0.20	0.5265	0.0400
9	33.650	−84.433	7	9	9	13.87	3.031	71	−0.24	0.7102	0.0500
9	33.650	−84.433	8	9	4	9.82	2.950	74	−0.12	0.5631	0.0500
9	33.650	−84.433	9	6	7	12.76	2.097	97	−0.13	0.4113	0.0300
9	33.650	−84.433	10	4	11	14.96	1.363	144	−0.04	0.5238	0.0300
9	33.650	−84.433	11	6	11	14.26	1.370	109	−0.07	0.7063	0.0400

(*continued*)

Table F.2. (*cont.*)

Station ID	Latitude (°N)	Longitude (°W)	Month	m_ν	m_{t_r} (h)	m_h (mm)	m_i (mm h^{-1})	m_{t_b} (h)	$\rho[i, t_r]$	κ_o	λ_o (mm^{-1})
9	33.650	−84.433	12	9	7	10.94	1.188	69	0.22	0.5126	0.0400
10	41.733	−87.767	1	9	6	4.89	0.806	70	−0.04	0.3346	0.0683
10	41.733	−87.767	2	8	6	4.82	0.956	72	−0.11	0.5264	0.1092
10	41.733	−87.767	3	11	6	6.50	1.080	60	−0.06	0.4965	0.0764
10	41.733	−87.767	4	10	6	8.97	1.478	60	−0.09	0.5164	0.0575
10	41.733	−87.767	5	9	5	8.43	1.659	73	−0.09	0.5218	0.0619
10	41.733	−87.767	6	9	4	10.35	2.518	71	−0.08	0.4397	0.0425
10	41.733	−87.767	7	9	3	10.80	2.949	77	−0.02	0.4677	0.0420
10	41.733	−87.767	8	7	4	11.14	2.771	90	−0.02	0.4677	0.0420
10	41.733	−87.767	9	6	7	10.93	1.886	97	−0.11	0.5837	0.0534
10	41.733	−87.767	10	6	10	10.60	1.131	114	−0.09	0.3364	0.0317
10	41.733	−87.767	11	7	10	9.06	0.970	89	−0.10	0.5487	0.0606
10	41.733	−87.767	12	6	20	11.32	0.749	104	−0.19	0.5432	0.0480
11	40.667	−89.683	1	7	9	5.72	0.542	88	0.11	0.2921	0.0511
11	40.667	−89.683	2	8	6	4.54	0.629	74	0.10	0.4362	0.0961
11	40.667	−89.683	3	11	6	6.84	0.928	60	0.10	0.4972	0.0727
11	40.667	−89.683	4	10	7	9.15	1.218	59	0.03	0.4355	0.0476
11	40.667	−89.683	5	10	6	9.33	1.729	65	−0.08	0.5113	0.0548
11	40.667	−89.683	6	9	4	10.76	2.339	71	−0.04	0.5177	0.0481
11	40.667	−89.683	7	9	3	10.85	2.886	71	−0.03	0.5713	0.0527
11	40.667	−89.683	8	8	4	9.55	2.359	82	−0.03	0.4696	0.0492
11	40.667	−89.683	9	7	8	12.50	1.648	91	−0.06	0.4441	0.0355
11	40.667	−89.683	10	6	9	10.40	1.168	110	−0.06	0.5162	0.0496
11	40.667	−89.683	11	7	9	8.33	0.868	89	0.00	0.5877	0.0706
11	40.667	−89.683	12	8	8	6.99	0.651	79	0.24	0.3822	0.0547
12	39.733	−86.267	1	10	7	6.37	0.640	60	0.22	0.2946	0.0462
12	39.733	−86.267	2	10	6	6.06	0.722	58	0.27	0.3877	0.0639
12	39.733	−86.267	3	6	32	15.00	0.629	90	−0.22	0.6949	0.0463
12	39.733	−86.267	4	11	7	7.99	1.317	54	−0.08	0.6044	0.0755
12	39.733	−86.267	5	10	7	9.39	1.591	62	−0.13	0.6097	0.0650
12	39.733	−86.267	6	8	6	11.66	2.284	75	−0.11	0.4491	0.0385
12	39.733	−86.267	7	11	3	9.71	2.812	62	0.04	0.4701	0.0484
12	39.733	−86.267	8	9	3	8.81	2.508	73	−0.04	0.4969	0.0564
12	39.733	−86.267	9	7	5	9.53	1.701	90	−0.01	0.5363	0.0563
12	39.733	−86.267	10	7	7	9.28	1.196	92	−0.01	0.4470	0.0482
12	39.733	−86.267	11	8	10	9.77	0.957	75	0.01	0.5721	0.0586
12	39.733	−86.267	12	11	9	7.42	0.674	58	0.19	0.4317	0.0581
13	41.533	−93.650	1	6	7	4.08	0.458	105	0.22	0.3087	0.0757
13	41.533	−93.650	2	7	5	3.90	0.548	80	0.23	0.3671	0.0940
13	41.533	−93.650	3	8	8	6.44	0.751	75	0.04	0.5002	0.0777
13	41.533	−93.650	4	11	6	7.39	1.172	58	0.01	0.4474	0.0606
13	41.533	−93.650	5	9	8	10.54	1.496	70	−0.11	0.6166	0.0585
13	41.533	−93.650	6	11	4	9.77	2.252	59	0.02	0.4264	0.0436
13	41.533	−93.650	7	9	4	9.81	2.379	76	0.00	0.4583	0.0467
13	41.533	−93.650	8	8	5	12.25	2.330	81	−0.04	0.3939	0.0322
13	41.533	−93.650	9	7	6	10.17	1.678	85	−0.07	0.5267	0.0518
13	41.533	−93.650	10	7	6	8.33	1.321	93	−0.04	0.4953	0.0595
13	41.533	−93.650	11	4	13	9.35	0.718	138	−0.02	0.5283	0.0565

Table F.2. (*cont.*)

Station ID	Latitude (°N)	Longitude (°W)	Month	m_ν	m_{t_r} (h)	m_h (mm)	m_i (mm h^{-1})	m_{t_b} (h)	$\rho[i, t_r]$	κ_0	λ_0 (mm^{-1})
13	41.533	−93.650	12	6	7	4.31	0.516	101	0.11	0.4788	0.1111
14	42.367	−71.033	1	10	8	9.32	0.828	60	0.41	0.5194	0.0558
14	42.367	−71.033	2	12	6	8.11	0.855	49	0.44	0.3813	0.0470
14	42.367	−71.033	3	12	7	8.51	0.583	53	0.45	0.3691	0.0434
14	42.367	−71.033	4	12	6	7.73	0.961	51	0.24	0.3476	0.0450
14	42.367	−71.033	5	12	6	7.24	0.983	54	0.15	0.2487	0.0343
14	42.367	−71.033	6	10	5	7.57	1.371	63	0.00	0.2864	0.0378
14	42.367	−71.033	7	10	4	7.19	1.691	68	0.03	0.4043	0.0562
14	42.367	−71.033	8	9	5	9.15	1.881	70	−0.04	0.2137	0.0233
14	42.367	−71.033	9	9	5	8.64	1.367	68	0.14	0.2829	0.0328
14	42.367	−71.033	10	8	6	9.68	1.145	78	0.22	0.3668	0.0379
14	42.367	−71.033	11	10	7	11.08	1.095	61	0.34	0.5068	0.0457
14	42.367	−71.033	12	16	4	6.63	0.864	41	0.48	0.3248	0.0490
15	46.833	−92.183	1	10	7	2.98	0.323	63	0.21	0.2924	0.0980
15	46.833	−92.183	2	10	5	2.09	0.404	57	0.02	0.2944	0.1409
15	46.833	−92.183	3	9	8	4.82	0.470	70	0.19	0.3262	0.0677
15	46.833	−92.183	4	10	6	5.40	0.941	60	−0.08	0.3967	0.0735
15	46.833	−92.183	5	10	10	8.44	0.933	64	−0.07	0.5099	0.0604
15	46.833	−92.183	6	12	5	8.61	1.801	53	−0.09	0.4826	0.0560
15	46.833	−92.183	7	14	2	7.41	2.506	50	0.00	0.4335	0.0585
15	46.833	−92.183	8	11	5	8.88	1.706	59	−0.01	0.4097	0.0461
15	46.833	−92.183	9	11	6	7.78	1.231	55	−0.01	0.3711	0.0477
15	46.833	−92.183	10	8	8	7.32	0.781	76	0.09	0.3449	0.0471
15	46.833	−92.183	11	7	14	6.64	0.438	88	0.06	0.3698	0.0557
15	46.833	−92.183	12	10	8	3.42	0.342	66	0.22	0.2360	0.0691
16	48.567	−93.383	1	10	6	2.06	0.349	63	−0.05	0.3921	0.1907
16	48.567	−93.383	2	8	5	2.05	0.356	72	0.02	0.4376	0.2133
16	48.567	−93.383	3	8	9	3.56	0.394	84	−0.01	0.3608	0.1013
16	48.567	−93.383	4	6	11	5.82	0.569	93	−0.07	0.4419	0.0759
16	48.567	−93.383	5	8	11	8.08	0.831	78	−0.14	0.5755	0.0713
16	48.567	−93.383	6	11	6	8.56	1.480	54	−0.09	0.4500	0.0526
16	48.567	−93.383	7	14	3	6.84	2.085	49	0.01	0.3581	0.0524
16	48.567	−93.383	8	14	3	5.44	1.647	48	0.07	0.4090	0.0752
16	48.567	−93.383	9	11	5	7.19	1.163	57	0.02	0.3600	0.0500
16	48.567	−93.383	10	9	7	5.48	0.653	74	0.08	0.3187	0.0582
16	48.567	−93.383	11	9	9	3.68	0.359	70	0.06	0.4001	0.1086
16	48.567	−93.383	12	10	7	2.17	0.302	65	−0.04	0.3766	0.1732
17	39.283	−114.850	1	4	20	4.74	0.386	165	−0.32	0.6604	0.1392
17	39.283	−114.850	2	5	7	2.93	0.427	114	−0.10	0.4488	0.1532
17	39.283	−114.850	3	5	13	4.32	0.462	114	−0.29	0.7384	0.1709
17	39.283	−114.850	4	4	15	4.88	0.444	138	−0.22	0.7710	0.1578
17	39.283	−114.850	5	5	11	4.99	0.539	119	−0.14	0.4679	0.0937
17	39.283	−114.850	6	3	13	5.72	0.593	209	−0.20	0.3712	0.0648
17	39.283	−114.850	7	3	18	4.90	0.634	187	−0.31	0.5802	0.1185
17	39.283	−114.850	8	4	6	3.89	0.947	149	−0.14	0.4918	0.1264
17	39.283	−114.850	9	3	9	6.57	0.862	194	−0.08	0.4002	0.0609
17	39.283	−114.850	10	3	14	6.30	0.593	219	−0.26	0.7992	0.1268
17	39.283	−114.850	11	3	11	4.63	0.443	183	−0.12	0.5858	0.1266

(*continued*)

Table F.2. (*cont.*)

Station ID	Latitude (°N)	Longitude (°W)	Month	m_v	m_{t_r} (h)	m_h (mm)	m_i (mm h^{-1})	m_{t_b} (h)	$\rho[i, t_r]$	κ_o	λ_o (mm^{-1})
17	39.283	−114.850	12	4	14	4.19	0.398	156	−0.23	0.5080	0.1214
18	44.267	−71.300	1	22	8	8.42	0.711	24	0.39	0.2722	0.0323
18	44.267	−71.300	2	20	8	10.20	0.814	24	0.32	0.1585	0.0155
18	44.267	−71.300	3	15	13	13.37	0.798	33	0.25	0.3592	0.0269
18	44.267	−71.300	4	16	10	11.59	0.892	32	0.22	0.3609	0.0311
18	44.267	−71.300	5	14	10	12.18	1.153	40	0.01	0.4734	0.0389
18	44.267	−71.300	6	14	9	12.90	1.590	40	−0.09	0.5341	0.0414
18	44.267	−71.300	7	17	6	10.26	1.767	35	−0.06	0.5079	0.0495
18	44.267	−71.300	8	15	7	12.78	1.807	39	−0.07	0.4919	0.0385
18	44.267	−71.300	9	16	7	11.49	1.453	36	0.05	0.4741	0.0413
18	44.267	−71.300	10	12	12	14.44	1.063	49	0.07	0.4760	0.0330
18	44.267	−71.300	11	17	11	13.75	0.977	30	0.24	0.3759	0.0273
18	44.267	−71.300	12	23	8	9.69	0.782	23	0.40	0.2590	0.0267
19	35.050	−106.617	1	2	12	3.76	0.471	252	−0.25	0.7335	0.1949
19	35.050	−106.617	2	3	8	3.25	0.525	187	−0.37	1.0852	0.3337
19	35.050	−106.617	3	4	4	3.19	0.605	170	0.09	0.4473	0.1404
19	35.050	−106.617	4	2	9	4.84	0.716	293	−0.27	0.5317	0.1099
19	35.050	−106.617	5	3	8	3.92	0.724	208	−0.21	0.5583	0.1423
19	35.050	−106.617	6	3	6	4.40	0.972	209	−0.13	0.4098	0.0931
19	35.050	−106.617	7	5	16	6.08	1.104	116	−0.29	0.6199	0.1019
19	35.050	−106.617	8	9	2	4.13	1.742	77	−0.07	0.4844	0.1173
19	35.050	−106.617	9	4	9	5.47	1.010	160	−0.26	0.6110	0.1117
19	35.050	−106.617	10	3	10	7.03	0.894	211	−0.13	0.4989	0.0709
19	35.050	−106.617	11	2	7	3.57	0.634	249	−0.25	0.9604	0.2692
19	35.050	−106.617	12	3	7	3.81	0.597	213	−0.16	0.5529	0.1451
20	35.867	−78.783	1	10	6	9.06	1.093	67	0.23	0.5361	0.0592
20	35.867	−78.783	2	10	6	9.20	1.263	60	0.13	0.6871	0.0747
20	35.867	−78.783	3	11	5	8.04	1.293	59	0.08	0.5771	0.0718
20	35.867	−78.783	4	7	6	9.36	1.351	86	0.01	0.5019	0.0537
20	35.867	−78.783	5	8	8	11.98	2.019	83	−0.20	0.6998	0.0584
20	35.867	−78.783	6	8	5	10.44	2.506	76	−0.11	0.4834	0.0463
20	35.867	−78.783	7	9	7	12.15	2.535	72	−0.21	0.6329	0.0521
20	35.867	−78.783	8	9	4	11.80	2.565	71	−0.04	0.5083	0.0431
20	35.867	−78.783	9	6	7	12.79	2.139	100	−0.10	0.5245	0.0410
20	35.867	−78.783	10	6	8	11.94	1.510	116	−0.04	0.5498	0.0460
20	35.867	−78.783	11	7	6	9.17	1.311	84	0.13	0.3849	0.0420
20	35.867	−78.783	12	8	7	9.48	1.232	81	0.04	0.7115	0.0750
21	40.650	−75.433	1	12	6	6.65	0.760	52	0.45	0.4135	0.0622
21	40.650	−75.433	2	11	6	7.01	0.856	53	0.42	0.5571	0.0795
21	40.650	−75.433	3	13	5	7.18	0.958	50	0.28	0.4412	0.0614
21	40.650	−75.433	4	10	7	9.23	1.115	59	0.09	0.5317	0.0576
21	40.650	−75.433	5	11	6	8.59	1.435	56	−0.04	0.4139	0.0482
21	40.650	−75.433	6	10	4	8.68	1.919	63	−0.06	0.4868	0.0561
21	40.650	−75.433	7	11	3	9.48	2.450	59	0.05	0.5047	0.0532
21	40.650	−75.433	8	10	4	11.04	2.641	67	−0.10	0.4065	0.0368
21	40.650	−75.433	9	9	5	11.36	1.865	72	0.04	0.3319	0.0292
21	40.650	−75.433	10	8	6	9.29	1.228	83	0.15	0.5084	0.0547
21	40.650	−75.433	11	10	7	9.88	1.110	65	0.28	0.4171	0.0422

Table F.2. *(cont.)*

Station ID	Latitude (°N)	Longitude (°W)	Month	m_v	m_{t_r} (h)	m_h (mm)	m_i (mm h^{-1})	m_{t_b} (h)	$\rho[i, t_r]$	κ_o	λ_o (mm^{-1})
21	40.650	−75.433	12	12	6	7.40	0.872	53	0.46	0.4400	0.0595
22	39.883	−75.433	1	11	6	7.32	0.888	56	0.41	0.4721	0.0645
22	39.883	−75.433	2	10	5	6.94	0.869	55	0.39	0.4842	0.0698
22	39.883	−75.433	3	12	5	7.17	0.978	54	0.32	0.5276	0.0736
22	39.883	−75.433	4	10	6	8.57	1.087	61	0.17	0.5234	0.0611
22	39.883	−75.433	5	10	6	8.93	1.302	64	0.02	0.5153	0.0577
22	39.883	−75.433	6	10	5	8.84	1.824	62	−0.05	0.4839	0.0547
22	39.883	−75.433	7	10	4	9.63	2.358	68	−0.05	0.4441	0.0461
22	39.883	−75.433	8	9	4	10.96	2.568	75	−0.03	0.4310	0.0393
22	39.883	−75.433	9	7	6	12.83	2.179	89	−0.08	0.4287	0.0334
22	39.883	−75.433	10	7	6	10.75	1.589	97	0.06	0.4073	0.0379
22	39.883	−75.433	11	7	7	10.25	1.207	89	0.20	0.4499	0.0439
22	39.883	−75.433	12	10	6	8.79	1.040	66	0.33	0.4676	0.0532
23	33.950	−81.117	1	9	8	11.74	1.176	68	0.23	0.5325	0.0453
23	33.950	−81.117	2	9	7	11.47	1.411	65	0.14	0.5656	0.0493
23	33.950	−81.117	3	11	5	11.32	1.723	62	0.13	0.5510	0.0487
23	33.950	−81.117	4	7	7	12.56	1.843	91	−0.06	0.5322	0.0424
23	33.950	−81.117	5	7	9	13.65	2.211	97	−0.19	0.5208	0.0381
23	33.950	−81.117	6	8	6	13.59	2.783	81	−0.16	0.4940	0.0363
23	33.950	−81.117	7	9	9	15.57	2.730	74	−0.18	0.5176	0.0332
23	33.950	−81.117	8	10	4	14.58	3.806	68	−0.08	0.4470	0.0306
23	33.950	−81.117	9	7	6	13.77	2.348	95	−0.03	0.3237	0.0235
23	33.950	−81.117	10	5	9	14.95	1.749	139	−0.07	0.4299	0.0288
23	33.950	−81.117	11	5	10	13.57	1.485	125	−0.11	0.7074	0.0521
23	33.950	−81.117	12	9	6	9.02	1.176	72	0.25	0.4808	0.0533
24	35.800	−84.000	1	13	7	9.38	1.111	50	0.20	0.4825	0.0514
24	35.800	−84.000	2	12	6	8.11	1.164	46	0.17	0.6287	0.0776
24	35.800	−84.000	3	14	5	8.95	1.369	46	0.19	0.3732	0.0417
24	35.800	−84.000	4	11	5	8.60	1.524	58	0.00	0.5153	0.0599
24	35.800	−84.000	5	9	8	10.40	1.616	69	−0.16	0.4837	0.0465
24	35.800	−84.000	6	9	6	10.77	2.303	73	−0.16	0.6561	0.0609
24	35.800	−84.000	7	11	5	9.83	2.353	59	−0.13	0.5641	0.0574
24	35.800	−84.000	8	9	4	8.50	2.307	73	−0.12	0.4879	0.0574
24	35.800	−84.000	9	7	8	10.25	1.611	93	−0.15	0.6787	0.0662
24	35.800	−84.000	10	7	7	10.41	1.353	94	0.07	0.5567	0.0535
24	35.800	−84.000	11	8	8	11.00	1.255	76	0.00	0.7062	0.0642
24	35.800	−84.000	12	13	5	8.26	1.130	49	0.40	0.4121	0.0499
25	32.433	−99.683	1	2	11	7.54	0.608	293	0.12	0.3127	0.0415
25	32.433	−99.683	2	2	11	7.12	0.756	313	−0.17	0.7838	0.1100
25	32.433	−99.683	3	2	9	8.67	1.321	270	−0.20	0.6534	0.0753
25	32.433	−99.683	4	3	4	7.74	2.078	213	−0.14	0.6789	0.0877
25	32.433	−99.683	5	3	6	10.95	2.089	197	−0.13	0.6414	0.0586
25	32.433	−99.683	6	2	11	14.43	2.257	247	−0.25	0.7804	0.0541
25	32.433	−99.683	7	2	14	15.79	2.118	303	−0.18	0.7481	0.0474
25	32.433	−99.683	8	3	6	13.53	2.333	217	−0.08	0.9387	0.0294
25	32.433	−99.683	9	3	9	14.31	2.063	213	−0.11	0.3538	0.0247
25	32.433	−99.683	10	2	10	14.65	1.652	246	−0.13	0.3803	0.0260
25	32.433	−99.683	11	2	7	11.85	1.659	243	−0.02	0.2649	0.0223

(continued)

Table F.2. (*cont.*)

Station ID	Latitude (°N)	Longitude (°W)	Month	m_ν	m_{t_r} (h)	m_h (mm)	m_i (mm h^{-1})	m_{t_b} (h)	$\rho[i, t_r]$	κ_o	λ_o (mm^{-1})
25	32.433	−99.683	12	2	11	9.78	0.926	334	−0.04	0.5119	0.0523
26	37.500	−77.333	1	12	5	6.83	0.927	55	0.27	0.4443	0.0651
26	37.500	−77.333	2	9	7	8.79	1.023	67	0.14	0.6709	0.0763
26	37.500	−77.333	3	13	5	6.76	1.133	50	0.15	0.4988	0.0738
26	37.500	−77.333	4	9	6	8.53	1.309	73	−0.03	0.5053	0.0592
26	37.500	−77.333	5	9	7	10.10	1.926	71	−0.17	0.6191	0.0613
26	37.500	−77.333	6	9	3	9.26	2.493	69	−0.02	0.3821	0.0413
26	37.500	−77.333	7	10	5	13.30	3.033	68	−0.11	0.4533	0.0341
26	37.500	−77.333	8	9	4	13.13	3.007	71	−0.07	0.3779	0.0288
26	37.500	−77.333	9	7	6	12.17	2.077	98	−0.07	0.4639	0.0381
26	37.500	−77.333	10	7	7	12.39	1.446	94	0.16	0.3501	0.0283
26	37.500	−77.333	11	7	7	10.76	1.263	85	0.10	0.4706	0.0437
26	37.500	−77.333	12	9	6	8.53	1.053	69	0.31	0.5116	0.0600
27	37.317	−79.967	1	9	7	7.78	0.814	75	0.29	0.5680	0.0730
27	37.317	−79.967	2	9	7	8.76	0.943	67	0.37	0.5071	0.0579
27	37.317	−79.967	3	6	14	12.43	0.893	93	−0.05	0.7957	0.0640
27	37.317	−79.967	4	9	6	8.94	1.233	68	0.07	0.2929	0.0328
27	37.317	−79.967	5	10	6	8.45	1.436	61	−0.06	0.4092	0.0484
27	37.317	−79.967	6	9	4	8.33	2.185	69	−0.10	0.4045	0.0486
27	37.317	−79.967	7	10	4	8.24	2.290	63	−0.15	0.5912	0.0718
27	37.317	−79.967	8	9	5	10.83	2.556	70	−0.14	0.4156	0.0384
27	37.317	−79.967	9	7	7	11.57	1.708	91	−0.10	0.3591	0.0310
27	37.317	−79.967	10	6	9	14.14	1.377	107	0.08	0.2621	0.0185
27	37.317	−79.967	11	7	7	8.72	0.956	85	0.21	0.5589	0.0641
27	37.317	−79.967	12	10	5	7.22	0.870	68	0.50	0.4759	0.0659
28	38.367	−81.600	1	19	5	4.64	1.135	32	−0.13	0.4118	0.0887
28	38.367	−81.600	2	15	6	5.19	1.060	36	−0.11	0.5255	0.1012
28	38.367	−81.600	3	17	5	5.38	1.226	36	−0.09	0.4672	0.0869
28	38.367	−81.600	4	15	5	5.57	1.169	40	−0.09	0.4856	0.0871
28	38.367	−81.600	5	12	7	8.15	1.381	53	−0.12	0.5674	0.0696
28	38.367	−81.600	6	9	7	9.51	1.701	67	−0.18	0.6814	0.0717
28	38.367	−81.600	7	15	3	8.65	2.807	45	−0.07	0.4503	0.0521
28	38.367	−81.600	8	12	3	8.23	2.863	56	−0.11	0.4648	0.0565
28	38.367	−81.600	9	8	7	9.42	1.433	77	−0.11	0.5137	0.0545
28	38.367	−81.600	10	8	8	8.37	0.992	80	−0.02	0.4342	0.0519
28	38.367	−81.600	11	9	9	8.50	0.826	62	0.04	0.4591	0.0540
28	38.367	−81.600	12	17	5	4.80	0.875	37	0.00	0.3370	0.0702
29	44.483	−88.133	1	9	7	3.37	0.398	74	0.11	0.4942	0.1467
29	44.483	−88.133	2	7	7	3.73	0.461	79	0.09	0.3896	0.1043
29	44.483	−88.133	3	9	7	5.09	0.609	69	0.13	0.5150	0.1011
29	44.483	−88.133	4	9	9	7.83	0.891	71	−0.03	0.7077	0.0904
29	44.483	−88.133	5	9	7	7.38	1.320	69	−0.06	0.4934	0.0669
29	44.483	−88.133	6	10	4	7.75	1.863	64	−0.08	0.5393	0.0696
29	44.483	−88.133	7	11	3	7.77	2.327	63	0.04	0.5078	0.0653
29	44.483	−88.133	8	10	4	8.51	1.914	67	−0.07	0.5438	0.0639
29	44.483	−88.133	9	9	7	9.06	1.367	70	−0.07	0.5215	0.0576
29	44.483	−88.133	10	8	6	6.73	0.980	81	0.09	0.3828	0.0569
29	44.483	−88.133	11	5	19	8.88	0.545	107	−0.16	0.7027	0.0792

Table F.2. (*cont.*)

Station ID	Latitude (°N)	Longitude (°W)	Month	m_ν	m_{t_r} (h)	m_h (mm)	m_i (mm h^{-1})	m_{t_b} (h)	$\rho[i, t_r]$	κ_0	λ_0 (mm^{-1})
29	44.483	−88.133	12	11	5	3.22	0.455	60	0.27	0.4206	0.1307
30	43.867	−91.250	1	9	4	2.52	0.477	76	0.28	0.3937	0.1561
30	43.867	−91.250	2	6	6	3.68	0.504	101	0.23	0.4721	0.1284
30	43.867	−91.250	3	8	9	6.54	0.700	81	0.02	0.5797	0.0886
30	43.867	−91.250	4	10	6	7.30	1.214	65	−0.03	0.5039	0.0690
30	43.867	−91.250	5	11	5	7.93	1.512	60	−0.10	0.6106	0.0770
30	43.867	−91.250	6	12	4	9.40	2.289	55	0.00	0.4571	0.0486
30	43.867	−91.250	7	10	3	9.45	2.472	66	0.04	0.4638	0.0491
30	43.867	−91.250	8	9	5	9.75	2.094	79	−0.12	0.4993	0.0512
30	43.867	−91.250	9	8	7	10.61	1.457	82	0.00	0.6594	0.0622
30	43.867	−91.250	10	6	6	7.48	1.355	106	−0.09	0.5019	0.0671
30	43.867	−91.250	11	5	10	6.91	0.710	132	−0.07	0.4436	0.0642
30	43.867	−91.250	12	9	5	2.82	0.558	78	−0.02	0.4971	0.1764
31	39.150	−84.517	1	9	9	8.79	0.837	62	0.06	0.3591	0.0408
31	39.150	−84.517	2	9	7	7.11	0.859	63	0.11	0.4996	0.0703
31	39.150	−84.517	3	12	6	7.28	1.092	49	0.08	0.3717	0.0511
31	39.150	−84.517	4	12	5	7.36	1.678	50	−0.10	0.5795	0.0787
31	39.150	−84.517	5	9	8	10.73	1.908	67	−0.13	0.4898	0.0456
31	39.150	−84.517	6	10	4	9.29	2.369	62	−0.09	0.5976	0.0643
31	39.150	−84.517	7	10	3	9.02	2.573	62	−0.05	0.5659	0.0628
31	39.150	−84.517	8	9	3	8.47	2.738	68	−0.09	0.5125	0.0605
31	39.150	−84.517	9	7	5	9.53	1.923	84	−0.07	0.4490	0.0471
31	39.150	−84.517	10	7	6	8.38	1.441	90	−0.06	0.5815	0.0694
31	39.150	−84.517	11	8	8	10.08	2.395	76	−0.09	0.3752	0.0372
31	39.150	−84.517	12	11	6	6.46	0.853	56	0.17	0.5538	0.0857
32	41.417	−81.867	1	19	5	3.40	0.459	33	0.22	0.3047	0.0896
32	41.417	−81.867	2	15	6	3.93	0.505	38	0.21	0.3009	0.0766
32	41.417	−81.867	3	16	6	4.88	0.667	39	0.08	0.4213	0.0864
32	41.417	−81.867	4	15	6	5.66	0.970	40	−0.05	0.5677	0.1003
32	41.417	−81.867	5	10	8	8.41	1.294	63	−0.18	0.5440	0.0647
32	41.417	−81.867	6	11	4	8.29	1.817	60	−0.03	0.5054	0.0610
32	41.417	−81.867	7	11	3	7.79	2.387	61	−0.02	0.5187	0.0666
32	41.417	−81.867	8	10	4	8.52	2.055	66	−0.07	0.5859	0.0687
32	41.417	−81.867	9	9	5	8.06	1.552	67	−0.08	0.6200	0.0769
32	41.417	−81.867	10	8	10	7.26	0.870	74	−0.09	0.5088	0.0700
32	41.417	−81.867	11	13	8	6.22	0.682	46	0.08	0.3869	0.0622
32	41.417	−81.867	12	18	6	4.02	0.503	33	0.19	0.3281	0.0816
33	34.650	−86.767	1	12	6	11.12	1.343	55	0.20	0.4046	0.0364
33	34.650	−86.767	2	11	6	10.49	1.563	55	0.12	0.5826	0.0555
33	34.650	−86.767	3	12	5	13.51	2.104	53	0.19	0.3710	0.0275
33	34.650	−86.767	4	9	6	14.00	2.048	73	0.01	0.6004	0.0429
33	34.650	−86.767	5	9	6	14.47	2.278	74	−0.04	0.4668	0.0323
33	34.650	−86.767	6	8	7	12.39	2.500	84	−0.14	0.5023	0.0405
33	34.650	−86.767	7	11	3	10.50	3.087	59	0.01	0.3690	0.0351
33	34.650	−86.767	8	9	4	10.11	2.517	78	−0.07	0.4181	0.0413
33	34.650	−86.767	9	7	8	15.00	2.359	97	−0.16	0.5679	0.0379
33	34.650	−86.767	10	7	5	12.67	1.856	102	0.15	0.4539	0.0358
33	34.650	−86.767	11	9	8	14.24	1.719	77	0.11	0.6612	0.0464

(*continued*)

Table F.2. (*cont.*)

Station ID	Latitude (°N)	Longitude (°W)	Month	m_v	m_{t_r} (h)	m_h (mm)	m_i (mm h^{-1})	m_{t_b} (h)	$\rho[i, t_r]$	κ_o	λ_o (mm^{-1})
33	34.650	−86.767	12	11	6	12.15	1.455	58	0.25	0.3404	0.0280
34	32.300	−86.400	1	10	6	10.47	1.657	64	−0.03	0.5326	0.0509
34	32.300	−86.400	2	10	6	12.84	2.079	60	0.01	0.4353	0.0339
34	32.300	−86.400	3	10	6	14.84	2.643	65	−0.09	0.5636	0.0380
34	32.300	−86.400	4	7	5	14.46	2.416	85	0.01	0.4851	0.0336
34	32.300	−86.400	5	7	7	14.15	2.589	97	−0.15	0.5105	0.0361
34	32.300	−86.400	6	7	6	11.90	2.910	86	−0.18	0.5533	0.0465
34	32.300	−86.400	7	11	4	11.98	3.634	62	−0.16	0.5586	0.0466
34	32.300	−86.400	8	10	2	8.42	3.183	70	−0.04	0.4115	0.0489
34	32.300	−86.400	9	6	9	19.61	3.004	108	−0.15	0.3600	0.0184
34	32.300	−86.400	10	4	8	12.85	1.866	148	−0.11	0.5670	0.0441
34	32.300	−86.400	11	6	7	13.64	2.085	104	−0.07	0.6234	0.0457
34	32.300	−86.400	12	9	6	12.97	1.790	69	0.05	0.5692	0.0439
35	43.567	−116.217	1	8	12	4.50	0.396	72	−0.11	0.3937	0.0875
35	43.567	−116.217	2	7	10	3.80	0.423	78	−0.15	0.5273	0.0875
35	43.567	−116.217	3	8	7	3.61	0.524	78	−0.12	0.6460	0.1791
35	43.567	−116.217	4	6	7	4.52	0.644	96	−0.06	0.4739	0.1047
35	43.567	−116.217	5	7	6	4.29	0.694	100	−0.03	0.4536	0.1057
35	43.567	−116.217	6	4	7	4.59	0.697	141	−0.16	0.4374	0.0953
35	43.567	−116.217	7	2	2	3.14	1.204	271	−0.03	0.4883	0.1557
35	43.567	−116.217	8	2	6	4.33	0.849	306	−0.12	0.4018	0.0929
35	43.567	−116.217	9	2	11	5.85	0.597	228	−0.17	0.6624	0.1133
35	43.567	−116.217	10	4	9	4.37	0.566	153	−0.22	0.9206	0.2107
35	43.567	−116.217	11	5	26	7.15	0.385	114	−0.34	0.8887	0.1243
35	43.567	−116.217	12	4	38	7.15	0.302	118	−0.36	0.5910	0.0827
36	45.800	−108.533	1	4	22	4.76	0.336	135	−0.31	0.6674	0.1401
36	45.800	−108.533	2	6	8	3.00	0.361	100	−0.09	0.5207	0.1736
36	45.800	−108.533	3	8	7	3.36	0.466	84	−0.04	0.5164	0.1537
36	45.800	−108.533	4	7	8	5.77	0.574	83	0.12	0.2836	0.0491
36	45.800	−108.533	5	9	8	6.68	0.873	70	−0.07	0.3631	0.0543
36	45.800	−108.533	6	9	5	5.30	1.106	68	−0.12	0.3969	0.0749
36	45.800	−108.533	7	7	3	3.56	1.527	100	−0.09	0.3287	0.0924
36	45.800	−108.533	8	6	3	3.81	1.157	104	−0.06	0.3810	0.1001
36	45.800	−108.533	9	6	8	5.62	0.759	108	−0.07	0.3930	0.0699
36	45.800	−108.533	10	5	6	5.02	0.631	123	0.17	0.4370	0.0871
36	45.800	−108.533	11	4	10	4.28	0.457	140	0.11	0.5000	0.1167
36	45.800	−108.533	12	4	12	3.94	0.402	141	−0.22	0.6144	0.1559
37	46.767	−100.767	1	7	7	1.86	0.279	99	−0.16	0.6162	0.3316
37	46.767	−100.767	2	6	6	1.92	0.299	102	−0.05	0.3733	0.1946
37	46.767	−100.767	3	7	6	2.82	0.371	95	0.17	0.2212	0.0783
37	46.767	−100.767	4	6	9	6.24	0.671	103	−0.02	0.4478	0.0718
37	46.767	−100.767	5	5	18	9.93	0.818	108	−0.20	0.5123	0.0516
37	46.767	−100.767	6	10	5	7.16	1.651	65	−0.13	0.5107	0.0713
37	46.767	−100.767	7	9	3	6.14	1.995	79	−0.10	0.5043	0.0821
37	46.767	−100.767	8	8	3	5.33	1.600	81	0.07	0.3948	0.0741
37	46.767	−100.767	9	6	5	5.16	1.002	100	−0.02	0.3586	0.0695
37	46.767	−100.767	10	4	10	5.58	0.503	169	0.03	0.3696	0.0663
37	46.767	−100.767	11	4	9	2.95	0.357	142	−0.10	0.4243	0.1437

Table F.2. (*cont.*)

Station ID	Latitude (°N)	Longitude (°W)	Month	m_v	m_{t_r} (h)	m_h (mm)	m_i (mm h^{-1})	m_{t_b} (h)	$\rho[i, t_r]$	κ_0	λ_0 (mm^{-1})
37	46.767	−100.767	12	5	14	2.32	0.233	121	−0.34	0.6865	0.2953
38	44.383	−98.217	1	4	8	3.21	0.306	148	−0.18	0.4705	0.2130
38	44.383	−98.217	2	4	8	3.67	0.392	130	0.19	0.2746	0.0748
38	44.383	−98.217	3	5	12	6.71	0.483	122	0.14	0.3071	0.0458
38	44.383	−98.217	4	7	8	6.70	0.770	88	0.05	0.4394	0.0656
38	44.383	−98.217	5	8	8	8.43	1.170	80	−0.09	0.4657	0.0552
38	44.383	−98.217	6	10	3	7.73	1.953	62	0.01	0.3728	0.0482
38	44.383	−98.217	7	9	2	6.90	2.312	79	0.01	0.3541	0.0513
38	44.383	−98.217	8	9	3	6.15	1.914	78	0.02	0.3684	0.0599
38	44.383	−98.217	9	7	4	5.84	1.158	93	0.11	0.4333	0.0742
38	44.383	−98.217	10	4	8	7.47	0.931	149	−0.02	0.3693	0.0494
38	44.383	−98.217	11	4	8	4.61	0.481	162	0.07	0.3967	0.0860
38	44.383	−98.217	12	5	6	2.78	0.355	140	0.17	0.3504	0.1261
39	47.633	−117.533	1	8	17	6.82	0.420	68	−0.08	0.7167	0.1051
39	47.633	−117.533	2	7	13	5.22	0.447	72	−0.20	0.8084	0.1550
39	47.633	−117.533	3	8	11	4.46	0.476	76	−0.18	0.7597	0.1705
39	47.633	−117.533	4	7	7	3.98	0.537	94	−0.07	0.6801	0.1708
39	47.633	−117.533	5	6	10	4.94	0.642	97	−0.16	0.6374	0.1290
39	47.633	−117.533	6	6	6	4.81	0.848	99	−0.10	0.5016	0.1043
39	47.633	−117.533	7	3	6	3.89	0.745	193	−0.10	0.5212	0.1339
39	47.633	−117.533	8	2	19	5.88	0.610	226	−0.39	0.7523	0.1280
39	47.633	−117.533	9	3	21	6.14	0.498	191	−0.32	0.7487	0.1219
39	47.633	−117.533	10	3	27	7.27	0.399	162	−0.36	0.6854	0.0943
39	47.633	−117.533	11	9	11	5.90	0.546	62	−0.06	0.6890	0.1167
39	47.633	−117.533	12	12	9	4.84	0.491	47	0.02	0.6380	0.1319
40	35.333	−94.367	1	5	10	10.20	0.956	121	0.02	0.4093	0.0401
40	35.333	−94.367	2	7	6	9.24	1.233	84	0.15	0.3675	0.0398
40	35.333	−94.367	3	9	5	10.68	1.725	75	0.06	0.4741	0.0444
40	35.333	−94.367	4	10	4	10.01	2.001	64	0.04	0.4573	0.0457
40	35.333	−94.367	5	9	6	13.93	2.506	74	−0.09	0.4854	0.0348
40	35.333	−94.367	6	7	6	12.63	2.533	94	−0.13	0.5710	0.0452
40	35.333	−94.367	7	6	5	12.86	2.538	107	−0.09	0.2998	0.0233
40	35.333	−94.367	8	6	4	11.08	2.479	106	0.01	0.5050	0.0456
40	35.333	−94.367	9	5	10	14.46	2.037	116	−0.23	0.7597	0.0525
40	35.333	−94.367	10	5	10	17.78	2.047	129	−0.10	0.5240	0.0295
40	35.333	−94.367	11	5	8	16.77	1.873	117	0.06	0.5430	0.0324
40	35.333	−94.367	12	5	10	12.73	1.147	119	0.11	0.4754	0.0373
41	41.980	−97.433	1	4	11	3.54	0.361	166	−0.17	0.4574	0.1291
41	41.980	−97.433	2	5	7	3.89	0.456	126	0.18	0.2444	0.0628
41	41.980	−97.433	3	6	9	7.23	0.723	104	0.07	0.3771	0.0521
41	41.980	−97.433	4	7	8	7.78	0.999	86	−0.05	0.6016	0.0773
41	41.980	−97.433	5	9	7	10.58	1.660	70	−0.13	0.5080	0.0480
41	41.980	−97.433	6	9	4	11.62	2.338	71	0.03	0.4478	0.0386
41	41.980	−97.433	7	9	3	9.19	2.783	74	−0.01	0.4700	0.0511
41	41.980	−97.433	8	9	2	6.67	2.176	73	0.07	0.5187	0.0778
41	41.980	−97.433	9	6	6	8.46	1.425	99	−0.07	0.4848	0.0573
41	41.980	−97.433	10	4	10	8.78	1.080	164	−0.13	0.4656	0.0530
41	41.980	−97.433	11	3	10	6.48	0.554	192	0.10	0.4211	0.0650

(*continued*)

Table F.2. (*cont.*)

Station ID	Latitude (°N)	Longitude (°W)	Month	m_ν	m_{t_r} (h)	m_h (mm)	m_i (mm h^{-1})	m_{t_b} (h)	$\rho[i, t_r]$	κ_o	λ_o (mm^{-1})
41	41.980	−97.433	12	4	7	3.73	0.495	153	0.07	0.4843	0.1298
42	39.567	−97.667	1	3	11	3.80	0.389	203	−0.13	0.5407	0.1424
42	39.567	−97.667	2	4	7	3.68	0.513	153	0.00	0.5373	0.1462
42	39.567	−97.667	3	5	8	6.98	0.696	120	0.21	0.3892	0.0557
42	39.567	−97.667	4	6	5	7.53	1.360	107	−0.03	0.3224	0.0428
42	39.567	−97.667	5	7	9	10.74	1.709	87	−0.16	0.5387	0.0502
42	39.567	−97.667	6	7	6	10.71	2.418	98	−0.16	0.6736	0.0629
42	39.567	−97.667	7	7	5	11.11	2.790	91	−0.17	0.5267	0.0474
42	39.567	−97.667	8	7	3	8.57	2.187	93	0.04	0.3934	0.0459
42	39.567	−97.667	9	6	4	7.78	1.794	112	−0.02	0.4101	0.0527
42	39.567	−97.667	10	3	10	10.05	1.181	203	−0.12	0.3248	0.0323
42	39.567	−97.667	11	3	8	5.38	0.682	198	−0.03	0.4447	0.0827
42	39.567	−97.667	12	3	6	3.83	0.523	188	0.10	0.3632	0.0949
43	35.400	−97.600	1	3	12	8.37	0.729	188	−0.07	0.4959	0.0593
43	35.400	−97.600	2	4	9	7.71	0.847	129	−0.04	0.4536	0.0589
43	35.400	−97.600	3	5	10	11.66	1.415	129	−0.18	0.6606	0.0566
43	35.400	−97.600	4	6	8	12.20	1.960	111	−0.20	0.6693	0.0549
43	35.400	−97.600	5	8	7	17.64	2.616	83	−0.07	0.5244	0.0297
43	35.400	−97.600	6	7	5	14.86	3.220	92	−0.09	0.5062	0.0341
43	35.400	−97.600	7	5	8	14.13	2.264	137	−0.17	0.4835	0.0342
43	35.400	−97.600	8	6	4	10.16	2.374	116	−0.07	0.6066	0.0597
43	35.400	−97.600	9	5	8	15.80	2.172	123	−0.10	0.4965	0.0314
43	35.400	−97.600	10	5	8	16.13	1.668	136	0.09	0.3112	0.0193
43	35.400	−97.600	11	3	15	12.96	0.882	199	−0.06	0.5799	0.0448
43	35.400	−97.600	12	4	9	8.02	0.899	163	−0.03	0.4209	0.0525
44	37.233	−93.383	1	6	10	7.66	0.660	110	0.07	0.4150	0.0542
44	37.233	−93.383	2	8	6	6.59	0.829	72	0.21	0.3064	0.0465
44	37.233	−93.383	3	9	7	9.59	1.330	72	−0.03	0.5404	0.0564
44	37.233	−93.383	4	11	4	8.99	1.857	56	−0.01	0.5405	0.0602
44	37.233	−93.383	5	11	5	10.27	2.085	61	−0.06	0.4968	0.0484
44	37.233	−93.383	6	9	6	13.71	2.700	73	−0.13	0.6009	0.0438
44	37.233	−93.383	7	6	6	12.98	2.624	103	−0.16	0.4049	0.0312
44	37.233	−93.383	8	7	5	10.92	2.332	90	−0.09	0.5337	0.0489
44	37.233	−93.383	9	6	8	16.54	2.105	100	−0.09	0.4748	0.0287
44	37.233	−93.383	10	7	7	12.40	1.485	90	0.10	0.4242	0.0342
44	37.233	−93.383	11	7	7	11.02	1.375	86	0.08	0.3454	0.0313
44	37.233	−93.383	12	7	9	10.71	0.942	97	0.16	0.4768	0.0445
45	30.533	−91.150	1	9	7	12.77	1.627	74	0.03	0.6483	0.0508
45	30.533	−91.150	2	8	6	15.60	2.194	71	0.02	0.5317	0.0341
45	30.533	−91.150	3	9	5	12.93	2.235	75	0.06	0.4883	0.0378
45	30.533	−91.150	4	7	5	19.10	2.765	95	0.17	0.2593	0.0136
45	30.533	−91.150	5	6	7	18.43	2.885	106	−0.11	0.4914	0.0267
45	30.533	−91.150	6	6	13	14.74	2.134	101	−0.27	0.5310	0.0360
45	30.533	−91.150	7	11	8	16.09	4.023	59	−0.24	0.6517	0.0405
45	30.533	−91.150	8	11	4	12.19	3.733	62	−0.13	0.4305	0.0353
45	30.533	−91.150	9	7	7	14.34	3.314	86	−0.12	0.4006	0.0279
45	30.533	−91.150	10	4	9	18.43	2.276	162	−0.10	0.3770	0.0205
45	30.533	−91.150	11	6	7	15.88	2.287	105	−0.03	0.5861	0.0369

Table F.2. (*cont.*)

Station ID	Latitude (°N)	Longitude (°W)	Month	m_ν	m_{t_r} (h)	m_h (mm)	m_i (mm h^{-1})	m_{t_b} (h)	$\rho[i, t_r]$	κ_o	λ_o (mm^{-1})
45	30.533	−91.150	12	9	7	15.19	1.885	75	0.12	0.4510	0.0297
46	44.767	−123.017	1	8	33	20.84	0.547	57	0.17	0.4031	0.0193
46	44.767	−123.017	2	8	28	15.53	0.519	56	0.04	0.4042	0.0260
46	44.767	−123.017	3	8	28	12.76	0.493	58	−0.05	0.4491	0.0352
46	44.767	−123.017	4	8	19	7.98	0.478	71	−0.13	0.4218	0.0529
46	44.767	−123.017	5	7	15	6.85	0.549	89	−0.17	0.5080	0.0742
46	44.767	−123.017	6	4	14	7.32	0.565	131	−0.12	0.5000	0.0683
46	44.767	−123.017	7	2	8	4.93	0.605	306	−0.07	0.4259	0.0864
46	44.767	−123.017	8	2	24	8.40	0.489	303	−0.28	0.3643	0.0434
46	44.767	−123.017	9	3	38	13.36	0.513	193	−0.27	0.5487	0.0411
46	44.767	−123.017	10	4	38	17.60	0.513	114	−0.11	0.5252	0.0298
46	44.767	−123.017	11	8	35	21.96	0.614	56	0.01	0.4480	0.0204
46	44.767	−123.017	12	9	26	16.92	0.571	49	0.11	0.3380	0.0200
47	44.767	−106.967	1	7	9	2.64	0.335	96	−0.20	0.6229	0.2362
47	44.767	−106.967	2	8	6	2.30	0.363	77	0.03	0.4478	0.1949
47	44.767	−106.967	3	9	6	2.88	0.424	69	−0.02	0.5918	0.2055
47	44.767	−106.967	4	9	7	4.90	0.553	67	0.11	0.2244	0.0458
47	44.767	−106.967	5	9	9	6.52	0.862	69	−0.07	0.3998	0.0613
47	44.767	−106.967	6	9	6	6.22	1.084	73	−0.11	0.3785	0.0609
47	44.767	−106.967	7	7	3	3.45	1.181	96	−0.01	0.2211	0.0642
47	44.767	−106.967	8	6	3	3.59	1.209	108	−0.07	0.4429	0.1235
47	44.767	−106.967	9	5	11	6.41	0.724	127	−0.15	0.6401	0.0998
47	44.767	−106.967	10	6	7	4.75	0.565	115	0.10	0.4720	0.0994
47	44.767	−106.967	11	6	8	3.36	0.435	106	−0.06	0.5128	0.1527
47	44.767	−106.967	12	7	7	2.25	0.353	94	−0.18	0.7894	0.3512
48	32.333	−88.750	1	9	7	13.13	1.833	69	−0.06	0.6107	0.0465
48	32.333	−88.750	2	10	5	12.69	2.074	60	0.06	0.4906	0.0387
48	32.333	−88.750	3	10	5	16.82	2.904	67	0.00	0.5900	0.0351
48	32.333	−88.750	4	8	5	15.43	2.884	76	0.01	0.5109	0.0331
48	32.333	−88.750	5	7	7	15.49	2.646	93	−0.17	0.5519	0.0356
48	32.333	−88.750	6	7	8	13.29	3.061	93	−0.25	0.8273	0.0623
48	32.333	−88.750	7	11	5	12.95	3.516	62	−0.17	0.5409	0.0418
48	32.333	−88.750	8	8	5	10.73	2.980	78	−0.17	0.5593	0.0521
48	32.333	−88.750	9	6	7	14.84	2.589	104	−0.20	0.4282	0.0289
48	32.333	−88.750	10	4	8	15.97	1.891	155	−0.02	0.4141	0.0259
48	32.333	−88.750	11	6	8	17.03	2.245	111	−0.14	0.8753	0.0514
48	32.333	−88.750	12	9	7	15.73	2.072	71	0.00	0.5807	0.0369
49	43.650	−70.317	1	12	7	7.76	0.790	51	0.44	0.4537	0.0585
49	43.650	−70.317	2	10	7	8.51	0.836	57	0.41	0.4005	0.0471
49	43.650	−70.317	3	12	7	8.71	0.821	54	0.42	0.4067	0.0467
49	43.650	−70.317	4	10	9	9.84	0.821	61	0.30	0.3737	0.0380
49	43.650	−70.317	5	12	6	7.18	0.974	53	0.09	0.3941	0.0549
49	43.650	−70.317	6	11	4	6.86	1.383	56	0.00	0.3436	0.0501
49	43.650	−70.317	7	11	3	6.61	1.744	60	0.01	0.4377	0.0663
49	43.650	−70.317	8	11	4	6.54	1.526	62	0.01	0.4368	0.0668
49	43.650	−70.317	9	9	5	8.80	1.330	70	0.18	0.2891	0.0329
49	43.650	−70.317	10	9	6	9.87	1.127	71	0.29	0.2880	0.0292
49	43.650	−70.317	11	10	8	11.90	1.049	58	0.36	0.4260	0.0358

(*continued*)

Table F.2. (*cont.*)

Station ID	Latitude (°N)	Longitude (°W)	Month	m_ν	m_{t_r} (h)	m_h (mm)	m_i (mm h^{-1})	m_{t_b} (h)	$\rho[i, t_r]$	κ_o	λ_o (mm^{-1})
49	43.650	−70.317	12	15	5	6.89	0.799	41	0.59	0.2822	0.0410
50	44.467	−73.150	1	16	5	3.05	0.416	40	0.32	0.2790	0.0915
50	44.467	−73.150	2	11	6	3.86	0.448	50	0.33	0.3105	0.0804
50	44.467	−73.150	3	11	8	4.85	0.552	55	0.08	0.4956	0.1022
50	44.467	−73.150	4	12	7	5.76	0.753	51	0.02	0.5678	0.0985
50	44.467	−73.150	5	11	8	6.78	0.949	54	−0.09	0.5571	0.0821
50	44.467	−73.150	6	12	6	7.20	1.392	52	−0.13	0.5931	0.0824
50	44.467	−73.150	7	15	3	6.05	1.774	46	0.04	0.4793	0.0792
50	44.467	−73.150	8	16	3	6.27	1.725	42	−0.02	0.4509	0.0719
50	44.467	−73.150	9	13	4	6.38	1.209	49	0.07	0.3998	0.0627
50	44.467	−73.150	10	10	7	6.65	0.840	60	0.02	0.5855	0.0881
50	44.467	−73.150	11	13	7	5.93	0.663	46	0.16	0.5427	0.0915
50	44.467	−73.150	12	19	4	3.18	0.481	33	0.41	0.2141	0.0672
51	41.167	−73.133	1	11	6	7.18	0.825	57	0.41	0.3740	0.0521
51	41.167	−73.133	2	10	6	7.27	0.884	57	0.40	0.4547	0.0625
51	41.167	−73.133	3	11	6	8.86	0.959	59	0.36	0.3770	0.0425
51	41.167	−73.133	4	10	7	9.43	1.092	64	0.18	0.4922	0.0522
51	41.167	−73.133	5	11	5	8.08	1.343	59	0.03	0.4588	0.0568
51	41.167	−73.133	6	8	5	8.93	1.541	79	−0.01	0.3103	0.0348
51	41.167	−73.133	7	8	4	10.95	2.664	79	−0.05	0.4286	0.0391
51	41.167	−73.133	8	10	3	8.89	2.169	67	0.02	0.4043	0.0455
51	41.167	−73.133	9	7	6	10.21	1.598	87	0.02	0.4095	0.0401
51	41.167	−73.133	10	7	6	11.69	1.537	100	0.12	0.4262	0.0365
51	41.167	−73.133	11	8	8	11.93	1.183	79	0.20	0.5859	0.0491
51	41.167	−73.133	12	12	5	7.26	0.923	53	0.41	0.4404	0.0606
52	37.100	−88.600	1	6	9	30.09	0.000	114	−0.16	0.1557	0.0052
52	37.100	−88.600	2	5	9	14.00	2.158	110	−0.18	0.4980	0.0356
52	37.100	−88.600	3	4	32	25.12	1.599	131	−0.30	0.6738	0.0268
52	37.100	−88.600	4	8	5	12.30	3.080	76	−0.16	0.6469	0.0526
52	37.100	−88.600	5	6	10	17.08	3.787	103	−0.13	0.6195	0.0363
52	37.100	−88.600	6	6	5	15.47	3.759	110	−0.19	0.7609	0.0492
52	37.100	−88.600	7	4	17	20.87	3.285	151	−0.27	0.7582	0.0363
52	37.100	−88.600	8	6	4	13.29	4.377	110	−0.18	0.7499	0.0564
52	37.100	−88.600	9	4	7	17.77	4.358	148	−0.16	0.5067	0.0285
52	37.100	−88.600	10	5	6	13.86	3.251	133	−0.11	0.6314	0.0456
52	37.100	−88.600	11	5	8	16.39	2.275	110	−0.19	0.6265	0.0382
52	37.100	−88.600	12	4	15	20.38	2.231	136	−0.13	0.8961	0.0440
53	42.933	−78.733	1	24	6	3.29	0.427	24	0.20	0.2820	0.0857
53	42.933	−78.733	2	17	7	3.68	0.426	31	0.20	0.2868	0.0780
53	42.933	−78.733	3	15	8	4.87	0.593	40	0.02	0.4599	0.0944
53	42.933	−78.733	4	15	6	5.03	0.785	41	0.00	0.5441	0.1081
53	42.933	−78.733	5	11	7	6.87	0.949	59	0.00	0.5058	0.0736
53	42.933	−78.733	6	10	5	7.65	1.581	66	−0.06	0.3746	0.0490
53	42.933	−78.733	7	11	3	6.87	1.916	64	0.02	0.4652	0.0677
53	42.933	−78.733	8	10	5	10.25	2.131	65	−0.12	0.5440	0.0531
53	42.933	−78.733	9	11	5	7.59	1.392	58	0.02	0.3569	0.0470
53	42.933	−78.733	10	9	11	8.61	0.820	71	−0.05	0.4997	0.0581
53	42.933	−78.733	11	14	9	6.96	0.701	39	0.08	0.4369	0.0628

Table F.2. (*cont.*)

Station ID	Latitude (°N)	Longitude (°W)	Month	m_v	m_{t_r} (h)	m_h (mm)	m_i (mm h^{-1})	m_{t_b} (h)	$\rho[i, t_r]$	κ_o	λ_o (mm^{-1})
53	42.933	−78.733	12	22	6	3.99	0.511	26	0.24	0.3857	0.0967
54	42.867	−100.550	1	3	7	1.95	0.312	189	−0.15	0.5401	0.2768
54	42.867	−100.550	2	3	9	3.33	0.350	168	0.01	0.4789	0.1438
54	42.867	−100.550	3	5	11	4.90	0.444	136	−0.01	0.4761	0.0971
54	42.867	−100.550	4	7	7	6.44	0.869	89	−0.03	0.4591	0.0713
54	42.867	−100.550	5	9	7	8.54	1.684	73	−0.15	0.4805	0.0563
54	42.867	−100.550	6	10	4	6.95	1.878	61	−0.07	0.5486	0.0790
54	42.867	−100.550	7	9	3	7.46	2.708	72	−0.06	0.4314	0.0578
54	42.867	−100.550	8	8	3	7.74	2.305	86	−0.07	0.4078	0.0527
54	42.867	−100.550	9	6	4	5.94	1.341	105	−0.07	0.5169	0.0870
54	42.867	−100.550	10	4	7	5.96	0.892	177	−0.04	0.6248	0.1049
54	42.867	−100.550	11	4	6	3.53	0.522	175	0.04	0.4692	0.1330
54	42.867	−100.550	12	4	7	2.41	0.343	181	−0.02	0.3875	0.1610
55	33.433	−112.017	1	2	17	7.22	0.647	297	−0.20	0.7697	0.1000
55	33.433	−112.017	2	2	12	6.88	0.684	279	−0.18	0.6539	0.0900
55	33.433	−112.017	3	2	12	8.37	0.781	288	−0.15	0.7176	0.0800
55	33.433	−112.017	4	1	6	4.28	0.761	548	−0.06	0.6963	0.1600
55	33.433	−112.017	5	1	2	3.07	1.101	643	0.08	0.5876	0.1900
55	33.433	−112.017	6	0	44	7.36	0.957	999	−0.27	0.4544	0.0600
55	33.433	−112.017	7	2	7	6.45	2.014	270	−0.23	0.3641	0.0500
55	33.433	−112.017	8	6	1	2.41	2.408	113	0.00	0.2939	0.1200
55	33.433	−112.017	9	2	6	7.22	1.606	308	−0.15	0.3472	0.0400
55	33.433	−112.017	10	2	8	7.83	0.815	341	0.07	0.4971	0.0600
55	33.433	−112.017	11	1	11	8.25	0.865	413	−0.25	1.1306	0.1300
55	33.433	−112.017	12	2	11	8.99	0.892	292	−0.11	0.5196	0.0500
56	32.133	−81.200	1	8	7	9.84	1.249	81	0.02	0.5346	0.0543
56	32.133	−81.200	2	11	4	7.43	1.341	55	0.19	0.4398	0.0592
56	32.133	−81.200	3	8	7	12.35	1.687	84	−0.02	0.5765	0.0467
56	32.133	−81.200	4	6	7	13.40	2.133	113	−0.11	0.5304	0.0396
56	32.133	−81.200	5	7	7	15.11	2.538	94	−0.14	0.5848	0.0387
56	32.133	−81.200	6	9	6	15.20	2.900	69	−0.14	0.4814	0.0317
56	32.133	−81.200	7	13	5	13.39	3.242	50	−0.11	0.4471	0.0334
56	32.133	−81.200	8	12	5	15.14	3.626	54	−0.13	0.4776	0.0315
56	32.133	−81.200	9	9	6	14.76	2.537	70	−0.07	0.2283	0.0155
56	32.133	−81.200	10	5	7	10.84	1.770	129	−0.10	0.3711	0.0342
56	32.133	−81.200	11	6	5	8.00	1.289	111	0.07	0.4571	0.0571
56	32.133	−81.200	12	8	6	8.38	1.237	82	0.14	0.5812	0.0694
57	40.783	−111.950	1	6	13	4.92	0.450	95	−0.19	0.7976	0.1622
57	40.783	−111.950	2	6	9	4.68	0.553	89	−0.14	0.7698	0.1646
57	40.783	−111.950	3	8	9	5.51	0.638	77	−0.08	0.8065	0.1464
57	40.783	−111.950	4	7	11	7.53	0.649	87	0.00	0.5895	0.0783
57	40.783	−111.950	5	6	11	6.96	0.674	104	−0.10	0.5915	0.0850
57	40.783	−111.950	6	4	11	4.65	0.628	166	−0.14	0.5078	0.0898
57	40.783	−111.950	7	3	7	5.48	1.192	193	−0.19	0.4158	0.0759
57	40.783	−111.950	8	4	8	5.60	0.969	165	−0.17	0.3827	0.0683
57	40.783	−111.950	9	3	13	7.55	0.685	173	−0.10	0.3023	0.0400
57	40.783	−111.950	10	4	12	7.90	0.675	160	−0.09	0.7614	0.0964
57	40.783	−111.950	11	4	18	7.17	0.515	133	−0.27	0.9865	0.1376

(*continued*)

Table F.2. (*cont.*)

Station ID	Latitude (°N)	Longitude (°W)	Month	m_ν	m_{t_r} (h)	m_h (mm)	m_i (mm h^{-1})	m_{t_b} (h)	$\rho[i, t_r]$	κ_o	λ_o (mm^{-1})
57	40.783	−111.950	12	6	14	5.56	0.488	108	−0.21	0.5987	0.1077
58	38.433	−113.017	1	4	14	3.88	0.363	151	−0.27	0.6891	0.1775
58	38.433	−113.017	2	5	7	3.19	0.479	113	−0.13	0.7285	0.2283
58	38.433	−113.017	3	5	18	5.49	0.452	124	−0.25	0.8829	0.1610
58	38.433	−113.017	4	4	15	5.77	0.553	161	−0.26	0.7614	0.1320
58	38.433	−113.017	5	4	10	4.67	0.751	169	−0.20	0.6035	0.1293
58	38.433	−113.017	6	2	6	4.06	0.750	253	−0.10	0.3939	0.0970
58	38.433	−113.017	7	3	19	5.77	0.887	214	−0.28	0.6894	0.1194
58	38.433	−113.017	8	5	4	3.87	1.115	129	−0.12	0.4478	0.1156
58	38.433	−113.017	9	3	5	5.53	0.956	189	0.01	0.4902	0.0886
58	38.433	−113.017	10	3	11	6.07	0.634	218	−0.22	0.8226	0.1356
58	38.433	−113.017	11	3	10	5.04	0.507	187	−0.10	0.7783	0.1546
58	38.433	−113.017	12	4	11	3.94	0.444	158	−0.21	0.5560	0.1411
59	31.800	−106.400	1	2	9	3.54	0.471	250	−0.20	0.8987	0.2542
59	31.800	−106.400	2	2	7	4.77	0.704	282	−0.07	0.7656	0.1604
59	31.800	−106.400	3	2	4	3.56	0.761	308	−0.06	0.5962	0.1675
59	31.800	−106.400	4	1	5	3.77	0.751	477	−0.05	0.3506	0.0930
59	31.800	−106.400	5	2	3	3.21	1.402	357	−0.11	0.4731	0.1473
59	31.800	−106.400	6	2	9	7.12	1.475	277	−0.22	0.4654	0.0654
59	31.800	−106.400	7	4	17	8.91	1.111	137	−0.27	0.3669	0.0412
59	31.800	−106.400	8	6	4	5.56	1.649	107	−0.10	0.3646	0.0656
59	31.800	−106.400	9	4	8	9.18	1.398	167	−0.11	0.3473	0.0378
59	31.800	−106.400	10	2	11	7.12	0.911	246	−0.15	0.4950	0.0695
59	31.800	−106.400	11	2	8	4.13	0.596	323	−0.14	0.4946	0.1199
59	31.800	−106.400	12	2	11	5.13	0.507	266	−0.12	0.4782	0.0932
60	27.767	−97.500	1	5	12	8.17	0.617	125	0.01	0.2161	0.0265
60	27.767	−97.500	2	4	13	10.48	0.752	133	0.03	0.3349	0.0320
60	27.767	−97.500	3	4	6	5.65	1.114	163	−0.08	0.2797	0.0495
60	27.767	−97.500	4	4	5	11.21	2.035	162	−0.04	0.2837	0.0253
60	27.767	−97.500	5	5	8	15.57	2.767	136	−0.14	0.4302	0.0276
60	27.767	−97.500	6	4	13	18.36	2.211	158	−0.21	0.3802	0.0207
60	27.767	−97.500	7	3	10	15.05	2.088	196	−0.11	0.2462	0.0164
60	27.767	−97.500	8	3	22	25.40	1.928	198	−0.18	0.2623	0.0103
60	27.767	−97.500	9	6	11	21.89	2.896	97	−0.14	0.2835	0.0129
60	27.767	−97.500	10	4	13	19.34	2.508	155	−0.21	0.4482	0.0232
60	27.767	−97.500	11	4	11	9.83	1.207	166	−0.17	0.4038	0.0411
60	27.767	−97.500	12	4	14	8.63	0.751	171	−0.12	0.3709	0.0430
61	44.117	−123.217	1	8	30	24.17	0.690	59	0.14	0.3578	0.0148
61	44.117	−123.217	2	7	26	18.39	0.674	61	0.02	0.4670	0.0254
61	44.117	−123.217	3	4	85	32.44	0.429	103	−0.15	0.6399	0.0197
61	44.117	−123.217	4	7	17	9.79	0.637	79	−0.16	0.4967	0.0507
61	44.117	−123.217	5	6	16	8.00	0.602	100	−0.18	0.5099	0.0637
61	44.117	−123.217	6	4	13	8.10	0.715	151	−0.12	0.5424	0.0670
61	44.117	−123.217	7	1	9	5.26	0.693	395	−0.20	0.3408	0.0648
61	44.117	−123.217	8	2	22	11.01	0.665	330	−0.17	0.4895	0.0444
61	44.117	−123.217	9	2	27	12.76	0.723	214	−0.25	0.5429	0.0425
61	44.117	−123.217	10	3	45	22.46	0.538	143	−0.09	0.4347	0.0194
61	44.117	−123.217	11	6	41	31.55	0.765	74	−0.01	0.4163	0.0132

Table F.2. *(cont.)*

Station ID	Latitude (°N)	Longitude (°W)	Month	m_{ν}	m_{t_r} (h)	m_h (mm)	m_i (mm h^{-1})	m_{t_b} (h)	$\rho[i, t_r]$	κ_o	λ_o (mm^{-1})
61	44.117	−123.217	12	6	47	32.41	0.589	71	0.17	0.3193	0.0099
62	46.150	−123.883	1	7	53	36.47	0.601	51	0.18	0.4755	0.0130
62	46.150	−123.883	2	8	36	23.83	0.627	44	0.05	0.5387	0.0226
62	46.150	−123.883	3	9	34	19.60	0.549	46	0.06	0.5257	0.0268
62	46.150	−123.883	4	12	17	10.23	0.572	42	0.01	0.4893	0.0478
62	46.150	−123.883	5	11	13	6.76	0.515	56	0.01	0.3231	0.0478
62	46.150	−123.883	6	7	18	8.54	0.468	75	−0.01	0.4612	0.0540
62	46.150	−123.883	7	5	13	5.76	0.459	133	−0.03	0.3293	0.0572
62	46.150	−123.883	8	5	15	7.02	0.438	133	0.02	0.3829	0.0545
62	46.150	−123.883	9	4	32	17.20	0.571	122	−0.06	0.4020	0.0234
62	46.150	−123.883	10	4	95	49.37	0.526	114	−0.02	0.3690	0.0075
62	46.150	−123.883	11	5	65	46.39	0.728	62	−0.03	0.5776	0.0125
62	46.150	−123.883	12	8	42	31.46	0.746	45	0.00	0.5662	0.0180
63	47.450	−122.300	1	10	22	13.52	0.579	49	0.04	0.5387	0.0399
63	47.450	−122.300	2	10	18	10.18	0.567	50	−0.02	0.5654	0.0555
63	47.450	−122.300	3	11	15	8.25	0.551	49	−0.07	0.5323	0.0645
63	47.450	−122.300	4	13	7	4.30	0.596	45	−0.01	0.5413	0.1258
63	47.450	−122.300	5	8	9	5.21	0.585	77	−0.10	0.4581	0.0879
63	47.450	−122.300	6	5	18	7.39	0.537	126	−0.24	0.6045	0.0818
63	47.450	−122.300	7	3	11	5.42	0.734	191	−0.27	0.7503	0.1384
63	47.450	−122.300	8	3	17	8.91	0.502	198	−0.01	0.4030	0.0453
63	47.450	−122.300	9	5	18	9.65	0.648	108	−0.20	0.5280	0.0547
63	47.450	−122.300	10	5	35	16.17	0.658	106	−0.25	0.4998	0.0309
63	47.450	−122.300	11	13	12	9.84	0.793	39	−0.02	0.5686	0.0578
63	47.450	−122.300	12	10	24	15.39	0.619	49	0.02	0.5003	0.0325
64	48.217	−106.617	1	5	12	1.68	0.234	120	−0.36	0.6256	0.3733
64	48.217	−106.617	2	6	6	1.19	0.254	106	−0.36	0.8175	0.6891
64	48.217	−106.617	3	6	6	1.65	0.337	111	−0.20	0.4260	0.2574
64	48.217	−106.617	4	6	6	2.74	0.471	110	−0.06	0.3622	0.1322
64	48.217	−106.617	5	7	11	6.30	0.645	95	−0.10	0.3774	0.0599
64	48.217	−106.617	6	10	4	5.64	1.398	65	−0.01	0.3658	0.0649
64	48.217	−106.617	7	7	4	5.81	1.657	97	−0.11	0.2924	0.0504
64	48.217	−106.617	8	5	5	5.73	1.037	119	−0.03	0.1860	0.0324
64	48.217	−106.617	9	5	7	4.79	0.638	127	−0.03	0.3063	0.0639
64	48.217	−106.617	10	5	3	2.84	0.556	130	0.32	0.3024	0.1064
64	48.217	−106.617	11	4	8	1.74	0.301	145	−0.29	0.6673	0.3830
64	48.217	−106.617	12	6	10	1.54	0.231	112	−0.49	0.6286	0.4091
65	48.183	−103.633	1	7	7	2.03	0.312	99	−0.21	0.6018	0.2965
65	48.183	−103.633	2	5	6	2.00	0.332	108	−0.10	0.5176	0.2583
65	48.183	−103.633	3	5	11	3.07	0.358	119	−0.21	0.4889	0.1593
65	48.183	−103.633	4	6	7	4.79	0.633	101	0.05	0.3817	0.0797
65	48.183	−103.633	5	5	15	7.99	0.682	112	−0.15	0.3871	0.0484
65	48.183	−103.633	6	9	6	7.09	1.474	71	−0.14	0.4678	0.0659
65	48.183	−103.633	7	9	2	5.59	1.872	77	0.02	0.3021	0.0540
65	48.183	−103.633	8	7	3	4.33	1.309	90	0.04	0.3681	0.0850
65	48.183	−103.633	9	6	7	5.78	0.721	109	0.10	0.3486	0.0603
65	48.183	−103.633	10	4	5	4.03	0.607	147	0.20	0.2594	0.0643
65	48.183	−103.633	11	4	9	2.82	0.373	150	−0.18	0.6543	0.2319

(continued)

Table F.2. (*cont.*)

Station ID	Latitude (°N)	Longitude (°W)	Month	m_ν	m_{t_r} (h)	m_h (mm)	m_i (mm h^{-1})	m_{t_b} (h)	$\rho[i, t_r]$	κ_o	λ_o (mm^{-1})
65	48.183	−103.633	12	5	14	2.77	0.276	130	−0.31	0.5592	0.2019
66	46.467	−84.367	1	14	12	4.09	0.309	38	0.04	0.5077	0.1240
66	46.467	−84.367	2	16	6	2.64	0.349	35	0.22	0.3325	0.1260
66	46.467	−84.367	3	12	8	4.60	0.491	53	0.10	0.4505	0.0980
66	46.467	−84.367	4	8	11	7.03	0.607	69	0.01	0.5449	0.0775
66	46.467	−84.367	5	10	7	6.98	0.963	65	−0.01	0.3898	0.0559
66	46.467	−84.367	6	12	5	6.95	1.278	53	0.03	0.4280	0.0615
66	46.467	−84.367	7	10	4	7.09	1.646	66	−0.04	0.5067	0.0715
66	46.467	−84.367	8	12	4	7.18	1.416	54	0.07	0.3753	0.0523
66	46.467	−84.367	9	15	5	6.28	1.108	40	0.07	0.4929	0.0785
66	46.467	−84.367	10	12	9	7.23	0.798	52	−0.03	0.2350	0.0325
66	46.467	−84.367	11	15	10	5.59	0.499	37	0.12	0.4734	0.0847
66	46.467	−84.367	12	18	9	3.85	0.373	31	0.13	0.4832	0.1256
67	30.683	−88.250	1	12	5	11.09	1.545	57	0.19	0.2966	0.0267
67	30.683	−88.250	2	9	6	14.33	1.991	63	0.08	0.5146	0.0359
67	30.683	−88.250	3	11	4	13.77	2.370	59	0.16	0.4634	0.0337
67	30.683	−88.250	4	6	5	17.68	2.720	100	0.12	0.3696	0.0209
67	30.683	−88.250	5	7	6	18.99	3.136	88	−0.05	0.4320	0.0227
67	30.683	−88.250	6	9	8	14.38	3.021	71	−0.19	0.3907	0.0272
67	30.683	−88.250	7	15	4	12.60	3.478	43	−0.12	0.3198	0.0254
67	30.683	−88.250	8	19	2	9.12	3.311	35	0.06	0.3710	0.0407
67	30.683	−88.250	9	9	7	17.53	2.848	71	−0.12	0.3323	0.0190
67	30.683	−88.250	10	5	8	15.86	2.096	137	−0.05	0.2703	0.0170
67	30.683	−88.250	11	6	8	14.59	1.860	101	−0.05	0.4078	0.0280
67	30.683	−88.250	12	9	7	14.68	1.794	74	0.05	0.5672	0.0386
68	24.550	−81.750	1	6	4	8.73	1.781	118	0.03	0.1100	0.0126
68	24.550	−81.750	2	5	5	8.48	1.658	115	−0.03	0.5399	0.0637
68	24.550	−81.750	3	5	3	8.70	2.414	135	0.02	0.3411	0.0392
68	24.550	−81.750	4	3	5	12.34	2.297	190	−0.01	0.2564	0.0208
68	24.550	−81.750	5	4	21	21.36	1.752	135	−0.14	0.2997	0.0140
68	24.550	−81.750	6	9	9	13.92	2.408	69	−0.15	0.2888	0.0207
68	24.550	−81.750	7	11	4	8.47	2.358	63	−0.13	0.4212	0.0497
68	24.550	−81.750	8	15	4	8.56	2.615	44	−0.10	0.3902	0.0456
68	24.550	−81.750	9	17	3	9.19	2.629	37	−0.04	0.2882	0.0313
68	24.550	−81.750	10	9	7	12.44	1.819	74	−0.04	0.2302	0.0185
68	24.550	−81.750	11	6	5	12.08	1.744	112	0.12	0.0590	0.0049
68	24.550	−81.750	12	5	6	9.91	1.595	126	−0.02	0.2123	0.0214
69	29.183	−81.050	1	7	4	7.80	1.453	91	0.18	0.2939	0.0377
69	29.183	−81.050	2	6	8	12.04	1.679	93	−0.11	0.4413	0.0366
69	29.183	−81.050	3	6	8	13.53	1.794	109	−0.08	0.4088	0.0302
69	29.183	−81.050	4	5	5	12.98	2.294	136	0.02	0.3937	0.0303
69	29.183	−81.050	5	5	12	14.05	1.807	113	−0.21	0.4715	0.0336
69	29.183	−81.050	6	8	12	18.38	2.669	72	−0.21	0.4715	0.0336
69	29.183	−81.050	7	11	7	13.45	2.884	59	−0.23	0.5419	0.0403
69	29.183	−81.050	8	12	6	13.71	3.380	55	−0.17	0.4347	0.0317
69	29.183	−81.050	9	11	7	14.95	2.475	53	−0.08	0.3721	0.0249
69	29.183	−81.050	10	8	10	15.16	1.785	81	−0.06	0.2587	0.0171
69	29.183	−81.050	11	6	5	9.50	1.740	100	0.02	0.2756	0.0290

Table F.2. (*cont.*)

Station ID	Latitude (°N)	Longitude (°W)	Month	m_v	m_{t_r} (h)	m_h (mm)	m_i (mm h^{-1})	m_{t_b} (h)	$\rho[i, t_r]$	κ_0	λ_0 (mm^{-1})
69	29.183	−81.050	12	6	6	9.47	1.394	107	0.05	0.3343	0.0353
70	43.117	−77.667	1	19	6	3.02	0.422	31	0.17	0.3433	0.1137
70	43.117	−77.667	2	16	6	3.46	0.441	33	0.28	0.2499	0.0722
70	43.117	−77.667	3	13	7	4.52	0.568	46	0.05	0.4672	0.1034
70	43.117	−77.667	4	10	11	6.84	0.688	61	−0.15	0.6874	0.1004
70	43.117	−77.667	5	10	7	6.38	0.973	65	−0.06	0.5354	0.0839
70	43.117	−77.667	6	9	5	7.60	1.500	70	−0.06	0.4860	0.0640
70	43.117	−77.667	7	9	4	6.80	1.938	71	−0.12	0.5347	0.0786
70	43.117	−77.667	8	10	4	8.30	1.965	66	−0.11	0.6045	0.0728
70	43.117	−77.667	9	11	4	6.40	1.225	58	0.06	0.4379	0.0684
70	43.117	−77.667	10	10	8	6.38	0.770	66	0.00	0.3554	0.0557
70	43.117	−77.667	11	14	7	5.17	0.626	42	0.11	0.3709	0.0718
70	43.117	−77.667	12	20	5	3.27	0.470	30	0.21	0.3522	0.1076
71	35.267	−75.550	1	9	6	11.28	1.472	69	0.24	0.3840	0.0340
71	35.267	−75.550	2	8	5	10.47	1.537	73	0.21	0.5524	0.0528
71	35.267	−75.550	3	10	4	8.01	1.469	68	0.23	0.3608	0.0450
71	35.267	−75.550	4	6	4	8.63	1.611	100	0.07	0.4726	0.0548
71	35.267	−75.550	5	6	9	13.51	1.811	112	−0.14	0.6349	0.0470
71	35.267	−75.550	6	7	5	11.52	2.269	99	−0.05	0.4030	0.0350
71	35.267	−75.550	7	9	6	12.48	2.455	77	−0.09	0.4020	0.0322
71	35.267	−75.550	8	9	5	12.68	2.286	75	0.03	0.3114	0.0246
71	35.267	−75.550	9	6	8	17.03	1.985	110	−0.01	0.2801	0.0164
71	35.267	−75.550	10	6	7	16.49	2.010	108	0.02	0.3508	0.0213
71	35.267	−75.550	11	6	6	14.29	1.683	98	0.21	0.3041	0.0213
71	35.267	−75.550	12	4	23	20.24	1.325	137	−0.23	0.8528	0.0421
72	36.900	−76.200	1	10	7	9.27	1.056	62	0.23	0.5538	0.0597
72	36.900	−76.200	2	10	6	8.37	1.083	56	0.18	0.6039	0.0722
72	36.900	−76.200	3	13	4	6.93	1.205	50	0.15	0.4239	0.0611
72	36.900	−76.200	4	9	5	8.13	1.422	70	−0.01	0.5064	0.0623
72	36.900	−76.200	5	9	7	10.78	1.921	75	−0.12	0.6055	0.0562
72	36.900	−76.200	6	9	4	9.86	2.363	72	−0.02	0.3392	0.0344
72	36.900	−76.200	7	10	6	12.90	2.843	67	−0.17	0.5016	0.0389
72	36.900	−76.200	8	10	4	12.53	2.780	67	−0.04	0.2721	0.0217
72	36.900	−76.200	9	7	7	15.88	2.015	95	0.03	0.3305	0.0208
72	36.900	−76.200	10	7	7	12.01	1.588	94	0.00	0.4691	0.0390
72	36.900	−76.200	11	7	7	9.51	1.216	91	0.07	0.6181	0.0650
72	36.900	−76.200	12	10	6	8.40	1.049	68	0.20	0.5402	0.0643
73	32.783	−79.933	1	9	6	8.03	1.175	70	0.10	0.5122	0.0638
73	32.783	−79.933	2	10	4	7.28	1.374	57	0.11	0.4737	0.0651
73	32.783	−79.933	3	9	6	11.88	1.801	74	0.03	0.5921	0.0498
73	32.783	−79.933	4	6	6	10.18	1.589	109	0.02	0.5264	0.0517
73	32.783	−79.933	5	6	8	13.48	2.067	107	−0.17	0.4948	0.0367
73	32.783	−79.933	6	7	11	18.12	2.181	88	−0.15	0.3361	0.0185
73	32.783	−79.933	7	11	5	13.46	3.890	59	−0.13	0.4733	0.0352
73	32.783	−79.933	8	10	7	15.41	2.913	64	−0.14	0.4360	0.0283
73	32.783	−79.933	9	9	6	15.22	2.832	71	−0.07	0.3004	0.0197
73	32.783	−79.933	10	4	9	14.54	1.620	145	−0.02	0.2809	0.0193
73	32.783	−79.933	11	6	5	7.86	1.372	103	−0.01	0.4435	0.0564

(*continued*)

Table F.2. (*cont.*)

Station ID	Latitude (°N)	Longitude (°W)	Month	m_v	m_{t_r} (h)	m_h (mm)	m_i (mm h^{-1})	m_{t_b} (h)	$\rho[i, t_r]$	κ_o	λ_o (mm^{-1})
73	32.783	−79.933	12	8	5	8.06	1.210	78	0.14	0.5212	0.0647
74	25.900	−97.433	1	1	12	8.95	0.670	631	0.05	0.2856	0.0319
74	25.900	−97.433	2	1	10	7.86	0.663	582	0.07	0.3644	0.0463
74	25.900	−97.433	3	1	5	4.17	1.040	705	−0.12	0.4284	0.1028
74	25.900	−97.433	4	0	20	10.43	1.762	1852	−0.31	0.5932	0.0569
74	25.900	−97.433	5	0	18	20.96	3.014	1278	−0.24	0.4471	0.0213
74	25.900	−97.433	6	0	17	20.69	3.679	828	−0.31	0.6234	0.0301
74	25.900	−97.433	7	1	8	16.35	3.472	644	−0.17	0.3133	0.0191
74	25.900	−97.433	8	1	6	10.63	2.513	736	−0.20	0.3819	0.0359
74	25.900	−97.433	9	1	18	29.15	2.573	643	−0.18	0.2353	0.0081
74	25.900	−97.433	10	1	8	16.76	2.314	570	−0.10	0.4454	0.0266
74	25.900	−97.433	11	1	10	15.41	1.677	705	−0.05	0.3929	0.0255
74	25.900	−97.433	12	0	38	14.96	0.463	953	−0.14	0.2853	0.0191
75	33.933	−118.400	1	2	31	26.32	0.945	219	−0.12	0.5117	0.0194
75	33.933	−118.400	2	2	20	21.00	1.141	189	−0.07	0.3897	0.0186
75	33.933	−118.400	3	3	14	12.16	1.079	179	−0.17	0.5856	0.0482
75	33.933	−118.400	4	2	14	9.61	0.817	287	−0.14	0.6370	0.0663
75	33.933	−118.400	5	1	4	2.82	0.589	568	0.01	0.1347	0.0477
75	33.933	−118.400	6	0	1	0.53	0.525	704	0.00	0.8433	1.6049
75	33.933	−118.400	7	0	1	0.41	0.412	1082	0.00	0.4555	1.1068
75	33.933	−118.400	8	1	1	1.52	1.521	582	0.00	0.3937	0.2588
75	33.933	−118.400	9	1	3	3.91	0.770	499	0.28	0.2991	0.0765
75	33.933	−118.400	10	1	2	3.08	1.115	334	0.14	0.2097	0.0680
75	33.933	−118.400	11	2	17	18.18	1.084	291	−0.05	0.4840	0.0266
75	33.933	−118.400	12	2	21	15.89	0.864	269	−0.10	0.4551	0.0286

G

Derivation of the G-function

In deriving the G-function *neglecting scattering* we follow the work of Ross and Nilson (1965) (see Ross, 1975, 1981). We refer to Fig. 3.13 for notation.

For the special case $\boldsymbol{r}_s = \boldsymbol{r}_z = z$, $h_\otimes = \frac{\pi}{2}$ and the angle being averaged is that between the vertical and the leaf normal giving

$$\mathrm{G}(h_\otimes) \equiv \mathrm{G}(\theta_\mathrm{L}) = \langle \cos(\theta_\mathrm{L}) \rangle \equiv \beta. \tag{G.1}$$

If $\mathrm{g}(\theta_\mathrm{L}, \phi_\mathrm{L})$ is the probability density function of foliage area orientation, its integral over the celestial hemisphere must satisfy

$$\frac{1}{2\pi} \int_0^{2\pi} \mathrm{g}(\theta_\mathrm{L}, \phi_\mathrm{L})\,\mathrm{d}\Omega \equiv 1, \tag{G.2}$$

in which an increment of the central solid angle, Ω, is $\Delta\Omega = \sin(\theta_\mathrm{L})\Delta\theta_\mathrm{L}\Delta\phi_\mathrm{L}$. It is unusual for leaves to be oriented with downward-facing upper leaf surfaces so we restrict the integration space of Eq. G.2 to the upper hemisphere. Equation G.2 is expanded as

$$\frac{1}{2\pi} \int_0^{2\pi} \mathrm{d}\phi_\mathrm{L} \int_0^{\frac{\pi}{2}} \mathrm{g}(\theta_\mathrm{L}, \phi_\mathrm{L}) \sin(\theta_\mathrm{L})\,\mathrm{d}\theta_\mathrm{L} \equiv 1. \tag{G.3}$$

Assuming azimuthal uniformity (i.e., axial symmetry) the dependence on ϕ vanishes, leaving $\mathrm{g}(\theta_\mathrm{L}, \phi_\mathrm{L}) = \mathrm{g}(\theta_\mathrm{L})$, whereupon Eq. G.3 can be written

$$\int_0^{\frac{\pi}{2}} \mathrm{g}(\theta_\mathrm{L}) \sin(\theta_\mathrm{L})\,\mathrm{d}\theta_\mathrm{L} = \int_0^{\frac{\pi}{2}} \tilde{\mathrm{g}}(\theta_\mathrm{L})\,\mathrm{d}\theta_\mathrm{L} \equiv 1, \tag{G.4}$$

in which, as in Chapter 2, $\tilde{\mathrm{g}}(\theta_\mathrm{L})$ is the pdf of leaf orientation angle.

With $\boldsymbol{r}$ in the Sun direction (i.e., $\boldsymbol{r} = \boldsymbol{r}_s$), the G-function can then be written

$$G(\theta) = \frac{1}{2\pi}\int_0^{2\pi} d\varphi_L \int_0^{\frac{\pi}{2}} \tilde{g}(\theta_L)\,|\cos\widehat{\boldsymbol{r}_S\boldsymbol{r}_L}\,|\,d\theta_L. \tag{G.5}$$

With $\theta \equiv$ solar zenith angle, the law of cosines gives

$$\cos\widehat{\boldsymbol{r}_S\boldsymbol{r}_L} = \cos\theta\cos\theta_L + \sin\theta\sin\theta_L\cos(\phi - \phi_L). \tag{G.6}$$

With azimuthal symmetry, the azimuth, ϕ, of the solar vector, $\boldsymbol{r}_s$, is arbitrary, so for convenience we will choose $\phi = 0$, and Eq. G.6 becomes

$$\cos\widehat{\boldsymbol{r}_S\boldsymbol{r}_L} = \cos\theta\cos\theta_L + \sin\theta\sin\theta_L\cos(\phi_L). \tag{G.7}$$

Since for $\boldsymbol{r} = \boldsymbol{r}_s$

$$\theta \equiv \frac{\pi}{2} - h_\otimes, \tag{G.8}$$

we can write Eq. G.7 in the alternative form

$$\cos\widehat{\boldsymbol{r}_S\boldsymbol{r}_L} = \sin(h_\otimes)\cos(\theta_L) + \cos(h_\otimes)\sin(\theta_L)\cos(\phi_L). \tag{G.9}$$

The notation $|\cos\widehat{\boldsymbol{r}_S\boldsymbol{r}_L}\,|$ in Eq. G.5 signifies the absolute value which is designed to avoid losing area by mathematical cancellation of contributions from opposite sides of the unit sphere. To assess its significance we rewrite Eq. G.9 as

$$|\cos\widehat{\boldsymbol{r}_S\boldsymbol{r}_L}\,| = \sin(h_\otimes)\cos(\theta_L)|[1 + \cot(h_\otimes)\tan(\theta_L)\cos(\phi_L)]|. \tag{G.10}$$

We can now see that

(1) For $\theta_L \le h_\otimes$, the bracketed term of Eq. G.10 remains positive for all ϕ_L, and Eq. G.6 becomes

$$G(h_\otimes) = \sin(h_\otimes)\int_0^{h_\otimes} \tilde{g}(\theta_L)\cos(\theta_L)\,d\theta_L, \quad \theta_L \le h_\otimes. \tag{G.11}$$

However, by definition

$$\beta \equiv \int_0^{\frac{\pi}{2}} \tilde{g}(\theta_L)\cos(\theta_L)\,d\theta_L, \tag{G.12}$$

or in this case, since $\theta_L \le h_\otimes$, $\tilde{g}(\theta_L) \equiv 0$ for $h_\otimes < \theta_L \le \frac{\pi}{2}$, and

$$\beta = \int_0^{h_\otimes} \tilde{g}(\theta_L)\cos(\theta_L)\,d\theta_L, \quad \theta_L \le h_\otimes. \tag{G.13}$$

Using Eq. 3.16, Eq. G.11 then becomes

$$\frac{G(h_\otimes)}{\sin(h_\otimes)} = \kappa_b = \beta, \quad \theta_L \le h_\otimes. \tag{G.14}$$

(2) For $\theta_L > h_\otimes$, $\cot(h_\otimes)\tan(\theta_L) > 1$, and the bracketed term of Eq. G.10 will be negative whenever

$$\cos(\phi_L) < -\tan(h_\otimes)\cot(\theta_L), \quad \theta_L > h_\otimes. \tag{G.15}$$

The value, ϕ_o, of the azimuthal angle at which the sign change takes place is then

$$\phi_o \equiv \cos^{-1}[-\tan(h_\otimes)\cot(\theta_L)], \tag{G.16}$$

and its supplement, $\phi_{os} \equiv \pi - \phi_o$, is

$$\phi_{os} \equiv \cos^{-1}[\tan(h_\otimes)\cot(\theta_L)]. \tag{G.17}$$

We can then write Eq. G.05 for $\theta_L > h_\otimes$ as

$$\begin{aligned} G(h_\otimes) = \frac{1}{\pi} \Bigg[& \sin(h_\otimes) \int_{h_\otimes}^{\frac{\pi}{2}} \tilde{g}(\theta_L)\cos(\theta_L) \int_0^{\varphi_o} \{1 + \cot(h_\otimes)\tan(\theta_L)\cos(\phi_L)\}\, d\phi_L\, d\theta_L \\ & - \sin(h_\otimes) \int_{h_\otimes}^{\frac{\pi}{2}} \tilde{g}(\theta_L)\cos(\theta_L) \int_{\varphi_o}^{\pi} \{1 + \cot(h_\otimes)\tan(\theta_L)\cos(\phi_L)\}\, d\phi_L\, d\theta_L \Bigg]. \end{aligned} \tag{G.18}$$

Integrating on ϕ_L gives

$$\begin{aligned} \frac{G(h_\otimes)}{\sin(h_\otimes)} = & \int_{h_\otimes}^{\frac{\pi}{2}} \tilde{g}(\theta_L)\cos(\theta_L)\left[\frac{2}{\pi}\phi_o - 1\right] d\theta_L \\ & + \frac{2}{\pi}\int_{h_\otimes}^{\frac{\pi}{2}} \tilde{g}(\theta_L)\cos(\theta_L)\cot(h_\otimes)\tan(\theta_L)\sin(\phi_o)\, d\theta_L. \end{aligned} \tag{G.19}$$

Replacing ϕ_o by $\pi - \phi_{os}$, noting that $\sin(\phi_o) = \sin(\phi_{os})$, and using Eq. G.17 we have

$$\frac{G(h_\otimes)}{\sin(h_\otimes)} = \int_{h_\otimes}^{\frac{\pi}{2}} \tilde{g}(\theta_L)\left\{1 + \frac{2}{\pi}[\tan(\phi_{os}) - \phi_{os}]\right\}\cos(\theta_L)\, d\theta_L, \quad \theta_L > h_\otimes. \tag{G.20}$$

Finally, Eqs. G.14 and G.20 can be combined to write the *general* relationship

$$\frac{G(h_\otimes)}{\sin(h_\otimes)} \equiv \kappa_b = \beta + \frac{2}{\pi}\int_{h_\otimes}^{\frac{\pi}{2}} \tilde{g}(\theta_L)\,[\tan(\phi_{os}) - \phi_{os}]\cos(\theta_L)\, d\theta_L, \quad 0 \le \theta_L \le \frac{\pi}{2}. \tag{G.21}$$

in which for $\theta_L \le h_\otimes$, the second term is identically zero. Remember that Eq. G.21 *neglects scattering*! We note that for $\theta_L > h_\otimes$, the second term is always positive and that it is dominated by the magnitude of $[\tan(\phi_{os}) - \phi_{os}]$ as given from Eq. G.17.

H

The canopy absorption index and compensation ratio as species constants

The Michaelis–Menten equation

A modified Michaelis–Menten equation of enzyme kinetics (White *et al.*, 1968) is commonly used to describe the relation between photosynthesis and light intensity (Monteith, 1963, p. 106; Gates, 1980, p. 50). For this application retain P_r to write, after Horn (1971, p. 68),

$$P_t = P + P_r = \frac{P_s \dfrac{I}{I_0}}{\dfrac{I}{I_0} + \dfrac{k}{I_0}}, \tag{H.1}$$

in which

P_t = total rate of photosynthesis of an isolated leaf,
P = net rate of photosynthesis of an isolated leaf,
P_s = light-saturated rate of photosynthesis of an isolated leaf,
= photosynthetic capacity of an isolated leaf,
P_r = respiration rate of an isolated leaf,
I = light intensity on an isolated leaf,
I_0 = intensity of "full sunlight" on an isolated leaf, and
k = empirical "binding constant measuring the effectiveness of an isolated leaf in getting and processing photons" (Horn, 1971, p. ix).

With $I = I_k$, the so-called "compensation" light intensity, the net rate of photosynthesis is zero by definition and Eq. H.1 becomes, for an isolated leaf

$$P_r = \frac{P_s \dfrac{I_k}{I_0}}{\dfrac{I_k}{I_0} + \dfrac{k}{I_0}}. \tag{H.2}$$

From the approximation in Fig. 8.5 we see that k is approximately equal to I_{SL} and $P_s = P_{sm}$ by definition, so that from Eq. H.2 for an isolated leaf

$$\frac{P_r}{P_{sm}} = \frac{\dfrac{I_k}{I_0}}{\dfrac{I_k}{I_0} + \dfrac{I_{SL}}{I_0}}. \tag{H.3}$$

Expanding to the whole canopy, we first assume that the compensation light intensity is the same for all leaves, i.e., $\widehat{I_k} = I_k$. Next, Eqs. 8.28 and 9.02 give $\widehat{I_{SL}} = I_{SL} = f_I(\beta L_t) I_0$, making Eq. H.3 for a climax canopy

$$\frac{P_r}{P_{sm}} = \frac{\dfrac{I_k}{I_0}}{\dfrac{I_k}{I_0} + f_I(\beta L_t)}, \tag{H.4}$$

and using Eq. 7.4 gives finally for the climax canopy

$$\frac{P_r}{P_{sm}} = \frac{1}{1 + f_I(\beta L_t) e^{\beta L_t}}. \tag{H.5}$$

Both P_r and P_{sm} are species constants which are separately temperature dependent (cf. Larcher, 1983, pp. 94, 97), but the temperature is fixed here by our optimization at the species constant, T_m. The ratio P_r/P_{sm} is thus a species constant at optimum temperature as is the canopy absorption index, βL_t, through Eq. H.5.

Finally, since $I_k/I_0 = \exp(-\beta L_t)$ from Eq. 7.4, I_k/I_0 is itself a species constant.

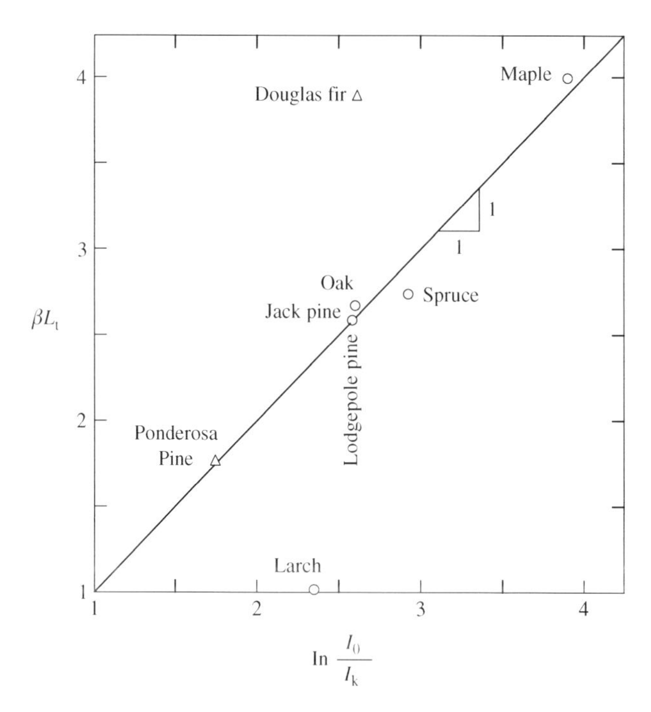

Fig. H.1. Compensation light intensity as a species constant fraction of ambient insolation. ○, $\beta L_t = a_w$ from Tables 2.2 and 2.3; △, $\beta L_t = \beta \overline{L_t}$ from Table 2.2; I_k/I_0 from Table 3.9.

Observational confirmation

For a few of Baker's (1950) genotypes listed in Table 3.9, observations of either the canopy absorption index, βL_t, or both β (or κ) and L_t, were found in the literature and are given in Tables 2.2 and 2.3. We use these to test the equivalence of βL_t and, using Eq. 3.53, Baker's I_k/I_0 as is shown in Fig. H.1. Essential agreement over the full range of Baker's species, from ponderosa pine to maple, lends observational support to our interpretation of the compensation intensities as a constant percentage of the ambient insolation, and therefore that κL_t is a constant for a given species. This result plays an important role in our work.

Glossary

Adiabat = state space path of constant heat

Advection = horizontal transport of a fluid quantity by virtue of the fluid motion

Albedo = ratio of light reflected from to that received by a surface

Amphistomatous = broadleaved or needle-leaved plant species having stomates on both leaf surfaces

Anabolic = pertaining to the synthesis, in living systems, of more complex substances from simpler ones

Arid climate = one in which the bare soil evaporation is under soil control, i.e., $E \leq 2/\pi$

Assimilation = the total process of plant nutrition including the uptake of external foods and the process of photosynthesis

Assimilation ratio = grams of organic matter produced per hour by a gram of chlorophyll under the light to which the system is adapted (Odum, 1975, p. 76)

Atmospheric emissivity $\equiv \varepsilon_a$ = efficiency of black body radiation from atmosphere

Atmospheric resistance $\equiv r_a$ = the time required for exchange of the sensible heat released in the condensation from 1 cm^3 of saturated air at one temperature in contact with 1 cm^2 of liquid water at a lower temperature

Azimuth = angular position about the stem

Bare soil evaporation effectiveness $\equiv E$ = independent climate–soil variable determining the bare soil evaporation efficiency

Bare soil evaporation efficiency $\equiv \beta_s$ = interstorm bare soil evaporation as a fraction of the potential value (Eq. 6.40)

Basal area = projected area on a horizontal plane

Basal leaf area index $\equiv \beta L_t$ = horizontal leaf area per unit of crown basal area

Beam radiation = radiation unmodified by the effects of intervening clouds (also called *direct* radiation)

Beam ratio = direct radiance fraction of total radiance

Big leaf model = representation of canopy by a single leaf having the equivalent properties of the full canopy

Biochemical assimilation capacity = the linear empirical function, $P_{sm} = \varepsilon I_{SL}$, relating maximum C_3 assimilation rate to the carbon-critical insolation

Biomass = weight of living plant material

Bowen ratio ≡ rate of sensible heat flux/rate of latent heat flux

Buffered tolerance = increased tolerance of unfavorable circumstances due to special mechanisms

Bulk drag parameter ≡ $\frac{u_*}{ku_0}$, and equal to the crown density (Eq. 4.35)

Canopy absorption index (or just "*canopy index*") ≡ horizontal leaf area index = βL_t (shown to be equal to κL_t) dimensionless index of exponential decay of either horizontal momentum (using β) or insolation (using κ)

Canopy conductance ≡ k_v = transpiration rate/potential rate of evaporation from a wet simple surface (Eq. 5.29)

Canopy cover ≡ M = fraction of substrate covered by vegetation

Canopy mass flux capacity ≡ $Mk_v^* E_{ps}$

Canopy resistance ≡ r_c = lumped resistance of all the leaves in the crown along with all the labyrinthine interleaf pathways to the canopy top

Canopy-scale non-homogeneity = condition in which crowns do not fill the canopy space

Canopy transpiration efficiency ≡ β_v = transpiration during the interstorm period as a fraction of the potential value (Eq. 6.47), dimensionless

Capillary rise ≡ w = rate at which water rises to a bare soil surface under the forces of capillarity (Eq. 6.51)

Carbon-critical insolation ≡ I_{SL} = insolation at which a leaf of a given species is photosynthetically saturated when water is non-limiting

Carbon demand function ≡ $f_D(\beta L_t)$ = the light-stimulated rate of crown carbon assimilation per unit of crown basal area assuming unlimited carbon supply to all leaves

Carryover storage = the difference, ΔS, between the stored soil moisture at the two ends of a partial-year growing season

Catabolic = pertaining to the degradation, in living organisms, of complex substances into their simpler constituents

Characteristic penetration depth = intersection of the asymptotes of the error function of the dimensionless diffusion depth (Appendix C-5)

Climate = long-term (at least 30-year) average of the local weather

Climate-controlled bare soil evaporation = evaporation under the saturating asymptote of β_s where $E \geq 2/\pi$

Climatic assimilation potential = the linear empirical relation, $P_{sm} = \varepsilon f_I(\beta L_t) I_0$, relating the maximum crown CO_2 assimilation rate to the climate and the species

Climatic climax = the globally optimal bioclimatic state $\widehat{I_l} = \widehat{I_{SL}} = \widehat{I_{SW}}$ for $M \leq 1$

Climatic insolation ≡ I_0 = daylight-hour average of the insolation during the growing season

Climax = the asymptotic growth state of a species; a stable stage in the evolution of a plant community supplantable only by change of the environment by some outside agency (Colinvaux, 1973); we use it in reference to the species that while fully exploiting the environment's productive potential is itself maximally productive, and hence stable. With this definition, the climax state is $\widehat{I_{SL}} = \widehat{I_l} = I_0$.

Closed canopy = canopy in which the horizontally projected leaf area equals the total substrate area thereby giving $M = 1$

Clustered needle varieties = trees with short, stiff needles clustered closely together on long twigs forming "shoots"; high needle drag interference causes the characteristic drag dimension to be that of the cluster

Compensation light intensity $\equiv I_k =$ insolation at which gross photosynthesis is just balanced by total respiration, making net photosynthesis equal to zero (Eq. 7.3)

Competing species = multiple species all meeting the feasibility conditions in the same environment

Critical light intensity = intensity at which water-equivalent of absorbed solar energy flux equals potential transpiration rate

Critical moisture state = soil moisture concentration at which stomatal closure begins

Crown = that portion of the tree above the lowest leaves

Crown density $\equiv (\frac{h-d_0}{h-h_s})\gamma L_t$ (cf. Eq. 4.35) governs the partition of drag between the substrate and the foliage elements for closed, sparse canopies (cf. Fig. 4.26)

Crown leaf area density = one-sided leaf area/crown volume

Crown-scale non-homogeneity = condition in which the leaf density is variable within the crown regardless of whether the crowns fill the canopy space

Cuticular resistance $\equiv r_{cu}$ = resistance impeding diffusion through the leaf cuticle

Dark respiration = the use of oxygen to break down substances in order to provide energy for plant metabolism

Deep percolation = downward movement of infiltrated water below the depth at which gravity becomes the dominant motive force

Dense canopy = canopy in which the momentum reaching the substrate is negligible and therefore the boundary shear stress may be neglected

Dense monolayer = monolayer with foliage density high enough that the canopy drag results primarily from the shape of the crown

Desiccation insolation $\equiv I_{SW}$ = insolation at which soil moisture reaches the critical value causing stomates to begin closing when light is non-limiting

Desorption = process of soil moisture reduction through capillary rise to the surface

Diffuse radiation = radiation arriving from every point of the sky

Diffusely distributed foliage = assumed infinite number of small, unconnected resistive elements behaving as a continuum-like sink for momentum

Diffusive penetration depth = depth to which moisture will penetrate a soil by simple diffusion during the average interstorm period

Direct solar radiation = parallel beams of the Sun's rays unmodified by intervening clouds (also called *beam* radiation)

Drag area density $\equiv a_D(z)$ = foliage element drag area in a unit vertical element of the crown-circumscribing circular cylinder

Drying power = the net capacity of the ambient atmosphere to augment (or diminish) the radiational forcing of evaporation through adiabatic transformation of excess sensible heat into additional latent heat (Eqs. 5.22, B.2, Fig. B.3)

Dry matter : radiation quotient $\equiv \varepsilon$ = grams of dry matter produced per megajoule of intercepted radiant energy

Dynamically active = contributing significant fluid drag

Ecotone = transition zone between different plant communities (Colinvaux, 1973)

Ecotype = same species raised in different habitats and retaining differences when cultivated under common conditions (Walter and Breckle, 1985, p. 198)

Eddy diffusivity = dimensional proportionality constant between turbulent flux density of fluid mass or sensible heat and the gradient (in the direction of transport) of the mean of this mass or sensible heat

Eddy viscosity (or *eddy momentum viscosity*) = dimensional proportionality constant between turbulent flux density of fluid momentum and the gradient (in the direction of transport) of the mean of this fluid momentum

Effective canopy conductance $\equiv \beta_v k_v^*$, dimensionless

Environmental insolation $\equiv I_0$ = insolation at the top of the crown

Equilibrium soil moisture $\equiv s_0$ = space–time average concentration of soil moistiure in the root zone of the soil, i.e., the "climatic" soil moisture, dimensionless

Equivalent atmospheric resistance = momentum concentration difference/ momentum flux

Equivalent interleaf resistance $\equiv r_{ci}$ = lumped resistance of the labyrinthine atmospheric pathway for fluxes through the crown

Equivalent leaf resistance $\equiv r_l$ = resistance of a unit leaf area including the leaf boundary layer

Equivalent leaf stomatal resistance $\equiv r_{ls}$ = resistance of a unit leaf area in which the density of individual stomates is accounted for

Erectophile = distribution of leaves tending toward the vertical (Eq. 2.13)

Evapotranspiration = the undifferentiated sum of evaporation (say from a bare soil substrate) and transpiration (from the vegetated component) as measured at screen height above the canopy (Eq. 6.49)

Exfiltration capacity $\equiv f_e^*$ = the rate at which, by virtue of its properties and moisture content, the soil can return moisture to the surface under given climatic conditions (Eq. 6.32)

Exfiltration rate = rate at which moisture leaves the soil at its surface under the force of capillarity less gravity

Extinction parameter = coefficient of exponential decay of a fluid quantity with vertical distance

Extremophile = distribution of leaves in which both horizontal and vertical tendencies predominate

Feasible habitat = habitat in which the species is unstressed

Flux–gradient relationship = linear relationship of turbulent flux density of fluid quantity to gradient in the mean of the quantity in the transport direction with proportionality coefficient being known as the kinematic eddy viscosity

Foliage area density $\equiv a_t(z)$ = one-sided foliage element area per unit horizontal area per unit of canopy depth (Eq. 4.15)

Foliage area index $\equiv L_t$ = total upper surface area of all foliage elements per unit of basal area (assumed approximately equal to *leaf area index*, *L*)

Foliage elements = leaves or needles, stems, shoots, pods, blossoms, fruit

Fractional relative humidity $\equiv S_r$

Free water surface evaporation = that which occurs from a thin film of water having insignificant heat storage (closely resembles potential evaporation from adequately watered "simple" natural surfaces such as individual leaves and bare soil)

Genotype = group or class sharing a specific genetic constitution: remains unaltered for a plant maintaining itself in a complex and changing environment (Sinnott, 1960, p. 303)

G-function $\equiv G(h_\otimes)$ = average over the canopy of the projection of a unit foliage area on a plane normal to the Sun direction; $G(h_{\otimes})/\sin(h_{\otimes})$ = shadow area per unit of foliage area (Eq. G.21)

Global radiation = combined direct solar radiation and diffuse radiation

Gross primary productivity ≡ GPP = NPP + respiration

Gross time = includes the hours of darkness

Growing season = period during which light, temperature and moisture conditions are conducive to plant growth

Habitat = the native environment of a plant species; we use it in reference to the environment in which a given species is maximally productive

Habitat heat proposition = Eq. 9.1, Fig. 9.4

Habitat light proposition: $\Pi_L = 1$ (Eq. 9.3, Figs. 9.5, 9.6)

Habitat water proposition; water-limited branch: $\Pi_W = 1$, $M \leq 1$ (Eq. 9.13, Fig. 9.7); light-limited branch: $\Pi_W < 1$, $M = 1$ (Eq. 9.14, Fig 9.7)

Heliotropism = the tendency of a plant to move toward or away from the light

Hemispherical distribution = *uniform distribution* = probability density function which is constant for all values of the variable (e.g., distribution of the normal direction to a unit element of the surface area of a hemisphere)

Homogeneity function ≡ $a(z) \equiv a_t(\xi)/L_t$, equal unity in homogeneous canopy

Homogeneous ≡ the same at all points

Homogeneous canopy = canopy of homogeneous cylindrical crowns with $M = 1$

Homogeneous crown = same foliage area density everywhere in crown

Horizontal leaf area index ≡ canopy absorption index = βL_t (shown to be equal to κL_t and to be a species constant) (Eq. 7.4)

Hypostomatous = broadleaved or needle-leaved plant species having stomates on only the underside of the leaves

Incipient illumination = the state of shadow in a canopy such that any increase in environmental insolation will reduce the "fullness" of that shadow

Individual needle varieties = trees with long needles which are separated to the point of relative drag independence

Infiltration capacity = infiltration rate as limited by soil properties and moisture content (Eq. 6.16)

Infiltration rate = rate at which moisture enters the soil under the forces of gravity and capillarity

Insolation = quantity of radiant solar energy arriving at a unit of horizontal surface during a given time

Insolation absorption index ≡ κL_t

Insolation extinction coefficient ≡ κ

Intercellular resistance ≡ r_{ic} = resistance controlling flow within the stomatal cavity

Interception = precipitation trapped on foliage surfaces

Interception capacity = depth of water which surface tension and gravity forces are capable of holding on the leaf surface

Interference function ≡ $s(\xi)$ = function describing the drag-modulating effects of neighboring foliage elements and/or crowns

Interleaf layer resistance ≡ r_i = average resistance of atmospheric path between adjacent leaf layers

Interstorm period = time between cessation of one rainstorm and the beginning of the next

Irradiance = the rate at which incident radiant energy is received per unit of surface area

Isothermal = constant temperature

Isotropic = insensitive to direction at a point

Leaf area density $\equiv a_L(\xi)$ = one-sided leaf area per unit horizontal area per unit of canopy depth

Leaf area index $\equiv L$ = total one-sided leaf area per unit of basal area (assumed approximately equal to the *foliage area index*, L_t)

Leaf boundary layer resistance $\equiv r_{la}$ = resistance restricting flow from the leaf surface out to the ambient atmosphere

Leaf inclination = *leaf angle* $\equiv \theta_L$ = angle between leaf surface and the horizontal = $\cos^{-1}\beta$

Leaf layer = horizontal crown layer containing leaves whose one-sided surface equals the plan area of the crown

Leaf layer stomatal resistance $\equiv r_{lls}$ = equivalent stomatal resistance of all leaves of a leaf layer

Leafy plants = plants with surfaces having a significant horizontal projection and for which the drag is primarily skin-friction

Light characteristic = asymptotic approximation of the leaf photosynthetic capacity curve (Fig. 8.7)

Light respiration = the release of CO_2 as a byproduct of plant photochemistry

Light-limited = the suboptimal bioclimatic state in which $\widehat{I_{SL}} > \widehat{I_l}$

Light-limited climax = the bioclimatic state $\widehat{I_l} = \widehat{I_{SL}} < \widehat{I_{SW}}$ with $M = 1$

Limiting soil moisture volume $\equiv V_e$ = necessary soil moisture volume at beginning of mean interstorm period such that at end of mean interstorm period the average soil moisture is the critical value, $s = s_5$ (Eq. 6.59)

Lumped canopy resistance = $\widehat{r_c}$ (Eq. 7.26)

Mat surface = dull surface

Matching conditions = the canopy top value of fluid quantities as determined separately by their different variation with vertical distance within and above the canopy but which must match at canopy top

Maximum canopy conductance = canopy conductance at minimum $\widehat{r_c/r_a}$ (Eq. 7.14)

Maximum canopy moisture flux $\equiv Mk_v^*$ (Eqs. 6.66, 7.36), dimensionless

Maximum crown moisture flux = flux at minimum $\widehat{r_c/r_a}$

Michaelis–Menten equation = photosynthetic rate as a function of insolation rate (Eq. 8.2)

Mixing length $\equiv l$ = average distance traversed by a fluctuating fluid element before it aquires the velocity of its new region

Momentum (horizontal) absorption index $\equiv \beta L_t$

Momentum (horizontal) extinction coefficient $\equiv \beta$

Monolayer = tree having the leaves concentrated in a thin layer of uniform density at the crown surface

Monolayer foliage volume density $\equiv \frac{\hat{d}}{t} L_t$ (Eq. 7.22)

Monsi–Saeki radiation extinction = $\frac{I(\xi)}{I_0} = \exp(-\kappa L_t \xi)$, (Eq. 7.1)

Multilayer = tree having the leaves distributed more-or-less homogeneously throughout the crown

Natural habitat = environment where the species has maximum stressless productivity

Net primary productivity = NPP = the sum of the accumulation of green plant biomass plus any net transfer of organic carbon from the green plant compartment to other compartments within the ecosystem (Bormann and Likens, 1979, p. 16)

Net radiation $\equiv R_n = q_i - q_b$ = the difference between incoming and back radiation

NIR = near infrared radiation

Open canopy = canopy in which the horizontally projected leaf area is less than the substrate area thereby giving $M < 1$

Optical optimality = the set of physical conditions producing maximum absorption of incident radiation by the canopy

Optimal bioclimatic state = $\widehat{I_{SW}} \geq \widehat{I_{SL}} = \widehat{I_l}$

Optimal canopy state = canopy with optimum foliage state and at critical moisture state (Fig. 7.10)

Optimal foliage state = the leaf area index and hence species for which, at given leaf angle and at optical optimality, the resistance ratio is minimum causing both the water vapor flux and nutrient flux to be maximum (Fig. 7.4)

PAR = photosynthetically active radiation

Penetration depth $\equiv z_i$ = average maximum penetration of the soil-wetting process, m

Penetration function $\equiv \exp(-\kappa L_t \xi)$ = decay of insolation with depth into the crown

Penman equation = classic expression for potential evaporation from a saturated simple surface (Eqs. 4.142, 5.20)

Penman–Monteith equation = classic expression for potential transpiration from a dry vegetated surface (Eq. 4.158)

Phenotype = the appearance of an organism resulting from the interaction of the genotype with the environment; may change as environment changes (Sinnott, 1960, p. 303)

Photosynthesis = the process by which absorbed solar radiation is used to construct energy-rich compounds in plants

Photosynthetic capacity curve = the net rate of CO_2 assimilation per unit of basal leaf area as a function of the insolation for a given species at optimum temperature (Fig. 8.9)

Phytotron = a chamber having controlled environment for plant growth experiments

Plagiophile = distribution of leaves in which the leaves tend toward a 45° inclination (Eq. 2.14)

Planophile = distribution of leaves in which the leaves tend toward the horizontal (Eq. 2.12)

Priestley–Taylor equation = isothermal relation for potential evaporation (Eq. 5.35)

Primary productivity = the rate at which energy is bound or organic material created by photosynthesis, per unit area, per unit time (Whittaker, 1975)

Potential rate of evaporation = the rate at which water will evaporate from a saturated simple surface under a particular fixed set of atmospheric conditions

Potential rate of transpiration = the rate at which water will transpire from a plant under given atmospheric conditions when the stomates are fully open

Potential water-limited photosynthesis = photosynthetic rate as limited by the available water if light were non-limiting

Primary production = the quantity of dry matter produced by vegetation covering an area

Productive efficiency = the factor by which the productive demand of the non-diffusive monolayer must be multiplied in a given climate to get the productive demand of the canopy (per unit basal area) (Eq. 10.44, Fig. 10.3)

Productive gain = the factor by which the productive demand of the groundcover monolayer $L_t = 1$, $\beta = 1$ must be multiplied in a given climate to get the productive demand of the canopy (per unit basal area) (Eq. 10.21, Fig. 10.1)

Productive potential = the light-saturated productivity of a species

Projected leaf area = the plan area of a leaf when it lies naturally (i.e., not flattened) on a horizontal surface

Quadratic soil = one where $c = d + 2$ so the water balance is quadratic in $s_* \equiv s_0^{1+d/2}$

Radiance = power emitted per unit solid angle per unit area of normal spherical surface

Radiant flux = rate of arrival of radiant energy at a normal surface

Radiant flux density = rate of arrival of radiant energy at a unit area of normal surface

Radiation absorption index $\equiv \kappa L_t$

Radiation extinction coefficient $\equiv \kappa$

Radiational forcing = the component of total evaporation produced solely by net radiation

Rainfall intensity $\equiv i$ = precipitation rate

Relative humidity = ratio of vapor density to saturated vapor density at a common temperature (when written as a fraction it is identical to the *saturation ratio*

Residual resistance $\equiv r_r$ = resistance governing flow across the cell walls into the stomatal cavity

Resistance ratio $\equiv r_c/r_a$ = lumped internal canopy resistance/lumped atmospheric resistance to screen height

Respiration = exudation of metabolically produced carbon dioxide (Monteith, 1973, p. 200)

Reynolds analogy = equality of eddy viscosities for the transfer of momentum and eddy diffusivities for the transfer of mass and heat in turbulent transfer in stable atmospheres (Eq. 4.121)

Riparian = of, pertaining to, or situated on the bank of a river or other body of water

Roughness density $\equiv \lambda$ = geometrical similarity parameter (cf. Eq. 4.106) governing the partition of drag between the substrate and the foliage elements for dense, open canopies (cf. Fig. 4.26)

Saturation insolation $\equiv I_{SL}$ = insolation at which net photosynthesis can be considered to have reached its limiting value, P_s

Saturation rate of assimilation $\equiv P_s$ = asymptotic assimilation rate as incident radiation increases

Saturation ratio = fractional relative humidity = vapor pressure/saturated vapor pressure, at constant temperature

Saturation vapor deficit = difference between saturated and actual vapor pressure at the actual temperature

Scattering = the sum of reflection and transmission of radiation

Screen height = 2 m = height above crown at which ambient environmental measurements are made

Selective instability = the species-limited situation in which $I_{SL} < I_o$ giving rise to pressure for productivity increase through species substitution

Shear velocity = fluid velocity at which the horizontal momentum per unit volume equals the canopy top shear stress, τ_0

Silhouette area = projection of crown surface area on a plane perpendicular to the radiation

Simple natural surface = surface supporting a film of water thin enough to be at constant temperature (and thus devoid of thermal convection) and freely replenished continuously

Soil-controlled bare soil evaporation = evaporation under the rising asymptote of β_s where $E < 2/\pi$

Solar altitude = angle, $h_{\otimes}$, between the observer's horizon plane and the plane containing the solar hour circle measured along that great circle on the celestial sphere containing the Sun, the observer, and the observer's zenith.

Sorption diffusivity $\equiv D_i$

Sparse monolayer = monolayer with foliage density low enough that the crown drag results primarily from drag on the individual foliage elements

Species = related individuals that resemble one another and are able to breed only among themselves

Species-limited = the suboptimal bioclimatic growth state in which $I_{SL} < I_o$

Specular reflection = sharply defined reflected beam produced by smooth surfaces such as polished metals, liquids, and mirrors (used here as an idealization of the reflection from leaf surfaces)

Stand = a homogeneous community of plants

State–space = the joint range of those variables describing the condition of the vegetation or the atmospheric moisture

Stemmy plants = plants with largely vertical surfaces and in which form drag predominates

Stomate = "pore" on leaf surface (usually the lower surface) through which the plant exchanges CO_2, O_2, and water vapor with the atmosphere

Stomatal cavity resistance $\equiv r_o$ = overall resistance of a single stomate from surrounding cells out through the leaf boundary layer

Stomatal resistance $\equiv r_s$ = resistance regulating plant water loss and CO_2 assimilation

Storm infiltration = storm depth less both storm surface runoff and storm surface retention

Storm rainfall excess $\equiv R^*_{sj}$ = the un-infiltrated storm rainfall (Eq. 6.22)

Storm surface retention $\equiv E_r$ = actual depth of surface retention in a storm (a part of storm rainfall excess)

Stress = exposure to extraordinarily unfavorable conditions (Larcher, 1983, p. 33) such as are created by water shortage and are signaled by stomatal closure

Substrate = ground surface

Surface emissivity $\equiv \varepsilon_o$ = efficiency of black body radiation from the surface

Surface retention capacity $\equiv h_o$ = depth of water that can be held on a surface without there being infiltration

Surface roughness length $\equiv z_0$ = vertical measure of the effective height of boundary roughness

Surface runoff $\equiv R_{sj}$ = storm rainfall excess less storm surface retention (Eq. 6.26)

Tapered crown = crown having an increasing cross-section with increasing depth, ξ

Time to stress = the time from cessation of rainfall until the stomates begin to close

Tolerance criterion = quantitative measure of the maximum survivable stress from a given source

Transpiration = evaporation of plant cellular moisture from the wet internal stomatal surfaces of the leaves

Trunk = the "stem" of a tree capped by the crown

Tundra = a generic term including vegetation types that range from tall shrub (up to 2 m high) to dwarf shrub heath, lichens, and mosses (Lewis and Callaghan, 1976, p. 400).

Turbulent flux capacity = the rate per unit area at which foliage-generated canopy turbulence is capable of vertical transport of momentum, mass and heat

Turgor = the degree of rigidity of plant cells resulting from internal pressure exerted by the cell contents

Uniform leaf distribution = "hemispherical" leaf orientation in that each element of solid angle contains the same foliage area (Eq. 2.15), i.e., foliage area is equally oriented in all directions

Uniform sky = diffuse radiation of uniform intensity from all directions

Unit vegetated area = a unit area covered completely by vegetation

Unstressed canopy conductance $\equiv k_v^* = E_{pv}/E_{ps}$

UV = ultraviolet radiation

Water balance = quantitative relation among the long-term averages of the partition of precipitation into evapotranspiration, surface runoff, and groundwater runoff (Eqs. 6.1, 6.65)

Water characteristic = asymptotic approximation of leaf transpiration rate as a function of incident radiation for fixed species, soil and precipitation (Fig. 8.15)

Water-critical insolation $\equiv I_{SW}$ = insolation at which stomates begin closure due to low soil moisture (remaining environmental components fixed)

Water demand rate = atmospheric potential for canopy transpiration at optimum foliage state

Water-limited climax = the bioclimatic state $\widehat{I_l} = \widehat{I_{SL}} = \widehat{I_{SW}}$ with $M < 1$

Water-limited community = a community having canopy cover $M < 1$

Water supply rate = average rate of exhaustion of available soil moisture

Water table = elevation, z_w, at which the soil is saturated from below

Wet-bulb depression = difference between the actual temperature and the saturation temperature at constant heat

Wet climate = one in which the bare soil evaporation is under climate control, i.e., $E > 2/\pi$

Wet season = rainy portion of a seasonal climate

Yield $\equiv Y_\tau$ = sum of surface and groundwater components of runoff

Zero-plane displacement height $\equiv d_0$ = similarity parameter of the velocity distribution over a rough surface (shown in Chapter 4 to be the level of action of the drag on the component parts of the canopy)

Zonal brightness = diffuse radiance on a horizontal surface

References

Abramowitz, M., and I.A. Stegun (editors), *Handbook of Mathematical Functions with Formulas, Graphs, and Mathematical Tables*, National Bureau of Standards Applied Mathematics Series 55, 9th Printing, U.S. Government Printing Office, Washington, D.C., 1964.

Addicott, F.T., and J.L. Lyon, Physiological ecology of abscission, in *Shedding of Plant Parts*, edited by T.T. Koslowski, pp. 85–124, Academic Press, New York, 1973.

Ahlgren, C.E., Phenological observations of nineteen native tree species in Northeastern Minnesota, *Ecology*, 38, 622–628, 1957.

Albrektson, A., *Biomass of Scots pine* (Pinus sylvestris *L.*). *Amount, Development, Methods of Mensuration*, Report No. 2, University of Agricultural Sciences, Department of Silviculture, Umea, Sweden, 1980a.

Albrektson, A., Relations between tree biomass fractions and conventional silvicultural measurements, in *Structure and Function of Northern Coniferous Forests: An Ecosystem Study*, edited by T. Persson, pp. 315–327, Ecological Bulletin 32, Swedish Natural Science Research Council (NFR), Stockholm, 1980b.

Allen, L.H., Turbulence and wind spectra within a Japanese Larch plantation, *J. Appl. Meteorol.*, 7, 73–78, 1968.

Allen, L.H., Jr., and E.R. Lemon, Carbon dioxide exchange and turbulence in a Costa Rican tropical rain forest, in *Vegetation and the Atmosphere*, vol. 2, *Case Studies*, edited by J.L. Monteith, pp. 265–308, Academic Press, New York, 1976.

Allen, L.H., C.S. Yocum, and E.R. Lemon, Photosynthesis under field conditions, VII. Radiant energy exchanges within a corn crop canopy and implications in water use efficiency, *Agron. J.*, 56(2), 253–259, 1964.

Amiro, B.D., Comparison of turbulence statistics within three boreal forest canopies, *Boundary-Layer Meteorol.*, 51, 99–121, 1990a.

Amiro, B.D., Drag coefficients and turbulence spectra within three boreal forest canopies, *Boundary-Layer Meteorol.*, 52, 227–246, 1990b.

Anderson, M.C., Stand structure and light penetration, II. A theoretical analysis, *J. Appl. Ecol.*, 3, 41–54, 1966.

Aronsson, A., and S. Elowson, Effects of irrigation and fertilization of mineral nutrients in Scots pine needles, in *Structure and Function of Northern Coniferous Forests: An Ecosystem Study*, edited by T. Persson, pp. 219–228, Ecological Bulletin 32, Swedish Natural Science Research Council (NFR), Stockholm, 1980.

Arris, L.L., and P.S. Eagleson, *A Physiological Explanation for Vegetation Ecotones in Eastern North America*, Ralph M. Parsons Laboratory Report 323, Department of Civil Engineering, Massachusetts Institute of Technology, Cambridge, MA, 1989a.

Arris, L.L., and P.S. Eagleson, Evidence of a physiological basis for the boreal–deciduous forest ecotone in North America, *Vegetatio*, 82, 55–58, 1989b.

Arris, L.L., and P.S. Eagleson, A water use model for locating the boreal/deciduous forest ecotone in Eastern North America, *Water Resources Res.*, 30(1), 1–9, 1994.

Aylor, D.E., Y. Wang, and D.R. Miller, Intermittent wind close to the ground within a grass canopy, *Boundary-Layer Meteorol.*, 66, 427–448, 1993.

Baker, F.S., *Principles of Silviculture*, McGraw-Hill, New York, 1950.

Baker, J.B., and O.G. Langdon, *Pinus taeda* L., in *Silvics of North America*, vol. 1, *Conifers*, coordinated by R.M. Burns and B.H. Honkala, pp. 497–512, Agriculture Handbook 654, U.S. Department of Agriculture, Washington, D.C., 1990.

Baker, M.B., Jr., *Hydrologic Regimes of Forested Areas in the Beaver Creek Watershed*, General Technical Report RM-90, U.S. Department of Agriculture Forest Service, Rocky Mountain Forest and Range Experiment Station, Fort Collins, CO, 1982.

Baker, M.B., Jr., Effects of ponderosa pine treatments on water yield in Arizona, *Water Resources Res.*, 22(1), 67–73, 1986.

Baldwin, H.I., The period of height growth in some northeastern conifers, *Ecology*, 12(4), 665–689, 1931.

Bartholomew, W.V., J. Meyer, and H. Laudelout, Mineral nutrient immobilization under forest and grass fallow in the Yangambi region (Belgian Congo), *Pub. Inst. Etude Agron. Congo Belge*, 57, 1953.

Billings, W.D., The historical development of physiological plant ecology, in *Physiological Ecology of North American Plant Communities*, edited by B.F. Chabot and H.A. Mooney, pp. 1–15, Chapman and Hall, New York, 1985.

Birkebak, R., and R. Birkebak, Solar radiation characteristics of tree leaves, *Ecology*, 45(3), 646–649, 1964.

Blum, B.M., *Picea rubens* Sarg., in *Silvics of North America*, vol. 1, *Conifers*, coordinated by R.M. Burns and B.H. Honkala, pp. 250–259, Agriculture Handbook 654, U.S. Department of Agriculture, Washington, D.C., 1990.

Bormann, F.H., and G.E. Likens, *Pattern and Process in a Forested Ecosystem*, Springer-Verlag, New York, 1979.

Boysen-Jensen, P., *Die Stoffproduktion der Pflanzen*, Gustav Fischer Verlag, Jena, Germany, 1932.

Brady, N.C., *The Nature and Properties of Soils*, 8th Edition, Macmillan, New York, 1974.

Bras, R.L., *Hydrology: An Introduction to Hydrologic Science*, Addison-Wesley, Reading, MA, 1990.

Bray, J.R., Primary consumption in three forest canopies, *Ecology*, 45, 165–167, 1964.

Briggs, L.J., Limiting negative pressure of water, *J. Appl. Phys.*, 21, 721–722, 1950.

Briscoe, P.V., Stomata and the plant environment, PhD thesis, University of Nottingham, U.K., 1969.

Britton, N.L., When the leaves appear, *Bull. Torrey Bot. Club*, 6, 235–237, 1878a.

Britton, N.L., When the leaves fall, *Bull. Torrey Bot. Club*, 6, 211–213, 1878b.

Brix, H., The effect of water stress on the rates of photosynthesis and respiration in tomato plants and loblolly pine seedlings, *Physiol. Plant.*, 15, 10–20, 1962.

Brooks, R.H., and A.T. Corey, Properties of porous media affecting fluid flow, *J. Irrig. Drain. Div., Am. Soc. Civil Eng., IR2*, 61–88, 1966.

Brown, H.P., Growth studies in forest trees, I. *Pinus rigida* Mill, *Bot. Gaz.*, 54, 386–402, 1912.

Brown, H.P., Growth studies in forest trees, II. *Pinus strobus* L, *Bot. Gaz.*, 59, 197–241, 1915.

Brown, K.W., and W. Covey, The energy-budget evaluation of the micro-meteorological transfer processes within a cornfield, *Agric. Meteorol.*, 3, 73–96, 1966.

Brutsaert, W.H., *Evaporation into the Atmosphere*, Reidel, Dordrecht, The Netherlands, 1982.

Budyko, M.I., *The Heat Balance of the Earth's Surface*, Leningrad, 1956 (in Russian). Translated by N.A. Stepanova, U.S. Weather Bureau Publication 131692, Department of Commerce, Washington, D.C., 1958.

Budyko, M.I., *Climate and Life*, 1971 (in Russian). English translation edited by D.H. Miller, Academic Press, New York, 1974.

Budyko, M.I. (editor), *Atlas of World Water Balance*, UNESCO, Paris, 1977.

Bunce, J.A., L.N. Miller, and B.F. Chabot, Competitive exploitation of soil water by five Eastern North American tree species, *Bot. Gaz.*, 138, 168–173, 1977.

Burdine, N.T., Relative permeability calculations from pore size distribution data, *Trans. AIME*, 198, 71–78, 1958.

Burns, R.M., and B.H. Honkala (technical coordinators), *Silvics of North America*, vol. 1, *Conifers*, Agriculture Handbook 654, U.S. Department of Agriculture, Washington, D.C., 1990a.

Burns, R.M., and B.H. Honkala (technical coordinators), *Silvics of North America*, vol. 2, *Hardwoods*, Agriculture Handbook 654, U.S. Department of Agriculture, Washington, D.C., 1990b.

Businger, J.A., J.C. Wyngaard, Y. Izumi, and E.F. Bradley, Flux–profile relationships in the atmospheric surface layer, *J. Atmos. Sci.*, 28, 181–189, 1971.

Cannell, M.G.R., *World Forest Biomass and Primary Production Data*, Academic Press, London, 1982.

Cannell, M.G.R., R. Milne, L.J. Sheppard, and M.H. Unsworth, Radiation interception and productivity of willow, *J. Appl. Ecol.*, 24, 261–278, 1987.

Carbon, B.A., G.A. Bartle, and A.M. Murray, Leaf area index of some eucalypt forests in south-west Australia, *Austral. Forest Res.*, 9, 323–326, 1979.

Carbon, B.A., G.A. Bartle, and A.M. Murray, Water stress, transpiration and leaf area index in eucalypt plantations in a bauxite mining area in south-west Australia, *Austral. J. Ecol.*, 6, 459–466, 1981.

Carslaw, H.S., and J.C. Jaeger, *Conduction of Heat in Solids*, 2nd Edition, Oxford University Press, New York, 1959.

Cauchy, A., Mémoire sur la rectification des courbes et la quadrature des surfaces courbes, presenté le 22 octobre, 1832,

Oevres complètes d'A. Cauchy, Ire série, vol. 2, pp. 167–177, Gauthier-Villars, Paris, 1908.

Chabot, B.F., and D.J. Hicks, The ecology of leaf life spans, *Annu. Rev. Ecol. Syst.*, 13, 229–259, 1982.

Chen, J.M., and T.A. Black, Defining leaf area index for non-flat leaves, *Plant, Cell Envir.*, 15, 421–429, 1992.

Christensen, N.L., Vegetation of the southeastern coastal plain, in *North American Terrestrial Vegetation*, edited by M.G. Barbour and W.D. Billings, pp. 317–363, Cambridge University Press, New York, 1988.

Cionco, R.M., A wind–profile index for canopy flow, *Boundary-Layer Meteorol.*, 3, 255–263, 1972.

Cionco, R.M., Analysis of canopy index values for various canopy densities, *Boundary-Layer Meteorol.*, 15, 81–93, 1978.

Clary, W.P., M.B. Baker, Jr., P.F. O'Connell, T.N. Johnsen, Jr., and R.E. Campbell, *Effects of Pinyon–Juniper Removal on Natural Resource Products and Uses in Arizona*, 28 pp., Research Paper RM-128, U.S. Department of Agriculture Forest Service, Rocky Mountain Forest and Range Experiment Station, Fort Collins, CO, 1974.

Colinvaux, P., *Introduction to Ecology*, John Wiley, New York, 1973.

Cooley, K.R., *Evaporation from Open Water Structures in Arizona*, Agricultural Experiment Station and Cooperative Extension Service Folder 159, University of Arizona, Tucson, 1970.

Cowan, I.R., The interception and absorption of radiation in plant stands, *J. Appl. Ecol.*, 5, 367–379, 1968.

Crank, J., *The Mathematics of Diffusion*, Oxford University Press, New York, 1956.

Culf, A.D., Equilibrium evaporation beneath a growing convective boundary layer, *Boundary-Layer Meteorol.*, 70, 37–49, 1994.

Daily, J.W., and D.R.F. Harleman, *Fluid Dynamics*, Addison-Wesley, Reading, MA, 1966.

Darwin, C., *On the Origin of Species by Means of Natural Selection*, John Murray, London, 1859.

Davies, J.A., and C.D. Allen, Equilibrium, potential and actual evaporation from cropped surfaces in Southern Ontario, *J. Appl. Meteorol.*, 12, 649–657, 1973.

Dawkins, R., *The Selfish Gene*, Oxford University Press, New York, 1976.

Deans, J.D., Fluctuations of the soil environment and fine root growth in a young Sitka spruce plantation, *Plant Soil*, 52, 195–208, 1979.

Deans, J.D., Dynamics of coarse root production in a young plantation of *Picea sitchensis, Forestry*, 54, 139–155, 1981.

de Bruin, H.A.R., A model for the Priestley–Taylor parameter α, *J. Climate Appl. Meteorol.*, 22, 572–578, 1983.

Decker, J.P., Effect of temperature on photosynthesis and respiration in red and loblolly pines, *Plant Physiol.*, 19, 679–688, 1944.

Denmead, O.T., Relative significance of soil and plant evaporation in estimating evapotranspiration, in *Plant Response to Climatic Factors*, edited by R.O. Slatyer, pp. 505–511, UNESCO, Paris, 1973.

Denmead, O.T., Temperate cereals, in *Vegetation and the Atmosphere*, vol. 2, *Case Studies*, edited by J.L. Monteith, pp. 1–31, Academic Press, New York, 1976.

Denmead, O.T., and E.F. Bradley, Flux–gradient relationships in a forest canopy, in *The Forest–Atmosphere Interaction*, edited by B.A. Hutchison and

B.B. Hicks, pp. 421–422, Reidel, Dordrecht, The Netherlands, 1985.

Denmead, O.T., and E.F. Bradley, On scalar transport in plant canopies, *Irrig. Sci.*, 8, 131–149, 1987.

Denmead, O.T., and I.C. McIlroy, Measurements of non-potential evaporation from wheat, *Agric. Meteorol.*, 7, 285–302, 1970.

Denmead, O.T., F.X. Dunin, S.C. Wong, and E.A.N. Greenwood, Measuring water use efficiency of eucalypt trees with chambers and micrometeorological techniques, *J. Hydrol.*, 150, 649–664, 1993.

Dickinson, R.E., A. Henderson-Sellers, C. Rosenzweig, and P.J. Sellers, Evapotranspiration models with canopy resistance for use in climate models: A review, *Agric. Forest Meteorol.*, 54, 373–388, 1991.

Doley, D., Photosynthetic productivity of forest canopies in relation to solar radiation and nitrogen cycling, *Austral. Forest Res.*, 12, 245–261, 1982.

Doorenbos, J., and W.O. Pruitt, *Crop Water Requirements*, FAO Irrigation and Drainage Paper 24, (revised 1977), United Nations, Rome, 1977.

Drake, B.G., M.A. Gonzàlez-Meler, and S.P. Long, More efficient plants: A consequence of rising atmospheric CO_2? *Annu. Rev. Plant Physiol. Plant Mol. Biol.*, 48, 609–639, 1997.

Drozdov, A.V., The productivity of zonal terrestrial plant communities and the moisture and heat parameters of an area, *Soviet Geography*, 12, 54–60, 1971.

Druilhet, A., A. Perrier, J. Fontan, and J.L. Laurent, Analysis of turbulent transfers in vegetation: Use of thoron for measuring the diffusivity profiles, *Boundary-Layer Meteorol.*, 2, 173–187, 1972.

Dunne, T., and R.D. Black, An experimental investigation of runoff prediction in permeable soils, *Water Resources Res.*, 6(2), 478–490, 1970.

Eagleson, P.S., *Dynamic Hydrology*, McGraw-Hill, New York, 1970.

Eagleson, P.S., Climate, soil, and vegetation, 1. Introduction to water balance dynamics, *Water Resources Res.*, 14(5), 705–712, 1978a.

Eagleson, P.S., Climate, soil, and vegetation, 2. The distribution of annual precipitation derived from observed storm sequences, *Water Resources Res.*, 14(5), 713–721, 1978b.

Eagleson, P.S., Climate, soil and vegetation, 3. A simplified model of soil moisture movement in the liquid phase, *Water Resources Res.*, 14(5), 722–730, 1978c.

Eagleson, P.S., Climate, soil and vegetation, 4. The expected value of annual evapotranspiration, *Water Resources Res.*, 14(5), 731–740, 1978d.

Eagleson, P.S., Climate, soil, and vegetation, 5. A derived distribution of storm surface runoff, *Water Resources Res.*, 14(5), 741–748, 1978e.

Eagleson, P.S., Climate, soil and vegetation, 6. Dynamics of the annual water balance, *Water Resources Res.*, 14(5), 749–764, 1978f.

Eagleson, P.S., Climate, soil, and vegetation, 7. A derived distribution of annual water yield, *Water Resources Res.*, 14(5), 765–776, 1978g.

Eagleson, P.S., Ecological optimality in water-limited natural soil–vegetation systems, 1. Theory and hypothesis, *Water Resources Res.*, 18(2), 325–340, 1982.

Eagleson, P.S., and T.E. Tellers, Ecological optimality in water-limited natural soil–vegetation systems, 2. Tests and applications, *Water Resources Res.*, 18(2), 341–354, 1982.

Ehleringer, J., Annuals and perennials of warm deserts, in *Physiological Ecology of North American Plant Communities*, edited by B.F. Chabot and H.A. Mooney, pp. 162–180, Chapman and Hall, New York, 1985.

Eshbach, O.W. (editor), *Handbook of Engineering Fundamentals*, John Wiley, New York, 1936.

Evans, G.C., and D.E. Coombe, Hemispherical and woodland canopy photography and the light climate, *J. Ecol.*, 47, 103–113, 1959.

Eyre, S.R., *Vegetation and Soils: A World Picture*, 2nd Edition, Edward Arnold, London, 1968.

Fakorede, M.A.B., and J.J. Mock, Leaf orientation and efficient utilization of solar energy by maize (*Zea mays* L.), in *Symposium on the Agrometeorology of the Maize Crop*, Iowa State University, 1976, pp. 207–230, WMO, Geneva, 1977.

Farnsworth, R.K., E.S. Thompson, and E.L. Peck, *Evaporation Atlas for the Contiguous 48 United States*, 26 pp. with maps, NOAA Technical Report NWS 33, Office of Hydrology, National Weather Service, Washington, D.C., 1982.

Fechner, G.H., *Picea pungens* Engelm., in *Silvics of North America*, vol. 1, *Conifers*, coordinated by R.M. Burns and B.H. Honkala, pp. 238–249, Agriculture Handbook 654, U.S. Department of Agriculture, Washington, D.C., 1990.

Federer, C.A., and D. Lash, *BROOK: A Hydrologic Simulation Model for Eastern Forests*, Water Resources Research Center Report 19, University of New Hampshire, Durham, NH, 1978.

Feller, M.C., Biomass and nutrient distribution in two eucalypt forest ecosystems, *Austral. J. Ecol.*, 5, 309–333, 1980.

Feynman, R.P., *The Feynman Lectures in Physics*, vol. 1, Addison-Wesley, Reading, MA, 1963.

Finnigan, J.J., Turbulent transport in flexible plant canopies, in *The Forest–Atmosphere Interaction*, edited by B.A. Hutchison and B.B. Hicks, pp. 443–480, Reidel, Dordrecht, The Netherlands, 1985.

Finnigan, J.J., and M.R. Raupach, Transfer processes in plant canopies in relation to stomatal characteristics, in *Stomatal Function*, edited by E. Zeiger, G.D. Farquhar, and I.R. Cowen, pp. 385–429, Stanford University Press, Stanford, CA, 1987.

Fogg, G.E. (editor), *The State and Movement of Water in Living Organisms*, Part II, *Water in the Plant*, Society for Experimental Biology, London, 1965.

Forrest, W.G., Biological and economic production in radiata pine plantations, *J. Appl. Ecol.*, 10, 259–267, 1973.

Forrest, W.G., and J.D. Ovington, Organic matter changes in an age series of *Pinus radiata* plantations, *J. Appl. Ecol.*, 7, 177–186, 1970.

Fraser, D.A., Vegetative and reproductive growth of black spruce (*Picea mariana* [Mill.] BSP.) at Chalk River, Ontario, Canada, *Can. J. Bot.*, 44, 567–580, 1966.

Fritschen, L.J., J. Hsia, and P. Doraiswamy, Evapotranspiration of a Douglas fir with a weighing lysimeter, *Water Resources Res.*, 13(1), 1977.

Fritz, S., Solar radiation during cloudless days, *Heat. Vent.*, 46, 69–74, 1949.

Fröhlich, C., and R.W. Brusa, Solar radiation and its variation in time, *Sol. Phys.*, 74, 209–215, 1981.

Gaastra, P., Climatic control of photosynthesis and respiration, in *Environmental Control of Plant Growth*, edited by L.T. Evans, pp. 113–140, Academic Press, New York, 1963.

Gail, F.W., The osmotic pressure of cell sap and its possible relation to winter killing and leaf fall, *Bot. Gaz.*, 81, 434–445, 1926.

Gardner, W.R., Dynamic aspects of water availability to plants, *Soil Sci.*, 89(2), 63–73, 1960.

Garratt, J.R., *Aerodynamic Roughness and Mean Monthly Surface Stress over Australia*, Division of Atmospheric Physics Technical Paper 29, CSIRO, Aspendale, Victoria, Australia, 1977.

Gash, J.H.C., and J.B. Stewart, The average surface resistance of a pine forest derived from Bowen ratio measurements, *Boundary-Layer Meteorol.*, 8, 453–464, 1975.

Gates, D.M., Energy, plants, and ecology, *Ecology*, 46(1,2), 1–13, 1965.

Gates, D.M., *Biophysical Ecology*, Springer-Verlag, New York, 1980.

Gates, D.M., R. Alderfer, and S.E. Taylor, Leaf temperatures of desert plants, *Science*, 159, 994–995, 1968.

Gee, G.W., and C.A. Federer, Stomatal resistance during senescence of hardwood leaves, *Water Resources Res.*, 8, 1456–1460, 1972.

George, M.G., M.J. Burke, H.M. Pellet, and A.G. Johnson, Low temperature exotherms and woody plant distribution, *Horticult. Sci.*, 9, 519–522, 1974.

Godman, R.M., H.W. Yawney, and C.H. Tubbs, *Acer saccharum* Marsh.: Sugar maple, in *Silvics of North America*, vol. 2, *Hardwoods*, coordinated by R.M. Burns and B.H. Honkala, pp. 78–91, Agricultural Handbook 654, U.S. Department of Agriculture, Washington, D.C., 1990.

Goel, N.S., Inversion of canopy reflectance models for estimation of biophysical parameters from reflectance data, in *Theory and Applications of Optical Remote Sensing*, edited by G. Asrar, pp. 205–251, John Wiley, New York, 1989.

Goldstein, S. (editor), *Modern Developments in Fluid Mechanics*, vols. 1 and 2, Oxford at the Clarendon Press, London, 1938.

Goodell, B.C., A reappraisal of precipitation interception by plants and attendant water loss, *J. Soil Water Conserv.*, 18, 231–234, 1963.

Gradshteyn, I.S., and I.M. Ryzhik, *Tables of Integrals, Series, and Products*, Academic Press, New York, 1980.

Grulois, J., Extinction du rayonnement global, tropismes et paramètres foliares, *Bull. Soc. Roy. Bot. Belgique*, 100, 315–335, 1967.

Gunderson, C.A., and S.D. Wullschleger, Photosynthetic acclimation in trees to rising CO_2: A broader perspective, *Photosynth. Res.*, 39, 369–388, 1994.

Hamon, W.R., Estimating potential evapo-transpiration, *J. Hydraul. Div., Am. Soc. Civil Eng.*, 87, HY-3, 107–120, 1961.

Harris, A.S., *Picea sitchensis* (Bong.) Carr., in *Silvics of North America*, vol. 1, *Conifers*, coordinated by R.M. Burns and B.H. Honkala, pp. 260–267, Agriculture Handbook 654, U.S. Department of Agriculture, Washington, D.C., 1990.

Hartog, G. den, and R.H. Shaw, A field study of atmospheric exchange processes within a vegetative canopy, in *Heat and Mass Transfer in the Biosphere, Pt. I, Transfer Processes in Plant Environment*, edited by D.A. deVries and N.H. Afgan, pp. 299–309, John Wiley, New York, 1975.

Havranek, W.M., and U. Benecke, Influence of soil moisture on water potential, transpiration and photosynthesis of conifer seedlings, *Plant Soil*, 49, 91–103, 1978.

Hawk, K.L., and P.S. Eagleson, *Climatology of Station Storm Rainfall in the Continental United States: Parameters of the Bartlett–Lewis and Poisson Rectangular Pulses Models*, with diskette, M.I.T. Department of Civil Engineering, R.M. Parsons Laboratory Report 336, Cambridge, MA, 1992.

Heath, M.C., and P.D. Hebblethwaite, Solar radiation intercepted by leafless, semi-leafless and leafed peas (*Pisum sativum*) under contrasting field conditions, *Ann. Appl. Biol.*, 107, 309–318, 1985.

Hendricks, D.W., and V.E. Hansen, Mechanics of evapo-transpiration, *J. Irrig. Drain. Div., Am. Soc. Civil Eng.*, 88, IR-2, 67–82, 1962.

Hicks, B.B., and C.M. Sheih, Some observations of eddy momentum fluxes within a maize canopy, *Boundary-Layer Meteorol.*, 11, 515–519, 1977.

Hicks, D.J., and B.F. Chabot, Deciduous forest, in *Physiological Ecology of North American Plant Communities*, edited by B.F. Chabot and H.A. Mooney, pp. 257–277, Chapman and Hall, New York, 1985.

Hinckley, T.M., M.O. Schroeder, J.E. Roberts, and D.N. Bruckeshoff, Effect of several environmental variables and xylem pressure potential on leaf surface resistance in white oak, *For. Sci.*, 21, 201 211, 1975.

Hinckley, T.M., R.G. Aslin, R.R. Aubuchon, C.L. Metcalf, and J.E. Roberts, Leaf conductance and photosynthesis in four species of the oak hickory forest type, *For. Sci.*, 24, 73–84, 1978.

Hoerner, S.F., *Fluid-Dynamic Drag*, Hoerner Fluid Dynamics, Brick Lawn, NJ, 1958.

Holbrook, N.M., and M.A. Zwieniecki, Embolism repair and xylem tension: Do we need a miracle?, *Plant Physiol.*, 120, 7–10, 1999.

Holmgren, P., P.G. Jarvis, and M.S. Jarvis, Resistances to carbon dioxide and water vapor transfer in leaves of different plant species, *Physiol. Plant.*, 18, 557–573, 1965.

Horn, H.S., *The Adaptive Geometry of Trees*, Princeton University Press, Princeton, NJ, 1971.

Horton, R.E., The role of infiltration in the hydrologic cycle, *Trans. Am. Geophys. U.*, 14, 446–460, 1933.

Hughes, G., J.D.H. Keatinge, and S.P. Scott, Pigeon peas as a dry season crop in Trinidad, West Indies, II. Interception and utilization of solar radiation, *Trop. Agric., Trinidad*, 58, 191–199, 1981.

Illick, J.S., *Pennsylvania Trees*, 4th Edition, Pennsylvania Commission of Forestry, Harrisburg, PA, 1923.

Inoue, E., On the turbulent structure of airflow within crop canopies, *J. Meteorol. Soc. Japan*, 41, 317–326, 1963.

Inoue, K., and Z. Uchijima, Experimental study of microstructure of wind turbulence in rice and maize canopies, *Bull. Nat. Inst. Agric. Sci. Tokyo, Japan, Ser. A*, 26, 1–88, 1979.

Iqbal, M., *An Introduction to Solar Radiation*, Academic Press, New York, 1983.

Jacquemin, B., and J. Noilhan, Sensitivity study and validation of a land surface parameterization using the HAPEX-MOBILHY data set, *Boundary-Layer Meteorol.*, 52, 92–134, 1990.

Jarvis, P.G., and J.W. Leverenz, Productivity of temperate, deciduous and evergreen forests, in *Encyclopedia of Plant Physiology*, New Series, vol. 12D, *Physiological Plant Ecology IV: Ecosystem Processes*, edited by O.L. Lange, P.S. Nobel, C.B. Osmond, and H. Ziegler, pp. 233–280, Springer-Verlag, New York, 1983.

Jarvis, P.G., and K.G. McNaughton, Stomatal control of transpiration: Scaling up from leaf to region, *Adv. Ecol. Res.*, 15, 1–49, 1986.

Jarvis, P.G., G.B. James, and J.J. Landsberg, Coniferous forest, in *Vegetation and the Atmosphere*, vol. 2, *Case Studies*, edited by J.L. Monteith, pp. 171–240, Academic Press, New York, 1976.

Jasinski, M.F., and P.S. Eagleson, Estimation of subpixel vegetation cover using red–infrared scattergrams, *IEEE Trans. Geosci. Remote Sens.*, 28(2), 253–267, 1990.

Jensen, M.E., and H.R. Haise, Estimating evapotranspiration from solar radiation, *J. Irrig. Drain. Div., Am. Soc. Civil Eng.*, IR-4, 15–41, 1963.

Jensén, P., and S. Pettersson, Nutrient uptake in roots of Scots pine, in *Structure and Function of Northern Coniferous Forests: An Ecosystem Study*, edited by T. Persson, pp. 229–237, Ecological Bulletin 32, Swedish Natural Science Research Council (NFR), Stockholm, 1980.

Jones, L.A., and H.R. Condit, Sunlight and skylight as determinants of photographic exposure, *J. Optic. Soc. Am.*, 38, 123–178, 1948.

Jordan, C.F., Derivation of leaf-area index from quality of light on the forest floor, *Ecology*, 50(4), 663–666, 1969.

Kaimal, J.C., and J.J. Finnigan, *Atmospheric Boundary Layer Flows*, Oxford University Press, New York, 1994.

Kármán, T. von, Mechanische Åhnlichkeit und Turbulenz, *Nachr. Ges. Wiss. Göttingen, Math.-Phys. Kl.*, 58–76, 1930.

Kármán, T. von, Turbulence and skin friction, *J. Aeronaut. Sci.*, 1, 1–20, 1934.

Kestemont, P., Production primaire de la strate aborée d'une hêtraie à fétuques, *Bull. Soc. Roy. Bot. Belgique*, 106, 305–316, 1973.

Kim, C.P., and D. Entekhabi, Examination of two methods for estimating regional evaporation using a coupled mixed layer and land surface model, *Water Resources Res.*, 33(9), 2109–2116, 1997.

Kimball, H.H., and I.F. Hand, Sky-brightness and daylight illumination measurements, *Monthly Weather Rev.*, 49, 481–487, 1921.

Klein, W.H., Calculation of solar radiation and the solar heat load on man, *J. Meteorol.*, 5, 119–129, 1948.

Knapp, A.K., and M.D. Smith, Variation among biomes in temporal dynamics of aboveground primary production, *Science*, 291, 481–484, 2001.

Kohler, M.A., T.J. Nordenson, and D.R. Baker, *Evaporation Maps for the United States*, Technical Paper no. 37, U.S. Weather Bureau, Washington, D.C., 1959.

Kondo, J., Relation between the roughness coefficient and other aerodynamic parameters, *J. Meteorol. Soc. Japan*, 49, 121–124, 1971.

Kondo, J., On a product of mixing length and coefficient of momentum absorption within plant canopies, *J. Meteorol. Soc. Japan*, 50, 487–488, 1972.

Kondo, J., and S. Akashi, Numerical studies on the two-dimensional flow in horizontally homogeneous canopy layers, *Boundary-Layer Meteorol.*, 10, 255–272, 1976.

Kondratyev, K. Ya., *Radiation in the Atmosphere*, Academic Press, New York, 1969.

Kozlowski, T.T. (editor), *Water Deficits and Plant Growth*, vol. 2, *Plant Water Consumption and Response*, Academic Press, New York, 1968.

Kozlowski, T.T., and F.X. Schumacher, Estimation of stomated foliar surface of pines, *Plant Physiol.*, 18, 122–127, 1943.

Kramer, P.J., *Plant and Soil Water Relationships: A Modern Synthesis*, McGraw-Hill, New York, 1969.

Kramer, P.J., and W.S. Clark, A comparison of photosynthesis in individual pine needles and entire seedlings at various light intensities, *Plant Physiol.*, 22, 51–57, 1947.

Kramer, P.J., and J.P. Decker, Relation between light intensity and rate of photosynthesis of loblolly pine and certain hardwoods, *Plant Physiol.*, 19, 350–358, 1944.

Kramer, P.J. and T.T. Kozlowski, *Physiology of Trees*, McGraw-Hill, New York, 1960.

Kriedeman, P.E., T.F. Neales, and D.H. Ashton, Photosynthesis in relation to leaf orientation and light interception, *Austral. J. Biol. Sci.*, 17, 591–600, 1964.

LaDeau, S.L., and J.S. Clark, Rising CO_2 levels and the fecundity of forest trees, *Science*, 292, 95–98, 2001.

Lamb, H., *Hydrodynamics*, 6th Edition, Dover, New York, 1932.

Landsberg, J.J., and P.G. Jarvis, A numerical investigation of the momentum balance of a spruce forest, *J. Appl. Ecol.*, 10, 645–655, 1973.

Landsberg, J.J., and A.S. Thom, Aerodynamic properties of a plant of complex structure, *Quart. J. Roy. Meteorol. Soc.*, 97, 565–570, 1971.

Lang, A.R.G., Leaf area and average leaf angle from transmission of direct sunlight, *Austral. J. Bot.*, 34, 349–355, 1986.

Lang, A.R.G., Application of some of Cauchy's theorems to estimation of surface areas of leaves, needles and branches of plants, and light transmittance, *Agric. Forest Meteorol.*, 55, 191–212, 1991.

Lang, A.R.G., and R.E. McMurtrie, Total leaf areas of single trees of *Eucalyptus grandis* estimated from transmittances of the sun's beam, *Agric. Forest Meteorol.*, 58, 79–92, 1992.

Lang, A.R.G., and Yueqin Xiang, Estimation of leaf area index from transmission of direct sunlight in discontinuous canopies, *Agric. Forest Meteorol.*, 37, 229–243, 1986.

Lang, A.R.G., Yueqin Xiang, and J.M. Norman, Crop structure and the penetration of direct sunlight, *Agric. Forest Meteorol.*, 35, 83–101, 1985.

Lang, A.R.G., R.E. McMurtrie, and M.L. Benson, Validity of surface area indices of *Pinus radiata* estimated from transmittance of the sun's beam, *Agric. Forest Meteorol.*, 57, 157–170, 1991.

Larcher, W., *Physiological Plant Ecology*, translated by M.A. Biederman-Thorson, Springer-Verlag, New York, 1975.

Larcher, W., *Physiological Plant Ecology*, corrected printing of the 2nd Edition, translated by M.A. Biederman-Thorson, Springer-Verlag, New York, 1983.

Larcher, W., *Physiological Plant Ecology*, 3rd Edition, translated by J. Wieser, Springer-Verlag, Berlin, 1995.

Lechowicz, M.J., Why do temperate deciduous trees leaf out at different times? Adaptation and ecology of forest communities, *Am. Nat.*, 124, 821–842, 1984.

Ledig, F.T., A.P. Drew, and J.G. Clark, Maintenance and constructive respiration, photosynthesis and net assimilation rate in seedlings of pitch pine (*Pinus rigida* Mill.), *Ann. Bot.*, 40, 289–300, 1976.

Lee, X., and T.A. Black, Atmospheric turbulence within and above a Douglas fir stand, I. Statistical properties of the velocity field, *Boundary-Layer Meteorol.*, 64, 149–174, 1993.

Lemon, E.R., Energy and water balance of plant communities, in *Environmental Control of Plant Growth*, edited by L.T. Evans, pp. 55–78, Academic Press, New York, 1963.

Lemon, E.R., Micrometeorology and the physiology of plants in their natural environment, in *Plant Physiology: A Treatise*, edited by F. C. Steward, pp. 203–227, Academic Press, New York, 1965.

Leonard, R.E., and C.A. Federer, Estimated and measured roughness parameters for a pine forest, *J. Appl. Meteorol.*, 12, 302–307, 1973.

Levy, E.B., and E.A. Madden, The point method of pasture analysis, *N. Z. J. Agric.*, 46, 267–279, 1933.

Lewis, M.C., and T.V. Callaghan, Tundra, in *Vegetation and the Atmosphere*, vol. 2, *Case Studies*, edited by J.L. Monteith, pp. 399–433, Academic Press, New York, 1976.

Lhomme, J.-P., A theoretical basis for the Priestley–Taylor coefficient, *Boundary-Layer Meteorol.*, 82, 179–191, 1997.

Lindeman, R.L., The trophic dynamic aspects of ecology, *Ecology*, 23, 399–418, 1942.

Linder, S., Potential and actual production in Australian forest stands, in *Research for Forest Management*, edited by J.J. Landsberg and W. Parson, pp. 11–51, CSIRO, Melbourne, 1985.

Lindroth, A., Canopy conductance of coniferous forest related to climate, *Water Resources Res.*, 21, 297–304, 1985.

Lindroth, A., Aerodynamic and canopy resistance of short-rotation forest in relation to leaf area index and climate, *Boundary-Layer Meteorol.*, 66, 265–279, 1993.

Liou, K.-N., *An Introduction to Atmospheric Radiation*, Academic Press, New York, 1980.

List, R.J., *Smithsonian Meteorological Tables*, 6th revised Edition, Smithsonian Institution, Washington, D.C., 1951.

Little, E.L., *Atlas of United States Trees*, vol. 1, *Conifers and Important Hardwoods*, U.S. Department of Agriculture Miscellaneous Publication 1141, U.S. Government Printing Office, Washington, D.C., 1971.

Lotka, A.J., Contribution to the energetics of evolution, *Proc. Nat. Acad. Sci., USA*, 8, 147–154, 1922.

Ludlow, M.M., and P.G. Jarvis, Photosynthesis in Sitka spruce, I. General characteristics, *J. Appl. Ecol.*, 8, 925–953, 1971.

MacMahon, J.A., Warm deserts, in *North American Terrestrial Vegetation*, edited by M.G. Barbour and W.D. Billings, pp. 232–264, Cambridge University Press, New York, 1988.

Mägi, H., and J. Ross, Phytometric characteristics and photosynthetic productivity of the barley stand, II. Growth dynamics of the assimilation area and increase of the dry-matter content. In *Photosynthetic Productivity of Plant Stand*, pp. 144–173, Institute of Physics and Astronomy, Academy of Sciences of the Estonian SSR, Tartu, 1969 (in Russian).

Mälkönen, E., *Annual Primary Production and Nutrient Cycle in some Scots Pine Stands*, Communicationes Instituti Forestalis Fenniae (Helsinki) 84, Finnish Forest Research Institute, Helsinki, 1974.

Marie-Victorin, Frère, *Les Gymnospermes du Québec*, Contributions du Laboratoire de Botanique de l'Université de Montréal 10, University of Montreal, 1927.

Martin, H.C., Average winds above and within a forest, *J. Appl. Meteorol.*, 10, 1132–1137, 1971.

McCaughey, J.H., and J.A. Davies, Diurnal variation in net radiation depletion within a corn crop, *Boundary-Layer Meteorol.*, 5, 505–511, 1974.

McNaughton, K.G., Evaporation and advection, I. Evaporation from extensive homogeneous surfaces, *Quart. J. Roy. Meteorol. Soc.*, 102, 181–191, 1976.

McNaughton, K.G., and P.G. Jarvis, Predicting the effects of vegetation changes on transpiration and evaporation, in *Water Deficits and Plant Growth*, vol. 2, edited by T.T. Kozlowski, pp. 1–47, Academic Press, London, 1983.

McNaughton, K.G., and W.T. Spriggs, An evaluation of the Priestley and Taylor equation and the complementary relationship using results from a mixed layer model of the convective boundary layer, in *Estimation of Areal Evaporation*, edited by T.A. Black, D.L. Spittlehouse, M.D. Novak, and D.T. Price, pp. 89–104, International Association of Hydrological Sciences, Gentbrugge, Belgium, 1989.

Mendham, N.J., P.A. Shipway, and R.K. Scott, The effects of delayed sowing and weather on growth, development and yield of winter oil-seed rape (*Brassica napus*), *J. Agric. Sci., Cambridge*, 96, 389–416, 1981.

Milburn, J.A., Cavitation: A review – past, present and future, in *Water Transport in Plants under Climatic Stress*, edited by M. Borghetti, J. Grace, and A. Raschi, pp. 14–26, Cambridge University Press, Cambridge, U.K., 1993.

Miller, H.G., J.D. Miller, and J.M. Cooper, Biomass and nutrient accumulation at different growth rates in thinned plantations of Corsican pine, *Forestry*, 53, 23–39, 1980.

Miller, P.C., Leaf temperatures, leaf orientation and energy exchange in quaking aspen (*Populus tremuloides*) and Gambell's oak (*Quercus gambellii*) in central Colorado, *Oecol. Plant.*, 2, 3, 241–270, 1967.

Mitchell, H.L., Trends in the nitrogen, phosphorus, potassium and calcium content of the leaves of some forest trees during the growing season, *Black Rock Forest Paper 1*, 30–44, 1936.

Monin, A.S., and A.M. Obukhov, The basic laws of turbulent mixing in the surface layer of the atmosphere, *Akad. Nauk, SSSR Trud. Geofiz. Inst.*, 24(151), 163–187, 1954.

Monin, A.S., and A.M. Yaglom, *Statistical Fluid Mechanics: Mechanics of Turbulence*, vol. 1, MIT Press, Cambridge, MA, 1971.

Monk, C.D., An ecological significance of evergreenness, *Ecology*, 47, 504–505, 1966.

Monsi, M., and T. Saeki, Über den Lichtfaktor in den Pflanzengesellschaften und seine Bedeutung für die Stoffproduktion, *Jap. J. Bot.*, 14(1), 22–52, 1953.

Monteith, J.L., Gas exchange in plant communities, in *Environmental Control of Plant Growth*, edited by L.T. Evans, pp. 95–112, Academic Press, New York, 1963.

Monteith, J.L., Evaporation and environment, *Symp. Soc. Exp. Biol.*, 19, 205–234, 1965.

Monteith, J.L., Light interception and radiative exchange in crop stands, in *Physiological Aspects of Crop Yield*, edited by J.D. Eastin, pp. 89–115, American Society of Agronomy, Madison, WI, 1969.

Monteith, J.L., *Principles of Environmental Physics*, Elsevier, New York, 1973.

Monteith, J.L., Climate and the efficiency of crop production in Britain, *Phil. Trans. Roy. Soc. Lond., B.*, 281, 277–294, 1977.

Monteith, J.L., Climatic variation and the growth of crops, *Quart. J. Roy. Meteorol. Soc.*, 107, 749–774, 1981.

Mooney, H.A., The carbon balance of plants, *Annu. Rev. Ecol. Syst.*, 3, 315–346, 1972.

Mooney, H.A., J. Kummerow, A.W. Johnson, D.J. Parsons, S. Keeley, A. Hoffmann, R.I. Hays, J. Giliberto, and C. Chu, The producers: their resources and adaptive responses, in *Convergent Evolution in Chile and California*, edited by H. Mooney, Dowden, Hutchinson & Ross, Stroudsburg, PA, 1977.

Mooney, H.A., P.J. Ferrar, and R.O. Slatyer, Photosynthetic capacity and carbon allocation patterns in diverse growth forms of Eucalyptus, *Oecologia (Berlin)*, 36, 103–111, 1978.

Moore, B., Some factors influencing the reproduction of red spruce, balsam fir and white pine, *J. Forestry*, 15, 842–844, 1917.

Morganstern, E.K., Range-wide genetic variation of black spruce, *Can. J. Forest Res.*, 8, 463–473, 1978.

Müller, D., Die Kohlensäureassimilation bei arktischen Pflanzen und die Abhängigkeit der Assimilation von der Temperatur, *Planta*, 6, 22–39, 1928.

Murphy, C.E., Jr., and K.R. Knoerr, Simultaneous determinations of the sensible and latent heat transfer coefficients for tree leaves, *Boundary-Layer Meteorol.*, 11, 223–241, 1977.

Nakashima, H., Über den Einfluss meteorologischen Faktoren auf den Baumzuwachs, II, *J. Fac. Agri. Hokkaido Imp. Uni. Sapporo, Japan*, 22, 301–327, 1929.

Nemeth, J.C., Dry matter production in young loblolly (*Pinus taeda* L.) and slash pine (*Pinus elliottii* Engelm.) plantations, *Ecol. Monogr.*, 43, 21–41, 1973.

Nichiporovich, A.A., Properties of plant crops as an optical system, *Sov. Plant Physiol.*, 8, 428–435, 1961a (in Russian).

Nichiporovich, A.A., On properties of plants as an optical system, *Sov. Plant Physiol.*, 8, 536–546, 1961b (in Russian).

Nilson, T., A theoretical analysis of the frequency of gaps in plant stands, *Agric. Meteorol.*, 8, 25–38, 1971.

Norman, J.M., and P.G. Jarvis, Photosynthesis in Sitka spruce (*Picea Sitchensis* (Bong.) Carr.), III. Measurements of canopy structure and interception of radiation, *J. Appl. Ecol.*, 11, 375–398, 1974.

Nye, P.H., Organic matter and nutrient cycles under moist tropical forest, *Plant and Soil*, 13, 333–346, 1961.

Odum, E.P., *Fundamentals of Ecology*, W.B. Saunders, Philadelphia, PA, 1959.

Odum, E.P., *Ecology*, 2nd Edition, Holt, Rinehart and Winston, New York, 1975.

Odum, H.T., B.J. Copeland, and R.Z. Brown, Direct and optical assay of leaf mass of the lower montane rain forest of Puerto Rico, *Proc. Nat. Acad. Sci., USA*, 49, 429–434, 1963.

O'Loughlin, E.M., Resistance to flow over boundaries with small roughness concentrations, PhD dissertation, University of Iowa, 1965.

Oliver, H.R., Wind profiles in and above a forest canopy, *Quart. J. Roy. Meteorol. Soc.*, 97, 548–553, 1971.

Osmond, C.B., K. Winter, and S.B. Powles, Adaptive significance of carbon dioxide cycling during photosynthesis in water-stressed plants, in *Adaptation of Plants to Water and High Temperature Stress*, edited by N. C. Turner and P. J. Kramer, pp. 139–154, John Wiley, New York, 1980.

Ovington, J.D., Dry matter production of *Pinus sylvestris* L., *Ann. Bot.*, N.S., 21, 287–316, 1957.

Ovington, J.D., Mineral content of plantations of *Pinus sylvestris* L., *Ann. Bot.*, N.S., 23, 75–88, 1959.

Ovington, J.D., Some aspects of energy flow in plantations of *Pinus sylvestris* L., *Ann. Bot.*, N.S., 25, 12–20, 1961.

Paeschke, W., Experimentelle Untersuchungen zum Rauhigkeits- und Stabilitätsproblem in der bodennahen

Luftschicht, *Beiträge z. Phys. d. freien Atmos.*, 24, 163–189, 1937.

Parkhurst, D.F., and O.L. Loucks, Optimal leaf size in relation to environment, *J. Ecol.*, 60, 505–537, 1972.

Pearson, G.A., The growing season of western yellow pine, *J. Agric. Res.*, 29, 203–204, 1924.

Penman, H.L., Natural evaporation from open water, bare soil, and grass, *Proc. Roy. Soc. Lond., A*, 193, 120–145, 1948.

Penning de Vries, F.W.T., A.H.M. Brunsting, and A.H. van Laar, Products, requirements and efficiency of biosynthesis, a quantitative approach, *J. Theoret. Biol.*, 45, 339–377, 1974.

Perrier, A., Land surfaces: vegetation, in *Land Surface Processes in Atmospheric General Circulation Models*, edited by P.S. Eagleson, pp. 395–448, Cambridge University Press, Cambridge, U.K., 1982.

Philip, J.R., The theory of infiltration, IV. Sorptivity and algebraic infiltration equations, *Soil Sci.*, 84, 257–264, 1957.

Philip, J.R., General method of exact solution of the concentration-dependent diffusion equation, *Austral. J. Phys.*, 13(1), 1–12, 1960.

Philip, J.R., The distribution of foliage density with foliage angle estimated from inclined point quadrat observations, *Austral. J. Bot.*, 13, 2, 357–366, 1965.

Philip, J.R., Plant–water relations: some physical aspects, *Annu. Rev. Plant Physiol.*, 17, 245–268, 1966.

Philip, J.R., The theory of infiltration, in *Advances in Hydroscience*, vol. 5, edited by V.T. Chow, pp. 215–296, Academic Press, New York, 1969.

Philip, J.R., A physical bound on the Bowen ratio, *J. Climate Appl. Meteorol.*, 26, 8, 1043–1045, 1987.

Pike, J.G., The estimation of annual runoff from meteorological data in a tropical climate, *J. Hydrol.*, 2, 116–123, 1964.

Põldmaa, V., On distribution of diffuse radiation over the sky, in *Investigations in Atmospheric Physics*, vol. 4, pp. 111–119, Institute of Physics and Astronomy, Academy of Sciences of the Estonian SSR, Tartu, 1963 (in Russian).

Prandtl, L., Ueber die ausgebildete Turbulenz, *Proc. 2nd Int. Cong. Appl. Mech.*, 62–75, 1926.

Prandtl, L., *Essentials of Fluid Mechanics*, Hafner, New York, 1952.

Priestley, C.H.B., and R.J. Taylor, On the assessment of surface heat flux and evaporation using large-scale parameters, *Monthly Weather Rev.*, 100(2), 81–92, 1972.

Pruitt, W.O., D.L. Morgan, and F.J. Lourence, Momentum and mass transfers in the surface boundary layer, *Quart. J. Roy. Meteorol. Soc.*, 99, 370–386, 1973.

Raison, R.J., B.J. Myers, and M.L. Benson, Dynamics of *Pinus radiata* foliage in relation to water and nitrogen stress, I. Needle production and properties, *Forest Ecol. Mgmt*, 52, 139–158, 1992.

Rauner, J.L., Deciduous forests, in *Vegetation and the Atmosphere*, vol. 2, *Case Studies*, edited by J.L. Monteith, pp. 241–264, Academic Press, New York, 1976.

Raupach, M.R., Canopy transport processes, in *Flow and Transport in the Natural Environment: Advances and Applications*, edited by W.L. Steffen and O.T. Denmead, pp. 95–127, Springer-Verlag, Berlin, 1988.

Raupach, M.R., Drag and drag partition on rough surfaces, *Boundary-Layer Meteorol.*, 60, 375–395, 1992.

Raupach, M.R., Simplified expressions for vegetation roughness length and zero-plane displacement as functions of canopy height and area index, *Boundary-Layer Meteorol.*, 71, 211–216, 1994.

Raupach, M.R., and J.J. Finnigan, 'Single-layer models of evaporation from plant canopies are incorrect but useful, whereas multilayer models are correct but useless': discuss, *Austr. J. Plant Physiol.*, 15, 705–716, 1988.

Raupach, M.R., and R.H. Shaw, Averaging procedures for flow within vegetation canopies, *Boundary-Layer Meteorol.*, 22, 79–90, 1982.

Raupach, M.R., and A.S. Thom, Turbulence in and above plant canopies, *Annu. Rev. Fluid Mech.*, 13, 97–129, 1981.

Raupach, M.R., A.S. Thom, and I. Edwards, A wind tunnel study of turbulent flow close to regularly arrayed rough surfaces, *Boundary-Layer Meteorol.*, 18, 373–397, 1980.

Raupach, M.R., R.A. Antonia, and S. Rajagopalan, Rough-wall turbulent boundary layers, *Appl. Mechs. Rev.*, 44, 1–25, 1991.

Reader, R., J.S. Radford, and H. Lieth, Modeling important phytophenological events in Eastern North America, in *Phenology and Seasonality Modeling*, edited by H. Lieth, pp. 329–343, Springer-Verlag, New York, 1974.

Rees, L.W., Growth studies in forest trees: *Picea rubra* Link., *J. Forest.*, 27, 384–403, 1929.

Relf, E.F., *Discussion of the Results of Measurements of the Resistance of Wires, with some Additional Tests on the Resistance of Wires of Small Diameter*, 5 pp., Aeronautics Research Council of Great Britain Reports and Memoranda No. 102, HMSO, London, 1914.

Richards, L.A., Capillary conduction of liquids through porous mediums, *Physics*, 1, 318–333, 1931.

Ripley, E.A., and R.E. Redmann, Grassland, in *Vegetation and the Atmosphere*, vol. 2, *Case Studies*, edited by J.L. Monteith, pp. 349–398, Academic Press, New York, 1976.

Rodríguez-Iturbe, I., D.R. Cox, and V. Isham, Some models for rainfall based on stochastic point processes, *Proc. Roy. Soc. Lond., A*, 410, 269–288, 1987.

Rodríguez-Iturbe, I., D.R. Cox, and V. Isham, A point process model for rainfall: further developments, *Proc. Roy. Soc. Lond., A*, 417, 283–298, 1988.

Rogers, R.W., and W.E. Westman, Seasonal nutrient dynamics of litter in a subtropical eucalypt forest, North Stradbroke Island, *Austral. J. Bot.*, 25, 47–58, 1977.

Rogers, R.W., and W.E. Westman, Growth rhythms and productivity of a coastal subtropical *Eucalyptus* forest, *Austral. J. Ecol.*, 6, 85–98, 1981.

Rogerson, T.L., Throughfall in pole-sized Loblolly pine as affected by stand density, in *International Symposium on Forest Hydrology*, edited by W.E. Sopper and H.W. Lull, pp. 187–190, Pergamon Press, New York, 1967.

Romell, L.G., Växttidsundersökningar a tall och gran, *Meddel. fran Statens Skogsförsöksanstalt*, 22, 45–124, 1925.

Rosen, R., *Optimality Principles in Biology*, Plenum Press, New York, 1967.

Rosenzweig, M.L., Net primary productivity of terrestrial communities: prediction from climatological data, *Am. Nat.*, 102, 67–74, 1968.

Ross, J., Radiative transfer in plant communities, in *Vegetation and the Atmosphere*, vol. 1, *Principles*, edited by J.L. Monteith, pp. 13–55, Academic Press, New York, 1975.

Ross, J., *The Radiation Regime and Architecture of Plant Stands*, Dr. W. Junk, Boston, MA, 1981.

Ross, J., and Nilson, T., The extinction of direct radiation in crops. In *Questions on Radiation Regime of Plant Stand*, pp. 25–64, Institute of Physics and Astronomy, Academy of Sciences of the Estonian SSR, Tartu, 1965 (in Russian).

Rouse, H., *Elementary Mechanics of Fluids*, John Wiley, New York, 1946.

Russell, G., P.G. Jarvis, and J.L. Monteith, Absorption of radiation by canopies and stand growth, in *Plant Canopies: Their Growth, Form and Function*, edited by G. Russell, B. Marshall, and P.G. Jarvis, pp. 21–39, Cambridge University Press, Cambridge, U.K., 1989a.

Russell, G., B. Marshall, and P.G. Jarvis (editors), *Plant Canopies: Their Growth, Form and Function*, Cambridge University Press, Cambridge, U.K., 1989b.

Rutter, A.J., Water consumption by forests, in *Water Deficits and Plant Growth*, vol. 2, *Plant Water Consumption and Response*, edited by T.T. Kozlowski, pp. 23–84, Academic Press, New York, 1968.

Sakai, A., Low temperature exotherms of winter buds of hardy conifers, *Plant Cell Physiol.*, 19, 1439–1446, 1978.

Sakai, A., Freezing avoidance mechanism of primordial shoots of conifer buds, *Plant Cell Physiol.*, 20, 1381–1390, 1979.

Sakai, A., and C.J. Weiser, Freezing resistance of trees in North America with reference to tree regions, *Ecology*, 54, 118–126, 1973.

Salvucci, G.D., A test of ecological optimality for semiarid vegetation, SM thesis, Department of Civil Engineering, MIT, 1992.

Salvucci, G.D., An approximate solution for steady vertical flux of moisture through an unsaturated homogeneous soil, *Water Resources Res.*, 29(11), 3749–3753, 1993.

Salvucci, G.D., and P.S. Eagleson, *A Test of Ecological Optimality for Semiarid Vegetation*, Ralph M. Parsons Laboratory Report No. 335, Department of Civil Engineering, MIT, Cambridge, MA, 1992.

Salvucci, G.D., and D. Entekhabi, Hillslope and climatic controls on hydrologic fluxes, *Water Resources Res.*, 31(7), 1725–1739, 1995; corrected in *Water Resources Res.*, 33(1), 277, 1997.

Saugier, B., Sunflower, in *Vegetation and the Atmosphere*, vol. 2, *Case Studies*, edited by J.L. Monteith, pp. 87–119, Academic Press, New York, 1976.

Schlichting, H., *Boundary Layer Theory*, translated by J. Kestin, McGraw-Hill, New York, 1955.

Schmidt, W., Der Massenaustausch bei der ungeordneten Strömung in freier Luft und seine Folgen, *Sitzber. Kais. Akad. Wissen. Wien [2a]*, 126, 757–804, 1917.

Scott, R.K., and E.J. Allen, Crop physiological aspects of importance to maximum yields in potatoes and sugar beet. In *Maximum Yield of Crops*, edited by J.K.R. Gasser and B. Wilkinson, pp. 25–30, HMSO, London, 1978.

Seginer, I., P.J. Mulhearn, E.F. Bradley, and J.J. Finnigan, Turbulent flow in a model plant canopy, *Boundary-Layer Meteorol.*, 10, 423–453, 1976.

Sellers, P. J., Vegetation–canopy spectral reflectance and biophysical processes, in *Theory and Applications of Optical Remote Sensing*, edited by G. Asrar, pp. 297–335, John Wiley, New York, 1989.

Sellers, P.J., Y. Mintz, Y.C. Sud, and A. Dalcher, A simple biosphere model (SiB) for use within general circulation models, *J. Atmos. Sci.*, 43(6), 505–531, 1986.

Sharkey, T.D., Photosynthesis in intact leaves of C_3 plants: Physics, physiology and rate limitations, *Bot. Rev.*, 51(1), 53–105, 1985.

Shaw, R.H., and A.R. Pereira, Aerodynamic roughness of a plant canopy: A numerical experiment, *Agric. Meteorol.*, 26, 51–65, 1982.

Shuttleworth, W.J., *Evaporation*, Institute of Hydrology Report 56, Institute of Hydrology, Wallingford, U.K., 1979.

Shuttleworth, W.J., Micrometeorology of temperate and tropical forest, *Phil. Trans. Roy. Soc. Lond. B*, 324, 299–334, 1989.

Shuttleworth, W.J., and I.R. Calder, Has the Priestley–Taylor equation any relevance to forest evaporation? *J. Appl. Meteorol.*, 18, 639–646, 1979.

Sinnott, E.W., *Plant Morphogenesis*, McGraw-Hill, New York, 1960.

Slatyer, R.O., The influence of progressive increases in total soil moisture stress on transpiration, growth and internal water relationships of plants, *Austral. J. Biol. Sci.*, 10, 320–336, 1957a.

Slatyer, R.O., Significance of the permanent wilting percentage in studies of plant and soil water relations, *Bot. Rev.*, 23, 585–636, 1957b.

Slatyer, R.O., *Plant–Water Relationships*, Academic Press, New York, 1967.

Slatyer, R.O., and I.C. McIlroy, *Practical Micrometeorology*, CSIRO, Melbourne, Australia, 1961.

Smith, J.W., Phenological dates and meteorological data recorded by Thomas Mikesell between 1873 and 1912 at Wauseon, Ohio, *Monthly Weather Review, Supplement 2, Part II*, W.B. No. 558, pp. 21–93, U.S. Government Printing Office, Washington, D.C., 1915.

Solot, S., Computation of depth of precipitable water in a column of air, *Monthly Weather Rev.*, 67, 100–103, 1939.

Spurr, S.H., and B.V. Barnes, *Forest Ecology*, 3rd Edition, John Wiley, New York, 1980.

Stanhill, G., A simple instrument for the field measurement of turbulent diffusion flux, *J. Appl. Meteorol.* 8, 509–513, 1969.

Stone, E.C., Dew as an ecological factor, *Ecology*, 38, 407–422, 1957.

Sun, R.J., and J.B. Weeks, *Bibliography of Regional Aquifer-System Analysis Program of the U.S. Geological Survey, 1978–91*, U.S. Geological Survey Water Resources Investigation Report 91-4122, U.S. Government Printing Office, Washington, D.C., 1991.

Swinbank, W.C., Long-wave radiation from clear skies, *Quart. J. Roy. Meteorol. Soc.*, 89, 339–348, 1963, with discussions 90, 488–493, 1964.

Szeicz, G., Solar radiation for plant growth, *J. Appl. Ecol.*, 11, 617–636, 1974.

Taylor, F.G., Jr., Phenodynamics of production in a mesic deciduous forest, in *Phenology and Seasonality Modeling, Ecological Studies: Analysis and Synthesis*, vol. 8, edited by H. Lieth, pp. 237–254, Springer-Verlag, New York, 1974.

Thom, A., *An Investigation of Fluid Flow in Two Dimensions*, Aeronautical Research Committee Reports and Memoranda 1194, HMSO, London, 1929.

Thom, A.S., Momentum absorption by vegetation, *Quart. J. Roy. Meteorol. Soc.*, 97, 414–428, 1971.

Thom, A.S., Momentum, mass and heat exchange of plant communities, in *Vegetation and the Atmosphere*, vol. 1, *Principles*, edited by J.L. Monteith, pp. 57–109, Academic Press, New York, 1975.

Thornthwaite, C.W., and J.R. Mather, Instructions and tables for computing potential evapotranspiration and the water balance, *Drexel Institute of Technology, Lab. Climatol., Pub. Climatol.*, 10, 181–311, 1957.

Tolsky, A.P., *Trans. For. Exp. St. Petersburg*, 47, 1913 (Abstract by R. Zon *For. Quart.*, 12, 277–278, 1914.)

Tranquillini, W., Die Bedeutung des Lichtes und der Temperature für die Kohlensäureassimilation von *Pinus cembre* Jungwachs an einem hochalpinen Standort, *Planta*, 46, 154–178, 1955.

Trewartha, G.T., *An Introduction to Climate*, 3rd Edition, McGraw-Hill, New York, 1954.

Tubbs, C.H., and D.R. Houston, *Fagus grandifolia* Ehrh., in *Silvics of North America*, vol. 2, *Hardwoods*, coordinated by R.M. Burns and B.H. Honkala, pp. 325–332, Agriculture Handbook 654, U.S. Department of Agriculture, Washington, D.C., 1990.

Uchijima, Z., Studies on the micro-climate within plant communities, I. On the turbulent transfer coefficient within plant layer, *J. Agric. Meteorol. (Nogyo Kisho) Japan*, 18, 1–9, 1962.

Uchijima, Z., Maize and rice, in *Vegetation and the Atmosphere*, vol. 2, *Case Studies*, edited by J.L. Monteith, pp. 33–64, Academic Press, New York, 1976.

Uchijima, Z., and J.L. Wright, An experimental study of air flow in a corn plant–air layer, *Bull. Nat. Inst. Agric. Sci. Japan (Nogyo Gijutsu Kenkyusho Hokoku)*, A11, 19–66, 1964.

Uchijima, Z., T. Udagawa, T. Horie, and K. Kobayashi, Studies of energy and gas exchange within crop canopies, VIII. Turbulent transfer coefficient and foliage exchange velocity within a corn canopy, *J. Agric. Meteorol. (Nogyo Kisho) Japan*, 25, 215–227, 1970.

Udagawa, T., Z. Uchijima, T. Horie, and K. Kobayashi, Studies of energy and gas exchange within crop canopies, III. Canopy structure of corn plants, *Proc. Crop Sci. Japan*, 37, 589–596, 1968.

U.S. Army Corps of Engineers, *Snow Hydrology*, North Pacific Division, Portland, OR, 1956. (Available from U.S. Department of Commerce, Clearinghouse for Federal Scientific and Technical Information as PB 151660.)

U.S. Weather Service, Normals, Means and Extremes – 1974, *Climatological Bulletin*, Washington, D.C., 1974.

Valentine, H.T., Budbreak and leaf growth functions for modelling herbivory in some gypsy moth hosts, *Forest Sci.*, 29, 607–617, 1983.

Van Bavel, C.H.M., Potential evaporation: The combination concept and its experimental verification, *Water Resources Res.*, 2(3), 455–467, 1966.

Van Keulen, H., J. Goudriaan, and N.G. Seligman, Modelling the effects of nitrogen on canopy development and crop growth, in *Plant Canopies: Their Growth, Form and Function*, edited by G. Russell, B. Marshall, and P.G. Jarvis, pp. 83–104, Cambridge University Press, Cambridge, U.K., 1989.

Vertessy, R.A., R. Benyon, and S.R. Haydon, Melbourne's forest catchments: Effect of age on water yield, *Water (Australian Water and Wastewater Association)*, 21(2), 17–20, 1994.

Walter, H., Die Wasservergorgung der Wustenpflanzen, *Scientia, Bologna*, 59, 1–7, 1962.

Walter, H., *Vegetation of the Earth in Relation to Climate and the Eco-Physiological Conditions*, Springer-Verlag, New York, 1973.

Walter, H., and S.-W. Breckle, *Ecological Systems of the Geobiosphere*, vol. 1, *Ecological Principles in Global Perspective*, translated by S. Gruber, Springer-Verlag, New York, 1985.

Walter, H., and S.-W. Breckle, *Ecological Systems of the Geobiosphere*, vol. 2,

Tropical and Subtropical Zonobiomes, translated by S. Gruber, Springer-Verlag, New York, 1986.

Waring, R.H., Estimating forest growth and efficiency in relation to canopy leaf area, in *Advances in Ecological Research*, vol. 13, edited by A. Macfadyen and E.D. Ford, pp. 327–354, Academic Press, New York, 1983.

Waring, R.H., and S.W. Running, Sapwood water storage: Its contribution to transpiration and effect upon water conductance through the stems of old growth Douglas fir, *Plant Cell Envir.*, 1, 131–140, 1978.

Waring, R.H., D. Whitehead, and P.G. Jarvis, The contribution of stored water to transpiration in Scots pine, *Plant Cell Envir.*, 2, 309–317, 1979.

Warren Wilson, J., Inclined point quadrats, *New Phytol.*, 59(1), 1–8, 1960.

Warren Wilson, J., Estimation of foliage denseness and foliage angle by inclined point quadrats, *Austral. J. Bot.*, 11(1), 95–105, 1963.

Warren Wilson, J., Stand structure and light penetration, I. Analysis by point quadrats, *J. Appl. Ecol.*, 2, 383–390, 1965.

Warren Wilson, J., Stand structure and light penetration, III. Sunlit foliage area, *J. Appl. Ecol.*, 4, 159–165, 1967.

Warren Wilson, J., and J.E. Reeve, Analysis of the spatial distribution of foliage by two-dimensional point quadrats, *New Phytol.*, 58, 92–101, 1959.

Warren Wilson, J., and J.E. Reeve, Inclined point quadrats, *New Phytol.*, 59, 1–8, 1960.

Westman, W.E., and R.W. Rogers, Biomass and structure of a subtropical eucalypt forest, North Stradbroke Island, *Austral. J. Bot.*, 25, 171–191, 1977a.

Westman, W.E., and R.W. Rogers, Nutrient stocks in a subtropical eucalypt forest, North Stradbroke Island, *Austral. J. Ecol.*, 2, 447–460, 1977b.

White, A., P. Handler, and E. Smith, *Principles of Biochemistry*, McGraw-Hill, New York, 1968.

Whitehead, D., F.M. Kelliher, P.M. Lane, amd D.S. Pollock, Seasonal partitioning of evaporation between trees and understorey in a widely-spaced *Pinus radiata* stand, *J. Appl. Ecol.*, 31, 528–542, 1994.

Whitford, K.R., I.J. Colquhoun, A.R.G. Lang, and B.M. Harper, Measuring leaf area index in a sparse eucalypt forest: A comparison of estimates from direct measurement, hemispherical photography, sunlight transmittance and allometric regression, *Agric. Forest Meteorol.*, 74, 237–249, 1995.

Whittaker, R.H., Net production of heath balds and forest heaths in the Great Smoky Mountains, *Ecology*, 44, 176–182, 1963.

Whittaker, R.H., Forest dimensions and production in the Great Smoky Mountains, *Ecology*, 47(1), 103–121, 1966.

Whittaker, R.H., *Communities and Ecosystems*, 2nd Edition, Macmillan, New York, 1975.

Whittaker, R.H., and G.E. Likens, The biosphere and man, in *Primary Productivity of the Biosphere*, edited by H. Lieth and R.H. Whittaker, pp. 305–328, Springer-Verlag, New York, 1975.

Whittaker, R.H., and P.L. Marks, Methods of assessing terrestrial productivity, in *Primary Productivity of the Biosphere*, edited by H. Lieth and R.H. Whittaker, pp. 55–118, Springer-Verlag, New York, 1975.

Wijk, W.R. van, *Physics of Plant Environment*, North Holland, Amsterdam, The Netherlands, 1963.

Williams, J.A., and T.C. Anderson, Jr., *Soil Survey of Beaver Creek Area, Arizona*, Report of the U.S. Department of

Agriculture Forest Service and Soil Conservation Service in cooperation with the Arizona Agricultural Experiment Station, U.S. Government Printing Office, Washington, D.C., 1967.

Wilson, C.C., Fog and atmospheric carbon dioxide as related to apparent photosynthetic rate of some broadleaf evergreens, *Ecology*, 29, 507–508, 1948.

Wilson, N.R., and R.H. Shaw, A higher order closure model for canopy flow, *J. Appl. Meteorol.*, 16, 1197–1205, 1977.

Wit, C.T. de, *Photosynthesis of Leaf Canopies*, 57 pp., Agricultural Research Report No. 663, Center for Agricultural Publications and Documentation, Wageningen, The Netherlands, 1965.

Wright, J.L., and K.W. Brown, Comparison of momentum and energy balance methods of computing vertical transfer within a crop, *Agron. J.*, 59, 427–432, 1967.

Zahner, R., Water deficits and growth of trees, in *Water Deficits and Plant Growth*, edited by K.K. Kozlowski, pp. 191–254, Academic Press, New York, 1968.

Zim, H.S., and A.C. Martin, *Trees: A Guide to Familiar American Trees*, Golden Press, New York, 1987.

Zwieniecki, M.A., and N.M. Holbrook, Diurnal variation in xylem hydraulic conductivity in white ash (*Fraxinus americana* L.), red maple (*Acer rubrum* L.) and red spruce (*Picea rubens* Sarg.), *Plant Cell Envir.*, 21, 1173–1180, 1998.

Zwieniecki, M.A., P.J. Melcher, and N.M. Holbrook, Hydrogel control of xylem hydraulic resistance in plants, *Science*, 291, 1059–1062, 2001.

Author index

Subject index

Page numbers given in bold type refer to plates; those in italics refer to figures.